Basic Principles of Forensic Chemistry

AF182400

JaVed I. Khan • Thomas J. Kennedy
Donnell R. Christian, Jr.

Basic Principles of Forensic Chemistry

Humana Press

JaVed I. Khan
U.S. Crime Laboratory
California Department of Justice
Riverside, CA, USA
Javed.Khan@doj.ca.gov

Thomas J. Kennedy
Department of Chemistry
Victor Valley Community College
Victorville, CA, USA
tkennedy34@verizon.net

Donnell R. Christian, Jr.
Director of Forensic Programs
Professional Business Solutions, Inc.
1000 Lake St. Louis Blvd, Suite 129
Lake St. Louis, MO, USA
donnell.christian@criminalist.us

ISBN 978-1-62703-892-8 ISBN 978-1-59745-437-7 (eBook)
DOI 10.1007/978-1-59745-437-7
Springer New York Dordrecht Heidelberg London

© Springer Science+Business Media, LLC 2012
Softcover reprint of the hardcover 1st edition 2012
All rights reserved. This work may not be translated or copied in whole or in part without the written permission of the publisher (Humana Press, c/o Springer Science+Business Media, LLC, 233 Spring Street, New York, NY 10013, USA), except for brief excerpts in connection with reviews or scholarly analysis. Use in connection with any form of information storage and retrieval, electronic adaptation, computer software, or by similar or dissimilar methodology now known or hereafter developed is forbidden.
The use in this publication of trade names, trademarks, service marks, and similar terms, even if they are not identified as such, is not to be taken as an expression of opinion as to whether or not they are subject to proprietary rights.

Printed on acid-free paper

Humana Press is part of Springer Science+Business Media (www.springer.com)

This book is dedicated to the students and teachers in the field of forensic sciences.

JaVed I. Khan

To my family: Tammy, Breanna, McKenna, and Holden. This work is dedicated to each of you for the sacrifices you have made. Thank you for enduring the long hours, you are my inspiration, and I love you all very much!!
For my mother, Nancy, and my sister, Susan. Thank you for your love, guidance, and support for so many years. It is the foundation that drives me to be a better man. I love you.

Thomas J. Kennedy

To my parents and my wife Stephanie.

Donnell R. Christian, Jr.

Foreword

Forensic chemistry was once the foundation of the crime laboratory. The modern forensic laboratory seems to be drifting away from its traditional roots, with the introduction of DNA evidence and the plethora of forensic shows on television. Emphasis in biology has replaced chemistry with serological and biological examinations. Degrees in forensic science have been created to address the demand created by the CSI culture. However, forensic chemistry remains the backbone of the modern forensic laboratory.

I was once asked how I would council a person seeking a degree in forensic science. I reflected on the words of my mentor, my own personal experience with a degree in criminalistics, and the employment prospects for new graduates with a degree in forensic science as I formulated my response. I responded that I would advise them against seeking a degree in forensic science. Unfortunately, that was not the response the selection committee wanted to hear.

This may seem an odd opening for a book foreword. However, the issues that factored into my response have been incorporated into this text.

My mentor believed that he was not a science teacher. He would teach me how to apply the science I knew to the analysis of physical evidence. He was not going to waste his time teaching me things I should have learned in college. He was a brilliant man and could teach anyone to do the analysis, given enough time. He knew that it requires a scientist to understand how the analysis functions. He wanted to develop an examiner's mind to be able to solve a problem, not train a technician to push buttons.

A background in science is essential to work as a forensic examiner. The minimum requirement for most entry-level forensic laboratory positions is a degree in a hard or physical science. It was not until recently that forensic science was added to the list of accepted degrees. Additionally, a demonstrable minimum number of credit hours in chemistry and physics is required as part of the applicant's course work. These requirements are in place to ensure that an entry-level person had a basic understanding of science to build a forensic scientist from.

Finally, my degree is in criminalistics. In the early 1980s, no one knew what a criminalist or a forensic scientist was. Because a degree in chemistry was a requirement, every job application had a letter from the Chemistry Department Chairman stating that I had the equivalent course work to an ACS-certified degree in chemistry. This was in addition to a copy of my college transcripts.

This rambling story does relate to *Basic Principles of Forensic Chemistry*. It has to do with the way the book is organized. *Basic Principles of Forensic Chemistry* is designed to develop the student's understanding of forensic chemistry in a sequential manner. Basic chemistry principles are established. Generic examination techniques are presented followed by specific applications. Each section builds on the information developed in the previous sections.

The focus of *Basic Principles of Forensic Chemistry* is on the analysis of controlled substances, specifically drugs of abuse. However, it provides all of the conceptual information used in any forensic chemistry section of a modern forensic laboratory. The science and the examination techniques discussed are as applicable to the analysis of drugs as they are to trace evidence.

Specific reagents may change or sample preparation techniques may be modified, but the concepts are interchangeable.

Part I lays the scientific foundation that the examiner needs to understand the science of analysis. The information in Part I reviews basic principles of chemistry beginning with atomic structure and expanding through molecules and into organic chemistry. The section is intended to be a review of chemistry basics, not a replacement for formal class work.

Part II discusses the tools used by all examiners in the forensic chemistry section. Chapters 5 and 6 establish the terminology and paperwork flow common to all forensic chemistry sections. The following chapters discuss analytical techniques. Each chapter begins with generic theory and follows it with drug-analysis applications. Chapters 7 through 9 discuss the use of nonspecific tests and sample preparation techniques that are used as part of the screening process. Each section presents the basis for the examinations followed by sections of practical application. Chapters 10 and 11 address the instrumentation frequently used by drug chemists to confirm the identity of the controlled substance indicated by the screening process. The following Chaps. 8 and 9 are theoretical discussions on practical application. Additionally, the strengths and weaknesses of each instrument are addressed.

Part III deals with the job at hand, specifically the different types of controlled substances encountered by forensic drug chemists. This section divides the controlled substances into generic categories based on structural similarities. Each chapter addresses the drugs most frequently encountered in the group under discussion. In turn, a brief history of topic drugs is provided along with pharmacological information and the analytical techniques used to identify them.

Part IV concentrates on the most challenging portion of a forensic drug chemist's job, clandestine laboratory operations. These operations force the forensic chemist out of the clinical analytical mindset. This type of analysis requires the chemist to utilize his knowledge of chemistry (Part I), combine it with analytical tools (Part II), and understand drugs of abuse (Part III). These examinations allow the drug chemist to use all the tools in his toolbox, along with deductive reasoning, to objectively examine and evaluate the data from evidence obtained from suspected clandestine drug labs.

As you can see, *Basic Principles of Forensic Chemistry* is a process. A foundation of chemical knowledge supports an analytical scheme. The tools from the analytical toolbox are used to initially identify a generic class of drug followed by a specific compound identification. Finally, the chemist's complete knowledge base and power of deductive reasoning are used to bring calm from the chaos of the evidence obtained from clandestine lab operations.

Basic Principles of Forensic Chemistry will not turn the reader into a forensic chemist. However, it will provide the fundamental knowledge required to begin a very rewarding journey. Good luck on your journey.

Chesterfield, VA Donnell R. Christian, Jr.

Preface

I have not reinvented the wheel on forensic chemistry in this book. This book is merely an effort to consolidate previously developed, yet scattered, forensic chemistry-related information under one umbrella. I used all reliable resources that my predecessors and contemporary experts in the field of forensic chemistry have developed. For this reason, I consider myself an editor rather than an author of this book. The material presented is very basic and is not intended or recommended for legislative use.

Primarily, this book is a milestone textbook toward teaching forensic chemistry at colleges and universities. Second, it is the first major, consolidated resource book for forensic laboratories throughout the country and overseas to train newly hired staff in controlled substance examination. I have developed a parallel laboratory manual with the book. The laboratory manual has 17 experiments that are exclusively designed to provide initial training to students and trainees of forensic chemistry. I also have developed an instructional PowerPoint presentation to assist instructors when teaching this course. This presentation is available to instructors at no extra cost.

Unlike many other chemistry books, most of the questions at the end of each chapter in this book pertain to court testimony. The answers to these questions affect many lives positively or negatively. For this reason, I wanted the students of forensic chemistry to learn to answer court testimony-related questions.

This book has room for improvement. I would like your suggestions, complaints, compliments, or concerns about the book. Please do not hesitate to drop me a line with suggestions for improvements to the future editions of this book.

Illustrations

I have used many original illustrations in this book.

I owe a special thanks to my forensic community at large for providing me hundreds of valuable original photos and illustrations for this book. If I ever requested one photo, I was given ten photos to choose from. Thank you once again.

In addition, GCMS and FTIR instrument manufacturers such as Varian Instruments, Avatar (Nicolet) Instruments, HP Instruments, and Agilent Instruments have generously permitted the use of photos of their instruments for this book. Thank you all.

I also used some photos from the public domain (citizen information centers) of US DOJ-DEA and other (forensic-related public domains) websites. Thank you for posting valuable photos on your websites for educational purposes. Microgram also permitted me to use illustrations.

I am thankful to all of you for permitting the use your material for this book. This book would not be in its current form without your generous contributions.

I always respected the copyrights and have avoided the use of any copyrighted material without permission in this book.

Riverside, CA JaVed I. Khan

Acknowledgments

Many selfless and silent individuals are behind a successful individual. Although I do not consider myself so successful, I do consider myself fortunate to be surrounded by the many silent, selfless, giving, and encouraging individuals of the forensic community.

I owe a special thanks to the forensic community at large for their support and assistance from various angles. Your contributions have made this book very valuable and resourceful.

California Department of Justice (CA-DOJ)–Division of Law Enforcement (DLE)–Bureau of Forensic Service (BFS) has invested and contributed in my professional growth and development for many years. Thank you.

Joseph Rynearson and John DeHaan of the BFS are my role models. Both are the authors of well-read books in the field of forensic science. My mentor, the late Alfred Moses, was also a very inspiring soul.

I thank the entire BFS staff for their support. Bureau chief Lance Gima and assistant chiefs Gill Spriggs and Eva Steinberger deserve my special thanks. My laboratory director, Gary Asbury, placed various encouraging posters all around the laboratory for me. My supervisor, Kristen Rager, provided me with unreserved support. Assistant laboratory directors Tom Nasser, Elisa Mayo, Steven Secofsky, Glen Owens, and Caroline Kim supported me as well. I thank Jerry Massetti of the California Criminalistic Institute (CCI) for his support and Waheed Jawadi of the CCI library for dispatching requested material in a timely manner. I thank my retired administrators Arthur Young, Mike White, Cecil Hider, and the late Jan Bashinsky.

I thank my colleagues Hillary Bantrup, Anatoly Zolatoryov, Chantalle Clement, Brian Reinarz, Jennifer Dernoncourt, Bronwyn Weis, Gina Williams, and Larry Joiner for permitting the use of their photos in this book, and Bertha Castro, Cosette Larsen, Christina Ramirez, Alicia Lomas-Gross, David Wu, Marla Richardson, Michele Merritt, Rich Takanaga, Marianne Stam, Paul Sham, Lourdes Peterson, Phil Palez, Jim Hall, Kim Kreuz, Lynn Melgoza, Donna Merrill, John Bowden, Trina Duke-Robinson, Greg Crew, Frank Shagoya, Martin Romero, Tom Abercrombie, Terry Fickies, Tory Johnson, Bill Matty, and Theresa Anderson at Riverside Crime Laboratory for support and assistance. I learned much from all of my past and current colleagues.

I thank my college colleagues Richard Collins, Lori Kildal, Camille Kraft, Richard Rowley, Pat Gummo, Angela Seavey, Richard LeGarra, E. Ozolin, Brianna Aliabdi, John Schuler, Nancy Politano, Jeff Splesky, Shelly Aguilar, and Randy Lim for support and assistance.

I thank my co-authors T.J. Kennedy and D.R. Christian who helped me improve the overall outcome of this book. We made an awesome team. Thank you for believing in me.

My parents would be proud of my efforts if they were alive. But I feel their prayers are with me forever.

Best of all, my family members deserve the utmost credit and thanks. Knowing my life's goals and ambitions, they never demanded any time from me. They, instead, created a calm home environment so I could work on the computer for hours without any disturbance or distraction. Thank you Fatima, Ahmed, and Sadia for unconditional support and love. Yes, I love you too.

JaVed I. Khan

My appreciation and respect to JaVed Khan for his leadership on this project. I would also like to thank those individuals who took time to contribute to this project.

Thomas J. Kennedy

I would like to thank Mr. Khan and Mr. Kennedy for the opportunity to participate in this project. I would also like to thank the people who researched the source documents used to reference this work.

Donnell R. Christian, Jr.

About the Authors

JaVed I. Khan

Mr. Khan holds a master of science degree in biochemistry from the University of California, Riverside, a master of science degree in biochemistry from the University of Agriculture, Faisalabad, Pakistan, and a bachelor of science in chemistry/biology from B.Z. University, Multan, Pakistan. He has more than 17 years of experience in the field of forensic sciences at the California State Department of Justice in the Riverside Crime Laboratory. He is an accomplished forensic scientist whose expertise expands in various fields, including forensic chemistry, forensic biology, and forensic toxicology. He has authored a number of publications on related topics.

Mr. Khan responded to hundreds of clandestine laboratory operations as a clandestine laboratory field expert. He examined the evidence from thousands of clandestine laboratory operations. Mr. Khan testified in various courts as a clandestine laboratory analyses expert witness. Mr. Khan is a recognized expert witness in other fields of forensics and has appeared more than 300 times in federal, state, and local county courts in this capacity.

Mr. Khan has more than 12 years of teaching experience as a part-time college instructor of forensic chemistry and biology. He has developed curriculum of forensic chemistry, forensic biology, and forensic toxicology courses for Mt. San Jacinto College in Riverside, California. He is a dedicated professional whose insight and tireless efforts provided the driving force for this project.

Thomas J. Kennedy

Mr. Kennedy holds a bachelor of science degree in chemistry from the University of Rochester, Rochester, NY and a master of science degree in chemistry from California State Polytechnic University, Pomona, CA. Mr. Kennedy is a former police officer and has been teaching chemistry at Victor Valley College since 1994. He has been chairman of the Department of Chemistry for the past eight years. Mr. Kennedy is a passionate teacher focused on student success.

Donnell R. Christian

Donnell R. Christian is a PhD candidate at the University of South Africa, holds a Masters in Criminal Justice from American Military University and Bachelor's degrees in Chemistry and Police Administration from Northern Arizona University. He is the author of *Forensic Investigation of Clandestine Laboratories* (2003). His companion book, *Field Guide to Clandestine Laboratory Identification and Investigation* (2004), provides a ready reference for police, fire, and emergency responders who potentially encounter clandestine labs in their daily activities. Mr. Christian has published articles on the analysis, and the clandestine manufacture and analysis, of controlled substances and has developed training programs for investigators, laboratory examiners, and attorneys involved in the investigation, examination, and prosecution of clandestine labs. He also authored a chapter concerning the analysis of controlled substances in *Forensic Science, An Introduction to Scientific and Investigative Techniques* (2002, 2005, 2009) and the forensic chemistry section in *The Forensic Laboratory Handbook, Procedures and Practices* (2005, 2011).

Mr. Christian is the director of Forensic Programs at Professional Business Solutions, Inc., and is the former Forensic Science Development Coordinator for the United States Department of Justice's International Criminal Investigative Training Assistance Program (ICITAP). With ICITAP, he has assisted in establishing forensic science programs in the developing democracies of Armenia, Azerbaijan, Bosnia, Bulgaria, Georgia, Haiti, Kazakhstan, Kyrgyzstan, Senegal, Turkmenistan, and Uzbekistan. Mr. Christian served as president and chairman of the Board of Directors for the Southwestern Association of Forensic Scientists (SWAFS). Additionally, he spent 15 years with the Arizona Department of Public Safety Crime Laboratory specializing in forensic chemistry and trace analysis, with emphasis in the clandestine manufacture of controlled substances (i.e., drugs and explosives). He has responded to hundreds of clandestine lab scenes, examined thousands of exhibits, and provided untold hours of testimony.

Contents

Part I Introduction to Forensic Chemistry

1 **Introduction** 3
  1.1 Forensic Chemistry 3
  1.2 Scientific Investigation 4
  1.3 Forensic Investigation 4
  1.4 Properties of Matter 5
  1.5 Physical Properties 5
  1.6 Chemical Properties 6
  1.7 Questions 7

2 **Atomic Structure** 9
  2.1 Introduction 9
  2.2 Periodic Table 9
  2.3 Atomic Structure 10
  2.4 Subatomic Particles 11
  2.5 The Arrangement of Electrons in an Atom 12
  2.6 Electron Configurations 13
  2.7 Periodic Trends: Understanding the Periodic Table 17
  2.8 Isotopes 18
  2.9 Radioactivity 18
  2.10 Types of Radioactive Decay 19
  2.11 Nuclear Radiation: Forensic Applications 19
  2.12 The Mole and Molar Mass 20
  2.13 Elements of Forensic Interest 20
  2.14 Questions 20
  Suggested Reading 21

3 **Molecules** 23
  3.1 Introduction 23
  3.2 Chemical Bonding 23
    3.2.1 Ionic Bonds 23
    3.2.2 Covalent Bonds 24
    3.2.3 Polar Bonds 25
    3.2.4 Hydrogen Bonding 26
    3.2.5 Multiple Bonds 26
  3.3 Predicting Bond Types 27
    3.3.1 Nonpolar Covalent Bonds 27
    3.3.2 Polar Covalent Bonds 27

3.3.3 Hydrogen Bonds 27
3.3.4 Ionic Bonds 27
3.4 Molar Mass 27
3.5 Molarity 28
3.6 Chemical Reactions 28
3.7 Questions 29

4 Organic Chemistry 31
4.1 Introduction 31
4.2 Classification of Organic Compounds: Functional Groups 31
4.2.1 Alkanes 32
4.2.2 Alkenes 37
4.2.3 Alkynes 39
4.2.4 Aromatic Compounds 40
4.2.5 Alcohols 41
4.2.6 Ketones 44
4.2.7 Aldehydes 46
4.2.8 Carboxylic Acids 47
4.2.9 Esters 49
4.2.10 Nitro Compounds 50
4.2.11 Amines 51
4.3 Methyl Group ($–CH_3$) 54
4.4 Compounds Containing Multiple Functional Groups 54
4.5 Chirality 55
4.6 Questions 56
Suggested Reading 57

Part II Tools of Forensic Chemistry

5 Forensic Language 61
5.1 Defining Drugs 61
5.2 Origin of Drugs (Narcotics) 61
5.2.1 Natural Drugs 61
5.2.2 Synthetic Drugs 61
5.2.3 Psychotropic Drugs (Mind Altering) 61
5.3 Dependence and Addiction 62
5.3.1 Physical Dependence 62
5.3.2 Psychological Dependence 62
5.4 Drug Abuse 63
5.5 Hazards of Drug Abuse 63
5.6 Structural Relationships 64
5.6.1 Analogs 64
5.6.2 Designer Drugs 64
5.6.3 Isomers 65
5.7 Controlled Substance Statutes 66
5.7.1 Controlled Substances Act 66
5.7.2 Controlled Substances Laws 66
5.7.3 Controlled Substance: Charges and Offenses 67
5.8 Controlled Substance Submission to Crime Laboratories 67
5.9 Drug Cases in Crime Laboratories 68
5.10 Examination of Controlled Substances 69
5.11 Usable Quantity 69
5.12 Court Testimony 69
5.13 Qualifications and Education 69
5.14 Questions 70
Suggested Reading 70

6 Forensic Documentation 71
6.1 Introduction 71
6.2 Chain of Custody 71
6.3 Case Notes 72
6.3.1 Types 72
6.3.2 Purpose 74
6.3.3 Content 74
6.3.4 Format 75
6.3.5 Dissemination 75
6.4 Case Report 75
6.4.1 Purpose 75
6.4.2 Format and Content 75
6.5 Examples 76
6.5.1 Example One 76
6.5.2 Example Two 76
6.6 Questions 76
Suggested Reading 77

7 Chemical Screening 79
7.1 Introduction 79
7.2 Chemistry of Color Formation 79
7.3 Limitations of Chemical Color Tests 81
7.4 Chemical Color-Test Methods 81
7.5 Documentation 82
7.6 Chemical Color Tests 82
7.6.1 Chen's Test 82
7.6.2 Dille–Koppanyi's Test 83
7.6.3 Mecke's Test 83
7.6.4 Marquis' Test 84
7.6.5 Nitric Acid Test 85
7.6.6 Primary Amine Test 86
7.6.7 Secondary Amine Test 86
7.6.8 Tertiary Amine Test 86
7.6.9 Van-Urk's Test 86
7.6.10 Duquenois–Levine Test 87
7.6.11 Froehde's Test 87
7.6.12 Janovsky Test 87
7.6.13 Weber Test 88
7.7 Summary of Chemical Color Tests 88
7.8 Questions 90
Suggested Reading 90

8 Microcrystal Techniques 91
8.1 Introduction 91
8.2 Advantages of Microcrystal Techniques 91
8.3 Disadvantages of Microcrystal Techniques 92
8.4 Documentation 92
8.5 Microcrystal Test Techniques 94
8.5.1 Aqueous Test Technique 94
8.5.2 Volatility Test Technique 95
8.5.3 Acid and Anionic Test Technique 95
8.6 Aqueous Test Reagents 95
8.6.1 Gold Chloride Test 95
8.6.2 Gold Chloride in Phosphoric Acid Test 95
8.6.3 Platinum Chloride Test 96

8.6.4 Mercuric Iodide Test ... 96
8.6.5 Mercuric Chloride Test ... 96
8.6.6 Potassium Permanganate Test ... 96
8.6.7 Sodium Acetate Test ... 96
8.7 Critical Considerations ... 96
8.8 Questions ... 97
Suggested Reading ... 97

9 Chemical Extractions and Sample Preparation ... 99
9.1 Introduction ... 99
9.2 Techniques ... 99
9.2.1 Solid–Liquid Extraction ... 99
9.2.2 Liquid–Liquid Extraction ... 100
9.2.3 Acid–Base Extraction ... 101
9.2.4 Neutral Compound Extraction ... 102
9.3 Sample Preparation ... 103
9.4 Gas Chromatography/Gas Chromatography Mass Spectrometry ... 103
9.5 Dry-Extraction Gas-Chromatography Modification ... 104
9.6 Acid–Base-Extraction Gas-Chromatography Modification ... 104
9.7 Infrared Spectroscopy ... 104
9.8 Acid–Base-Extraction Infrared-Modification-I ... 105
9.9 Acid–Base-Extraction Infrared-Modification-II ... 105
9.10 Methanol Extraction ... 105
9.11 Questions ... 106
Selected Reading ... 106

10 Chromatography and Mass Spectrometry ... 107
10.1 Introduction ... 107
10.2 Chromatographic Techniques ... 107
10.2.1 Paper Chromatography ... 107
10.2.2 Thin-Layer Chromatography ... 109
10.2.3 Column Chromatography ... 110
10.2.4 Ion-Exchange Chromatography ... 110
10.2.5 High-Performance Liquid Chromatography ... 111
10.2.6 Gas Chromatography ... 112
10.2.7 Chromatography: Limitations ... 115
10.2.8 Interpretation of GC Chromatograms ... 115
10.3 Mass Spectrometry ... 116
10.3.1 Ionization ... 117
10.3.2 Electron Impact ... 117
10.3.3 Chemical Ionization ... 118
10.3.4 Mass Spectral Fragmentation ... 118
10.3.5 Mass Analyzers (Filters) ... 119
10.3.6 Quadrupole Mass Analyzers ... 119
10.3.7 Magnetic Sector Mass Analyzers ... 123
10.3.8 Ion Trap Mass Analyzers ... 123
10.4 Advantages of Gas Chromatography Mass Spectrometry ... 124
10.5 Disadvantages of Gas Chromatography Mass Spectrometry ... 124
10.6 Questions ... 125
Suggested Reading ... 125

11 Infrared Spectroscopy ... 127
11.1 Introduction ... 127
11.2 Theory of Infrared Spectroscopy ... 127
11.3 Infrared Spectrum ... 129

11.4 Instrumentation ... 129
11.4.1 Dispersive Infrared Spectrometer ... 129
11.4.2 Spectrometer Components ... 129
11.4.3 Spectrometer Design ... 131
11.4.4 Limitations of Dispersive Infrared ... 132
11.5 Fourier Transform Infrared Spectrometer ... 132
11.5.1 Spectrometer Components ... 133
11.5.2 Spectrometer Design ... 133
11.5.3 Advantages of Fourier Transform Infrared Spectrometers ... 134
11.5.4 Fourier Transform Infrared Sample Preparation Techniques ... 134
11.6 Sampling Techniques ... 136
11.6.1 Nujol Mull ... 136
11.6.2 Cast Film A ... 136
11.6.3 Cast Film B ... 136
11.6.4 Pellets ... 136
11.6.5 Synthetic Membrane Sample Cards ... 137
11.7 Reflectance ... 137
11.8 Fourier Transform Infrared Spectroscopy ... 137
11.9 Advantages of Fourier Transform Infrared Spectroscopy ... 137
11.10 Disadvantages of Fourier Transform Infrared Spectroscopy ... 137
11.11 Instrument Selection for Forensic Identification ... 137
11.12 Inorganic Analysis ... 138
11.13 Organic Analysis ... 139
11.14 Questions ... 141
Suggested Reading ... 141

Part III Examination of Drugs/Narcotics

12 Cannabis ... 145
12.1 Introduction ... 145
12.2 History ... 145
12.3 Packaging for Forensic Examination ... 147
12.4 Forms of Cannabis ... 147
12.5 Psychoactive Ingredient ... 147
12.6 Forensic Identification of Marijuana ... 149
12.6.1 Botanical Identification ... 149
12.6.2 Macroscopic Properties ... 149
12.6.3 Microscopic Identification ... 151
12.6.4 Chemical Identification (Duquenois–Levine Test) ... 151
12.6.5 Thin-Layer Chromatography ... 153
12.6.6 Gas Chromatography Mass Spectrometry ... 154
12.7 Documentation ... 154
12.8 Questions ... 156
Suggested Reading ... 156

13 Phenethylamines ... 157
13.1 Introduction ... 157
13.2 Methyl Derivatives ... 157
13.2.1 Amphetamine ... 158
13.2.2 Methamphetamine ... 159
13.2.3 Phentermine ... 161
13.3 Hydroxyl Derivatives ... 162
13.3.1 Phenylpropanolamine ... 162
13.3.2 Ephedrine/Pseudoephedrine ... 163
13.3.3 Ephedra Plant: Introduction and History ... 164

13.4 Ketone Derivatives 164
13.4.1 Cathinone 164
13.4.2 Methcathinone 165
13.4.3 Khat 165
13.5 Methylenedioxy Derivatives 166
13.5.1 3,4-Methylenedioxyamphetamine 167
13.5.2 3,4-Methylenedioxymethamphetamine 167
13.6 Methoxy Derivatives 168
13.6.1 Mescaline 168
13.7 Analytical Methods 170
13.7.1 Visual Inspection 170
13.7.2 Chemical Screening 170
13.7.3 Microcrystal Tests 170
13.7.4 Extraction Techniques 171
13.7.5 Extraction of Mescaline from Peyote 172
13.7.6 Confirmatory Examination 172
13.8 Questions 175
Suggested Reading 175

14 Tertiary Amines 177
14.1 Introduction 177
14.2 Natural Tertiary Amines 177
14.2.1 Cocaine 177
14.2.2 Opiates 179
14.3 Synthetic Tertiary Amines 182
14.3.1 Phenylcyclohexylpiperidine 182
14.4 Analytical Methods 183
14.4.1 Visual Inspections 183
14.4.2 Chemical Screening of Tertiary Amines 183
14.4.3 Confirmatory Examination 185
14.5 Questions 190
Suggested Reading 190

15 Tryptamines 191
15.1 Introduction 191
15.2 Natural Tryptamines 192
15.2.1 Psilocin and Psilocybin (Psychoactive Mushrooms) 192
15.2.2 Bufotenin 193
15.2.3 Methoxy Derivatives 195
15.3 Synthetic Tryptamines 197
15.4 Analytical Methods 197
15.4.1 Visual Identification 197
15.4.2 Chemical Screening Tests 198
15.4.3 Extraction of Psilocin and Psilocybin from Mushrooms 199
15.4.4 Thin-Layer Chromatography 199
15.4.5 Gas-Chromatography Mass Spectrometry 199
15.5 Questions 205
Suggested Reading 206

16 Anabolic Steroids 207
16.1 Introduction and History 207
16.2 Naturally Occurring Steroid Hormones 208
16.3 Anabolic Steroids 210
16.3.1 General Structure 210
16.3.2 Physical and Psychological Effects 211

16.3.3 Methods of Administration ... 211
16.3.4 Nomenclature of Anabolic Steroids ... 211
16.3.5 Frequently Encountered Steroids ... 212
16.4 Analytical Methods ... 213
16.4.1 Visual Inspections ... 213
16.4.2 Gas Chromatography Mass Spectrometry ... 213
16.4.3 Mass Spectra of Commonly Encountered Steroids ... 214
16.5 Questions ... 222
Suggested Reading ... 222

17 Miscellaneous Controlled Substances ... 223
17.1 Introduction ... 223
17.2 Barbiturates ... 223
17.3 Fentanyl ... 225
17.4 *Gamma*-Hydroxybutyric Acid: γ-Hydroxybutyric Acid ... 226
17.5 Ketamine ... 227
17.6 Lysergic Acid Diethylamide ... 228
17.7 Analytical Methods ... 229
17.7.1 Visual Identification ... 229
17.7.2 Chemical Screening Tests ... 230
17.7.3 Gas-Chromatography Mass Spectrometry ... 230
17.8 Questions ... 237
Suggested Reading ... 237

Part IV Clandestine Laboratory Operations

18 Clandestine Operations: Synthetic Methods, Hazards, and Safety ... 241
18.1 Introduction ... 241
18.2 Clandestine Operations ... 241
18.2.1 Synthesis of Cocaine ... 241
18.2.2 Synthesis of Fentanyl ... 242
18.2.3 Synthesis of γ-Hydroxybutyric Acid ... 242
18.2.4 Synthesis of Heroin ... 244
18.2.5 Synthesis of Lysergic Acid Diethylamide ... 245
18.2.6 Synthesis of 3,4-Methylenedioxymethamphetamine ... 246
18.2.7 Synthesis of Methcathinone ... 247
18.2.8 Synthesis of Phencyclidine ... 248
18.2.9 Synthesis of *N,N*-Dimethyltryptamine ... 248
18.3 Synthesis of Methamphetamine: The Clandestine Operation of Choice ... 249
18.3.1 Cold Method ... 249
18.3.2 Hot Method ... 251
18.4 Potential Hazards Associated with Clandestine Operations ... 253
18.5 Safety Considerations ... 253
18.6 Role of the Forensic Chemist at Clandestine Lab Sites ... 253
18.6.1 Advisory ... 254
18.6.2 Evidence Collection ... 254
18.7 Questions ... 254
Suggested Reading ... 254

19 Evidence Identification and Collection ... 257
19.1 Clandestine Operations: A Forensic Analogy ... 257
19.2 Signs of Clandestine Operations ... 257
19.3 Identification of Related Evidence ... 258
19.4 Solutions Frequently Encountered at Clandestine Sites ... 260

19.5 Clandestine Production of Methamphetamine ... 261
 19.5.1 Extraction of Pseudoephedrine from Cold Tablets (Step I) ... 261
 19.5.2 Manufacturing of Methamphetamine (Step II) ... 262
 19.5.3 Processing of Methamphetamine (Step III) ... 263
 19.5.4 Icing of Methamphetamine (Step IV) ... 263
19.6 Collection of Evidence ... 263
19.7 Collection of Washes ... 266
19.8 Questions ... 267
Suggested Reading ... 267

20 Examination of Clandestine Evidence ... 269
20.1 Introduction ... 269
20.2 Examination of Evidence to Prove Extraction (Step I) ... 269
 20.2.1 Evidence Type ... 269
 20.2.2 Examination ... 270
 20.3 Examination of Evidence to Prove Manufacturing of Methamphetamine (Step II) ... 270
 20.3.1 Evidence Type ... 270
 20.3.2 Examination ... 271
 20.3.3 Confirmatory Examination ... 274
20.4 Examination of Evidence to Prove Processing of Methamphetamine (Step III) ... 274
 20.4.1 Evidence Type ... 274
 20.4.2 Examination of Biphasic Solutions ... 275
20.5 Examination of Evidence to Prove Icing (Step IV) ... 275
20.6 Examination of Stains ... 277
20.7 Examination of Washes ... 278
20.8 Determining Methods of Methamphetamine Production ... 279
20.9 Questions ... 282
Suggested Reading ... 282

Laboratory Manual ... 283
Experiment # 2 ... 286
Experiment # 3 ... 288
Experiment # 4 ... 290
Experiment # 5 ... 292
Experiment # 6 ... 298
Experiment # 7 ... 300
Experiment # 8 ... 304
Experiment # 9 ... 310
Experiment # 10 ... 312
Experiment # 11 ... 314
Experiment # 12 ... 316
Experiment # 13 ... 320
Experiment # 14 ... 323
Experiment # 15 ... 325
Experiment # 16 ... 328
Experiment # 17 ... 332
Experiment # 18 ... 335
Experiment # 19 ... 338
Experiment # 20 ... 341

Index ... 345

Part I

Introduction to Forensic Chemistry

1 Introduction

1.1 Forensic Chemistry

Forensic science is the application of scientific principles to matters involving the law. This area of science is generally considered quite fascinating and it continues to experience growing popularity. Many would agree that the current public interest in forensics is a direct result of CSI-related television programming. These weekly shows have brought a once relatively unknown area of science to the forefront of public mainstream. Viewers are captivated and intrigued by well-informed scientists working in spotless labs with ominous lighting and a modern music background. The use of cutting-edge technology provides last-minute revelations culminating in the solution of a complex crime. These programs are entertaining and have certainly increased public awareness to the field of forensics; but alas, television is not reality. Although it is true that forensic science has experienced tremendous growth, few would (or should) believe this to be the result of fictional television programming.

Media coverage of high-profile cases has increased over the last decade in both numbers and content. Crime-scene investigation and forensic analysis have been brought out of the lab and into the public's "scrutinizing eye." Forensic science, once a broad field, has become segregated into highly specialized disciplines. For example, forensic chemistry, forensic pathology, forensic dentistry, forensic entomology, and forensic DNA analysis have evolved into independent fields of forensics. It seems more appropriate – and clearly more realistic – to attribute the unprecedented popularity of forensic investigation to enhanced public awareness and an increase in the availability of career opportunities.

Chemistry is the study of the composition of matter and the changes it undergoes. Forensic chemistry is a specialized area of forensic science involving the application of chemical principles and techniques to the field of forensic investigation. The role of forensic chemistry in criminal investigations is vast and ranges from techniques used to collect and preserve evidence, to complex chemical procedures used to identify elements and compounds. Identification procedures are highly reliable and are frequently based on the chemical and physical properties of the substance supported by data obtained from analytical analysis. Most chemical techniques used for isolation, purification, and identification are valid forensic techniques; however, chemical analysis differs from forensic chemical analysis in two ways: regulatory and judiciary.

The results of forensic investigation may have a serious impact on lives. Therefore, techniques performed during forensic analysis must be closely regulated to ensure the accuracy and integrity of experimental results. Forensic laboratories must develop two operating manuals designed to meet the specific needs of each laboratory. The technical procedures manual outlines the step-by-step details of all procedures used in forensic examinations. The quality-control manual is designed to maintain the highest standards of reliability and integrity of work done by scientists in the lab. Adherence to both the technical procedures manual and lab quality manual is a crucial part of any analysis and is strictly enforced both internally and externally. Internal quality control includes, but is not limited to, periodic instrument calibration, checking reagents for expiration, and performance evaluations on scientists working in the laboratory. In addition, a detailed record is kept of all internal quality procedures performed. Outside regulatory agencies are responsible for external quality control and these agencies may vary from state to state in the US. The American Society of Crime Laboratory Directors (ASCLD) has recently accepted the painstaking task of regulating various fields within forensic science worldwide. This includes the forensic chemistry section in the United States. ASCLD is the regulatory organization responsible for supervising, evaluating, and directing all laboratories within its membership. Their designated inspectors evaluate technical staff and conduct periodic site inspections to ensure the highest standards of quality and technical performance. The efforts of ASCLD have helped to streamline and

J.I. Khan et al., *Basic Principles of Forensic Chemistry*, DOI 10.1007/978-1-59745-437-7_1,
© Springer Science+Business Media, LLC 2012

standardize forensic analytical techniques worldwide. In addition, ASCLD provides direction and qualified solutions to potential issues facing member laboratories.

Courtroom presentation of scientific principles and techniques used during forensic examination is the judiciary responsibility of the forensic chemist. Forensic chemists are often called upon to describe complex chemical procedures to individuals who have a limited understanding of scientific principles. This responsibility can present a variety of challenges to the forensic chemist as an expert witness. Courtroom testimony is carefully prepared using common terminology and the presentation must be in a clear, simple manner that avoids confusion and misinterpretation. To achieve this, forensic chemists often use common analogies to describe complex chemical and analytical techniques. For example, a gas chromatograph is an instrument used to separate a gaseous mixture into individual components based on size and/or charge. The description of how a gas chromatograph functions may contain a reference to coin-separating machines frequently found in local grocery stores. A coin machine separates the mixture of coins based on size, and totals each pile based on weights. This analogy would illustrate how a gas chromatograph functions and may help members of a jury be more comfortable with testimony about this complex instrument.

Similar analogies will be used in the following chapters to describe complex chemical procedures and analytical techniques frequently used in forensic chemistry. These analogies are designed to promote an understanding of the topic under discussion while adding clarity and continuity to the subject.

1.2 Scientific Investigation

Imagine yourself in a classroom for an extended period of time without the ability to see outside. When you exit the building, you immediately notice that the ground is wet. Your first thought is that it rained while you were inside. To confirm this, you look to the sky to identify rain clouds. If the sky is cloudy, you are reasonably sure that it rained. If the sky is clear, you consider another possibility – perhaps sprinklers wet the ground. To confirm this, you look for sprinklers in the immediate area. If they are found, you are reasonably sure of why the ground is wet. If no sprinklers are found, you consider another possibility and the cycle repeats. Each time you consider a possible cause, you search for supporting evidence to confirm that cause. You accept or reject a possibility based on the presence or absence of supporting evidence. In the above scenario, you observe a water truck spraying an adjacent construction site. You are now reasonably sure of how the ground became wet. The wet ground was your *observation*. The possibility of rain was your first *hypothesis*. Searching the sky for clouds was your *experimentation*. The absence of clouds in the sky caused you to *reject your hypothesis*. Other hypotheses were considered and subsequently rejected based on a lack of supporting evidence. Finally, the water truck hypothesis was confirmed when you saw the truck in the immediate area. Your determination that the water truck wet the ground is your *conclusion or theory*. This deductive procedure is termed the *scientific method*: *the process used to form theories*. It begins with an *observation*: the discovery and recognition of some type of unexplained phenomenon. The observation is followed by the *hypothesis*: the proposal of a possible cause of the observation. The hypothesis is tested during the *experimentation* phase using experiments specifically designed to prove the hypothesis. If experimental results do not support the hypothesis, another possibility is considered and tested. If the experiments are successful and repeatable, the hypothesis becomes a *theory* and is presented to the scientific community.

1.3 Forensic Investigation

Imagine a distant planet, similar to earth, with diverse climates and distinctly different environments across its surface. Now imagine that four space programs on earth send their astronauts to the new planet that, by chance, land in different regions characterized as a desert landscape, a tropical rainforest, a frozen landscape, and mountainous landscape. The astronauts explore their regions collecting samples, data, and video from their distinctly different environments. They return to their respective countries with a description of the planet supported by evidence collected during exploration. Each space program presents their information to the world, but the views are conflicting. Each country defends their position and accuses the others of presenting false or misleading information. Whom do you believe? Intuitively, you trust your astronauts and reject the other three despite the fact that, in reality, each is truthful and correct. It is not uncommon for different forensic scientists to arrive at different conclusions after examining the same piece of evidence. This is acceptable, if not expected, in the field of forensic investigation. The results of forensic examinations must never be accepted or rejected because you know or trust one scientist more than another. You must keep an unbiased, open mind, knowing that two or more scientists may present different perspectives when evaluating the same piece of evidence.

An unfortunate aspect of forensic investigation is that the results of your examination will always have a negative impact on one party. If the evidence supports the suspect's innocence, the victim is unhappy; if it supports the suspect's guilt, the suspect is unhappy. This is both unfortunate and unavoidable; however, it is the duty of the forensic chemist to present the unbiased story of the evidence.

1.4 Properties of Matter

Matter is anything that has *mass and occupies space.* It is difficult to imagine something that has mass that does not occupy space, or something that occupies space that does not have mass. Do not spend too much time pondering the previous, I cannot think of anything either (perhaps something on the previously referenced imaginary planet). Despite the apparent redundancy in the definition of matter, it must satisfy the two parameters. There is a difference between the mass of an object and its weight. Weight is a *force resulting from the pull of gravity on a given mass.* Mass is defined as a *specific quantity of matter* and is not affected by the pull of gravity. The weight of an object on earth will be different from its weight on the moon because the force of gravity is different. The mass of an object will be constant at these locations despite the differences in gravitational field strength. For this reason, the term "mass" should always be used in any area of science when referring to "weight." There are three states (or phases) of matter: solid, liquid, and gas. Solids have a *defined volume and a fixed shape*; liquids have a *defined volume* and *undefined shape* – they conform to the shape of their container; and gases have an *undefined volume and undefined shape* – they take the shape and volume of the container holding the gas. Elements are the *fundamental building blocks of all matter.* The symbols used to identify all known elements can be found on the periodic table, an *arrangement of the elements based on atomic properties.* For example, "H" represents the element hydrogen and "O" represents the element oxygen. Compounds are formed through the *combination of two or more elements.* Chemical formulas are used to represent compounds. They specify the identity and relative number of each atom present using symbols from the periodic table and subscripts attached to each symbol. For example, the chemical formula for water is H_2O, a compound containing two atoms of the element hydrogen (note subscript 2 attached to H) and one atom of the element oxygen. Elements and compounds may exist as pure substances or as mixtures. Pure substances contain only *one component* and have the *same composition* throughout, for example, pure gold, pure sugar, and pure water. Mixtures contain *two or more pure substances* and may be homogeneous or heterogeneous. Homogeneous mixtures have the *same composition and properties throughout.* They are not pure substances because they contain more than one component. For example, pure sugar water is a homogeneous mixture containing sugar and water. It has the same sweetness throughout; however, evaporating one component (the water) will produce the other (sugar crystals). Heterogeneous mixtures have *distinctly different properties within the mixture*; water and sand would be an example. The sand and water are easily identified, regardless of the degree of mixing.

There are fundamental properties associated with all forms of matter. These *distinguishing characteristics* may be physical or chemical in nature and are frequently used to identify and classify a particular substance.

1.5 Physical Properties

Physical properties such as eye and hair color, skin tone, and general build are features used to distinguish individuals. Generally, physical properties are of genetic origin and are therefore highly reliable and difficult to conceal. In chemistry, a physical property is anything that can be *measured or observed without changing the chemical composition of the substance.* The melting point and boiling point of water are both physical properties of water. These temperatures can be measured without changing the chemical composition of water. When water freezes or boils it does not change composition, it merely changes states; ice, liquid water, and steam are all still water. Although melting points and boiling points are physical properties of a substance, the process of melting and boiling results in a physical change: *a change in the state of matter, but not its chemical composition.* Other physical properties often used in the forensic identification of elements and compounds are color, odor, density, solubility, conductivity, and sublimation. The sublimation of iodine crystals (*changing from solid phase directly to gas phase*) produces a yellow gas that can stain packaging material and certain types of fabric. This is a distinguishing characteristic of elemental iodine that does not change composition; it is, therefore, a physical property (Fig. 1.1).

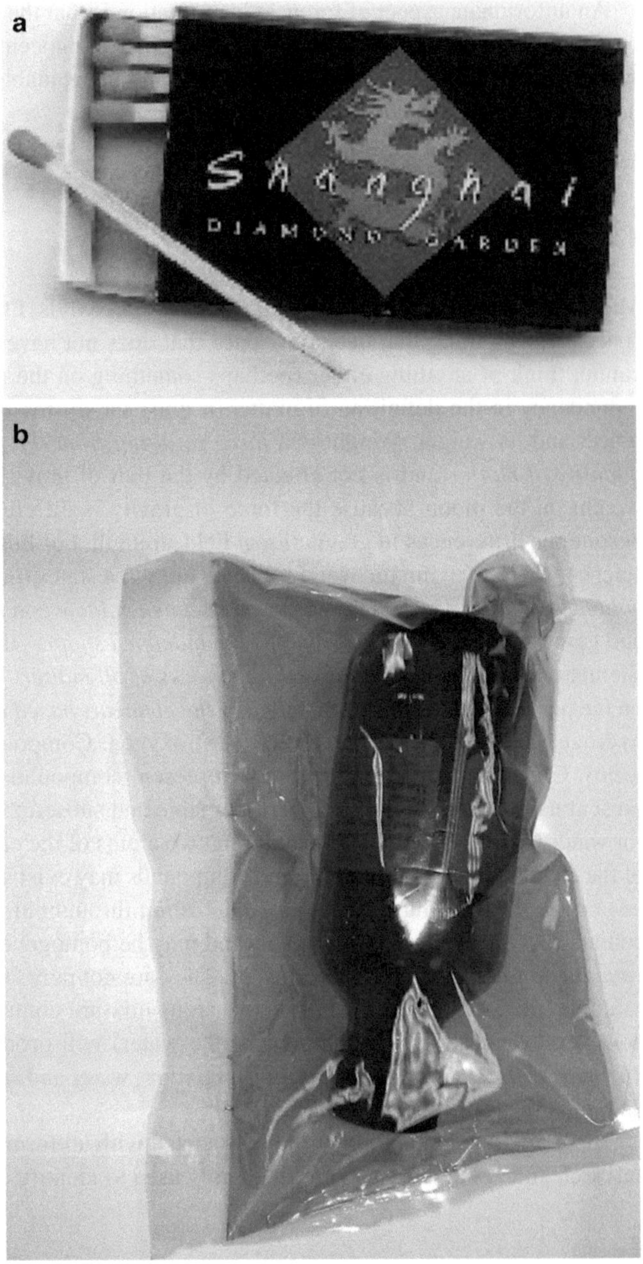

Fig. 1.1 The production of gas during sublimation is visible inside a sealed package containing iodine. A substance that has a small temperature range between melting and boiling points will sublime and change from a solid directly to a gas. Note the absence of a liquid phase.

1.6 Chemical Properties

Individuals exposed to the same situation will generally respond differently because most are unique in society. A particular response or action is based on a person's interpretation or perception of the event, which is often influenced by their morals, ethics, beliefs, etc. How individuals "react" in similar situations may be a distinguishing characteristic. Criminals have no regard for the laws of society and their behavior makes them unique among normal law-abiding citizens. Similarly, how elements or compounds react under carefully controlled conditions may be a distinguishing characteristic. A chemical reaction produces products that are different from the starting material. Chemical properties are a measure of the *ability of a substance to produce new substances*, or, more simply stated, a measure of the *reactivity of a substance*. Burning paper produces ash, silver tarnishes, iron rusts; these are all chemical reactions that produce chemical changes. The *products are chemically and physically different from the starting material* in each case.

Paper can burn, silver can tarnish, and iron can rust; these are all chemical properties because they relate to the reactivity of the substance. A matchstick produces fire when struck on the phosphorus-coated side (Fig. 1.1a). This is a chemical property

Fig. 1.2 The formation of a solid in solution is observable evidence that a chemical reaction has taken place. Chemical properties may be used to detect the presence or absence of specific ions in solution. The different solids indicate the presence of different ions.

of elemental phosphorus. A solution of silver nitrate will produce a white precipitate (solid) in the presence of chloride ions and a yellow precipitate in the presence of iodide ions. These reactions illustrate the unique response of two different ions (chloride and iodide) exposed to the same chemical environment (silver nitrate solution) (Fig. 1.2).

Physical and chemical properties are frequently used to identify elements and compounds in the field of forensic chemistry. Therefore, these properties may be used to support or reject specific parts of an investigation. For example, burning is a chemical property associated with heat. A variety of materials will burn; however, they may react differently when exposed to heat. A wall from a house fire would show several distinct burn marks: the wood frame, the wiring in the wall, the paint on the wall, and even the nails in the wood. Burn marks are distinguishing characteristics that provide information beyond the chemical property of combustion. They may be used to determine duration of contact, depth of burn, degree of heat intensity, and source of burn. Fire, hot objects, strong acids, and strong bases will all burn skin; however, each would leave a distinctly different burn mark that would allow identification of a single source from all others. Forensic scientists would use this knowledge to answer specific questions that arise during an investigation.

1.7 Questions

1. Define forensic chemistry.
2. Discuss the difference between chemical analysis and forensic chemical analysis.
3. Use the scientific method to develop a theory based on an observation of your choice.
4. Explain to the members of the jury why two highly qualified scientists presented conflicting testimony when examining the same piece of evidence.
5. Describe the three phases of matter to members of the jury.
6. Explain the difference between a pure substance and a pure mixture.
7. Discuss the difference between a homogeneous and heterogeneous mixture.
8. Identify each of the following mixtures as homogeneous or heterogeneous:
   (a) Italian dressing
   (b) Saltwater
   (c) Steel
   (d) Soil
   (e) Spoiled milk
9. Define a physical property.
10. Discuss how physical properties are related to physical changes.
11. Explain to members of the jury the difference between a physical property and a chemical property.
12. Identify each of the following as a chemical change or a physical change:
    (a) Burning paper
    (b) Melting butter
    (c) An explosion
    (d) Sublimation of dry ice

2 Atomic Structure

2.1 Introduction

At first glance, the periodic table may appear quite intimidating – a collection of seemingly unrelated numbers and symbols thrown together with little thought given to organization, continuity, or relevance. It may be comforting to know that the information contained in the periodic table is simply presented in a manner unfamiliar to most. A few basic skills are all that is required to appreciate the periodic table as a highly organized source of invaluable information.

2.2 Periodic Table

In 1869, a Russian chemist named Dmitri I. Mendeleev (1834–1907) discovered that elements exhibit a repeating pattern of properties when organized in order of increasing atomic mass. He called this observation the *periodic law*. Mendeleyev arranged the elements in different ways to determine if a relationship among the elements could be established. Frustrated and without success, he finally organized the elements into rows, beginning a new row with each repeating cycle, and the "primitive" periodic table was born. The German chemist Lothar Meyer also observed this "periodic" relationship around the same time, but as fate would have it, Mendeleyev is credited with the discovery because he was the first to organize the elements. The current form of the periodic table was later discovered by Henry Moseley who observed that elements organized by increasing atomic numbers created a more uniform arrangement.

The periodic table contains detailed information on all the known elements. The elements are represented by symbols that are translated into names. The symbols contain one-, two-, or three-letter designations that are case sensitive; the first letter is *always uppercase* with subsequent letters, if present, *always lowercase*. The names of the elements must be memorized along with their symbols. This task is somewhat simplified by the fact that many of the symbols share a common letter with the name. For example, N is the symbol for nitrogen and O is the symbol for oxygen. Unfortunately, this is not always the case, and some elements appear to have symbols that are completely unrelated to their names (Fig. 2.1). For example, Au is the symbol for gold and Fe is a symbol for iron, etc. In these cases, the symbol is derived from the element's name in Latin, German, or Greek; for example, Au is derived from the Latin word *aurum* meaning gold (see Appendix 1 for symbols and names of common elements). The elements are organized into vertical columns called groups and horizontal rows called periods. The group numbers appear at the top of each column in Roman numerals and range from I to VIII (one to eight). Elements in the same group (vertical column) have similar chemical and physical properties and a few groups are given characteristic names that you should become familiar with: Group IA elements are called alkali metals, group IIA are called alkaline earth metals, group VIIA the halogens, and group VIIIA the noble (or inert) gases. The A and B designations associated with the group numbers have no definitive meaning; they are simply used (in the US) to differentiate the main group elements (also called representative elements) from the transition metals. This practice can vary with table suppliers, particularly those from European countries. Nonetheless, the taller group columns located on each side of the table ("A" designations above) are called the main group elements and the middle groups ("B" designations above) are called the transition metals. The periods are numbered on the left side of the table from 1 to 7 downward, beginning with hydrogen.

J.I. Khan et al., *Basic Principles of Forensic Chemistry*, DOI 10.1007/978-1-59745-437-7_2,
© Springer Science+Business Media, LLC 2012

	1 IA	2 IIA	3 IIIB	4 IVB	5 VB	6 VIB	7 VIIB	8 VIIIB	9 VIIIB	10 VIIIB	11 IB	12 IIB	13 IIIA	14 IVA	15 VA	16 VIA	17 VIIA	18 VIIIA
1	1 H 1.01																	2 He 4.00
2	3 Li 6.94	4 Be 9.01											**5 B 10.81**	6 C 12.01	7 N 14.01	8 O 16.00	9 F 19.00	10 Ne 20.18
3	11 Na 22.99	12 Mg 24.31											13 Al 26.98	**14 Si 28.09**	15 P 30.97	16 S 32.06	17 Cl 35.45	18 Ar 39.95
4	19 K 39.10	20 Ca 40.08	21 Sc 44.96	22 Ti 47.88	23 V 50.94	24 Cr 52.00	25 Mn 54.94	26 Fe 55.85	27 Co 58.93	28 Ni 58.71	29 Cu 63.55	30 Zn 65.38	31 Ga 69.72	**32 Ge 72.59**	**33 As 74.92**	34 Se 78.96	35 Br 79.90	36 Kr 83.80
5	37 Rb 85.47	38 Sr 87.62	39 Y 88.91	40 Zr 91.22	41 Nb 92.91	42 Mo 95.94	43 Tc 98.91	44 Ru 101.1	45 Rh 102.9	46 Pd 106.4	47 Ag 107.87	48 Cd 112.4	49 In 114.8	50 Sn 118.7	**51 Sb 121.8**	**52 Te 127.6**	53 I 126.90	54 Xe 131.3
6	55 Cs 132.9	56 Ba 137.33	71 Lu* 175.0	72 Hf 178.5	73 Ta 180.9	74 W 183.9	75 Re 186.2	76 Os 190.2	77 Ir 192.2	78 Pt 195.1	79 Au 196.97	80 Hg 200.59	81 Tl 204.4	82 Pb 207.2	83 Bi 209.0	84 Po (209)	**85 At (210)**	86 Rn (222)
7	87 Fr (223)	88 Ra (226.0)	103 Lr** (262)	104 Rf (261)	105 Db (262)	106 Sq (263)	107 Bh (262)	108 Hs (265)	109 Mt (266)									

***Lanthanides**	57 La 138.9	58 Ce 140.1	59 Pr 140.9	60 Nd 144.2	61 Pm (147)	62 Sm 150.4	63 Eu 152.0	64 Gd 157.3	65 Tb 158.9	66 Dy 162.5	67 Ho 164.9	68 Er 167.3	69 Tm 168.9	70 Yb 173.0	71 Lu 175.0
****Actinides**	89 Ac 227.0	90 Th 232.0	91 Pa (231)	92 U 238.0	93 Np (237)	94 Pu (244)	95 Am (243)	96 Cm (247)	97 Bk (247)	98 Cf (251)	99 Es (252)	100 Fm (257)	101 Md (258)	102 No (259)	103 Lr (260)

Fig. 2.1 Periodic table of the elements (reprinted with permission from Dr. Thomas Basiri, Professor of Chemistry, Victor Valley College. Copyright 2008).

The periods extend left to right across the entire table. All periodic tables have a distinct line of separation running step-wise from boron/aluminum (B/Al) to polonium/astatine (Po/At). This line separates the metals to the left and the nonmetals to the right. The shaded elements bordering the line are called *metalloids* or *semi-metallics* because they possess both metallic (metal) and nonmetallic (nonmetal) properties. Metallic character decreases as you move left to right across a given period. Group IA and IIA elements are the most metallic and group VIIA and VIIIA are the most nonmetallic. A transition from metallic to nonmetallic character occurs through the group B elements and is the reason they are termed transition metals. The number above each symbol is called the atomic number and the number below each symbol is called the mass number. These numbers and their significance will be discussed below.

2.3 Atomic Structure

Elements are the *fundamental building blocks of matter.* Atoms are the *smallest, indivisible unit of an element that retains all chemical and physical properties* of the element. For example, a single atom of gold has the same physical and chemical properties as 10 tons of gold. Atoms are composed of three subatomic particles: protons, neutrons, and electrons. The protons and neutrons are located at the center of the atom in a region termed the nucleus. The mass number represents the *mass of the nucleus* or the total *mass of protons and neutrons.* The electrons are located in three-dimensional regions around the nucleus called *orbitals.* A large portion of any atom is empty space. The nucleus is surrounded by electrons in regions (orbitals) that are separated by great distances on a relative scale. Electrons located furthest from the nucleus, the *outermost electrons*, are called valence electrons and determine the chemical and physical properties of each element (Fig. 2.2).

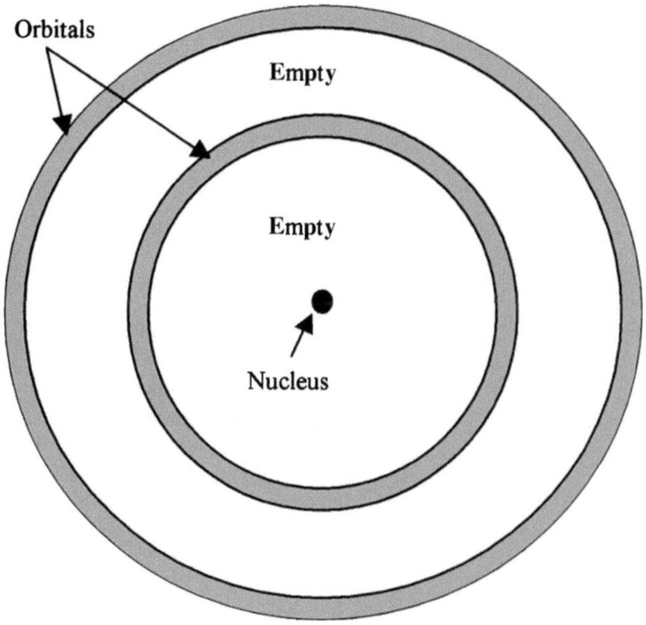

Fig. 2.2 The nucleus of subatomic particles is surrounded by electrons in orbital regions.

2.4 Subatomic Particles

Protons (p^+) are positively charged subatomic particles located in the nucleus. The *total number of protons* in the nucleus is given on the periodic table by the *atomic number* and *positively* identifies the element. A change in the number of protons changes the identity of the element; therefore, different elements *must* have different atomic numbers. A proton has a mass of one atomic mass unit (1 a.m.u $= 1.66 \times 10^{-24}$ g) and is comparable in size to a neutron.

Neutrons (n) are neutral subatomic particles also located in the nucleus. They carry no charge and therefore do not affect nuclear charge or the number of electrons in a neutral atom. Neutrons do contribute significantly to the mass of the atom (nuclear mass) because their mass is about the same as a proton. A change in the number of neutrons will change the mass number, but not the identity of the element. Atoms that contain the same number of protons but have different mass numbers are called *isotopes*. The number of neutrons contained in any nucleus is determined by subtracting the atomic number from the mass number.

Mass number = # of neutrons + # protons (atomic number)
Rearrange: # of neutrons = mass number – # protons (atomic number)

Electrons (e^-) are negatively charged subatomic particles located around the nucleus in predictable regions called *orbitals*. In a neutral atom, the number of electrons (negatives) is equal to the number of protons (positives). A change in the number of electrons in a neutral atom creates an ion, *an electrically charged atom*. Ions may carry a positive or negative charge depending on the number of electrons relative to the number of protons. The mass of an electron is approximately 2,000 times smaller than that of a proton or a neutron. As a result, electrons do not contribute significantly to the overall mass of the atom.

An atom is composed of two distinct regions – the nucleus and the region immediately surrounding the nucleus. The nucleus contains only protons and neutrons and is therefore positively charged. The region immediately surrounding the nucleus contains only electrons and is negatively charged. Protons and electrons are the only subatomic particles that are electrically charged. The charges are of equal magnitude despite the extreme disparity in mass, that is, an electron's charge will cancel a proton's charge. The overall charge on an atom is determined by comparing the number of protons to the number of electrons. If the number of electrons is *greater* than the number of protons (atomic number), the atom carries a *negative* charge; if the number of electrons is *less*, the atom carries a *positive* charge; if they are *equal*, the atom is *neutral*. The net charge on an atom in its standard state (natural conditions) is always zero (Fig. 2.3).

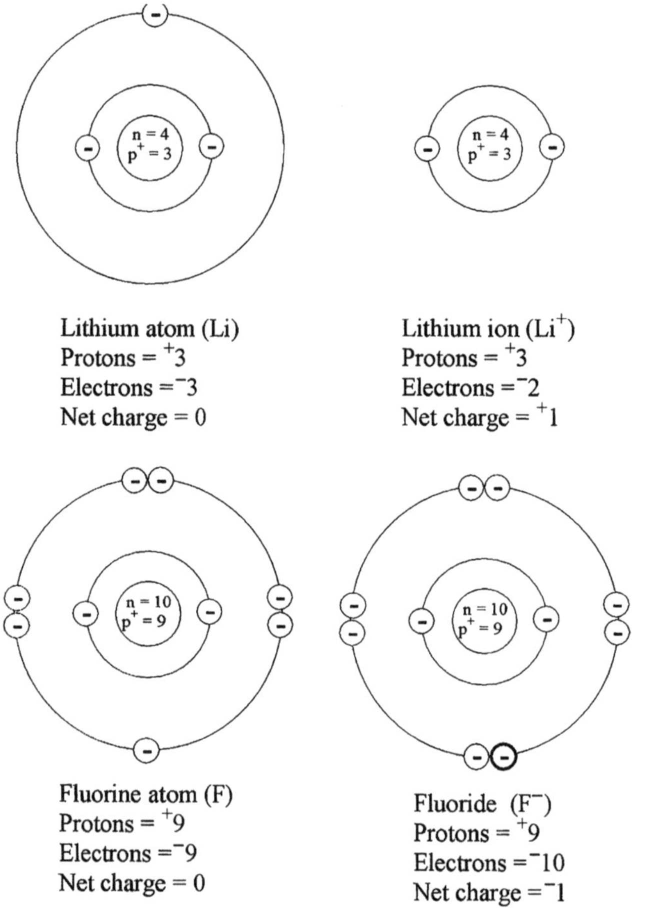

Fig. 2.3 Note how the net charge of protons and electrons determines if an atom's charge is neutral (0), positive (+), or negative (–).

2.5 The Arrangement of Electrons in an Atom

I grew up in a small town in upstate New York. On hot summer nights, our yard would fill with lightning bugs – interesting little flying creatures that periodically emit high-intensity light. When we were young, my sister and I would catch these elusive flying insects and put them in jars. We would punch out some small air holes in the lids, maybe throw in a piece of lettuce, and proudly display our new pets. Unfortunately, few ever saw the next night, a crushing reality to two small children. Imagine that you catch a single lightning bug and starve him for a few hours (incidentally, you did catch a male). You release your famished pet near a tree where you have graciously prepared a gourmet lightning bug dinner (you do, of course, know the diet of nocturnal insects!). You direct a camera at the food and take a time-lapsed photo over the remainder of the night (Fig. 2.4).

You would be very confident that, at any given time, there would be a high probability of finding the lightning bug in the region near the food. This statement is based on knowledge you possess, specifically, you know he is hungry. Your photo would most likely support your statement. You would see a region, or distribution, of light from the insect around the food. It is possible to find the lightning bug far from the food, but not likely given his current state of hunger. Electrons behave in a similar manner around the nucleus. It is not possible to know exactly where the electrons are; however, we can define regions where there is a high probability of finding them. Electrons are not randomly distributed around the nucleus; they are confined to specific energy levels called orbitals. At the subatomic level, it is not practical to use common units of length to measure distances, that is, it is not commonly stated that a specific electron may be 2 nm from the nucleus. Instead, we use energy to define distances. The electrons fill outward from the nucleus with the lowest energy, most stable electrons occupying regions close to the nucleus. The arrangement is very similar to an onion except great distances separate the

Fig. 2.4 The author's childhood experiment with lightning bugs shows a similarity with the relationship between electrons and the nucleus of an atom.

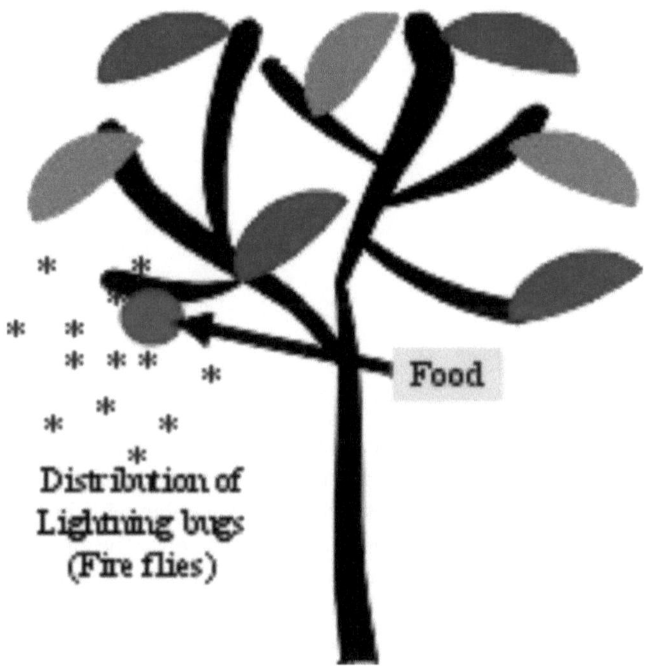

individual "peels." Thus, in comparing two electrons of different energy, it would be stated that electrons of higher energy are located further from the nucleus.

The energy of an electron is well defined and only certain energies are allowed (we call this quantized). If you have trouble with this statement, consider the musical tones created when you blow across the top of a bottle containing a fixed volume of water. It is not possible to create a full musical scale under these conditions. You hear only octaves or specific, allowable musical tones. The same can be said of an electron; it can have only specific, allowable energy values. This energy determines the location of the electron around the nucleus (remember, energy equates to distance; therefore, specific energy values translate to specific distances). The first levels of electron arrangement are "energy shells" called principal energy levels. They are given by the *period numbers* from the periodic table; for example, H and He represent *period 1* and have electrons in *principal energy level 1*, Li, Be, B, C, N, O, F, and Ne represent *period 2* and have electrons in *principal energy level 2*, and so forth. These regions (shells) increase in both size and energy as their distance from the nucleus increases. Principal energy level 1 is the lowest energy level (and smallest); therefore, electrons occupying this region are closest to the nucleus. Electrons occupying successively higher levels possess greater energy and are further from the nucleus. The *principal energy levels* are further *divided into sublevels*, or *orbitals*, designated *s*, *p*, *d*, and *f*. The orbitals are *regions around the nucleus where there is a high probability of finding an electron of specific energy*. Electrons occupy orbitals within principal energy levels and the energy of the electron determines which orbital it resides in and therefore its location around the nucleus. It may be helpful to compare electron arrangement to people staying at a hotel (I know…but bear with me!). First, hotels vary in size; some can have several floors with many rooms, while others have only a few floors with a small number of rooms (different atoms). Different people (the electrons) stay in different rooms (the orbitals) on different floors (the principal energy levels). Also, people are obligated to stay in only one room and generally cannot roam around and stay in any room they choose (specific energy of the electron).

2.6 Electron Configurations

Electron configurations illustrate the *arrangement of electrons around the nucleus of an atom*. The *aufbau principle* is used to construct electron configurations for ground-state (neutral) atoms and ions (*aufbau*: German for "build-up"). To determine the order that electrons fill around the nucleus, we must first construct the aufbau triangle (Fig. 2.5).

Fig. 2.5 An example of an aufbau triangle.

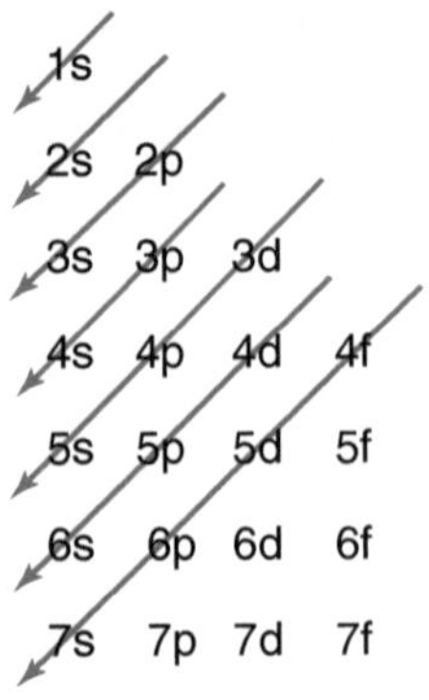

Draw the triangle as shown in Fig. 2.5; notice all rows contain the same number and all columns contain the same orbital (row three contains all 3's and the first column contains all *s*-orbitals). Draw parallel lines through the orbital designations as shown. Follow the arrows tail to head, beginning with 1s, and write the sequence: 1s2s2p3s3p4s3d…, etc. This is the sequence used to fill electrons around a nucleus. The electrons are located in orbitals (*s*, *p*, *d*, *f*) within principal energy levels (1, 2, 3, 4, etc.). To write an electron configuration, the number of electrons contained in the atom or ion must be calculated. Recall that protons and electrons are the only subatomic particles that carry a charge. In a neutral atom, the net charge is zero. This neutral state exists only when the number of electrons (negative charges) equals the number of protons (positive charges). Therefore, the number of electrons around the nucleus of a neutral atom is given by the atomic number. The following example illustrates how to write electron configurations. We will limit our discussion to *main group elements* only, that is, no *transition metals*.

Example: Write the electron configuration for Na.

First, determine the number of electrons in a neutral sodium atom. The periodic table gives an atomic number of 11 for Na. This means that there are 11 protons, or positive charges, in the nucleus of a sodium atom. Because no charge is written on the atom in our example, the number of electrons (negative charges) must also be 11.

Next, write a segment of the aufbau sequence.

$$1s2s2p3s3p4s3d$$

The number of electrons in each orbital is shown as a superscript attached to the orbital designation. The maximum number of electrons in an orbital: *s-orbital* is 2, *p-orbital* is 6, *d-orbital* is 10, and *f-orbital* is 14. The orbitals fill from *lowest to highest* energy and you *cannot* add electrons to higher levels until the preceding level is *full*. Revisiting our hotel analogy, you must fill the first floor before adding to the second; fill the second before adding to the third, etc. We start with the 1s orbital, principal energy level 1 containing a single s-orbital. We fill the orbital by placing a superscript 2 on the 1s designation (s-orbitals have a maximum occupancy of two electrons). Principal energy level 2 is filled next and contains a 2s and 2p orbital. The 2s orbital is filled in a manner similar to the 1s. The 2p orbital is filled using a superscript 6 attached (p-orbitals have a maximum occupancy of six electrons). We add the superscripts and find we have accounted for ten electrons. We have 11 electrons in total; so, the next orbital in our sequence (the 3s) will contain a single electron shown as a superscript of 1. The complete electron configuration for neutral Na is shown below.

$$1s^2 2s^2 2p^6 3s^1 3p4s3d$$

Our segment is too long so we simply erase the unused orbitals. If our segment was too short, we would add more orbitals to our sequence from the aufbau triangle.

The electron configuration for Na would be written as:

$$Na - 1s^2 2s^2 2p^6 3s^1$$

We would "read" this, 1s two, 2s two, 2p six, 3s one. The superscript on the last orbital depends on the number of electrons required to complete the configuration. It can be any number up to the maximum allowed in the orbital but can *never* exceed the maximum. Valence electrons are e*lectrons in the outermost principal energy level*. This may or may not be the last orbital written.

$$Na\left(11e^-\right) - 1s^2 2s^2 2p^6 3s^1$$

The outermost principal energy level containing electrons is level 3. Counting all electrons in level 3, we have 1 valence electron. In this case, the outermost level was the last one written.

$$O\left(8e^-\right) - 1s^2 2s^2 2p^4$$

If we examine the electron configuration for oxygen, we see that the outermost principal energy level is 2. We have two orbitals in level 2 containing a total of 6 valence electrons, 2 in the 2s and 4 in the 2p (add the superscripts). In this case, the outermost level included the last two orbitals written. Care must be taken when determining the number of valence electrons; do not immediately jump to the last orbital and use its superscript. By chance, sometimes it is the last orbital, but sometimes it is not. Valence electrons occupy the highest principal energy level (level...level), not orbital (I think that I'll stop beating that horse now!).

The valence electrons determine the chemical and physical properties of the element. If two atoms were brought together during the course of a chemical reaction, their first point of contact would be the electrons in the outermost levels. If we know the electron configuration, and, therefore, the valence configuration, we can make predictions on properties and reactivity. Core electrons are located in levels below the valence electrons and generally do not influence reactivity.

The electron configurations for group I elements Li, Na, and K are shown below.

$$Li\left(3e^-\right) - 1s^2 2s^1$$
$$Na\left(11e^-\right) - 1s^2 2s^2 2p^6 3s^1$$
$$K\left(19e^-\right) - 1s^2 2s^2 2p^6 3s^2 3p^6 4s^1$$

If we look closely at the configurations, we see that each ends with s^1, a single electron in the outermost energy level. It is no coincidence that all the above elements are members of group I. The group numbers on the periodic table represent the number of valence electrons for each member of the group. Group I elements have 1 valence electron, group II elements have 2, group III elements have 3, etc.

Some members of group II:

$$Be\left(4e^-\right) - 1s^2 2s^2$$
$$Mg\left(12e^-\right) - 1s^2 2s^2 2p^6 3s^2$$
$$Ca\left(20e^-\right) - 1s^2 2s^2 2p^6 3s^2 3p^6 4s^2$$

Some members of group VII:

$$F\left(9e^-\right) - 1s^2 2s^2 2p^5$$
$$Cl\left(17e^-\right) - 1s^2 2s^2 2p^6 3s^2 3p^5$$

Electron configurations for ions are constructed in a similar manner except the charge must be considered in determining the total number of electrons. For ions, the number of electrons is calculated by subtracting the charge on the ion (with its sign) from the atomic number.

$$\#e^-\ (\text{for ions}) = \text{atomic number}\ (\#\ \text{of protons}) - (\text{charge})$$

Using the above relationship, we can calculate the total number of electrons contained in each of the following ions:

$$N^{3-}\ \text{N has atomic number 7;}\ \#e^- = 7 - (^-3)\ \text{or}\ 10e^-$$
$$S^{2-}\ \text{S has atomic number 16;}\ \#e^- = 16 - (^-2)\ \text{or}\ 18e^-$$
$$Br^-\ \text{Br has atomic number 35;}\ \#e^- = 35 - (^-1)\ \text{or}\ 36e^-$$
$$Na^+\ \text{Na has atomic number 11;}\ \#e^- = 11 - (^+1)\ \text{or}\ 10e^-$$
$$Ca^{2+}\ \text{Ca has atomic number 20;}\ \#e^- = 20 - (^+2)\ \text{or}\ 18e^-$$

Notice that negative ions have more electrons than protons and positive ions have fewer electrons than protons. Ions are created by changing the number of electrons relative to the number of protons. It is worth noting that moving around electrons is not rocket science, if you have ever rubbed a balloon on your head and stuck it on a wall you have accomplished this miraculous feat. Below are the electron configurations for N^{3-}, O^{2-}, F^-, Ne, Na^+, and Mg^{2+}.

$$N^{3-}\left(10e^-\right)-1s^2 2s^2 2p^6$$
$$O^{2-}\left(10e^-\right)-1s^2 2s^2 2p^6$$
$$F^-\left(10e^-\right)-1s^2 2s^2 2p^6$$
$$Ne\left(10e^-\right)-1s^2 2s^2 2p^6$$
$$Na^+\left(10e^-\right)-1s^2 2s^2 2p^6$$
$$Mg^{2+}\left(10e^-\right)-1s^2 2s^2 2p^6$$

The above configurations are identical and each contains a total of 10 electrons with eight valence electrons (remember our dead horse; valence electrons occupy the highest principal energy level, which is 2 in this case). Although the configurations are identical, the charges on the ions vary and only one atom in our group is neutral. Neon is a "noble gas" or "inert gas," names given to all group VIIIA elements. Chemical reactivity is a stability-driven process – if products are more stable than reactants (starting material), the reaction occurs. The noble gases are extremely stable and show very little reactivity because of filled outer shell configurations. In the above examples, principal energy level 2 (the outer level or shell) is full; principal energy level 2 has only s and p orbitals. The addition of a single extra electron to any of the above configurations would require occupancy in the next higher energy shell, specifically the 3s orbital. A filled outer shell configuration is achieved when all orbitals in the outermost level are full. Generally, this is achieved with eight valence electrons. *The tendency of atoms to gain or lose electrons to obtain electron configurations similar to group VIII elements is called the* octet rule. The configurations above, all satisfy the octet rule and represent the elements in their most stable forms. The charges on the atoms result from a gain or loss of electrons in order to achieve a configuration identical to Ne. There are exceptions to the octet rule. For example, helium (He) is a group VIIIA element that has an electron configuration of $1s^2$. Principal energy level one contains only the 1s-orbital and requires only two electrons to fill. Helium satisfies this condition and has therefore achieved its octet.

There is a difference between the ground state of an element and its most stable state. The ground state is how the atom is most often found in nature and occurs when the atom is neutral. The most stable state is when the atom has achieved an octet, or filled outer shell configuration. This will generally require the atom to carry an overall net charge as a result of gaining or losing electrons. The group numbers provide us with important information on both states. We know that the number of valence electrons for a particular atom is given by the group number. However, what may have escaped our attention is that the group number gives the valence electrons for neutral atoms only; group I elements have 1 valence electron, group II have two, etc. Let us revisit our configurations above for N^{3-}, O^{2-}, F^-, Ne, Na^+, and Mg^{2+}; all contain eight valence electrons regardless of their group number, but none are neutral except Ne (you guessed it, a group VIIIA element). Consider the following examples, paying close attention to the differences in electron configuration and how this relates to charge in the most stable state:

Natural state of sodium – Na ($11e^-$) – $1s^2 2s^2 2p^6 3s^1$
Most stable state of sodium – Na^+ ($10e^-$) – $1s^2 2s^2 2p^6$
Natural state of magnesium – Mg ($12e^-$) – $1s^2 2s^2 2p^6 3s^2$
Most stable state of magnesium – Mg^{2+} ($10e^-$) – $1s^2 2s^2 2p^6$
Natural state of aluminum – Al ($13e^-$) – $1s^2 2s^2 2p^6 3s^2 3p^1$
Most stable state of aluminum – Al^{3+} ($10e^-$) – $1s^2 2s^2 2p^6$
Natural state of nitrogen – N ($7e^-$) – $1s^2 2s^2 2p^3$
Most stable state of nitrogen – N^{3-} ($10e^-$) – $1s^2 2s^2 2p^6$
Natural state of atomic oxygen – O ($8e^-$) – $1s^2 2s^2 2p^4$
Most stable state of atomic oxygen – O^{2-} ($10e^-$) – $1s^2 2s^2 2p^6$
Natural state of atomic fluorine – F ($9e^-$) – $1s^2 2s^2 2p^5$
Most stable state of atomic fluorine – F^- ($10e^-$) – $1s^2 2s^2 2p^6$

Relating the group number to the octet rule can provide information on the most stable state of group members. Group I elements have one valence electron and can obtain an octet (eight electrons in the outermost principal energy level) in two ways: lose one electron and take a $^+1$ charge or gain seven electrons and take $^-7$ charge. The more favorable choice would be to lose one electron and become $^+1$. Groups IA, IIA, and IIIA lose electrons to achieve their octets and take charges of $^+1$, $^+2$, and $^+3$, respectively, in their most stable forms. Group IV is unusual and will not be discussed at this point. Groups VA, VIA, and VIIA will gain electrons to complete their octets, taking charges of $^-3$, $^-2$, and $^-1$, respectively, in their most stable forms. We can summarize these observations by stating: *to satisfy the octet rule and achieve stable electron configurations, metals must lose electrons and nonmetals must gain electrons.*

Table 2.1 Orbitals within each principal energy level and the maximum number of electrons contained in each

Orbital		Maximum number of electrons	
s		2	
p		6	
d		10	
f		14	
Principal energy level	Number of orbitals	Orbital designation	Maximum number of electrons
1	1	1s	2
Maximum number of electrons in principal energy level 1 is 2			
2	2	2s	2
		2p	6
Maximum number of electrons in principal energy level 2 is 8			
3	3	3s	2
		3p	6
		3d	10
Maximum number of electrons in principal energy level 3 is 18			
4	4	4s	2
		4p	6
		4d	10
		4f	14

Table 2.2 Electron distribution in atomic orbitals

Principal energy level (n)	Formula $2n^2$	Maximum number of electrons
1	$2\times(1^2)$	2
2	$2\times(2^2)$	8
3	$2\times(3^2)$	18
4	$2\times(4^2)$	32

Table 2.1 shows the orbitals within each principal energy level and the maximum number of electrons contained in each. Notice that not all principal energy levels contain all orbitals.

The maximum number of electrons contained in a principal energy level can be calculated using the formula $2n^2$, where n is the principal energy level (period number from the periodic table) (Table 2.2).

2.7 Periodic Trends: Understanding the Periodic Table

The periodic table is an arrangement of the elements based on similarities in atomic properties. It can therefore be used to predict the chemical and physical properties of elements. Electronegativity is a measure of an atom's *desire for electrons* or the ability of an atom to draw electrons toward it. Fluorine (F) is the most electronegative element on the periodic table and, in general, the periodic trend is that the closer an element is to fluorine, the greater is its electronegativity. If asked which element, Br or Ca, has the greater ability to draw electrons, you would respond Br because it is closer to fluorine. The nonmetals are grouped near fluorine on the periodic table and must gain electrons to achieve octets; as a result, they have high electronegativities. Metallic elements, particularly groups IA and IIA, are not near fluorine on the periodic table. They must lose electrons to achieve octets and therefore have low electronegativities. The importance of electronegativity in chemical bonding cannot be overstated and a good rule of thumb is "the closer an element is to fluorine on the periodic table, the greater is its electronegativity"

Theoretically, the atomic radius of an atom is the distance from the center of the nucleus to the outer boundary of the atom. However, regions containing the valence electrons (outermost) do not have distinct boundaries. Not to worry; when two atoms of the same element are bound together, the centers of their nuclei are separated by a measurable distance. Therefore, the atomic radius of an atom is *defined as half the distance between the centers of two bonded atoms of the same element*; for example, the atomic radius of a single hydrogen atom is equal to half the distance from the centers of two bonded hydrogen atoms. Atomic radii increase down a given group of elements and, in general, decrease left to right across a period (Fig. 2.6).

Atomic radius decreases →

Atomic radius decreases ↑ (down the left side) — Electronegativity increases ↑ (up the right side)

Period	1 IA	2 IIA	3 IIIB	4 IVB	5 VB	6 VIB	7 VIIB	8 VIIIB	9 VIIIB	10 VIIIB	11 IB	12 IIB	13 IIIA	14 IVA	15 VA	16 VIA	17 VIIA	18 VIIIA
1	1 H 1.01																	2 He 4.00
2	3 Li 6.94	4 Be 9.01											5 B 10.81	6 C 12.01	7 N 14.01	8 O 16.00	9 F 19.00	10 Ne 20.18
3	11 Na 22.99	12 Mg 24.31											13 Al 26.98	14 Si 28.09	15 P 30.97	16 S 32.06	17 Cl 35.45	18 Ar 39.95
4	19 K 39.10	20 Ca 40.08	21 Sc 44.96	22 Ti 47.88	23 V 50.94	24 Cr 52.00	25 Mn 54.94	26 Fe 55.85	27 Co 58.93	28 Ni 58.71	29 Cu 63.55	30 Zn 65.38	31 Ga 69.72	32 Ge 72.59	33 As 74.92	34 Se 78.96	35 Br 79.90	36 Kr 83.80
5	37 Rb 85.47	38 Sr 87.62	39 Y 88.91	40 Zr 91.22	41 Nb 92.91	42 Mo 95.94	43 Tc 98.91	44 Ru 101.1	45 Rh 102.9	46 Pd 106.4	47 Ag 107.87	48 Cd 112.4	49 In 114.8	50 Sn 118.7	51 Sb 121.8	52 Te 127.6	53 I 126.90	54 Xe 131.3
6	55 Cs 132.9	56 Ba 137.33	57 La* 138.9	72 Hf 178.5	73 Ta 180.9	74 W 183.9	75 Re 186.2	76 Os 190.2	77 Ir 192.2	78 Pt 195.1	79 Au 196.97	80 Hg 200.59	81 Tl 204.4	82 Pb 207.2	83 Bi 209.0	84 Po (209)	85 At (210)	86 Rn (222)
7	87 Fr (223)	88 Ra (226.0)	89 Ac** (227)	104 Rf (261)	105 Db (262)	106 Sq (263)	107 Bh (262)	108 Hs (265)	109 Mt (266)									

Electronegativity increases →

Fig. 2.6 Understanding periodic trends.

2.8 Isotopes

Atoms that contain the same number of protons but have different mass numbers are called isotopes. Isotopes of an element differ only in the number of neutrons contained in the nucleus. Typically, atomic nuclei are most stable when they contain a certain number of protons and neutrons. The addition of neutrons to the nucleus increases the mass of the atom and creates instability. For example, the most abundant form of hydrogen contains a single proton in the nucleus. The addition of a neutron to a hydrogen nucleus creates an isotope of hydrogen called deuterium. Deuterium is a heavier and more energetic form of hydrogen, and is therefore less stable. The addition of a second neutron creates a third isotope of hydrogen called tritium, the most unstable and active form of hydrogen. The instability of isotopes is a direct result of increased nuclear mass and is detected through a release of energy called nuclear radiation (Fig. 2.7).

2.9 Radioactivity

Dig a ditch in 100° or take a nap on the couch; well, there is a tough choice. The relationship between energy and stability is a common thread that unifies most areas of science (and based on your response to my question, it appears that it extends into our daily lives as well). High energy translates to instability and there is always a natural tendency toward lowest energy and greatest stability (the nap on the couch). This is inescapable; however, I would not try this argument the next time your asked to do yard work, it does not work, believe me. The response of an atom to high energy is not much different from our own. It will not remain unstable indefinitely; eventually, the nucleus will emit energy in an effort to regain stability (its version of a nap). The *spontaneous emission of high-energy nuclear radiation from an unstable nucleus* is termed radioactivity (or radioactive decay). Atoms that exhibit this property are said to be radioactive and most elements with an atomic number of 90 or greater have radioactive isotopes. Early experiments identified three types of nuclear radiation: alpha (α), beta (β), and gamma (γ) rays. A sample of radioactive material is placed between the positive and negative poles of a magnet and emitted radiation is detected using a piece of X-ray film placed at the top of the apparatus (Fig. 2.8).

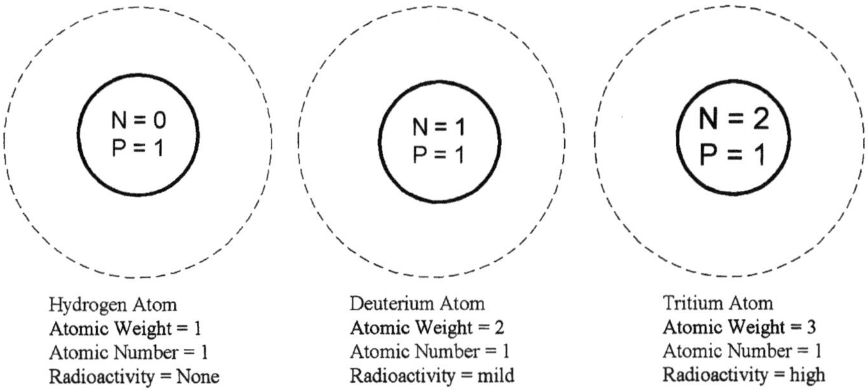

Fig. 2.7 As the nuclear mass of an element increases, so does the isotopes instability.

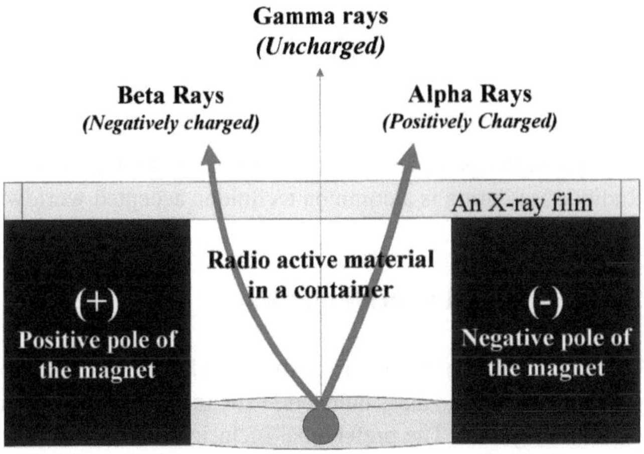

Fig. 2.8 A magnet is used to measure the emission of radioactivity from sample material.

Three spots were observed at different locations on the film. Two spots were deflected toward opposite poles of the magnet, whereas the third passed straight through, apparently unaffected. This implies that two of the particles are electrically charged and the third is neutral. Alpha (α) rays are positively charged particles that are deflected toward the negative pole and beta (β) rays are negatively charged particles deflected toward the positive pole. Gamma (γ) rays have no detectable charge (or mass) and therefore passed straight through.

2.10 Types of Radioactive Decay

The release of a helium nucleus (He^{2+}) during radioactive decay is called α-decay (alpha decay). This type of decay is a low-energy emission of positively charged particles. A thin sheet of paper will provide adequate protection against this type of radiation. The release of electrons (e^-) during radioactive decay is called β-decay (beta decay). This type of decay produces negatively charged particles of medium energy. A few hundred sheets of paper are required to provide adequate protection against this type of radiation. The release of electromagnetic radiation during radioactive decay is called γ-decay (gamma decay). This type of decay produces high-energy, neutral radiation capable of penetrating a 1-inch-thick wall of lead. This is the most dangerous and destructive form of radioactive decay.

2.11 Nuclear Radiation: Forensic Applications

Radioactive isotopes will lose intensity (gain stability) over time because of α-, β-, or γ-decay. The amount of time required for radioactive intensity to decrease by half is called the half-life. Carbon-14 is a radioactive isotope of carbon with a half-life of 5,720 years. A 100-g sample of radioactive carbon-14 will contain 50 g of active carbon-14 in 5,720 years, 25 g after an additional 5,720 years, and so on. Half-lives can range from fractions of a second to millions of years, depending on the isotope.

Table 2.3 The atomic number, atomic mass, and molar mass of selected elements

Element	Atomic number	Mass of 1 Atom (atomic mass)	Mass of one mole (molar mass)
H (hydrogen)	1	1.01 a.m.u.	1.01 g
He (helium)	2	4.00 a.m.u.	4.00 g
Li (lithium)	3	6.94 a.m.u	6.94 g
C (carbon)	6	12.01 a.m.u.	12.01 g
N (nitrogen)	7	14.01 a.m.u.	14.01 g
O (oxygen)	8	16.00 a.m.u.	16.00 g
F (fluorine)	9	19.00 a.m.u.	19.00 g
Na (sodium)	11	22.99 a.m.u.	22.99 g
P (phosphorous)	15	30.97 a.m.u.	30.97 g
Cl (chlorine)	17	35.45 a.m.u.	35.45 g
I (iodine)	53	126.90 a.m.u.	126.90 g

Forensic anthropologists use this information to determine the age of ancient artifacts, mummies, bones, and other material. Radioactive dating is a common technique accepted worldwide.

2.12 The Mole and Molar Mass

The atomic mass of carbon from the periodic table is 12.01, but 12.01 of "what"? Curiously, no "mass" units are given on the periodic table with "mass numbers." The reason is that mass numbers can have two equally important units: *atomic mass units* (a.m.u.) or *grams*. The preferential inclusion on the table of one unit over the other would undoubtedly spark a never-ending debate, dividing educators and authors worldwide. To avoid this debacle, and the certain demise of the modern world, no units are given; after all, the last thing we need is another source of debate. I momentarily digress, let us return to carbon: 12.01 amu's of carbon represents the mass of *one carbon atom*, 12.01 g of carbon represents the mass of 6.02×10^{23} *atoms* of carbon. The mass number in amu's of any element represents the mass of one atom of the element, whereas the mass number in grams represents the mass of 6.02×10^{23} atoms of the element. The mass "numbers" are the same; it is the units that distinguish the difference. If you were asked how many pencils are in a dozen pencils, you would reply 12. We associate the word "dozen" with the number "12" and define a dozen as anything that contains 12 "things." The same is true of a mole, an extremely important quantity used in chemistry. A mole is defined as anything that contains 6.02×10^{23} particles or "things." We associate the number 6.02×10^{23} with the word "mole." We can simplify our example above by stating: *the atomic mass of any element, in grams, contains* 6.02×10^{23} *atoms of the element, or one mole of the element, and is called the molar mass*. The quantity that defines a mole, 6.02×10^{23}, is called Avogadro's number in honor of its founder, the nineteenth-century Italian scientist Amadeo Avogadro.

2.13 Elements of Forensic Interest

See Table 2.3 for the elements of forensic interest.

2.14 Questions

1. Write the names of the elements represented by the following symbols:
   (a) I
   (b) P
   (c) Na

2. Write the symbols for the following elements:
   (a) Potassium
   (b) Nickel
   (c) Manganese
   (d) Magnesium
3. Name the three types of subatomic particles and give their location in the atom.
4. Provide the mass that contains:
   (a) One atom of carbon
   (b) One mole of magnesium
   (c) 6.02×10^{23} atoms of Li
   (d) 3.01×10^{23} atoms of Ca
5. Please explain to the members of the jury how two atoms of the same element can have different mass numbers.
6. Define radioactivity and the three types of nuclear radiation.
7. Cite a few examples of the application of radioactive decay to forensic investigation.
8. Explain the aufbau principle.
9. Give the maximum number of electrons in:
   (a) Principal energy level 2
   (b) A *p*-orbital
   (c) Principal energy level 4
   (d) The 4f-orbital
   (e) The 1s-orbital
10. Briefly explain to the members of the jury the difference between a neutral atom and an ion. How are ions formed?
11. Write the electron configuration for each of the following:
    (a) Na
    (b) F^-
    (c) Mg^{2+}
    (d) Li^+
    (e) Ar
12. Explain why the electron configurations for N^{3-}, O^{2-}, F^-, Ne, Na^+, and Mg^{2+} are identical.
13. What information does the group numbers of the periodic table give?
14. Describe the difference between the natural state of an atom and its most stable state.
15. Describe the periodic trends of electronegativity and atomic radius.

Suggested Reading

Jones, L.; Atkins, P. *Chemistry: Molecules, Matter, and Change*, 4th ed.; W.H. Freeman and Company: New York, 2002; pp 298–299, pp 959–964.

3 Molecules

3.1 Introduction

Compounds are formed through the combination of two or more elements held together by chemical bonds. The number and identity of each atom present in the compound are given by the chemical formula. Symbols from the periodic table are used to identify atoms, and the relative number of each atom present is indicated using a subscript attached to the symbol. Subscripts are used only when two or more atoms of the same element appear in the formula. The symbol without a subscript is used to represent the presence of a single atom in the formula. For example, H_2O is the chemical formula for water, a compound containing *one* atom of *oxygen* bound to *two* atoms of *hydrogen*. Compounds are electrically neutral and divided into two broad classes based on the type of chemical bond present: *ionic bonds* form ionic compounds and *covalent bonds* form covalent compounds. It is important to note that pure ionic and pure covalent represent the extremes of chemical bonding and rarely exist. The vast majority of chemical bonds contain both ionic and covalent character and classification is based on the type present in the highest percentage. For example, a bond that contains a higher percentage of ionic character is termed ionic; however, this does not mean that the bond contains no covalent character. Characterizing chemical bonds as ionic or covalent is a common, reliable practice that is universally accepted. This convenient language will be used to study bonding and structural properties in this chapter.

3.2 Chemical Bonding

Valency is *the number of bonds that a particular atom must form to achieve a neutral state*. It is directly related to the octet rule and measures an atom's ability to gain, lose, or share electron(s) when forming chemical bonds. We have already established a relationship between group number, valence electrons, and an atom's ability to gain or lose electrons (the octet rule). We may therefore confidently predict that a relationship between group number and valency exists; we are correct indeed. Although the valency of many elements is considered fixed, there are exceptions. These cases rarely have applications in forensic chemistry and, therefore, will not be discussed.

3.2.1 Ionic Bonds

Ionic bonds are *electrostatic forces of attraction between two ions resulting from a transfer of electrons*. This is an elaborate way of saying "opposites attract." This type of chemical bond is very similar to the attractive forces that hold the opposite poles of two magnets together. Ionic bonds are usually formed between *metals* and *nonmetals*, that is, between elements with low electronegativities and elements with high electronegativities. The metal will transfer electrons to the nonmetal resulting in a net charge on both as the ions achieve octets. For example, the chemical formula for common table salt is NaCl. The periodic table shows Na (a metal) in group IA and Cl (a nonmetal) in group VIIA; recall the division line on the periodic table separating metals to the left and nonmetals to the right. The Na atom transfers its single valence electron to Cl and takes a charge of positive one ($^{+}1$); the Cl atom accepts the electron and takes a charge of negative one ($^{-}1$). The two ions are attracted (because opposites attract) and an ionic bond is formed. In the mutually beneficial transfer, the octets of both atoms are satisfied.

J.I. Khan et al., *Basic Principles of Forensic Chemistry*, DOI 10.1007/978-1-59745-437-7_3,
© Springer Science+Business Media, LLC 2012

In reality, Na has no choice. A "conversation" between the two atoms might contain the following dialog: the highly electronegative Cl atom says, "I'm stronger than you, I want your electron and there's not much you can do about it." Na atom responds, "go ahead and take that electron, it's so far from the nucleus I have trouble keeping track of it anyway." The Cl atom steals or snatches the electron and both ions achieve an octet as a result. The electron transfer is illustrated in Fig. 3.1 (dots represent valence electrons).

Ionic compounds exist as crystal lattice structures containing ions packed together and held by ionic bonds. The ratio of ions in the crystal is given by the chemical formula. For example, common table salt (NaCl) exists as a crystal containing a large number of Na^+ and Cl^- ions. The total number of ions may vary from crystal to crystal; however, the ratio of Na^+ to Cl^- is given by the chemical formula and is one to one. In this model, the existence of a single molecule of NaCl is somewhat obscure. Therefore, it is generally improper and incorrect to refer to any compound containing a metal as a molecule. Accordingly, chemical formulas for ionic compounds represent formula units, not molecules. There are a few exceptions; but they are rare indeed. We shall see that the term "*molecule*" is reserved for compounds containing only *nonmetals*.

Ionic bonds are attractive forces between oppositely charged ions. The charges result from a transfer of electrons from elements of low electronegativity to elements of high electronegativity. A transfer of electrons is not the only method used by atoms to achieve an octet.

3.2.2 Covalent Bonds

Covalent bonds are formed when two nuclei *share electrons*. This type of chemical bond is typically found in compounds containing only *nonmetals*. Consider the formation of a hydrogen molecule, H_2 (Fig. 3.2). Hydrogen is *not* a metal despite

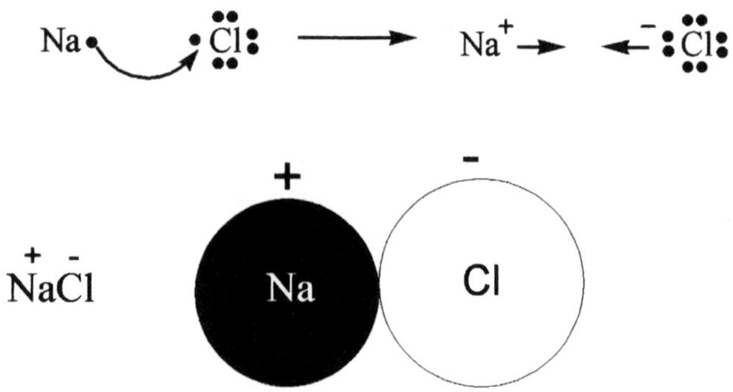

Fig. 3.1 Electron transfer in the formation of an ionic bond and the resulting formula unit. The ions are attracted by magnetic forces and arranged symmetrically into a crystal.

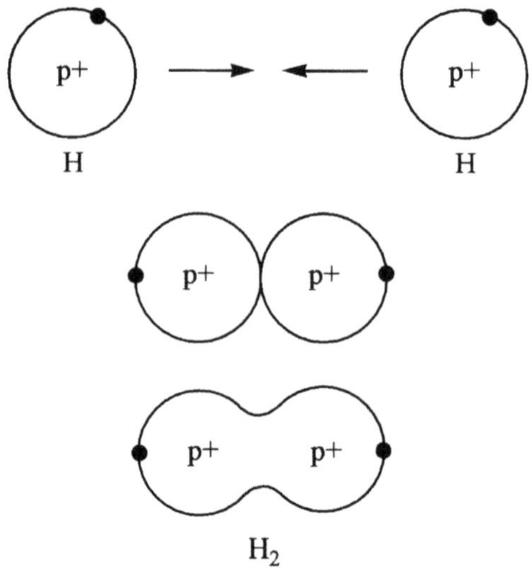

Fig. 3.2 Valence electrons from separate atoms are attracted into the internuclear space forming a molecule. The equally shared electrons form a covalent bond linking the two nuclei.

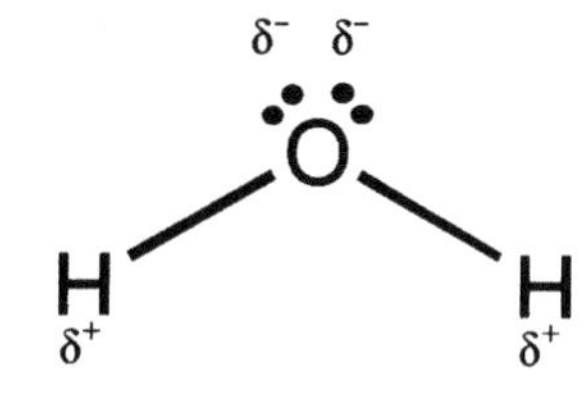

Fig. 3.3 The polarity of water results from the magnetic properties of the two polar covalent bonds between hydrogen and oxygen. The nonbonding electrons on oxygen (*dots*) repel the bonding electrons between hydrogen and oxygen (*lines*) producing the observed bent geometry.

its location on the periodic table. It is a member of group I simply because it has an electron configuration similar to group I elements ($1s^1$).

As the two hydrogen atoms approach one another, the valence electron from one atom begins to "feel" the positive force from the other nucleus and vice versa. The electronegativity of the two H atoms is the same; therefore, they each pull on the electrons with the same force. Each nucleus is not strong enough to pull the other's electron away, nor will it give up its own. The atoms continue to approach each other until the electrons orbit both nuclei. The electrons, which were once single valence electrons to individual nuclei, are now valence electrons to both nuclei; they are *co-valence*, and a *covalent bond* is created. Sharing electrons has satisfied the octets of both nuclei and the H_2 molecule is more stable than the individual hydrogen atoms alone. *Covalent compounds* exist as *molecules* and therefore have *molecular formulas*. They are easily distinguished from ionic compounds because no metal is present in the chemical formula.

The models of H_2 and NaCl represent extremes in chemical bonding and, in truth, most chemical bonds possess characteristics of both. Different elements have different electronegativities; therefore, no two elements on the periodic table have the same "desire for electrons." In a covalent bond between two different nonmetals, the atoms will not pull on the electrons with equal force. The tug of war will be won by the more electronegative element and will result in a slight distortion of the electron's path around the two nuclei. The shared electrons are pulled toward the more electronegative element, creating a region of slight negative charge. As a result, the region toward the less electronegative element will be slightly positive. The difference in electronegativity is not sufficient to cause an actual transfer of electrons; it merely creates a distorted path resulting in more electron density around the more electronegative element. This is called a polar covalent bond and is very similar to a weak bar magnet.

3.2.3 Polar Bonds

The term "polar covalent" is very simple to justify. The bond is very similar to a weak bar magnet; it has a "north and south pole," thus the term "polar." The bond results from a sharing of electrons, or, more specifically, an unequal sharing, thus the term "covalent." It is often helpful to associate the common characteristics of a simple magnet with words such as polar, polarity, dipole, and dipole moment.

Water contains a particularly common example of bond polarity (Fig. 3.3). Water contains two *polar covalent bonds* between oxygen and hydrogen. The bonding electrons are *not* shared equally between the two nuclei and are pulled toward the more electronegative oxygen (oxygen is closer to fluorine). This creates a slightly negative region on the oxygen and a slightly positive region on the hydrogen. The regions are represented above by the Greek letter δ (lowercase delta), meaning "slightly." It is important to note that δ^+ and δ^- are not fully developed $^+1$ and $^-1$ charges like those found in ionic bonds. The electronegativity difference between oxygen and hydrogen is not sufficient to cause a complete transfer of electrons. The distribution of the electrons between the two nuclei is simply distorted more toward the oxygen and less toward hydrogen. The result is a polar covalent bond.

The unequal sharing of bonding electrons in polar bonds may create polar molecules. The permanent dipoles in polar molecules can form weak bonds between adjacent molecules. These intermolecular bonds ("between molecules") are noncovalent (no sharing) in nature and may be quite extensive. One type of this interaction is termed hydrogen bonding due to the involvement of polar bonds containing hydrogen.

3.2.4 Hydrogen Bonding

A common example of hydrogen bonding is the association of water molecules in solution. The slightly negative oxygen on one molecule is weakly attracted to the slightly positive hydrogen on another (Fig. 3.4). Hydrogen bonding is responsible for many of the unusual properties of water, i.e.; high surface tension, high heat capacity, high boiling point, etc.

3.2.5 Multiple Bonds

Covalent bonds may be single, double, or triple, depending on the number of shared electrons. A single bond is formed when only two electrons are shared between the two bonded nuclei. The electrons may be donated from each atom or both from a single atom. Regardless of the source, they are shared equally by each nucleus as octets are achieved in both atoms. Single bonds are generally longer and weaker compared with double or triple bonds.

The structures in Fig. 3.5 illustrate single bonds. A pair of electrons (pair of dots) between two symbols represents a single covalent bond. It is also common to use a solid line to represent a covalent bond. The two structures above for methane (CH_4) and water (H_2O) are different representations of the same molecule. In methane, four pairs of electrons (or the four solid lines) represent the single covalent bonds formed between one carbon and four hydrogen atoms. The same is true for water, two pairs of electrons (or two solid lines) between oxygen and both hydrogen atoms illustrate the bonding arrangement. The remaining electron pairs on oxygen are termed *nonbonding pairs*; they are not located between two nuclei and therefore do not participate in chemical bonding.

Double bonds are formed when four electrons are shared between two combining atoms. Double bonds are shorter and stronger than single bonds and examples are found in oxygen (O_2) and carbon dioxide (CO_2) (Fig. 3.6).

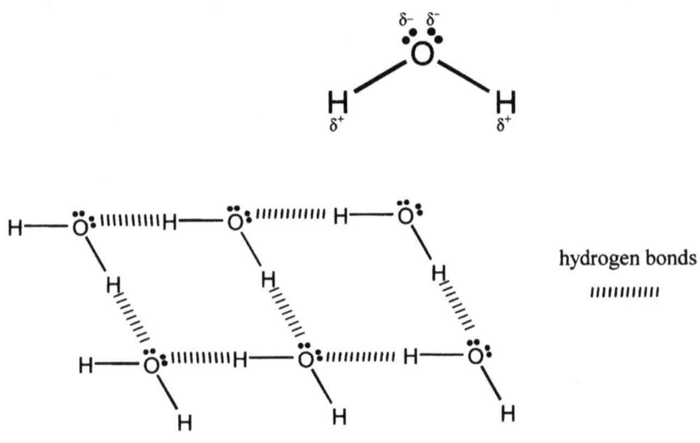

Fig. 3.4 Hydrogen bonding is a weak, noncovalent interaction between adjacent water molecules. Each molecule is capable of forming four intermolecular bonds: one at each hydrogen and one at each nonbonding electron pair on oxygen. The intermolecular bond length varies with the physical state of water.

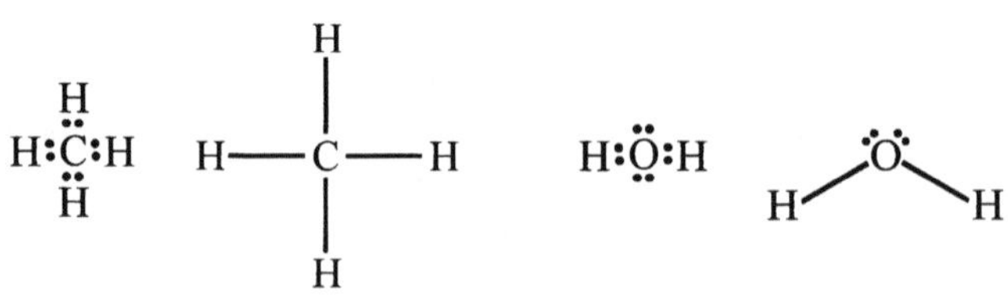

Fig. 3.5 Covalent bonds are commonly illustrated using *dots* (electrons) or *lines*. The use of lines provides insight into molecular geometry with each line representing a pair of bonding electrons.

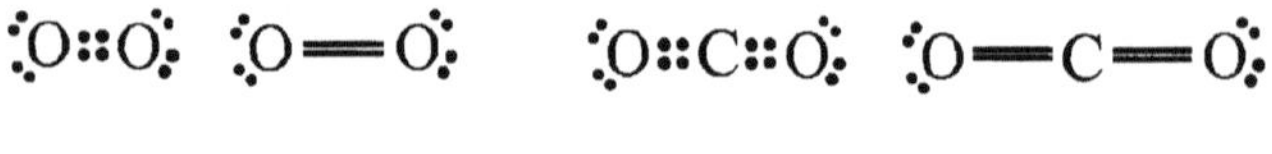

Fig. 3.6 Two atoms can share more than one pair of electrons to achieve an octet for each. A double bond results when two pairs are shared. The number of atoms with the capacity to form double bonds is small and generally limited to carbon (C), nitrogen (N), oxygen (O), sulfur (S), and phosphorus (P).

Fig. 3.7 Triple bonds contain the highest density of bonding electrons between two nuclei. The three pairs of shared electrons represent the upper limit on internuclear occupancy. Bonds containing four pairs of shared electrons between two nuclei do not exist (or have yet to be discovered).

:N:::N: :N≡N:

Triple bonds are formed when six electrons are shared between two combining atoms. These bonds are the shortest and strongest of the three types and found in compounds such as nitrogen (N_2) (Fig. 3.7).

3.3 Predicting Bond Types

3.3.1 Nonpolar Covalent Bonds

1. Bonds formed between two nonmetals that are the same element.
2. Diatomic molecules are the only examples of pure nonpolar covalent bonds: H_2, N_2, O_2, F_2, Cl_2, Br_2, and I_2.
3. The electrons are symmetrically distributed between the two nuclei and therefore no magnetic "poles" exist in the bond.

3.3.2 Polar Covalent Bonds

1. Bonds formed between two different nonmetals.
2. These bonds have permanent dipoles (poles) because the electrons are not symmetrically distributed between the two nuclei.
3. The more electronegative element (the one closest to fluorine) will bear a slight negative charge, and the less electronegative element will be slightly positive.
4. These bonds have the same properties and characteristics as weak bar magnets.

3.3.3 Hydrogen Bonds

1. Weak, intermolecular (between molecules) forces of attraction between molecules containing polar covalent bonds.
2. The polar covalent bonds must contain hydrogen.
3. They are noncovalent (no sharing of electrons) and electrostatic in nature (opposites attract).
4. They bridge adjacent molecules and can influence chemical and physical properties.

3.3.4 Ionic Bonds

1. Bonds formed between metals and nonmetals.
2. There is an actual transfer of electrons resulting in ion formation.
3. They are electrostatic forces of attraction (opposites attract).
4. The strongest ionic bonds are formed between group I and group VII elements.

3.4 Molar Mass

In our study of atoms, it was determined that the molar mass of an element is simply the atomic mass in grams. For compounds, the molar mass is calculated from the formula mass: the *sum total of the individual atomic masses of all elements present in the chemical formula*. For example, the formula mass of H_2O is 18, the sum of the atomic masses of one oxygen (16) and two hydrogens (2). The formula mass of NaCl is 58.45, the sum total of one Na atom (23) and one Cl atom (35.45). The formula mass can have units of amu's or grams; *18 amu's of water* is the mass of *one molecule of water* and *18 g of water* is the mass of 6.02×10^{23} molecules, or *one mole of water*. The chemical formula for NaCl contains a metal; it is therefore an ionic compound and does not exist as a molecule. Although the concept of the mole is consistent, we must use slightly different terminology to describe it: *58.45 amu's of NaCl* represents the mass of *one formula unit*, whereas *58.45 g* represents the mass of 6.02×10^{23} formula units, or *one mole of NaCl*. Ionic compounds are described in terms of formula units and covalent compounds in terms of molecules. This terminology, although technically correct, can be a source of confusion. But all is not lost, because the molar mass does not distinguish between ionic and covalent compounds; it is simply the *formula mass in units of grams*.

3.5 Molarity

A solution of saltwater is called a binary solution because it contains only two components – salt and water. In any binary solution, the component present in the greatest amount is termed the solvent, while the component present in the least amount is termed the solute. A solution of saltwater contains water as the solvent and salt as the solute. What quantity of salt is in our solution? Did we dissolve one teaspoon, two, or perhaps three? Often, it is important to know the exact concentration of solute in solution. Molar concentration or molarity (M) *is a unit of concentration defined as the number of moles of solute per liter of solution*. The molar mass of common table salt is 58.45 g/mol. A 1-M (one molar) salt solution would be prepared by dissolving 58.45 g of table salt in enough water to make one liter of total solution. *We do not add one liter of water* because a solution contains both the solvent and solute. The solid table salt will take up space in solution, so we add only the amount of water required to make one liter of total solution.

3.6 Chemical Reactions

A chemical reaction is *any process that results in a chemical change*. Reactions are represented by balanced chemical equations that *illustrate the quantitative relationship between starting materials (reactants) and products*.

$$\text{Reactants} \rightarrow \text{Products}$$

By convention, an arrow separates the two sides of a chemical equation; the *reactants* are written on the *left side* and the *products* on the *right*. The arrow always points from the reactants to the products and indicates that a reaction has taken place. It may be helpful to interpret the arrow as "react to form" or "forms." Heating elemental magnesium in the presence of oxygen gas forms solid magnesium oxide. This reaction is represented by the following chemical equation:

$$\text{magnesium(solid)} + \text{oxygen(gas)} \xrightarrow{\text{heat}} \text{magnesiumoxide(solid)}$$

Substituting formulas, we have the chemical equation:

$$Mg_{(s)} + O_{2(g)} \xrightarrow{\text{heat}} MgO_{(s)}$$

The law of conservation of mass states that *mass (atoms) cannot be created or destroyed during the course of a chemical reaction*. Accordingly, all chemical equations must be "mass balanced." This means that the *number and type of each atom on the reactant side of the arrow must equal the number and type on the product side*. The above equation is not balanced because there are two oxygen atoms on the reactant side (O_2) and only one on the product side (MgO). To balance oxygen, a *coefficient* of 2 is placed in front of MgO.

$$Mg_{(s)} + O_{2(g)} \xrightarrow{\text{heat}} 2MgO_{(s)}$$

The coefficient multiplies the entire chemical formula; therefore, 2 MgO means MgO + MgO, or two Mg atoms and two O atoms. In the process of balancing oxygen, we "unbalanced" Mg, so we must now insert a *coefficient* of two in front of Mg on the reactant side.

$$2Mg_{(s)} + O_{2(g)} \xrightarrow{\text{heat}} 2MgO_{(s)}$$

This is the balanced chemical equation for the reaction of elemental magnesium with oxygen. The coefficients required to balance chemical equations are termed stoichiometric coefficients and represent the *quantitative relationship between reactants and products*. The above equation is "read": *2 atoms of elemental magnesium* react with *one molecule of oxygen* to produce *2 formula units of magnesium oxide*. A coefficient of one is never written, because it is understood that the formula (alone) represents one. As stated previously, all chemical equations must be balanced; however, this does not mean that we must always add coefficients. Hydrochloric acid and sodium hydroxide react to form sodium chloride and water. This reaction is represented by the following chemical equation:

$$HCl + NaOH \rightarrow NaCl + H_2O$$

Although the elements have switched partners during the reaction, this equation is balanced. Verify that the number of each element on the left is equal to the number on the right.

The application of chemical principles to forensic investigation may not always center on the basic concepts of chemistry. For example, the concept of "balanced equations" may be lost in reactions involving the complex structures of drugs and controlled substances. Nonetheless, a sound foundation in the basic principles of chemical reactivity will enhance the understanding of applications in forensic investigation.

3.7 Questions

1. Describe a nonpolar covalent bond, a polar covalent bond, and an ionic bond.
2. Characterize each of the following as covalent or ionic compounds: CH_4, LiBr, CaO, CO_2, CCl_4, NaCl, and P_4O_{10}.
3. Explain the difference between covalent and ionic bonds.
4. Which groups of elements usually form polar covalent bonds? Explain.
5. Which groups of elements usually form ionic bonds? Why?
6. Which elements form nonpolar covalent bonds? Why?
7. Discuss the similarities and differences between atoms and molecules?
8. Explain the difference between valency and valence electrons.
9. What elements can form double bonds?
10. What elements do not form triple bonds?
11. Discuss the similarities and differences between nonpolar covalent and polar covalent bonds?
12. Discuss similarities and differences between ionic and polar covalent bonds.
13. Which elements will never form ionic bonds?
14. Which elements will never form covalent bonds?
15. What type of bond is formed between a metal and a nonmetal?
16. What type of bond is formed between a nonmetal and a nonmetal?
17. What type of compound (ionic or covalent) results from the bonding of four hydrogen atoms to a single carbon atom?
18. Please explain to the jury chemical bonding and the bonding ability of atoms.
19. Describe the "mole concept" to members of the jury and provide examples.
20. Calculate the molar masses of the following compounds: NaCl, CO_2, Li_2O, H_3PO_4, KOH, HCl, $H_2C_2O_4$, and Al_2O_3.
21. Balance the following chemical equations:

$$HNO_3 + KOH \rightarrow KNO_3 + H_2O$$
$$BaSO_4 + NaBr \rightarrow Na_2SO4 + BaBr_2$$
$$Li_3PO_4 + MgCl_2 \rightarrow LiCl + Mg_3(PO_4)_2$$
$$AlCl_3 + K_2O \rightarrow KCl + Al_2O_3$$

4 Organic Chemistry

4.1 Introduction

Historically, organic chemistry was once defined as the study of the structure and function of molecules originating from living organisms. It was once believed that the unexplained differences between organic compounds, those from living organisms, and inorganic compounds, those from mineral sources, were attributed to an unknown "life force" within the organic compounds. In today's world of science, advancements in research and technology have laid the mythical "life force" to rest. The complex chemical behavior exhibited by organic compounds is now explained in terms of reaction mechanisms, structural analysis, and thermodynamics. It is now possible to synthesize and manipulate organic molecules in a laboratory environment, and with this knowledge has come a modern definition of organic chemistry. Organic chemistry is the study of the properties, structure, and function of compounds containing carbon. Arguably the most complex and mysterious of the specialized areas of chemistry, organic chemistry often requires years of both practical and theoretical study to master. This chapter is designed to provide a basic survey of the concepts and principles of organic chemistry; reactions will be limited to applications in forensics and structural analysis will concentrate on the basic recognition of functional groups.

The defining element in organic molecules is clearly carbon; however, it is universally accepted, and common practice, to define them by the obligate presence of both carbon *and* hydrogen. This does not mean that organic molecules contain *only* carbon and hydrogen; elements such as nitrogen, oxygen, sulfur, phosphorus, and chlorine may also be present. The study of organic chemistry does not involve the individual study of the vast number of organic compounds, numbering in the hundreds of thousands and quite possibly millions. Instead, organic compounds are divided into broad classes based on the presence of a *common structural feature* termed a functional group. Functional groups classify the compound and also allow prediction of chemical behavior. For example, all organic *compounds containing a double bond* belong to a class termed alkenes. The double bond will undergo predictive chemistry that, in most cases, dominates the reactivity of the molecule. Therefore, through the study of the structure and chemical behavior of double bonds, knowledge is gained on an entire class of organic compounds. The study of functional groups is the most effective and efficient approach to the study of organic chemistry.

4.2 Classification of Organic Compounds: Functional Groups

Individuals possess unique characteristics or features that separate them from the general population. These characteristics may classify them into a particular group; for example, Hispanics, Caucasians, African-Americans, and Asians have an identity within a group that is usually based on physical characteristics. Organic molecules also have unique characteristics or features called *functional groups*.

Functional groups are atoms, groups of atoms, or common structural features used to classify organic molecules (Table 4.1). In general, functional groups will react in a unique, predictive manner and this chemical behavior is similar in all compounds containing a specific group. A few functional groups share a common structural feature; aldehydes, ketones, and carboxylic acids, for example, all contain a carbon–oxygen double bond called a *carbonyl group*. The type of atom or atoms bound to the carbonyl distinguishes them. It is possible, and quite common indeed, to have more than one functional group on a single

J.I. Khan et al., *Basic Principles of Forensic Chemistry*, DOI 10.1007/978-1-59745-437-7_4,
© Springer Science+Business Media, LLC 2012

Table 4.1 Common functional groups found in organic molecules

Class	Functional group	IUPAC ending
Alkane	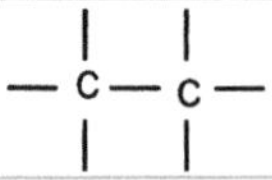	"-ane"
Alkene	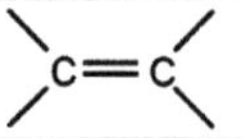	"-ene"
Alkyne	—C≡C—	"-yne"
Alcohol	R—OH	"-ol"
Ketone	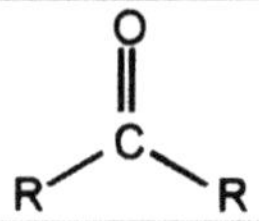	"-one"
Aldehyde	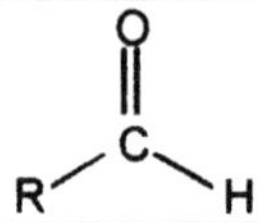	"-al"
Carboxylic acid	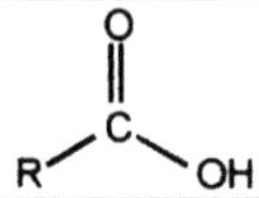	"-oic acid"
Ester	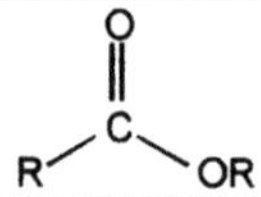	
Nitro compounds	R—NO_2	
Amines	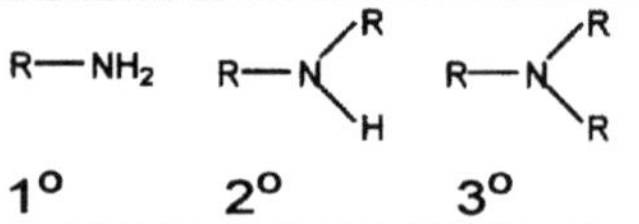	"-amine"

molecule. In these cases, the molecule will exhibit chemical and physical properties of all groups present. A wide range of functional groups can be found on different types of controlled substances. Underground chemists convert noncontrolled substances into illegal drugs, controlled substances, and designer drugs using functional-group reactivity, often by simply converting one functional group into another.

4.2.1 Alkanes

Alkanes are saturated hydrocarbons with a general formula C_nH_{2n+2}. There are three requirements in the definition of an alkane. First, the compound must be saturated; it must *contain only carbon–carbon single bonds*. Second, it must be a hydrocarbon; *contain only the elements carbon and hydrogen*. Lastly, the chemical formula must satisfy the general formula for an alkane, C_nH_{2n+2}. The names and chemical formulas for the first ten alkanes are shown in Table 4.2 and should be memorized.

The names of alkanes always end in "-ane." If we examine the chemical formulas, we see that each contains only carbon and hydrogen and each satisfies the general formula for an alkane. For example, butane has a chemical formula C_4H_{10}. In this case, $n=4$ and the general formula requires a number of H's equal to $2n+2$ or 10.

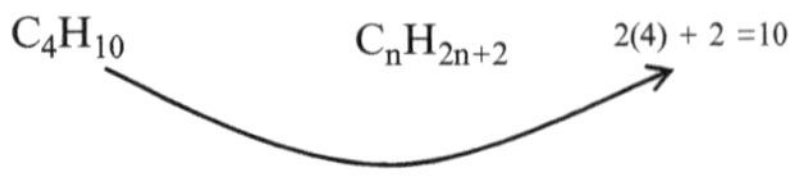

Any hydrocarbon that contains the number of carbons and hydrogens specified by C_nH_{2n+2} will contain only single bonds (are saturated). There are several methods used to represent organic compounds and each has advantages and disadvantages.

Table 4.2 The first ten alkenes

Name	Chemical formula
Methane	CH_4
Ethane	C_2H_6
Propane	C_3H_8
Butane	C_4H_{10}
Pentane	C_5H_{12}
Hexane	C_6H_{14}
Heptane	C_7H_{16}
Octane	C_8H_{18}
Nonane	C_9H_{20}
Decane	$C_{10}H_{22}$

The most common are *chemical formulas*, *structural formulas*, *condensed structural formulas*, and *skeletal* (or line) *formula*.

We are already familiar with chemical formulas; Table 4.2 contains several examples. Generally, these representations are the easiest and most convenient to write, but provide no information on the geometry of the molecule. Structural formulas are a *detailed representation of the bonding arrangement of atoms in the compound.* Typically, these formulas are the most tedious to draw. Structural formulas for straight-chain alkanes, commonly termed *n*-alkanes (*n* for *n*ormal), are drawn by connecting all the carbon atoms in a straight line using single bonds (single lines). The tetravalency (4-bonds) of carbon is maintained using hydrogen. This means that each carbon will have four total bonds (lines) and the bonds other than carbon–carbon bonds will be to hydrogen. The procedure for drawing the structural formula for butane, chemical formula C_4H_{10}, is given below.

First, draw four carbons in a straight, continuous chain connected by single lines (the single bonds).

Diagram 4.1

Next, maintain the tetravalency of carbon by ensuring all carbons have a total of four lines (bonds).

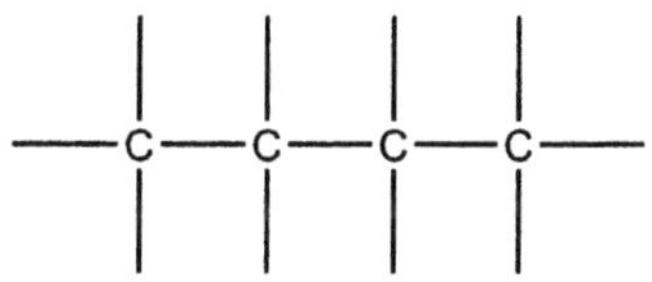

Diagram 4.2

Lastly, insert hydrogens at the end of each line (vacant bonds) to obtain the structural formula for butane. Verify the structure contains the number of carbons and hydrogens specified in the chemical formula for butane, C_4H_{10}.

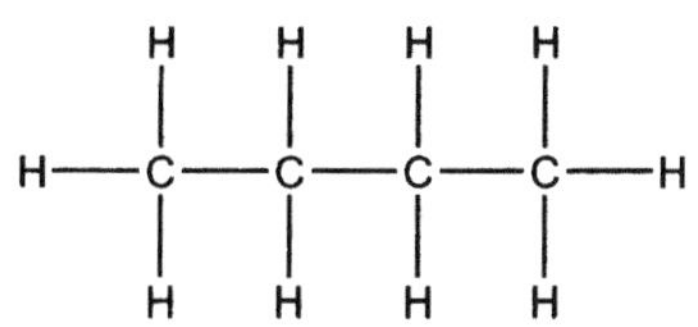

Diagram 4.3

A condensed structural formula can be easily obtained from the structural formula.

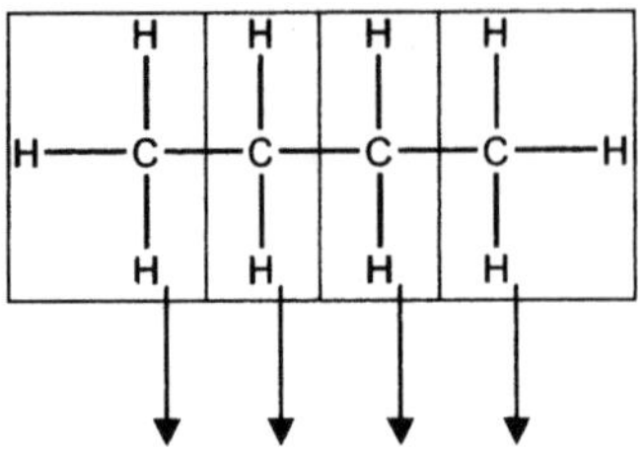

CH_3–CH_2–CH_2–CH_3 or alternatively $CH_3CH_2CH_2CH_3$

Diagram 4.4

When condensed structural formulas are used, it is understood that hydrogens are bound to the carbons they follow in the chain.

Skeletal (or line) formulas for *n*-alkanes show only the *carbon–carbon bonds, but not carbon or hydrogen atoms*. These structures illustrate an overall molecular geometry by showing realistic carbon–carbon bond angles. Butane is shown below using this method.

Diagram 4.5

In skeletal formulas, it is understood that carbons reside at both terminals (ends) and at each vertex, the points where the line changes direction (the peaks and valleys). There are enough hydrogens at each carbon to fill its tetravalency, but they are never written in this method. Skeletal formulas are the method of choice used to represent most complex organic compounds and will be used extensively in the following chapters. The next time you need a prescription, open the insert material and you will commonly see the structure of the drug represented using this method.

Slight variations to structural formulas can provide a *three-dimensional view* of the molecule. This *stereochemistry* is illustrated through the use of wedges; a solid wedge represents a bond extending out toward the viewer, and a dashed wedge represents a bond extending back from the viewer. Methane, for example, has a chemical formula of CH_4 and a tetrahedral geometry. A typical structural formula is shown below.

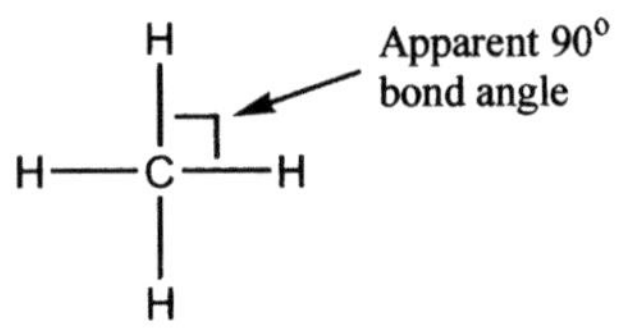

Diagram 4.6

At first glance, the structure of methane appears to be flat, with all bonds in the same plane with apparent H–C–H bond angles of 90°. The use of wedges adds depth to the structure and illustrates a more realistic view of the actual tetrahedral geometry which contains H–C–H bond angles of 109.5°.

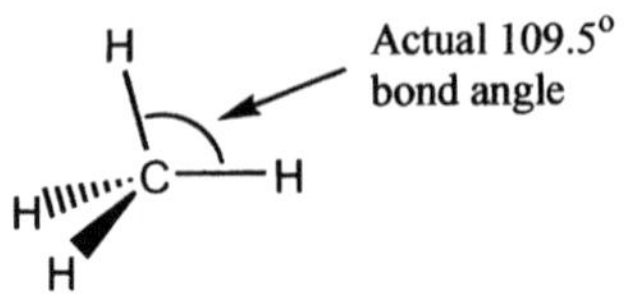

Diagram 4.7

The use of wedges to show stereochemistry (a three-dimensional view) is required in complex molecules where an illustration of bond depth is essential.

4.2.1.1 Naming Alkanes

Naming alkanes is a bit more involved than simply memorizing the first ten members, although this is an excellent start. Many organic molecules, including alkanes, contain substituted groups attached to a parent chain. The names of these compounds must incorporate the substituted groups, their location, and the parent chain. The rules for naming organic compounds are determined by an organization called the International Union of Pure and Applied Chemistry (IUPAC). Their goal is to maintain consistency in naming to ensure worldwide recognition of organic compounds.

4.2.1.2 Rules for Naming Alkanes

1. Determine the parent chain – the longest, continuous chain of carbons.
2. Name all substituted groups attached to the parent chain.
3. Number the parent chain in such a manner that the lowest number falls on the carbon containing the first substitution.
4. Locate the substituted groups on the parent chain using the carbon number containing the group. In cases with multiple substitutions, alphabetize the groups.
5. Name the alkane.
   Examples:

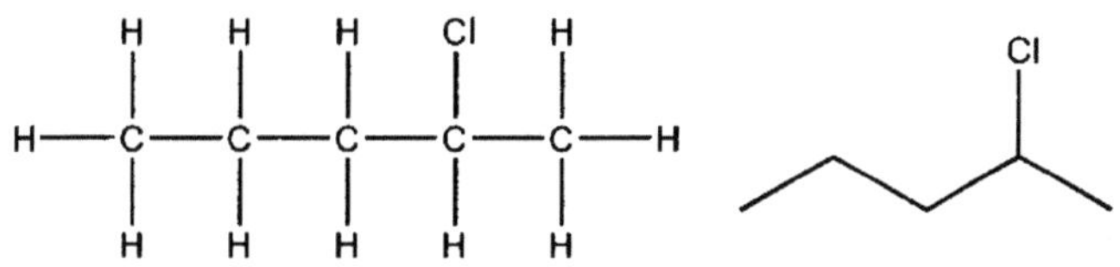

Diagram 4.8

The above structures are different representations of the same compound.

1. Verify the longest, continuous chain of carbons is five; the parent chain is pentane and this compound is a pentane derivative (C_5 from our table of alkanes).
2. Verify only one substituted group, a chlorine.
3. Numbering the chain left to right places the chlorine at carbon #4, numbering right to left places it at carbon #2. We chose right to left because it puts the lowest number on the first (and only) point of substitution.
4. This compound is 2-chloropentane, a 5-carbon parent (pentane) containing a chlorine atom at carbon #2. A hyphen always separates the carbon number from the substituted group attached at the carbon. Note the name of the substituted group is not its elemental name; chlorine (Cl) becomes chloro when it is attached to a parent chain. Group VIIA elements (halogens) are commonly found in organic compounds and their names as substituted groups are worth memorizing: F-fluoro, Cl-chloro, Br-bromo, and I-iodo.
   Verify the names of the following compounds:

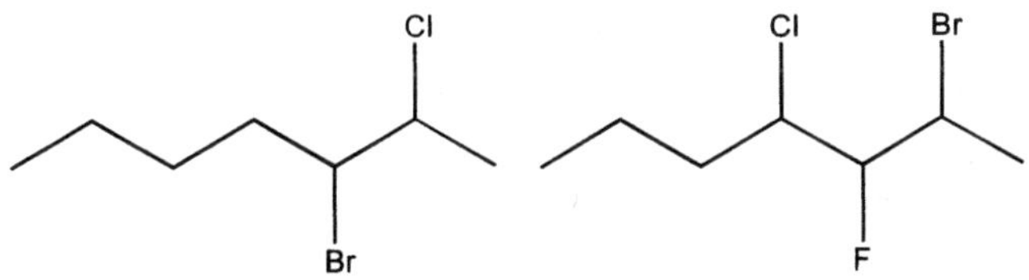

3-bromo-2-chloroheptane 2-bromo-4-chloro-3-fluoroheptane

Diagram 4.9

The above examples illustrate a few important principles in naming – alphabetizing takes priority over numbering, carbons containing substituted groups are not arranged in any particular ascending or descending order, and hyphens always separate substituted groups when multiple substitutions are present.

It is quite common to have small chain alkanes substituted on larger parent chains. In these cases, a single hydrogen atom must be removed from the shorter chain to create a point of attachment. *The removal of a single hydrogen from an alkane creates an alkyl group.* The names of alkyl groups end in "yl" and are determined by dropping the "ane" ending from the alkane name and adding the suffix "yl." Alkyl groups are represented using the letter "R" in cases requiring a "generic" hydrocarbon; accordingly, alkyl groups are termed R-groups. This terminology will be used extensively in our study of functional groups.

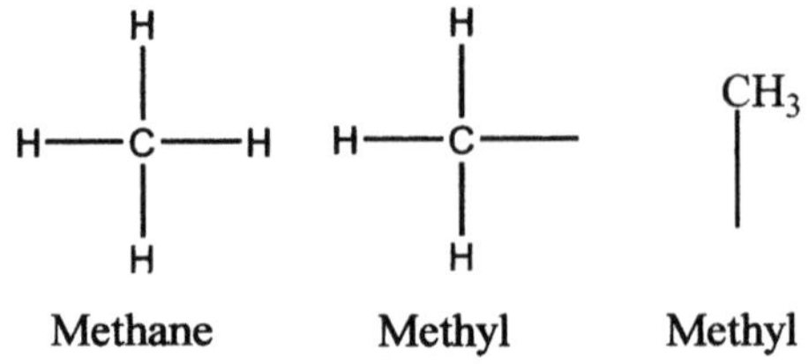

Diagram 4.10

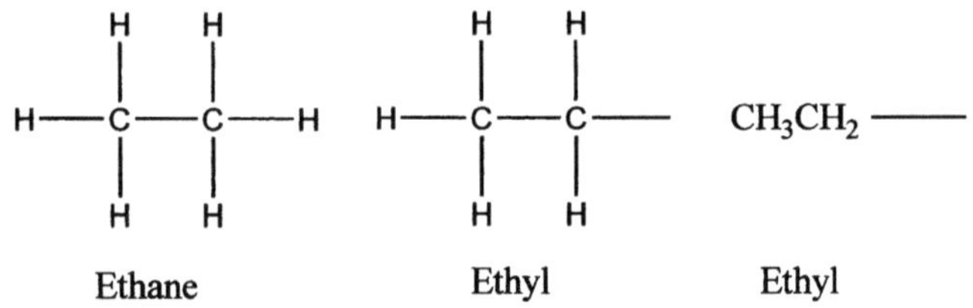

Diagram 4.11

In naming, alkyl groups are treated simply as substituted groups.

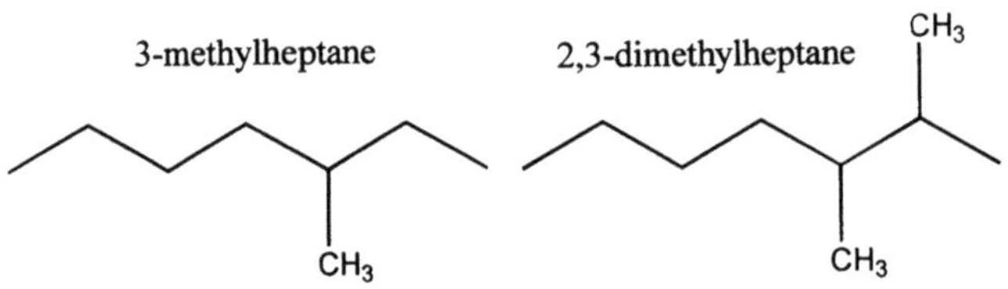

Diagram 4.12

Physical properties of alkanes, such as boiling points and melting points, are affected by the chain length or size of the alkane. In general, alkane boiling points and melting points increase with increasing chain length. For example, methane (CH_4) has a boiling point of −164°C and a melting point of −182°C, while decane ($C_{10}H_{22}$) has a boiling point of 174°C and a melting point of −30°C.

4.2.1.3 Cycloalkanes

Alkanes can also exist in closed ring structures called *cycloalkanes*. They are hydrocarbons that have a general formula C_nH_{2n}. Notice the number of hydrogens specified in the general formula is two less than that required for alkanes (C_nH_{2n+2}). Hydrocarbons containing a number of hydrogens less than the number required by its alkane counterpart are termed unsaturated. For this reason, cycloalkanes are classified as *unsaturated* compounds. Cycloalkanes are named using the prefix "cyclo" attached to the alkane name.

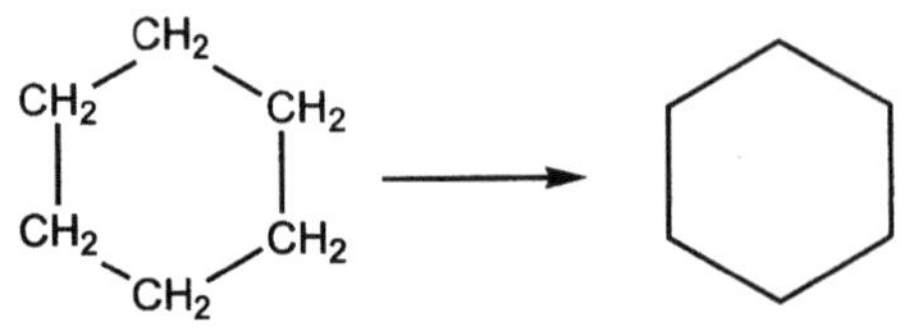

Diagram 4.13

The number of carbons contained in the above rings is six. The alkane containing six carbons is hexane and requires 14 hydrogens (C_nH_{2n+2}, where $n=6$), but the ring structures show a formula of C_6H_{12} (C_nH_{2n}, where $n=6$). The above structures are different representations of *cyclohexane*, an unsaturated compound. Cycloalkanes are almost exclusively represented using skeletal formulas and, in general, have the same chemical and physical properties as alkanes.

4.2.2 Alkenes

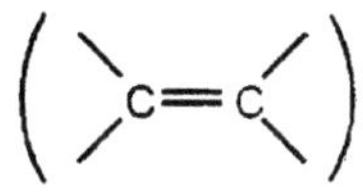

Diagram 4.14

Alkenes are unsaturated hydrocarbons with a general formula C_nH_{2n}. Note the general formula for alkenes is identical to that of cycloalkanes. There are similarities between alkane and alkene definitions; both are *hydrocarbons* and the number of carbon and hydrogen atoms *must satisfy a general formula.* The major difference, aside from the slightly different general formulas, is alkenes must be unsaturated; that is, they must *contain at least one carbon–carbon double bond.* It is worth noting that *carbon–carbon double bonds* are often termed *points of unsaturation.* Any compound that is an unsaturated hydrocarbon satisfying the general formula C_nH_{2n} and is not a ring belongs to the alkene class of organic compounds. The names and chemical formulas for the first nine alkenes are shown in Table 4.3. Why nine and not ten? You must have at least two carbons to form a double bond.

Alkene names end in "-ene," indicating the presence of at least one carbon–carbon double bond in the compound. Verify that the chemical formulas in Table 4.3 contain only carbon and hydrogen and each satisfies the general formula C_nH_{2n}.

Methods for drawing structural formulas, condensed structural formulas, and skeletal structures for alkenes are similar to those used for alkanes. The notable difference is the use of a "double line" to represent the double bond.

C_2H_4 Chemical Formula

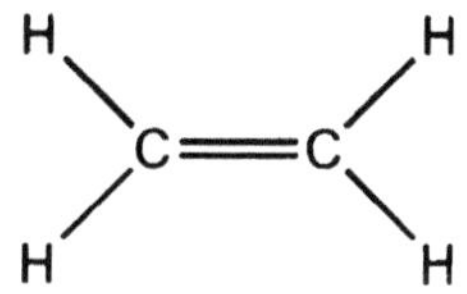

Structural Formula

CH_2CH_2 Condensed Structural Formula

Diagram 4.15

Table 4.3 The first nine alkenes

Name	Chemical formula
Ethene	C_2H_4
Propene	C_3H_6
Butene	C_4H_8
Pentene	C_5H_{10}
Hexene	C_6H_{12}
Heptene	C_7H_{14}
Octene	C_8H_{16}
Nonene	C_9H_{18}
Decene	$C_{10}H_{20}$

The location of the double bond must be specified when naming alkenes containing four or more carbons. In these cases, the number of the first carbon involved in the double bond is included in the name. Consider the following skeletal formulas for butene.

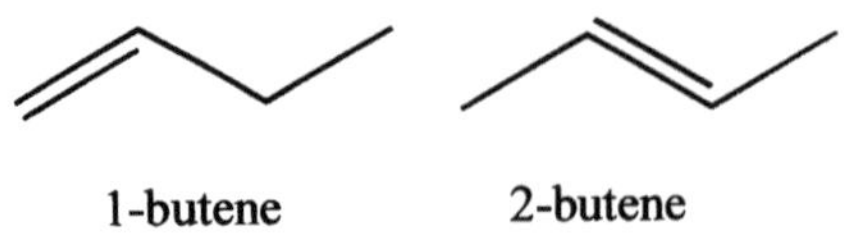

Diagram 4.16

Verify the number of carbons in each of the above structures is four. The alkane containing four carbons is butane (C_4H_{10}), but the above structures are both butenes because a double bond is present in each. The location of the double bond is clearly different and the carbons in the parent chain are always numbered in a manner that places the lowest number on the first carbon contained in the double bond. In the first example above, we can number the carbons left to right or right to left. If we number left to right, we find the double bond begins at carbon #1, but numbering right to left, we find the double bond begins at carbon #3. The chain is numbered left to right. This is 1-butene (*not* 3-butene); indicating the "ene" (double bond) begins at carbon #1. The other structure is 2-butene using similar reasoning. Can you justify the following names?

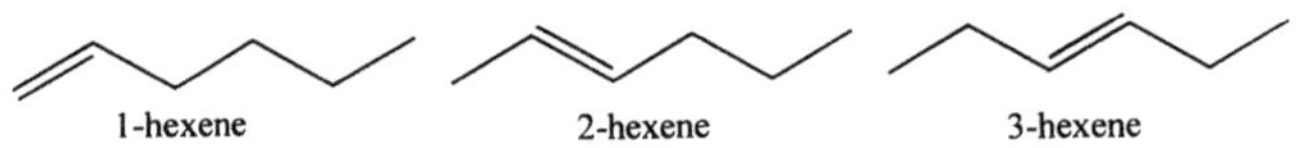

Diagram 4.17

4.2.2.1 Cycloalkenes

Alkenes can also exist in ring structures called cycloalkenes. The chemical formulas for cycloalkenes vary according to the number of double bonds present in the structure.

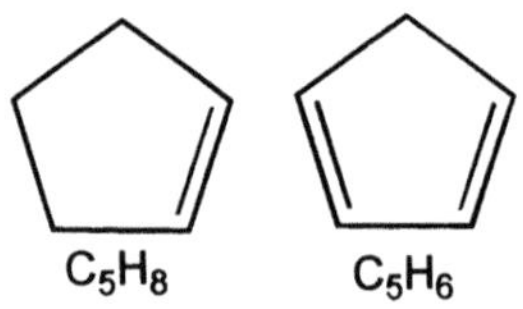

Diagram 4.18

Recall that skeletal formulas contain carbons at each vertex (change of direction) and the tetravalency of carbon is maintained with bonds to hydrogen that are never shown. Each of the above cycloalkenes contains five carbons. Next, determine the number of bonds to hydrogen required to total four bonds on each carbon. Verify the above formulas for each structure.

Naming cycloalkenes also (like cycloalkanes) requires the addition of the prefix "cyclo" to the parent alkene name.

Diagram 4.19

It is not necessary to locate the double bond in cycloalkenes containing only one point of unsaturation (double bond). However, if more than one double bond is present, the locations of all double bonds are specified using the first carbon in each carbon–carbon double bond. You may start with any carbon–carbon double in the structure, but you must number in the direction of the double bond and in such a way that the lowest number falls on the first carbon of the next double bond. In addition, the prefixes, di, tri, etc., must be added to the "ene" portion of the name.

1,4-cyclohexadiene	1,3-cyclohexadiene	1,3,5-cyclohexatriene

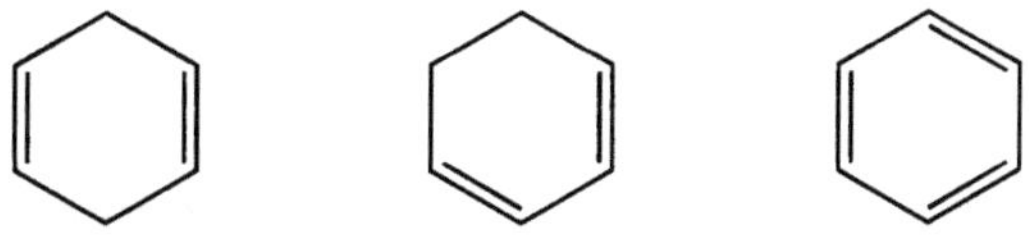

Not 1,3-cyclohexadiene	Not 1,5-cyclohexadiene	Also called benzene

Diagram 4.20

Many controlled substances contain cycloalkenes or cycloalkene derivatives. Cyclopentene, for example, is frequently used in clandestine laboratories to produce phencyclidine (PCP).

4.2.3 Alkynes

$$\left(-C \equiv C- \right)$$

Diagram 4.21

Alkynes are unsaturated hydrocarbons with a general formula C_nH_{2n-2}. This class of organic compound contains at least one carbon–carbon triple bond. Notice the general formula specifies four less Hs than that required for alkanes (C_nH_{2n+2}). The loss of two Hs from an alkane produced an additional bond (one point of unsaturation) and the alkene class. The loss of four Hs from an alkane produces two additional bonds (two points of unsaturation) and the alkynes. We may conclude that alkanes are "saturated" with hydrogens, and the loss of any hydrogens from an alkane produces an "unsaturated" compound at a rate of two Hs per additional bond (point of unsaturation). The names and chemical formulas for the first nine alkynes are shown in Table 4.4.

Alkyne names end in "-yne" with one notable exception: C_2H_2 is rarely named ethyne, it is almost exclusively called acetylene. Despite the "ene" ending in acetylene, it is not an alkene and does not contain a carbon–carbon double bond.

Chemical Formula	Structural Formula
C_2H_2	$H-C\equiv C-H$

Diagram 4.22

Notice the structure of acetylene is linear (a straight line). This is the geometry of all carbon–carbon triple bonds and results from the orientation of the combining orbitals on the carbons involved in the triple bond, a process termed hybridization.

Methods for drawing structural formulas, condensed structural formulas, and skeletal structures for alkynes are similar to those used for alkanes and alkenes. The triple bond is represented using a "triple line."

Table 4.4 The first nine alkynes

Name	Chemical formula
Ethyne	C_2H_2
Propyne	C_3H_4
Butyne	C_4H_6
Pentyne	C_5H_8
Hexyne	C_6H_{10}
Heptyne	C_7H_{12}
Octyne	C_8H_{14}
Nonyne	C_9H_{16}
Decyne	$C_{10}H_{18}$

Chemical Formula

C_4H_6

Condensed Structural Formula

CH_3CCCH_3

Structural Formula

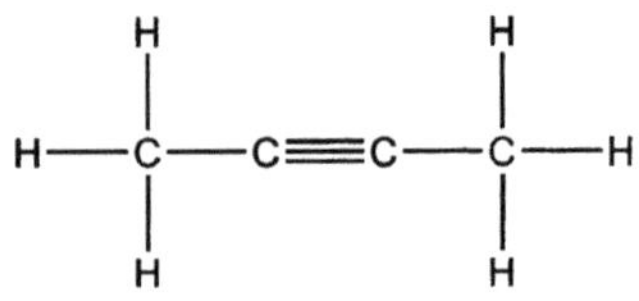

Diagram 4.23

The location of the triple bond must be specified when naming alkynes containing four or more carbons. Verify the names for each structure below.

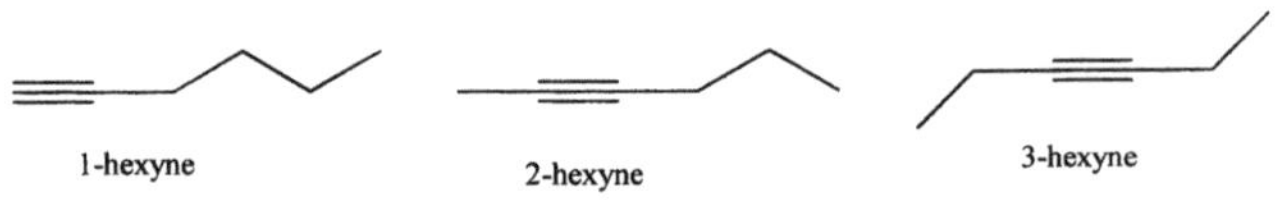

Diagram 4.24

4.2.3.1 Cycloalkynes

Cycloalkynes containing eight or more carbons are common. The linear geometry of carbon–carbon triple bonds introduces severe strain in small rings where bond angles deviate significantly from 180°. As ring size increases, the bond angles between adjacent carbons approach the favorable linear geometry of alkynes.

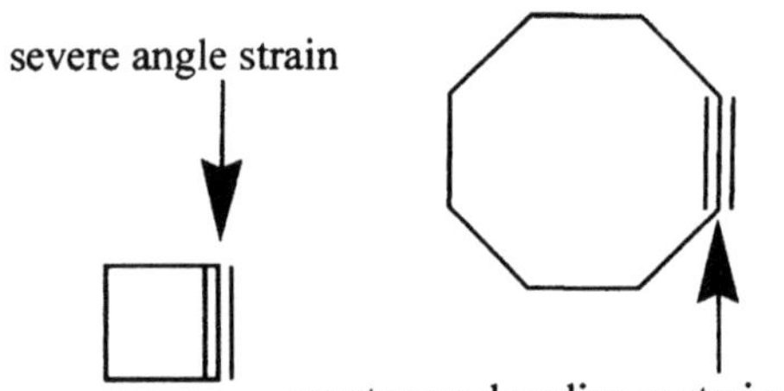

Diagram 4.25

4.2.4 Aromatic Compounds

Benzene is the common name for 1,3,5-cyclohexatriene, a unique member of the cycloalkene class. It is a flat ring with a chemical formula of C_6H_6.

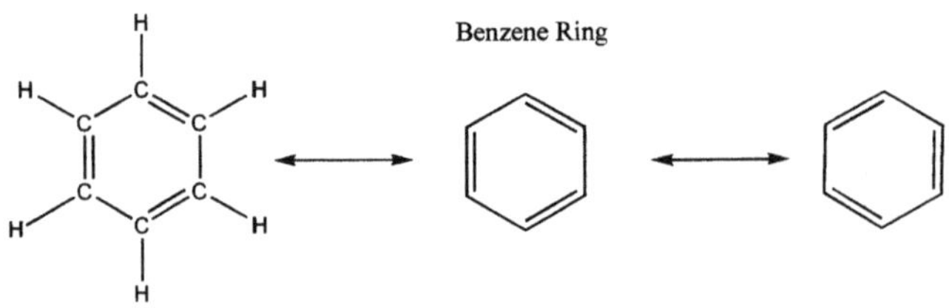

Diagram 4.26

Benzene is unusually stable and does not undergo reactions typical of alkenes. Surprisingly, structural analysis of benzene reveals six identical carbon–carbon bonds, not three carbon–carbon double bonds and three carbon–carbon single bonds as shown in the above structural formulas. The distance between adjacent carbons in benzene is longer than a carbon–carbon double bond, but shorter than a carbon–carbon single bond; in fact, the distance is almost exactly midway between the two. For this reason, benzene is frequently represented as a hexagon with an inscribed circle representing the six identical carbon–carbon "one and a half" bonds.

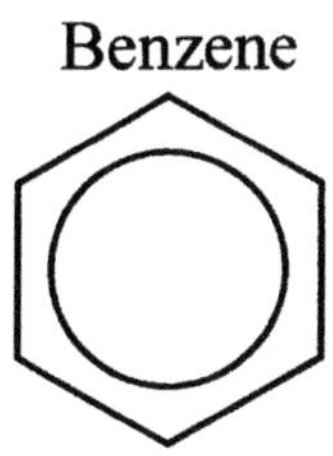

Diagram 4.27

The unusual stability exhibited by benzene is attributed to the fact that:

1. It is a ring
2. It is planar (flat)
3. It is conjugated
4. It satisfies the Huckel rule

A detailed explanation of the above conditions is beyond the scope of this text; however, any compound that satisfies the above will be aromatic and exhibit aromatic character. Benzene and derivatives of benzene are aromatic compounds, a class of organic molecules marked by unusual stability. In addition, aromatic compounds frequently have strong, pungent (often unpleasant) odors, a characteristic indicated by the term "aromatic," which is derived from aroma (to smell). Naphthalene, for example, is an aromatic compound used in the production of mothballs and is responsible for their distinct odor. Structural formulas of some aromatic compounds are shown below; notice the presence of benzene or "benzene-like" structures.

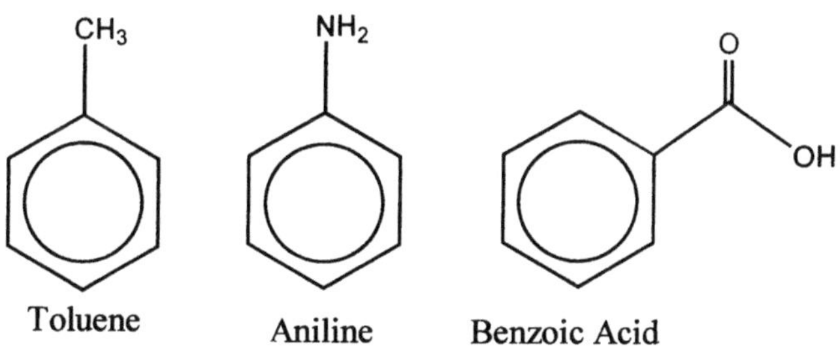

Diagram 4.28

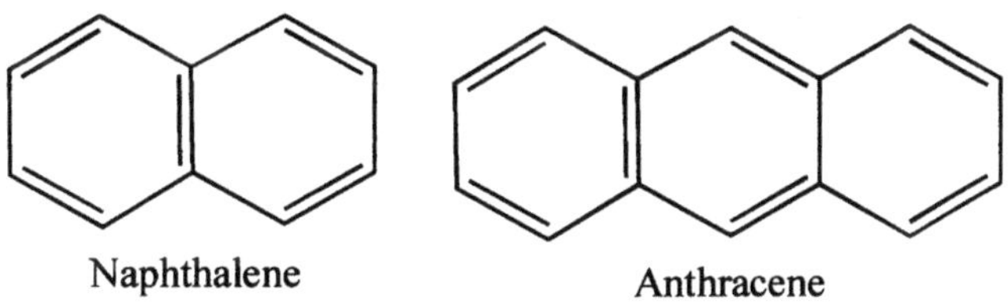

Diagram 4.29

4.2.5 Alcohols

$$(R—OH)$$

Diagram 4.30

Alcohols are organic compounds that contain the hydroxyl functional group (–OH). The names of alcohols end in "-ol," indicating the presence of the hydroxyl group, for example, methanol (fuel), ethanol (drinking alcohol), and isopropanol (rubbing alcohol). When naming alcohols, the "-e" is dropped from the alkane containing the "–OH" group and replaced with the suffix "-ol."

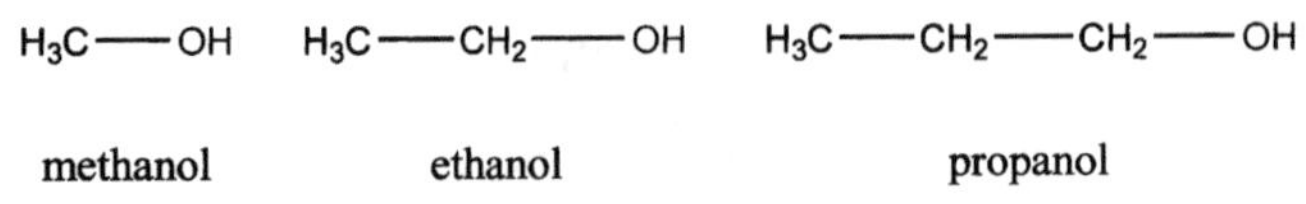

Diagram 4.31

Justify the names above; for example, in propanol, the hydroxyl group is attached to a three-carbon chain (propane), dropping the "-e" and adding "ol" gives propanol.

The location of the hydroxyl group must be specified in alcohols containing four or more carbons. The chain is numbered in a manner that places the lowest number on the carbon containing the –OH group.

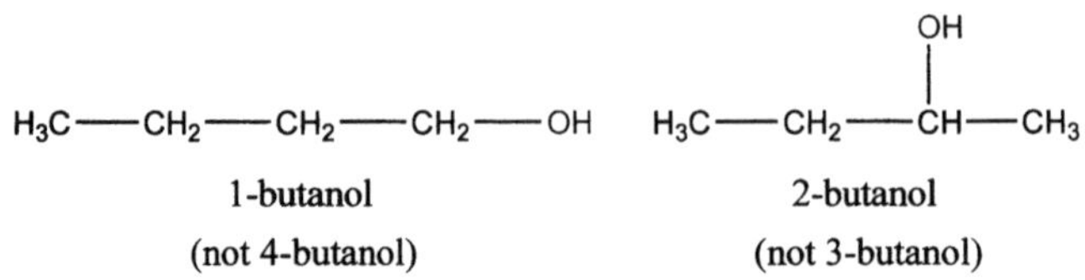

Diagram 4.32

The above examples both have an –OH attached to a four-carbon chain. The location of the –OH must be reflected in the name. Can you justify each name above? (Hint: the chains, in both cases, must be numbered right to left.)

Alcohol derivatives of propane represent an interesting case in which it is possible to name a single structure using two different names. The structures below are almost exclusively identified using the names under each, but the alternative names in parenthesis are also technically correct.

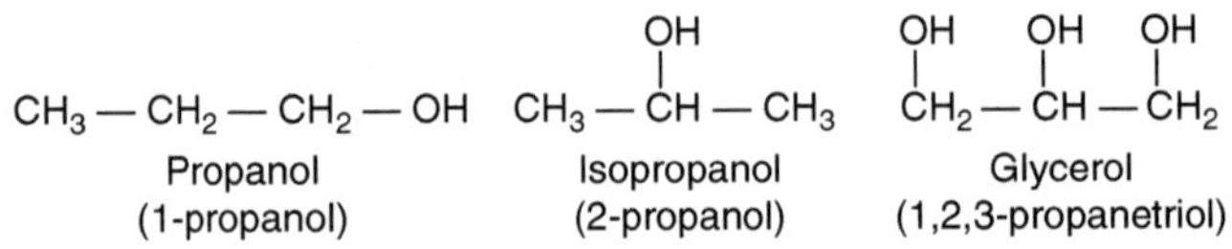

Diagram 4.33

Alcohols can also be classified based on the number of carbons attached to the carbon containing the hydroxyl group. A primary alcohol (1°) has one carbon attached to the carbon containing the hydroxyl group, secondary alcohols (2°) have two carbons, and tertiary alcohols (3°) have three. Identify the carbon containing the OH group in each example below. Determine the number of carbons bound to the OH containing carbon to verify each classification.

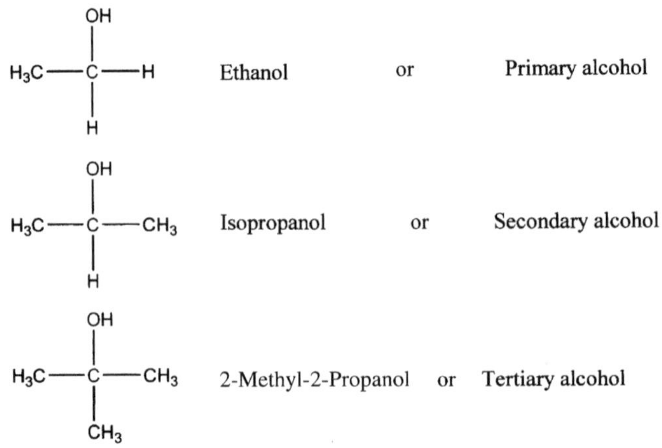

Diagram 4.34

Alcohols are polar organic solvents that exist in liquid form at room temperature (25°C). The solubility of alcohols (ability to mix) in other polar liquids (e.g., water) is dependent on the size of the carbon chain containing the –OH group. As the size of the chain increases, the alcohol tends to exhibit the nonpolar character of the hydrocarbon chain. This translates to a reduced solubility in water. Small-chain alcohols such as methanol, ethanol, and propanol will readily mix with water, while butanol, pentanol, and hexanol are slightly soluble. Octanol and larger alcohols are insoluble and form two layers when mixed with water. Alcohols do not exhibit acidic or basic character and generally have a pH comparable to water. They are moderately toxic if ingested, a characteristic exhibited by the most common member, ethanol (drinking alcohol). Methanol, ethanol, and isopropanol are common solvents used in the clandestine manufacturing of controlled substances (Fig. 4.1).

Designer drugs are produced through the chemical modification of illegal drugs or controlled substances. Designer drug manufacturing distinguishes itself from other types of drug manufacturing by requiring starting material that is already regulated by law. In some cases, production may be as simple as removing an –OH group. For example, bufotenine and psilocin are naturally occurring *positional isomers* that *differ only in the location* of the hydroxyl group. Both belong to the family of hallucinogenic drugs derived from dimethyltryptamine (DMT).

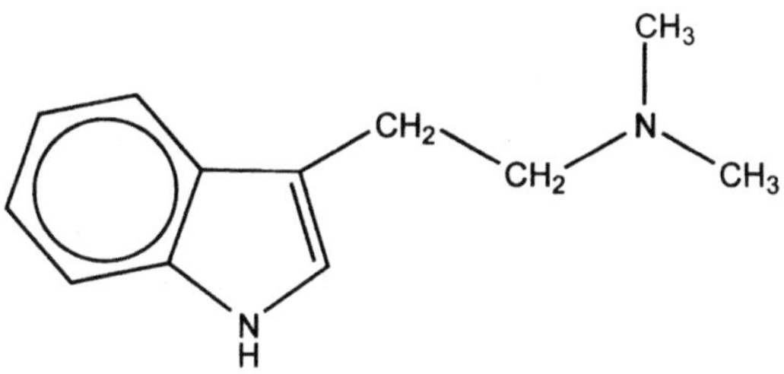

DMT

N,N-dimethyltryptamine

Diagram 4.35

Addition of a hydroxyl group at carbon four of the benzene ring on DMT produces psilocin, a hallucinogenic drug. Relocation of the hydroxyl group from carbon four on psilocin to carbon five generates bufotenine, another hallucinogenic drug.

Fig. 4.1 Examples of packaged denatured alcohols that are used as solvents in manufacturing controlled substances.

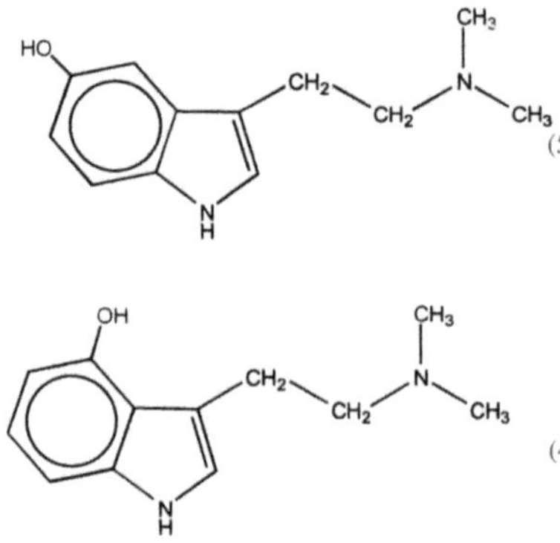

Bufotenine (5-hydroxy-N,N-dimethyltryptamine)

Psilocin (4-hydroxy-N,N-dimethyltryptamine)

Diagram 4.36

Bufotenine, psilocin, and DMT are derivatives of the indole family of organic compounds and are easily recognized by the characteristic fused five- and six-membered ring systems. It is often helpful to recognize fused ring systems that frequently occur in different types of organic compounds. Most steroids, for example, are immediately identified by the presence of the classical fused ring system in cholesterol, a common precursor in their production.

Ephedrine and pseudoephedrine also contain the hydroxyl functional group. They differ only in the special orientation of the –OH group and both are used to treat symptoms of hay fever, asthma, and nasal congestion. Unfortunately, they are also precursors in the production of methamphetamine, a controlled substance that stimulates the central nervous system. Ephedrine and pseudoephedrine are not currently controlled, but their distribution and sale are closely regulated. Recall the wedge convention; a solid wedge indicates a bond extending out toward the viewer and a dashed wedge indicates a bond extending back from the viewer.

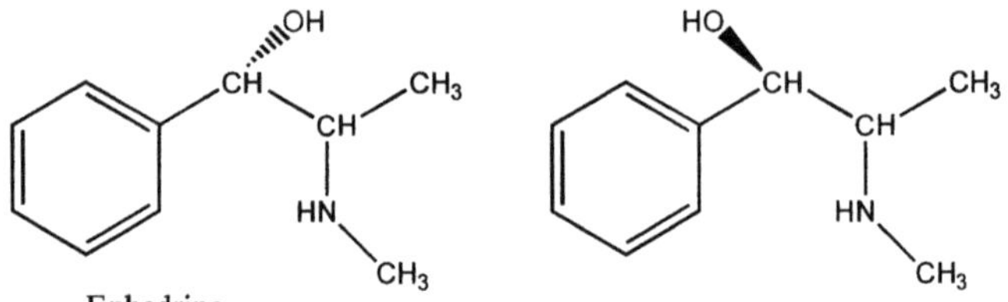

Ephedrine

Pseudoephedrine

Diagram 4.37

Pseudoephedrine is one of the most frequently used precursors in the illicit production of methamphetamine. The synthetic reaction is called a reduction reaction and results in the loss of the hydroxyl group.

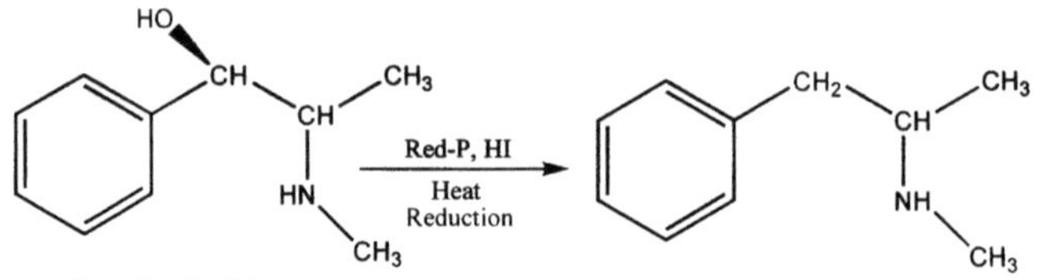

Pseudoephedrine

Methamphetamine

Diagram 4.38

4.2.6 Ketones

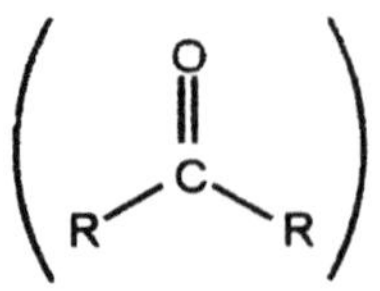

Diagram 4.39

Ketones are organic compounds that contain a carbon–oxygen double bond. This structural feature is called a carbonyl group (R–CO–R).

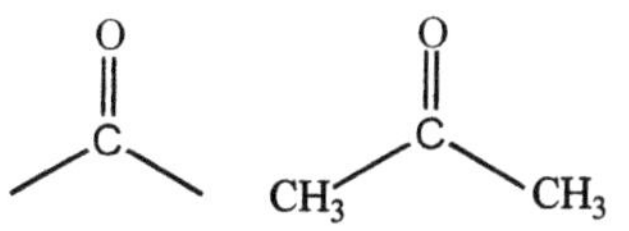

Carbonyl Ketone

Diagram 4.40

The names of ketones end in "-one," indicating the presence of the carbonyl group, i.e.; acetone, butanone, and pentanone. When naming ketones, the "-e" is dropped from the alkane containing the carbonyl group and replaced with the suffix "-one." The location of the carbonyl group must be specified in ketones containing five or more carbons. The chain is numbered in a manner that places the lowest number on the carbon containing the double bond to oxygen. It is worth noting that the carbonyl group can never appear on the end carbons (terminal carbons) in ketones. Carbonyl groups on terminal carbons will be covered shortly in our study of aldehydes and carboxylic acids.

Acetone (propanone) Butanone

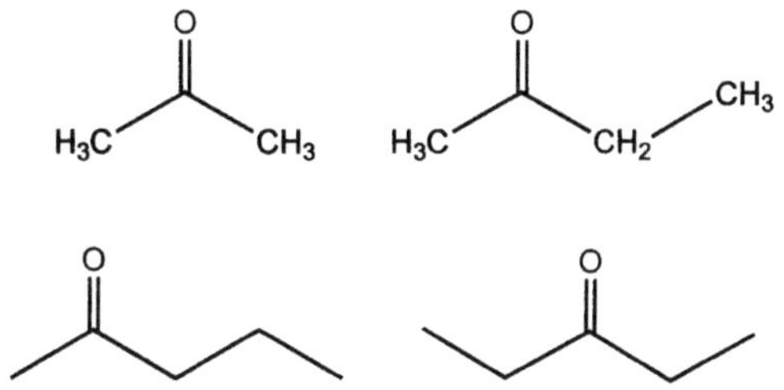

2-Pentanone 3-Pentanone

Diagram 4.41

Notice that the carbonyl carbons are not labeled in the top two structures above (see below for explanation) and the bottom two examples are complete skeletal formulas. Skeletal formulas will be used extensively in our study of forensic chemistry. We must begin to develop a familiarity with "interpreting" structures presented in this manner.

Can you justify the name of each ketone above? Acetone is the smallest member of the ketone family. A carbonyl group cannot appear on terminal carbons in the ketone class and a two-carbon chain would require the carbonyl to appear at a terminal carbon. Why is the second structure above butanone and not 2-butanone? In your mind, move the carbonyl to the other interior carbon and verify that it is the same structure.

In general, ketones are toxic and possess very strong, pungent odors. They are soluble in water as well as a variety of organic solvents. The carbonyl group is polar and therefore capable of hydrogen bonding in water. Similar to alcohols, ketone solubility depends on the length of the hydrocarbon chain. Acetone is a very common organic solvent that is used extensively in clandestine labs. It is generally the solvent of choice for the "icing" stage of methamphetamine production. Ketones are very prominent in all stages of drug production representing solvents, precursors, and even controlled substances.

Phenyl-2-propanone (P2P) is a precursor to methamphetamine. Barbiturates contain multiple carbonyl groups, ie; phenobarbital. Cathinone and methcathinone are well-known controlled substances containing ketone functional groups.

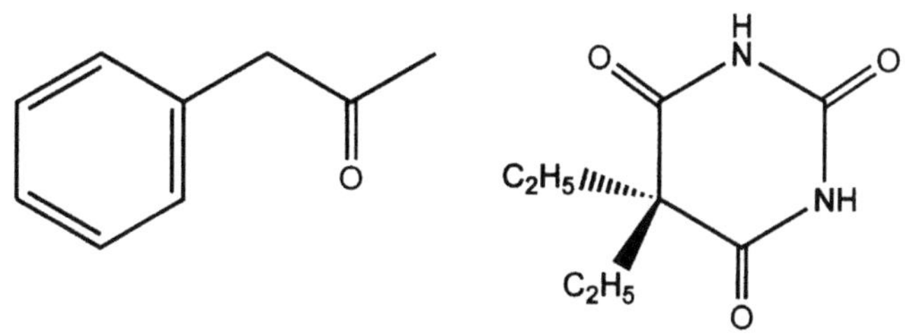

P2P (phenyl-2-propanone) Phenobarbital

Diagram 4.42

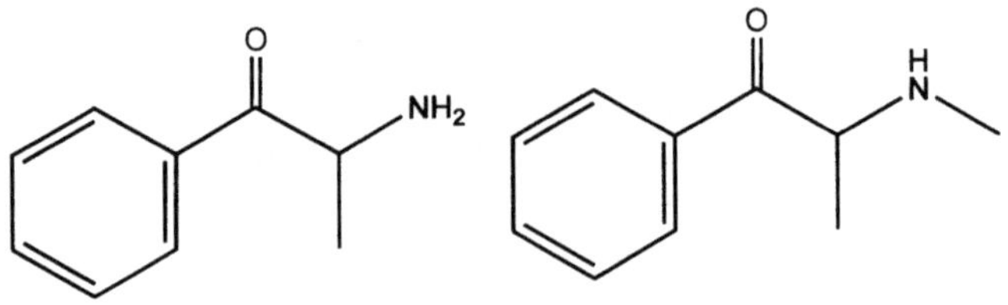

Cathinone Methcathinone

Diagram 4.43

Unfortunately, the chemical procedures used in the clandestine production of drugs are surprisingly simple and generally involve converting one functional group into another. In some cases, the ease with which this can be accomplished would astound, and perhaps frighten, any law-abiding citizen. Pseudoephedrine, a closely regulated but easily attainable drug, can be converted to methcathinone (ephedrone "Jeff") under relatively mild conditions using readily available chemicals.

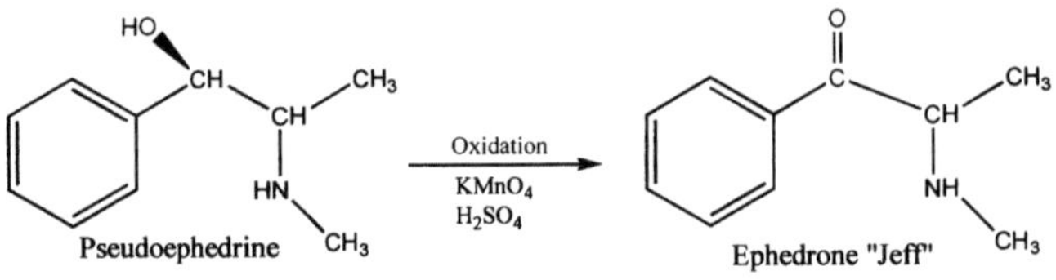

Diagram 4.44

4.2.7 Aldehydes

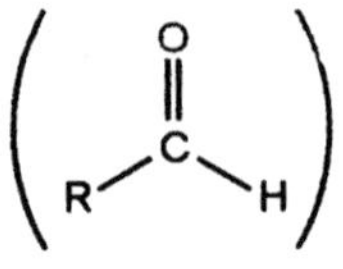

Diagram 4.45

Aldehydes are organic compounds that contain a carbonyl group bound to at least one hydrogen atom. These bonding conditions can only be satisfied if the carbonyl is located on a terminal carbon. In the aldehyde arrangement, a carbon must contain two bonds to oxygen (carbonyl carbon) and a single bond to hydrogen. This leaves only one vacant bonding position.

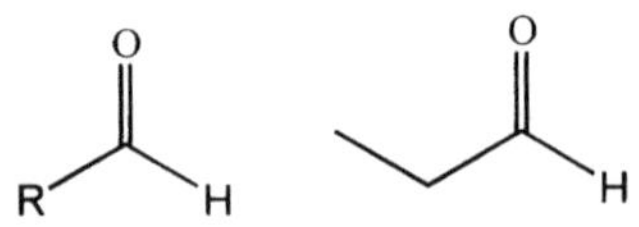

Aldehyde Group Aldehyde

Diagram 4.46

The names of aldehydes end in "-al," indicating the presence of the aldehyde group, i.e.; propanal, butanal, and pentanal. When naming aldehydes, the "-e" is dropped from the alkane containing the aldehyde group and replaced with the suffix "-al." The chain is numbered from the aldehyde group that is always located on the first carbon. Unlike other functional groups, the location of the aldehyde group is *not* included in the name because it can only appear at carbon #1.

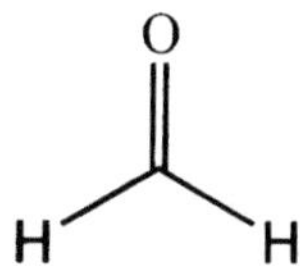

Formaldehyde or Methanal

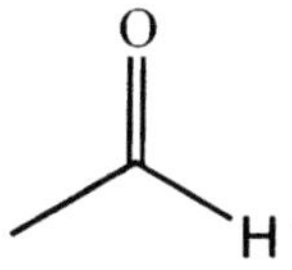

Acetaldehyde or Ethanal

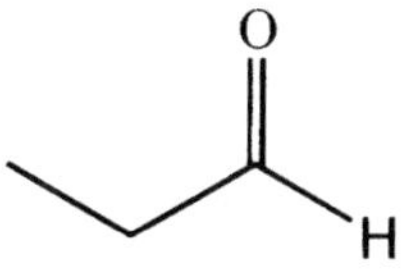

Propanal

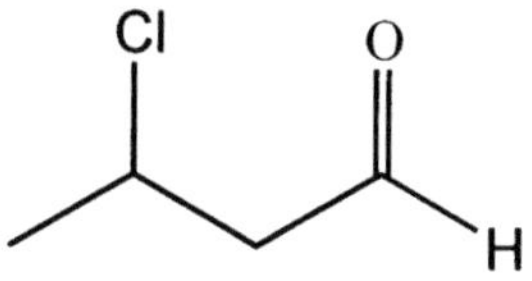

3-chlorobutanal

Diagram 4.47

A common mistake in aldehyde naming is the omission of the carbonyl carbon when determining chain length. The carbonyl carbon is *always the first carbon* in the parent chain and must be included.

The chemical and physical properties of aldehydes are similar to those of ketones. This comes as no surprise given the structural similarities between the two functional groups. In general, aldehydes are toxic and possess very strong, pungent odors. They are soluble in both water and organic solvents, depending on the length of the hydrocarbon chain. The polar carbonyl group in aldehydes is capable of hydrogen bonding in water.

Aldehyde functional groups are not usually found in the structures of illegal drugs or controlled substances; however, this group plays an important role in their identification. For example, formaldehyde is used in the Marquis test for initial screening of opium and phenethylamine families of controlled substances. Acetaldehyde is used in chemical-screening tests to detect marijuana, as well as screening secondary amines containing phenethylamine.

4.2.8 Carboxylic Acids

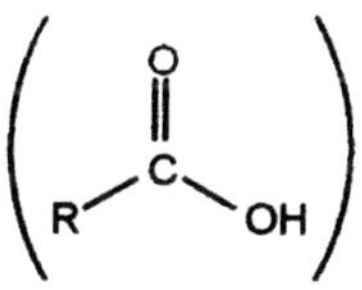

Diagram 4.48

Carboxylic acids are organic molecules characterized by the presence of a carbon containing both a double bond to oxygen (carbonyl group) and a single bond to a hydroxyl group (–OH). They represent a diverse class of organic molecules that can act as both acids (donate H^+) and bases (accept H^+), depending on the pH of the solution.

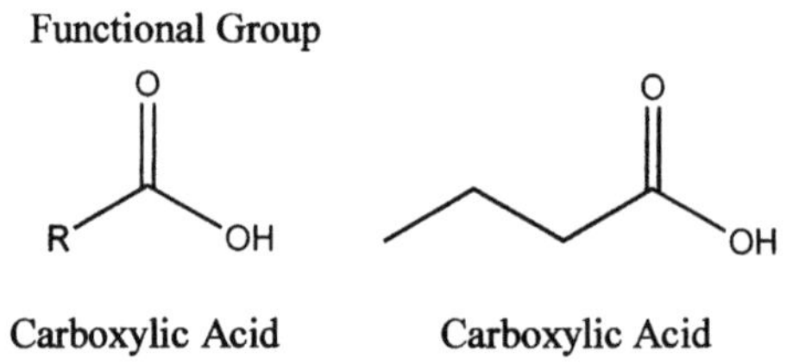

Diagram 4.49

The names of carboxylic acids end in "-oic acid," indicating the presence of the carboxyl group, i.e.; propanoic acid, butanoic acid, and pentanoic acid. When naming carboxylic acids, the "-e" is dropped from the alkane containing the carboxyl group and replaced with "-oic acid." The chain is numbered from the carboxylic acid group which is always located on the first carbon. Again, the location of the functional group is *not* included in the name.

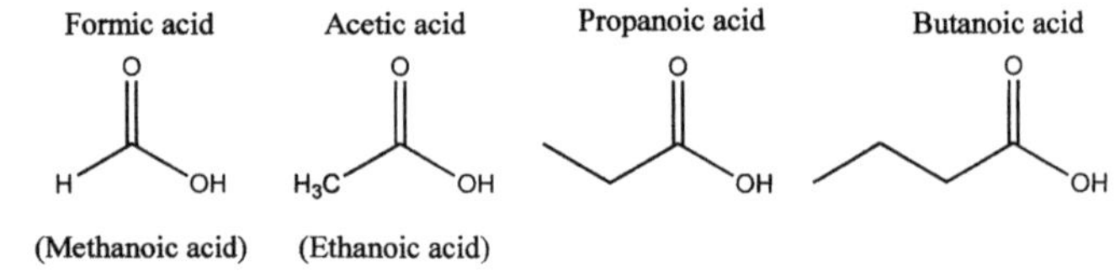

Diagram 4.50

Several members of this class are widely known by their common names and, unlike other functional groups, they can exist in two forms. The forms are distinguished by name and differ only in the presence or absence of hydrogen on the hydroxyl group. In general, the protinated forms (H is present) end in "-ic acid" and the deprotinated forms (H is absent) end in "-ate." This practice is limited to common names and is not routinely applied to systematic naming. Several examples are provided below.

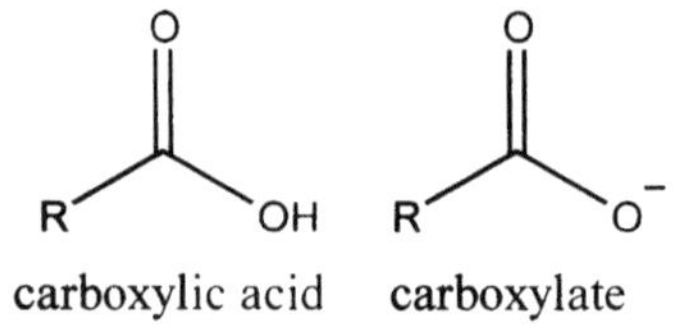

Diagram 4.51

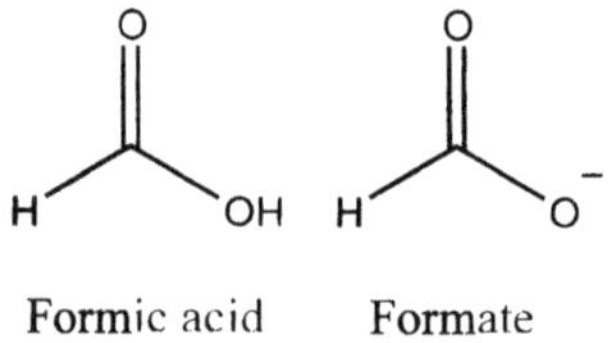

Diagram 4.52

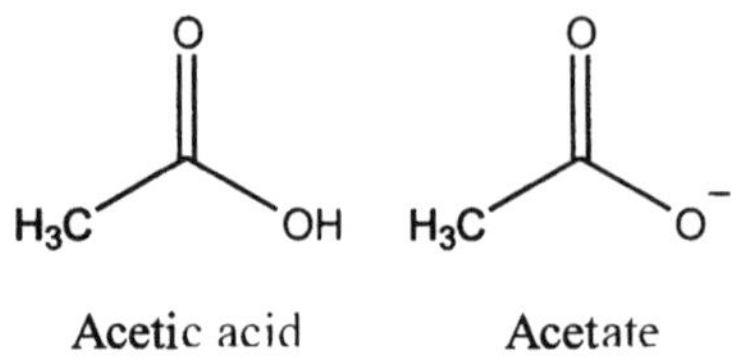

Diagram 4.53

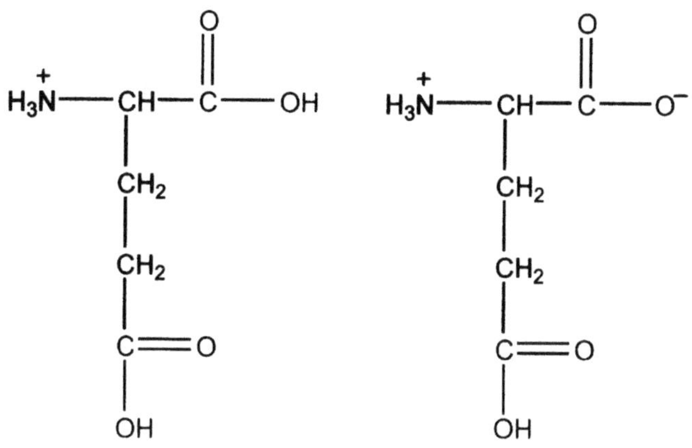

Glutamic Acid Glutamate

Diagram 4.54

Many controlled substances contain this functional group. The most famous is gamma-hydroxybutyric acid (GHB), which gained notoriety by its street name "the date rape drug."

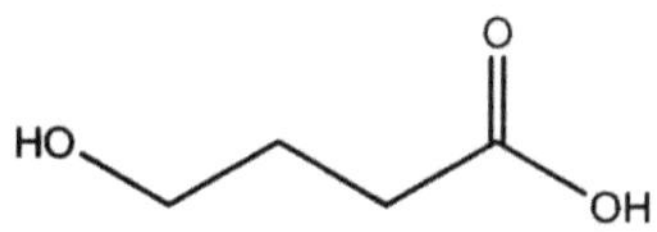

Gamma-hydroxybutyric acid (GHB)

Diagram 4.55

4.2.9 Esters

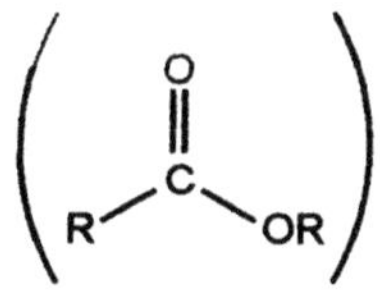

Diagram 4.56

Esters are organic compounds containing a carbonyl bound to an alkoxy group (–OR). The alkoxy group is oxygen bound to a hydrocarbon chain of varying length.

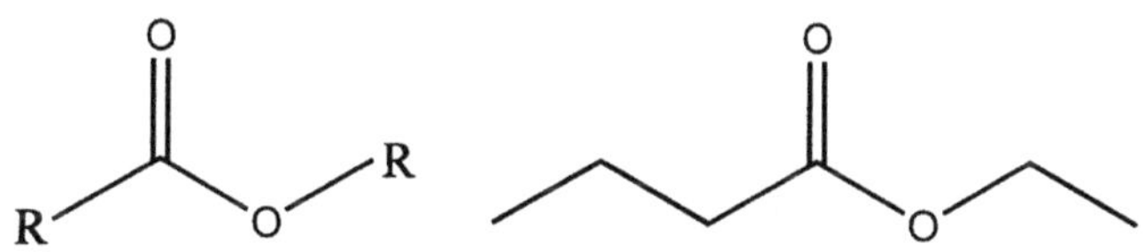

Ester Group An Ester

Diagram 4.57

The systematic naming of esters, although important, is not relevant to applications in forensic chemistry and therefore will not be discussed. Although some of the most widely recognizable drugs contain ester groups, they are not named as ester derivatives. Nonetheless, it is important to recognize this functional group as part of the structure of various drugs and other controlled substances. The consistency with which ester groups are found with amino groups in some of the most dangerous drugs in existence is somewhat surprising.

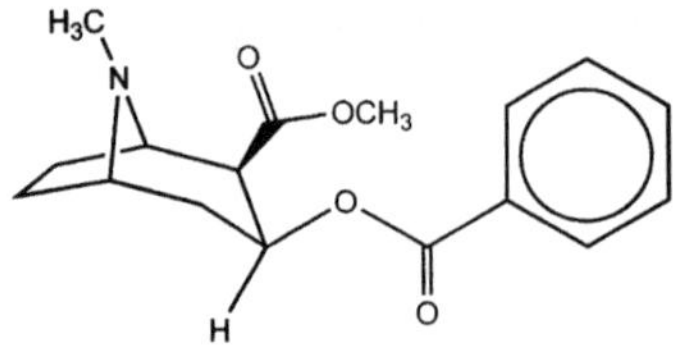

Cocaine

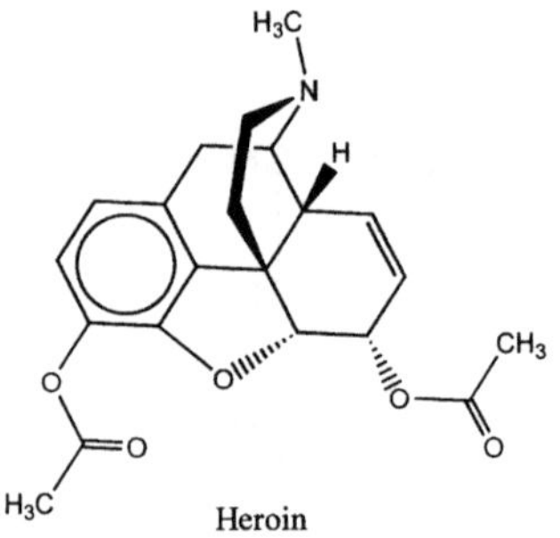

Heroin

Diagram 4.58

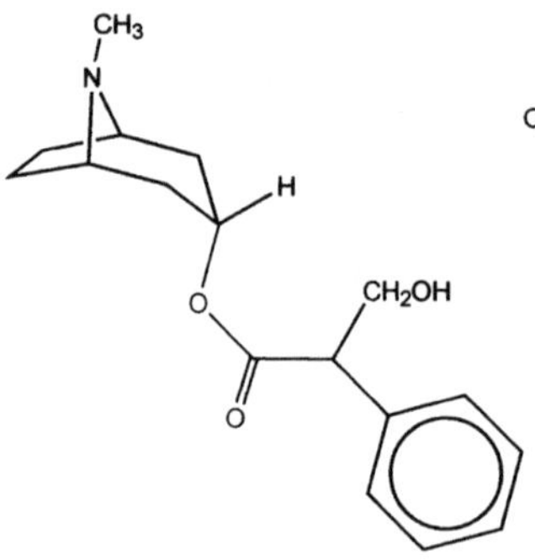

Atropine

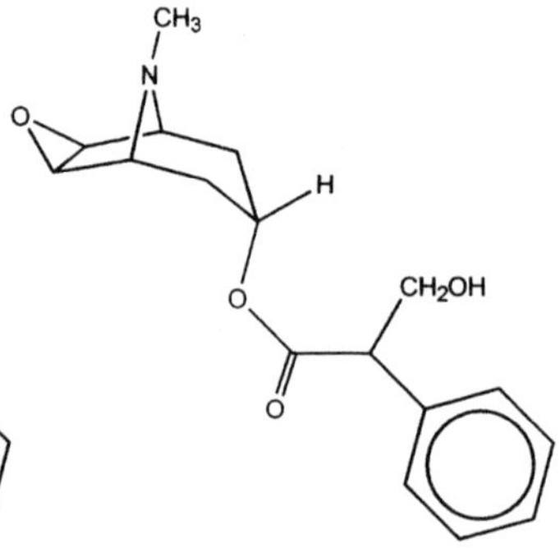

Scopolamine
(truth serum)

Diagram 4.59

4.2.10 Nitro Compounds

$$(R-NO_2)$$

Diagram 4.60

Nitro compounds are distinguished by the presence of the highly polar nitro functional group – a nitrogen atom bound to two oxygen atoms ($-NO_2$). The chemical formula shows that the group does not carry an overall net charge; however, the structural formula reveals the highly polar nature of this functional group.

CH_3NO_2

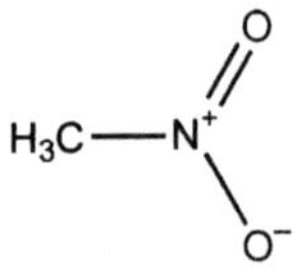

Nitromethane Nitromethane

Diagram 4.61

Nitro groups are not typically found in the structures of illegal drugs; however, they are an important part of forensic investigation. Nitro compounds are primarily used to test and detect a variety of functional groups frequently found in illegal drugs and controlled substances (Fig. 4.2).

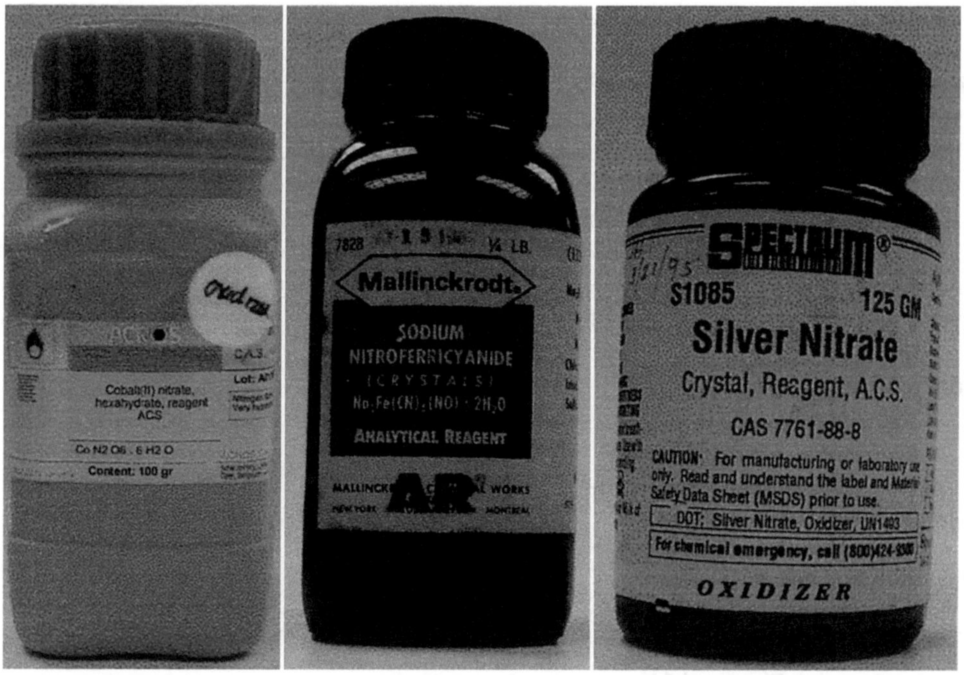

Fig. 4.2 Examples of nitro compounds used to detect a number of groups found in illegal drugs and controlled substances. (*Left*) Cobalt (1) nitrate is used to detect tertiary amines. (*Center*) Sodium nitroferricyanide is used to detect secondary amines; (*Right*) Silver nitrate is used to detect acids.

4.2.11 Amines

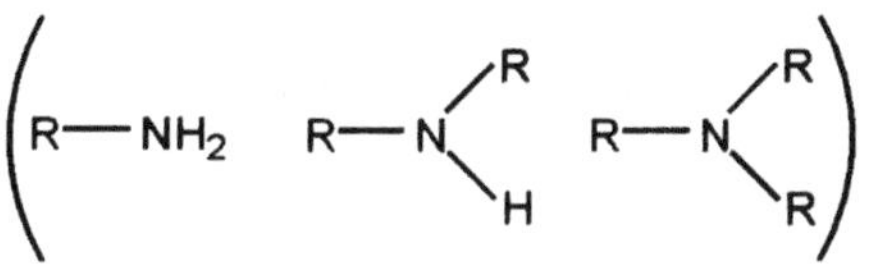

Diagram 4.62

Amines are organic compounds that contain an amino group – a nitrogen bound to one, two, or three hydrocarbon groups. All amines are organic derivatives of ammonia (NH_3) formed by replacing hydrogen atoms with hydrocarbons. Amines are classified as primary (1°), secondary (2°), or tertiary (3°), depending on the number of hydrocarbon substitutions;

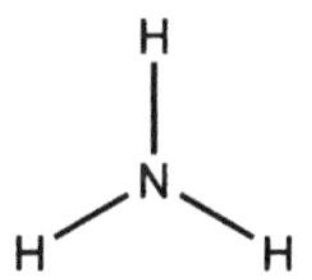

ammonia

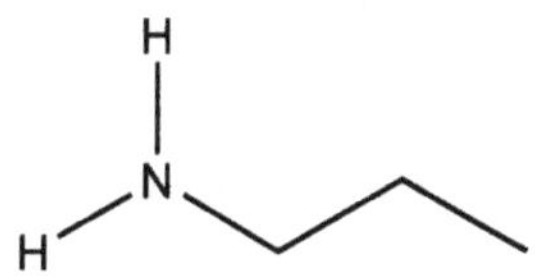

propylamine
(Primary amine)

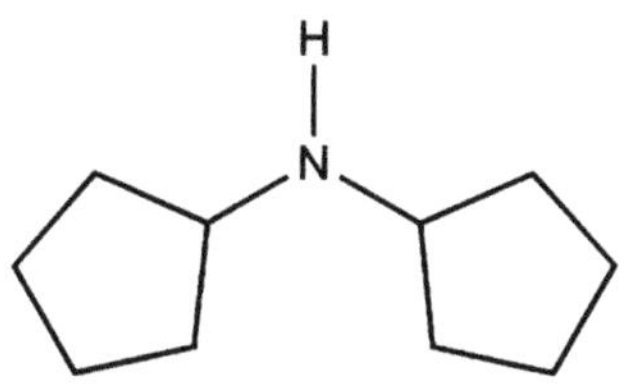

dicyclopentanamine
(Secondary amine)

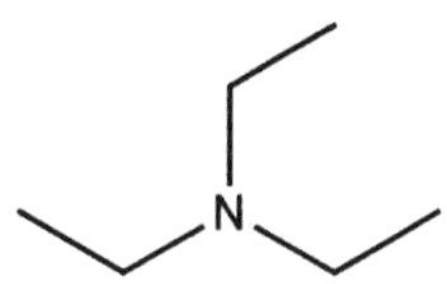

triethylamine
(Tertiary amine)

Diagram 4.63

primary amines contain *one hydrocarbon substitution*, *secondary amines* contain *two substitutions*, and *tertiary amines* contain *three*. Although several methods are used to name compounds in this class, it is common for the names to incorporate the substituted groups and end with the suffix "-amine." It is important to note that no constraints are placed on the hydrocarbon substitutions; they may be identical, different, chains, rings, etc.

In the field of forensic investigation, amines are arguably the most important class of organic compounds. Amino groups are found in the structures of some of the most destructive and addictive substances known. Heroin, cocaine, phencyclidine (PCP), lysergic acid diethylamide (LSD), and morphine are tertiary amines, as is the poison strychnine. Methamphetamine, ephedrine, pseudoephedrine, and ketamine are secondary amines. Amphetamine, tryptamine, and 3,4-methylenedioxyamphetamine (MDA) are examples of primary amines. The high occurrence of amino groups in the structures of drugs and controlled substances has led to the development of a variety of tests and screening methods to detect amines.

4.2.11.1 Primary Amines

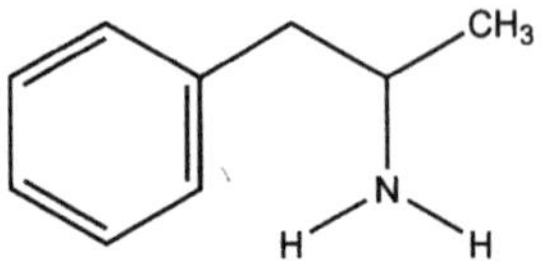

Amphetamine

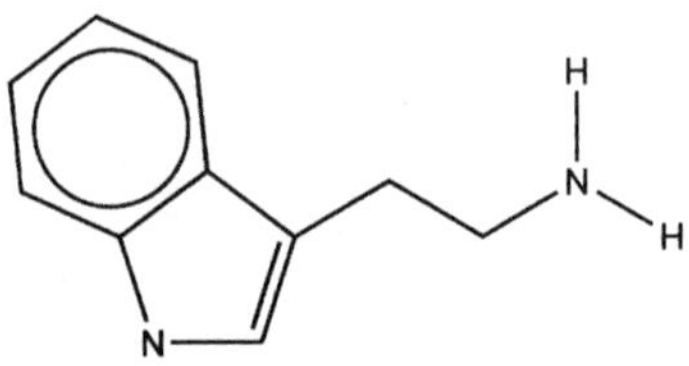

Tryptamine

Diagram 4.64

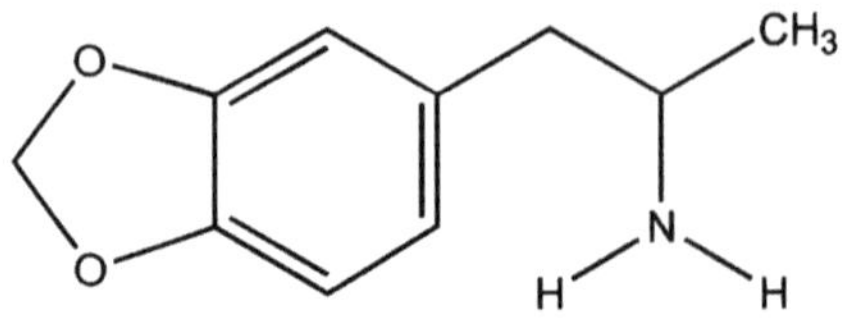

3,4-Methylenedioxyamphetamine

Diagram 4.65

4.2.11.2 Secondary Amines

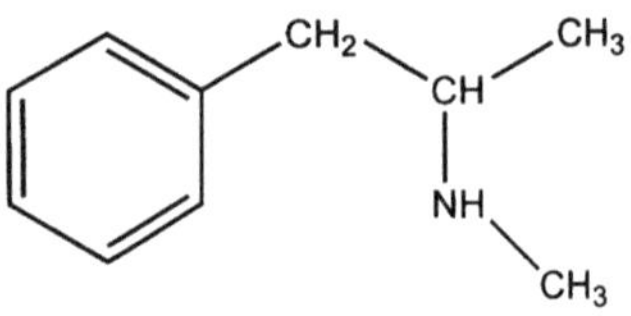

Methamphetamine

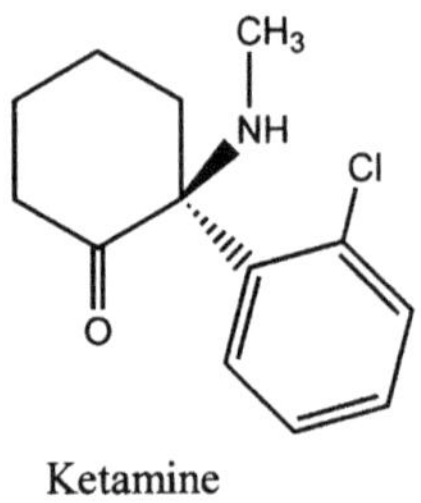

Ketamine

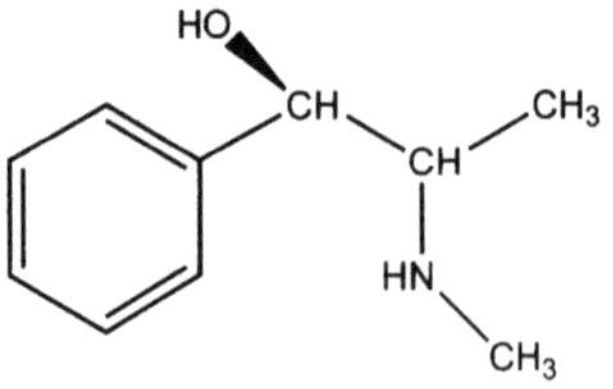

Pseudoephedrine

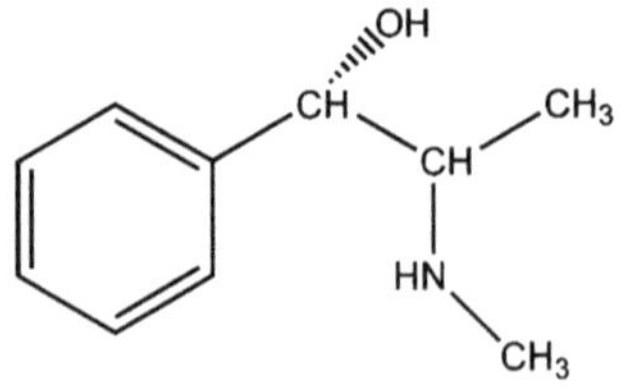

Ephedrine

Diagram 4.66

In general, chemical screening tests for secondary amines are very reliable. There are notable exceptions with ketamine, ephedrine, and pseudoephedrine where intramolecular (within molecule) interactions "mask" the secondary amine functional group. The hydroxyl group is the culprit with ephedrine and pseudoephedrine. The hydrogen atom on hydroxyl forms a hydrogen bond with the amino nitrogen and disguises the secondary amine. Pseudoephedrine is shown below for illustration.

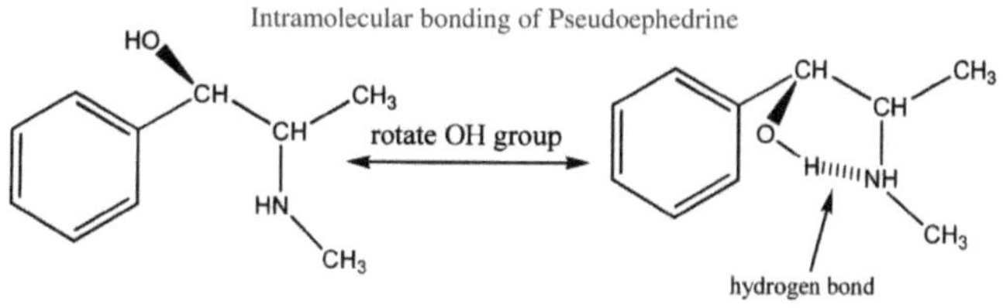

Diagram 4.67

Ketamine is slightly more complicated; it is a secondary amine that tests positive using tertiary amine chemical screening. This results from an intermolecular rearrangement involving the chlorine atom in ketamine.

4.2.11.3 Tertiary Amines

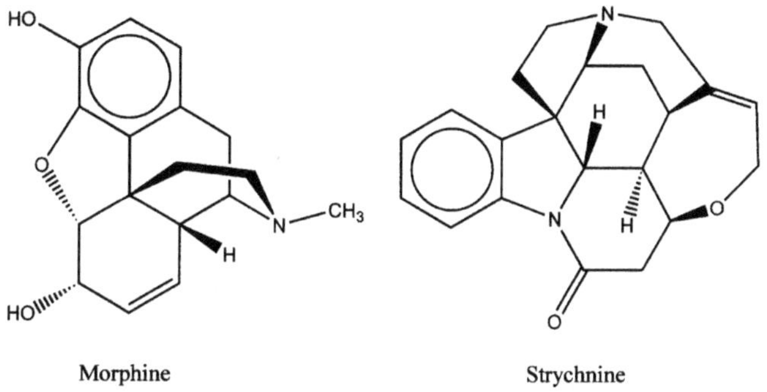

Diagram 4.68

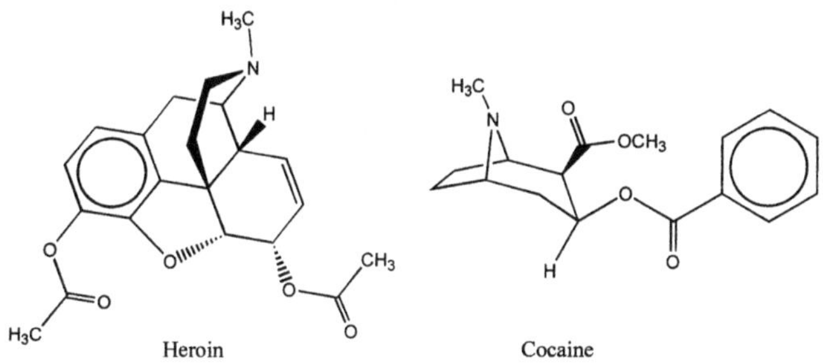

Diagram 4.69

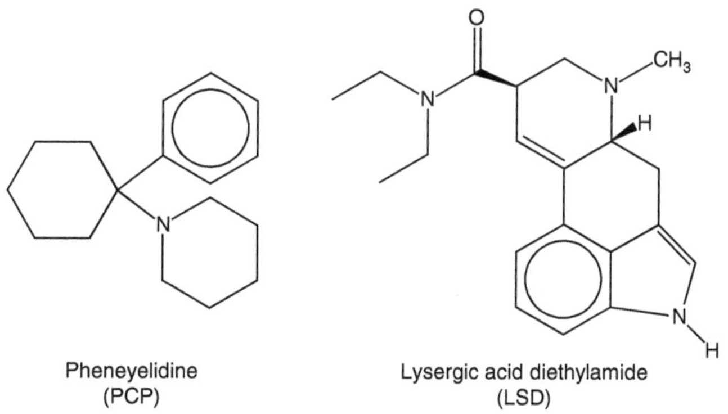

Diagram 4.70

LSD does not test positive using tertiary amine chemical screening. This behavior is the result of two factors: intermolecular rearrangements and structural symmetry of the amino group. All is not lost, because LSD does test positive using Van Urk's reagents. See Chap. 7, for more information.

4.3 Methyl Group (–CH_3)

The methyl group is not a functional group and is not used to classify organic molecules. This group does not represent a dominant reactive site that may be used to predict chemical behavior. However, the importance of methyl groups in the production of designer drugs and controlled substances warrants consideration. In some cases, there may be extreme differences in the biological activity of two structures differing only in the presence or absence of a single methyl group. Under these circumstances, it may be tempting to label the methyl group as a functional group. A close examination of the structure suggests that it may be more appropriate to attribute the observed differences in activity to a modification of existing functional groups. For example, amphetamine and methamphetamine differ only in the added methyl group seen in methamphetamine. The difference in activity between the two is more likely the result of the conversion of a primary amine into a secondary amine, and not based on the reactivity of the added methyl group alone.

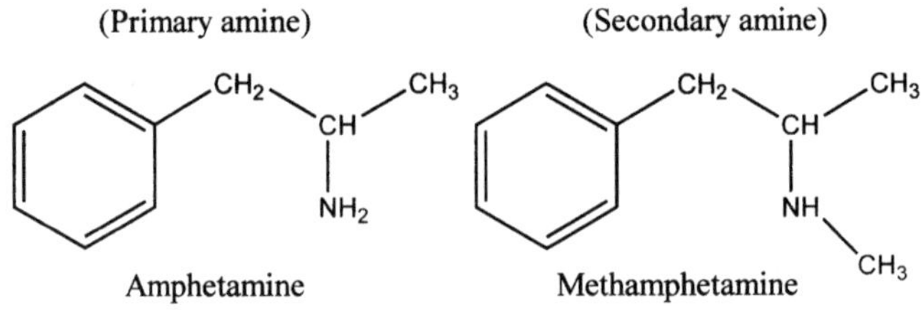

Diagram 4.71

It is also possible to produce illegal substances by simply changing the position of a methyl group on a single molecule. DMT and α,N-DMT (α,N-dimethyltryptamine) differ only in the position of a single methyl group. Molecules that exhibit this relationship are termed positional isomers. The production of positional isomers is often done to avoid legal restrictions or regulations imposed on one of the isomers and the conversion is generally a simple task.

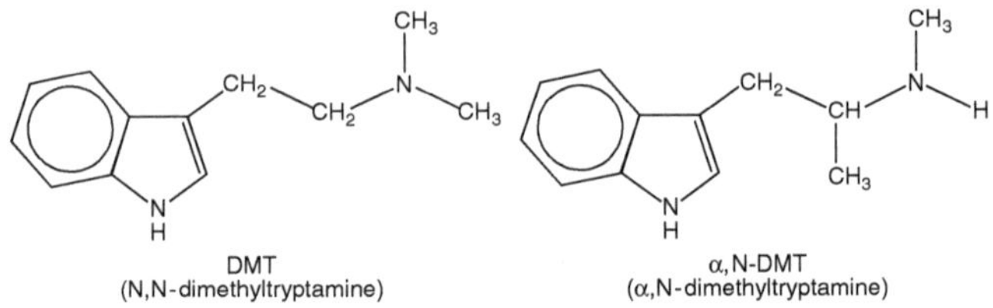

Diagram 4.72

4.4 Compounds Containing Multiple Functional Groups

Functional groups are atoms, groups of atoms, or structural features that undergo predictive chemistry. The chemical behavior of a specific functional group is generally used to detect, identify, and classify both legal and illegal chemicals. A vast majority of drugs, designer drugs, and controlled substances contain multiple functional groups. Although their classification is usually based on one of the groups, all functional groups have some influence on the reactivity of the compound. For this reason, a variety of tests are usually required to conclusively identify a particular substance. Examples of compounds containing multiple functional groups are shown below.

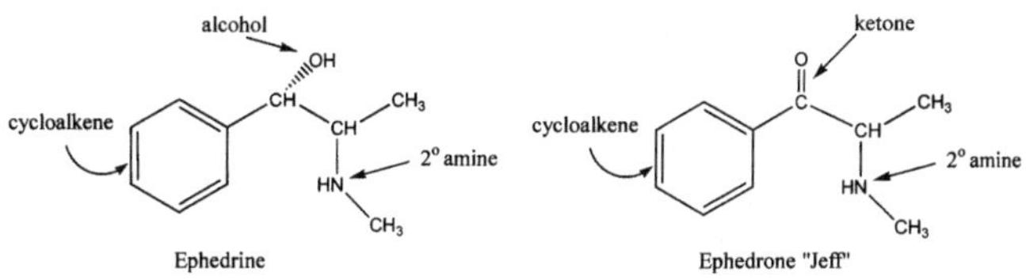

Diagram 4.73

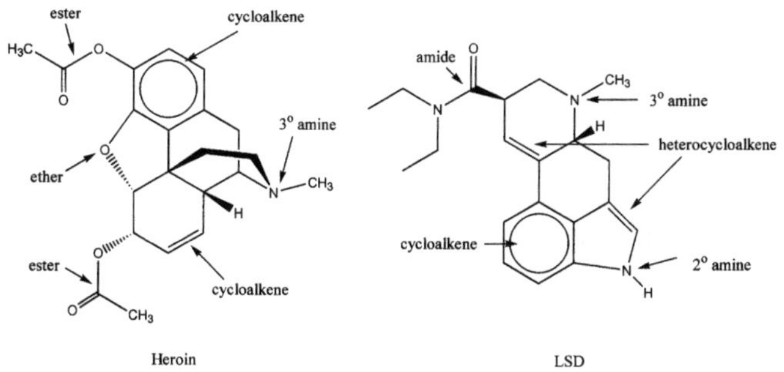

Diagram 4.74

Lysergic acid diethylamide (LSD) is named as an amide derivative, a functional group we did not discuss in our study of organic chemistry.

4.5 Chirality

Take a moment and compare your right hand to your left. They are virtually identical in most respects, but a right glove will not fit on your left hand. Your hands are mirror images that are not superimposable on one another, a fact you may have never considered, nor had cause to, until now. This property of "*handedness*," or chirality (*chiral*: Greek-"*handed*"), is characteristic of organic compounds containing chiral carbons – any *carbon bound to four different groups*. The *nonsuperimposable mirror images of chiral molecules* are stereoisomers called enantiomers. Stereoisomers have the *same chemical formula but differ in the spatial arrangement of atoms*. Enantiomers have almost identical physical and chemical properties; for example, same melting points, boiling points, and solubility. Your hands, which themselves are enantiomers, are clearly different from one another; by analogy, there must be a difference that distinguishes the enantiomers of chiral molecules. The search for this difference leads us to an unlikely candidate, the electromagnetic spectrum, or light. Enantiomers exhibit optical activity by *rotating plane-polarized light* in equal, but opposite directions. One enantiomer will rotate light *clockwise* (*right*) and is said to be *dextrorotary* (or also *dextrorotatory*), whereas the other will rotate light *counterclockwise* (*left*) and is referred to as *levorotary* (or also *levorotatory*). The direction of rotation is indicated in the names of optically active molecules using a "(*d*)" or "(+)" for dextrorotary, for example, (*d*)-morphine or (+)-morphine, and "(*l*)" or "(–)" for levorotary, for example, (*l*)-morphine or (–)-morphine. The optical activity of a specific molecule is determined using a *polarimeter* to measure the rotation of polarized light. The R/S convention is a quick, reliable method used to distinguish enantiomers without the necessity of polarimetry experiments. The details of this method are beyond the scope of this text; however, a basic overview is presented because this convention has relevance in our study of forensic chemistry.

The four different groups attached to chiral carbons are assigned priorities using the Cahn–Ingold–Prelog system (not discussed). The structure is oriented in a manner that allows the viewer to look down the carbon bond containing the lowest priority group, usually a carbon–hydrogen bond. The remaining groups are arranged in a circular configuration facing the viewer. One simply counts the remaining groups 1, 2, 3; if counting is in a *clockwise* direction, the *R-isomer* is present (*rectus*: Latin-"*right*"). If counting is *counterclockwise*, the *S-isomer* is present (*sinister*: Latin-"*left*"). It is worth noting that the R/S designation does *not* indicate optical rotation; it is simply a quick, easy method used to differentiate enantiomers. For example, (S)-glyceraldehyde is levorotary whereas (S)-alanine is dextrorotary. An understanding that optically active isomers exist, and the basic recognition of conventions used to differentiate them, is far more valuable than knowing the exact direction of rotation, which is usually of little consequence.

A sound foundation in organic chemistry is an absolute necessity in forensic investigation. Forensic scientists must be skilled in the interpretation of complex chemical procedures and results. In addition, they must be able to communicate this knowledge to members of a jury using common terminology. This ability is specific to forensics and generally not found in other areas of science. It is developed through knowledge, training, and experience and distinguishes forensic scientists as some of the most versatile and diverse in the scientific community.

4.6 Questions

1. Define organic chemistry.
2. Please define the term functional group to members of the jury.
3. Name the first ten alkanes with their chemical formulas.
4. Draw the structural formula and condensed structural formula for heptane.
5. Draw methane using the wedge convention.
6. Draw the following:
   (a) 2,2-dichlorohexane
   (b) 1,2-dibromo-4-fluorononane
   (c) 2,2,3-trimethyl-3-chloroheptane
   (d) Cyclopentane
7. Name the first four alkenes.
8. Draw the structures with a chemical formula C_4H_8 (there are three), and name each.
9. Draw the following:
   (a) 2-hexene
   (b) 3-chloro-2-heptene
   (c) Cyclohexene
10. Explain why acetylene is not a member of the alkene class.
11. Define aromatic character.
12. Name the following:

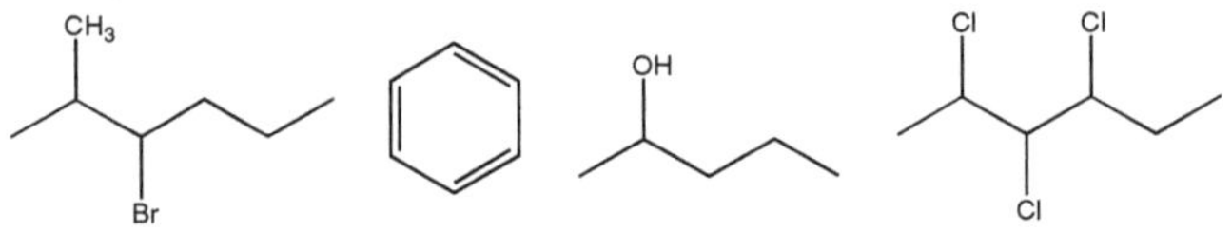

13. Draw the following alcohols and classify each as primary, secondary, or tertiary.
   (a) Isopropanol
   (b) 2-methyl-2-propanol
   (c) 1-hexanol
   (d) 3-heptanol
14. Discuss alcohol solubility in water.
15. Why is acetone the smallest member of the ketone class?
16. Draw 3-hexanone and butanone.
17. Name the following:

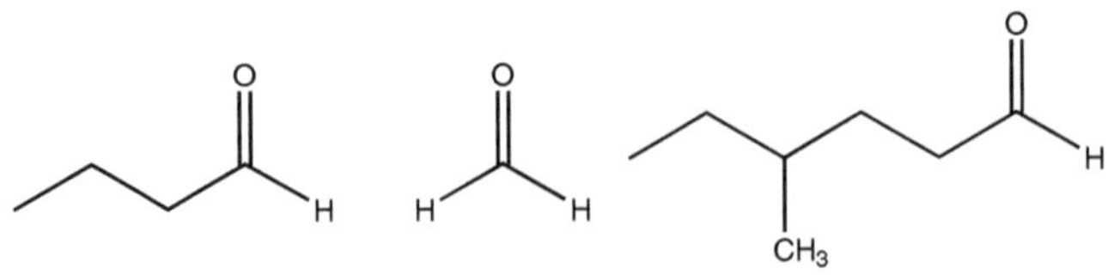

18. Draw the following:
   (a) 2-chlorobutanoic acid
   (b) Hexanoic acid
   (c) Acetic acid
19. Classify the following amines as primary, secondary, or tertiary.
   (a) Amphetamine

(b) 3,4-methylenedioxamphetamine
(c) PCP
(d) Cocaine
(e) Ketamine
(f) Ephedrine

20. Identify all functional groups in the following:
    (a) Heroine
    (b) LSD
    (c) Ephedrine
    (d) Cocaine
    (e) PCP

Suggested Reading

Jones, M. *Organic Chemistry*, 3rd ed.;W.W. Norton & Company: New York, 2004, (Chapter 21).

Part II

Tools of Forensic Chemistry

5 Forensic Language

5.1 Defining Drugs

Drugs, narcotics, and controlled substances may be defined as:

- Any substance that causes dependency in humans.
- Any substance intended for use in the diagnosis, cure, mitigation, treatment, or prevention of disease(s).
- Any substance that alters the mind, senses, mood, or thoughts.
- Any substance listed in the Official United States Pharmacopeia, the Official Homeopathic Pharmacopeia of the United States, or the Official National Formulary.

5.2 Origin of Drugs (Narcotics)

5.2.1 Natural Drugs

Heroin, cocaine, tetrahydrocannabinol (THC), bufotenine, psilocin, and psilocybin are narcotics derived from either plants or animals. Heroin is extracted from the poppy plant using a relatively simple procedure, and cocaine is easily isolated from coca plants. THC can be extracted from marijuana plants or its effects can be obtained from direct use of the dried plant (i.e., smoking). LSD is extracted from the ergot of rye and psilocin/psilocybin is isolated from mushrooms. Bufotenine is collected from the skin glands of the Bufo toad or from toadstool mushrooms.

5.2.2 Synthetic Drugs

Synthetic drugs are derived from mineral sources using a wide range of chemical processes. Barbiturates (produced from pyrimidine), phenethylamine analogs (except drugs in Khat or peyote plants), and the tryptamine family of drugs represent common examples of synthetic drugs.

5.2.3 Psychotropic Drugs (Mind Altering)

"Excitantia": Stimulants, such as caffeine and amphetamine
"Inebriantia": Intoxicants, such as ethanol and nitrous oxide
"Hypnotica": Hypnotics, such as methaqualone and mecloqualone
"Euphorica": Analgesics/tranquilizers, such as morphine and heroin
"Phantastica": Hallucinogens, such as psilocyn and mescaline
Psycoanaleptique: Cocaine and amphetamine

J.I. Khan et al., *Basic Principles of Forensic Chemistry*, DOI 10.1007/978-1-59745-437-7_5,
© Springer Science+Business Media, LLC 2012

Psycholeptiques: Morphine and heroin
"Entactogens": MDMA (3,4-methylenedioxymethamphetamine, ecstacy) and MDA (3,4-methylenedioxyamphetamine)
Psychodysleptiques: Mescaline and LSD (lysergic acid diethylamide)

5.3 Dependence and Addiction

5.3.1 Physical Dependence

Physical dependency is a condition resulting from chronic drug use that is characterized by the physiological side effects of tolerance and withdrawal.

Tolerance is the need to ingest progressively larger amounts of a drug to maintain a desired effect. It is a condition characterized by a marked decrease in both time duration and intensity of analgesia, euphoria, and sedation associated with a specific dosage. Tolerance development is inconsistent and unpredictable; toxic side effects often accompany increased tolerance.

Withdrawal is a term used to describe the unfavorable physical symptoms that result if drug use is suddenly stopped or dosage is drastically reduced. Withdrawal symptoms can range from mildly unpleasant to life threatening and their severity depends on a number of factors, i.e.: age, sex, type of drug, frequency and duration of abuse, daily dosage, route of administration, and concurrent abuse of other drugs. In general, short-term narcotics tend to produce shorter, more intense withdrawal symptoms, while longer-acting narcotics produce withdrawal symptoms that are protracted but less severe. The symptoms associated with heroin/morphine addiction are usually experienced just prior to the next scheduled dose. Initial symptoms may include watery eyes, runny nose, yawning, and sweating, followed by restlessness, irritability, loss of appetite, nausea, and tremors. The advanced stages are marked by severe depression, vomiting, elevated heart rate and blood pressure, muscle spasms, chills, excessive sweating, and pain in the bones, back muscles, and extremities. A suitable narcotic can be administered at any stage of withdrawal that will dramatically reduce the symptoms. Without intervention, the effects of withdrawal will slowly subside and most of the physical symptoms will disappear in 7–10 days.

Drug users often abuse a specific or preferred drug, and it is not uncommon to substitute drugs that produce similar effects (often the same drug class). Drugs within a class are often compared using terms, such as potency and efficacy. *Potency* defines the amount (dosage) of a drug that must be taken in order to produce a desired effect. *Efficacy* is the capacity of a drug to produce a given (desired) effect, regardless of dose.

The physical effects produced by any drug can vary significantly and are largely dependent on the dose, route of administration, and individual sensitivity to the drug. Concurrent use of several types of drugs may either enhance or block specific effects. As a result, abusers often take more than one drug in an effort to increase the desired effects, while minimizing unwanted side effects. This practice can dramatically increase the risks associated with drug abuse because the overall effects cannot be accurately predicted. This suggests that a genetic component may exist that predisposes individuals to either drug toxicity or addictive behavior.

5.3.2 Psychological Dependence

Psychological dependency is a perceived "need" or "desire" for a drug and is commonly associated with addiction. Individuals who are psychologically dependent often feel (or believe) that they cannot function normally without continuous drug use. Psychological dependency can last much longer than physical dependency and is a primary reason for relapse after a period of either treatment or abstinence.

The psychological dependence associated with narcotic addiction is complex and protracted. Addicts often ponder drug use long after the physical need for the drug has subsided. They may feel either uneasy or overwhelmed while performing daily activities without the influence of drugs (sober). There is a high probability of relapse after narcotic withdrawal, if changes are not made to either the addict's physical environment or behavioral motivators (associates).

There are two major patterns of narcotic abuse or dependence observed in the United States. One involves individuals whose drug use was initiated within the context of medical treatment. These individuals escalate use by obtaining fraudulent prescriptions or elicit drugs. The other, more common pattern is initiated outside the clinical setting through either the experimental or recreational use of narcotics. The majority of individuals in this category may abuse narcotics periodically for either months or even years. Although they may not become addicts, the social, medical, and legal consequences of their behavior are very serious. Some experimental users will escalate narcotic use to the point of physical and psychological dependency. Individuals initiating drug use at an early age are more likely to progress from casual use to dependence and addiction.

5.4 Drug Abuse

Narcotics are used therapeutically to treat pain, suppress cough, alleviate diarrhea, and induce anesthesia. They may be administered either orally, by injection, in suppository form, or through the skin (skin patches). Aside from their curative effects, narcotics produce a general sense of well-being by reducing tension, anxiety, and aggression. Although these effects are helpful in a clinical setting, they are also a contributing factor in their abuse. Narcotics are often smoked, sniffed, or injected when abused (Fig. 5.1).

A wide range of side effects (varying in severity), such as drowsiness, inability to concentrate, apathy, decreased physical activity, constriction of the pupils, dilation of subcutaneous blood vessels (causes flushing), constipation, nausea, vomiting, and respiratory depression, may accompany excessive and prolonged use of narcotics. The therapeutic and toxic effects become more pronounced with increased dosage. Apart from acute intoxication, there is generally no loss of motor coordination or slurred speech with narcotic use. This is in contrast to the general effects of most depressants.

5.5 Hazards of Drug Abuse

The health hazards of illicit drug use include an increased risk of infection, disease, and overdose (Fig. 5.2). Pharmaceutical products are manufactured in a controlled environment under strict regulatory control. The production process is well documented as is the concentration and purity of the final product. By contrast, street drugs produced in clandestine labs have

Fig. 5.1 Common vehicles used for narcotic abuse: smoking pipe, cigarette containing narcotics, injection needle, water pipe, equipment, and container used to make narcotic cigarettes.

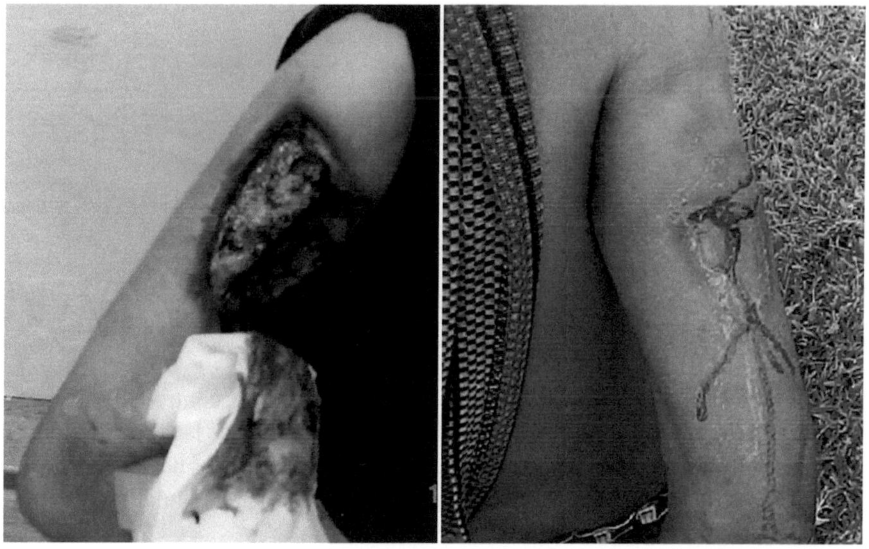

Fig. 5.2 The destructive nature of drug abuse is illustrated in the arms of the two addicts. Infection (*left*) and scars and track marks from injection (*right*).

unknown compositions. Health issues and medical complications common to most narcotic abusers are the result of adulterants, contaminants, and/or nonsterile injection practices.

Skin, lung, and brain abscesses, endocarditis (inflammation of the lining of the heart), hepatitis, and AIDS are prevalent among narcotic abusers. The composition and purity of illegally manufactured drugs is always in question. Therefore, the physiological effects are inconsistent and unpredictable and can be fatal in some cases. Physical signs of narcotic overdose include constricted (pinpoint) pupils, cold clammy skin, confusion, convulsions, severe drowsiness, and respiratory distress.

5.6 Structural Relationships

5.6.1 Analogs

The question of analogs is often encountered in trials involving either drugs or drug-related crimes. It is often asked if a specific chemical is an analog of a scheduled controlled substance. A substance is considered an analog of another substance if:

1. Both substances are structurally similar.
2. Both substances have the same chemical reactivity.
3. Both substances have similar physiological effects.
4. Both substances have similar toxicological effects.
5. Both substances have similar addictive natures.

The two substances must satisfy at least two of the above conditions to be considered analogs. For example, pseudoephedrine is structurally similar to methamphetamine and is used as a precursor in its production. However, pseudoephedrine does not have the same reactivity or physiological and toxicological effects as methamphetamine. Thus, pseudoephedrine is not an analog of methamphetamine.

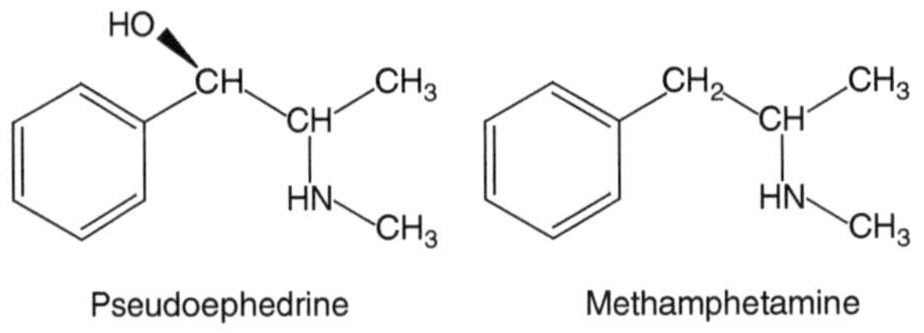

Diagram 5.1

Bufotenine and psilocin are analogs; note the similarity in structures.

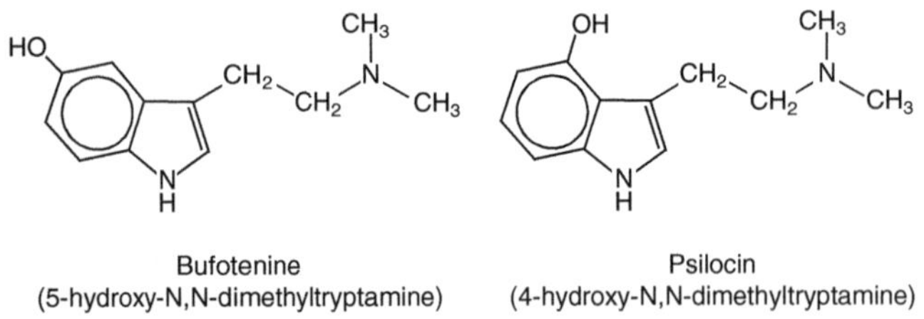

Diagram 5.2

5.6.2 Designer Drugs

The classification of a particular substance as an illegal drug, controlled substance, or designer drug is based on its molecular structure (structural formula). In an effort to circumvent these legal restrictions, underground chemists modify the structure of existing drugs (legal and illegal) to produce analogs known as designer drugs. For example, MDMA (street name: ecstasy) was "designed" from MDA in an effort to evade the laws and regulations controlling MDA.

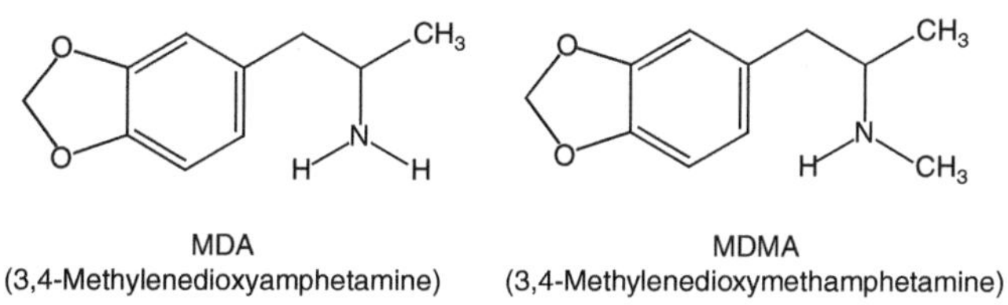

Diagram 5.3

The most common types of designer-drug analogs are phencyclidine (PCP), fentanyl and meperidine (synthetic analgesics), and the stimulants/hallucinogens amphetamine and methamphetamine. Designer drugs are often much stronger and more toxic than their precursor counterparts, and brain damage is possible with only a single dose. The federal government has passed legislation regulating all chemicals that are structurally similar to controlled drugs. These laws are designed to stop observed patterns of designer-drug production and usage.

5.6.3 Isomers

Isomers are compounds that have the same number and types of atoms (same chemical formula) but differ in the structural arrangement of the atoms. Isomers have different chemical and physical properties. There are several types of isomers, and the differences are indicated in the name of the isomer.

Structural: Compounds with the same molecular formula but a different connectivity of atoms (different structures).

Positional: Compounds with the same molecular formula but differing in the position/location of an atom or functional group.

Stereoisomers: Compounds with the same molecular formula and same connectivity but differing in the arrangement of atoms in the three-dimensional space.

Diastereomers: Compounds with the same molecular formula that are not mirror images of each other.

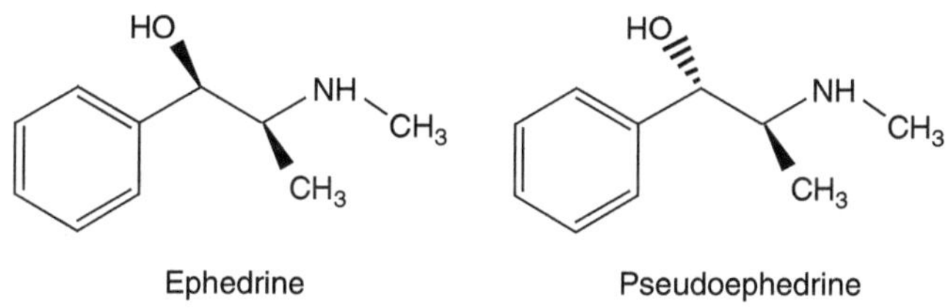

Diagram 5.4

Enantiomers: Compounds with the same molecular formula that are nonsuperimposable mirror images of one another. They must contain at least one chiral carbon, a carbon bound to four different atoms or groups. The nonsuperimposable mirror images are differentiated by their ability to rotate plane-polarized light and designated as *d* or *l*. The *d*- and *l*-forms of methamphetamine are an example.

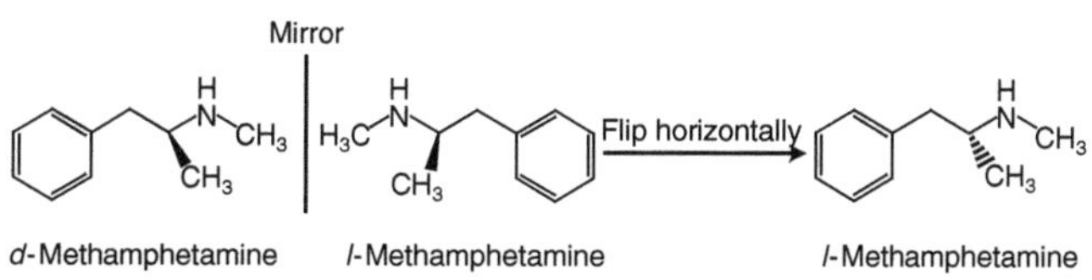

Diagram 5.5

5.7 Controlled Substance Statutes

5.7.1 Controlled Substances Act

The Controlled Substances Act (CSA) regulates five classes of drugs in the United States: narcotics, depressants, stimulants, hallucinogens, and anabolic steroids. Each class has distinguishing properties, and the drugs within each class often produce similar effects. However, all controlled substances, regardless of class, share a number of common features.

With the exception of anabolic steroids, most alter mood, thought, and feelings by targeting the central nervous system (brain and spinal cord). Drug abuse occurs when drugs are used in a manner (or amount) inconsistent with either the medical or social patterns of use in a particular culture. In legal terms, the nonsanctioned use of substances controlled in Schedules I through V of the CSA is considered drug abuse. Drugs classified in the CSA can be used legally for medical treatment under the supervision of a licensed physician; however, the use of these same drugs (or intended use) outside the clinical setting is drug abuse.

5.7.2 Controlled Substances Laws

The production, transportation, distribution, prescription, possession, and consumption of all narcotics, depressants, stimulants, hallucinogens, and anabolic steroids are strictly controlled and regulated. The controlled substances laws impose mandatory penalties (fines or imprisonment) on any individual or entity engaged in illegal activity relating to drugs classified in the CSA. The severity of the penalty depends on the quantity and type of controlled substances and may vary from state to state. A few of the drugs regulated by CSA are listed below under the appropriate schedule.

5.7.2.1 Schedule I

Drugs with a high potential for abuse and little to no medicinal value fall under schedule I (typically hallucinogens).

1. Cocaine base
2. Codeine base
3. GHB
4. Heroin and other opiates not listed in schedule II
5. LSD
6. Marijuana or its psychoactive ingredient (tetrahydrocannabinol)
7. Methaqualone
8. Mescaline, including peyote plant and its components
9. Morphine derivatives
10. PCP
11. Tryptamines
    (a) Synthetic analogs: All:
    (b) Natural:
    - Bufotenine: Pure form or present in:
      - Yopo seeds
      - Toadstool mushrooms
      - Bufo toads
    - Psilocin/psilocybin: Pure form or in mushrooms.

5.7.2.2 Schedule II

Drugs with an equal potential for abuse and medicinal use fall under schedule II (generally stimulants).

1. Barbiturates (most analogs)Cocaine: all salts
2. Levo-methorphan
3. Methadone
4. Morphine
5. Opium (raw, extracts, derivatives, and any opiates not listed in schedule I)
6. PCC (a precursor to PCP)

7. Phenethylamine family:
   (a) Amphetamine, its salts and isomers
   (b) Methamphetamine, its salts and isomers
   (c) *N,N*-dimethylamphetamine, its salts and isomer
   (d) Phenyl-2-propanone (P2P)
   (e) MDA
   (f) All analogs of phenethylamine

5.7.2.3 Schedule III

Drugs with less potential for abuse and more for medicinal use fall under schedule III (most anabolic steroids and depressants, some stimulants and prescription drugs).

1. Gamma-hydroxybutyric acid (GHB) and salts
2. Ketamine
3. Lysergic acid
4. Most anabolic steroids

5.7.2.4 Schedule IV

Drugs with a low potential for abuse and a high potential for medicinal use fall under schedule IV (typically depressants not listed in schedule III). Many prescription drugs are included in this category, as are precursors used in the manufacturing of controlled substances.

1. Atropine sulfate
2. Barbital
3. Clonazepam
4. Diazepam
5. Phenteramine

5.7.2.5 Schedule V

This schedule contains chemicals and precursors typically used in the manufacturing of controlled substances. There is considerable overlap between this category and regulated substances.

1. Barbituric acid, a precursor to barbiturates
2. Phenylpropanolamine, a precursor to amphetamine
3. Piperidine, a precursor to PCC
4. D-Lysergic acid, a precursor to LSD
5. Ephedrine/pseudoephedrine and its salts, precursors to methamphetamine
6. Safrol, a chemical used in the manufacturing of methamphetamine
7. Gamma hydroxybutyrolactone (GBL), a precursor to GHB
8. Hydroiodic acid, a chemical used in the manufacturing of methamphetamine

5.7.3 Controlled Substance: Charges and Offenses

The charges listed in Table 5.1 are taken from the State of California Health and Safety Codes.

5.8 Controlled Substance Submission to Crime Laboratories

Controlled substances are submitted to forensic laboratories from law-enforcement agencies in the local vicinity. Specially trained law-enforcement personnel follow the case-submission policies of the forensic laboratory. In general, all case evidence is sent to the laboratory in sealed and labeled bags or envelopes. The bags or envelopes will contain all agency-related case information, such as the suspect's name, agency name and case number, date of crime, and the quantity of suspected substance.

Table 5.1 Controlled substance: charges and offenses

Charge	Offense
11350	Possession of specified controlled substance (schedule I)
11351	Possession or purchase for sale of specified controlled substance (schedule I)
11351.5	Possession or purchase for sale of cocaine base
11352	Importing, selling, transporting, and or furnishing controlled substance (schedule I)
11353	Adult including minor to violate provisions (schedule I)
11353.1	Penalty enhancement for adult soliciting minor (schedule I)
11353.5	Adult providing controlled substance in a specified area (schedule I)
11353.6	Drug trafficking on or within 1,000 ft. of school ground – penalty enhancement (schedule I)
11355	Substances provided in lieu of controlled substance
11357	Unauthorized possession of marijuana
11358	Unauthorized planting of marijuana
11359	Unauthorized possession of marijuana for sale
11360	Unauthorized selling, transporting, importing, selling, furnishing of marijuana
11361	Adult employing minor or selling to minor
11362	Felony offense for bigger quantity of marijuana
11362.5	Medical use of marijuana
11362.9	California Marijuana Research Program
11363	Planting, cultivating, processing peyote
11364	Possession of devices (paraphernalia) for injecting or smoking controlled substance
11364.5	Drug paraphernalia for sale
11364.7	Unlawfully providing drug paraphernalia
11365	Presence where controlled substance unlawfully smoked or used
11377	Possession of specified controlled substance (schedule II)
11378	Possession or purchase for sale of specified controlled substance (schedule II)
11378.5	Possession for sale of PCP
11379	Importing, selling, transporting, and/or furnishing controlled substance (schedule II)
11379.2	Possession for sale of ketamine
11379.5	Importing, selling, transporting, furnishing PCP
11379.6	Manufacturing or processing controlled substance
11379.7	Penalty enhancement for manufacturing or processing controlled substance
11379.8	Penalty enhancement for possessing controlled substance
11380.5	Drug violation in public parks and beaches
11383	Possession of chemicals, precursors, solvents, and/or glassware for manufacturing of a controlled substance

5.9 Drug Cases in Crime Laboratories

The property section of the crime laboratory receives drug-related case information from client agencies via mail, UPS, or personal courier. They are responsible for logging case information into a database that generates a laboratory case number. All evidence is labeled with the designated laboratory case number and stored in a case-examination locker.

The case-examination locker is accessible to all analysts in the forensic-chemistry section. Prior to either examination or opening of a case, the analyst compares the computer entries to the information on the evidence package. This extra step minimizes data-entry errors and ensures that the analyst has the correct case evidence.

Following the data-entry review, the analysts open a case. The initial activity of the forensic chemist is strictly inventory. All evidence is carefully recovered from its packaging, weighed, and visually inspected. If a case contains more than one item, the analyst examines one item at a time to avoid cross-contamination. The forensic chemist then performs testing on suspected controlled substances. A detailed record of all activities is kept at each stage of the examination. When the examination of the evidence is complete, the analyst reseals the evidence, initials and dates the seal, and places the evidence in the outgoing evidence locker for release to the submitting agency. Forensic laboratories do not store evidence, including controlled substances, after completing their examination.

Table 5.2 Standard usable quantities of controlled substances

Substances	Dosage	Substances	Dosage
Amphetamine	10 mg	Barbiturates	100 mg
Cocaine base	10 mg	Cocaine HCl	10 mg
Codeine	60 mg	DET/DMT	60 mg
Dihydrocodeinone	5 mg	DMT and other analogs	50 mg
Hashish	35 mg	Hash oil	21 mg
Heroin	5 mg	Hydromorphine	2 mg
LSD	0.05 mg	Marijuana	500 mg
MDA/MDMA	100 mg	Mescaline	500 mg
Methamphetamine	5 mg	Morphine	10 mg
Opium	100 mg	Hydrocodone	5 mg
PCP	5 mg	Peyote	10 mg
Psilocin/psilocybin:	10 mg	STP	3 mg
THC	3 mg		

LSD Lysergic acid, *MDA* 3,4-Methylenedioxyamphetamine, *MDMA* 3,4-Methylenedioxymethamphetamin (ecstacy), *PCP* Phenylcyclohexylpiperidine (commonly shortened to phencyclidine), *THC* Tetrahydrocannabinol, *DET* *N*,*N*-diethyltryptamine, *DMT* *N*,*N*-dimethyltryptamine, *STP* 2,5-dimethoxy-4-methylamphetamine (*STP* serenity, tranquility, peace, also called DOM)

5.10 Examination of Controlled Substances

Examination of a controlled substance is a three-stage process. The first stage is a visual inspection of the suspected substance. The analyst relies on experience and the results of the visual inspection to determine the course of the second stage. The second stage consists of a series of color-screening examinations to determine the most suitable confirmatory method to use on the substance. Finally, a confirmatory examination is performed to identify the substance. The analyst reports the collective results from the color-screening- and confirmation examinations and submits a conclusion.

5.11 Usable Quantity

The forensic chemist is not qualified to give testimony on the physiological effect(s) of a given quantity of controlled substance. This is the responsibility of a forensic toxicologist. However, the Federal Drug Enforcement Agency (DEA) has developed a list of controlled substances containing standard dosages. The data listed in Table 5.2 was obtained from this list.

5.12 Court Testimony

An expert witness can testify in a court of law once their area of expertise has been established and recognized by the court. Witnesses are frequently asked questions about their education, qualifications, and relevant experience to determine if they qualify as an expert. The court will render a decision based on the answers and either accept or reject the witness as an expert. If accepted, the witness is qualified to give expert testimony relevant to the case. The list below contains representative questions that could be used to establish expertise.

5.13 Qualifications and Education

1. What is your educational background?
2. What do you do for a living?
3. How long have you been working in this field?
4. What are your job responsibilities?

5. What is your practical training and experience in this field?
6. What factors qualify you as an expert in the field of controlled substances?
7. Are you familiar with procedures used to test this substance?
8. How many times have you tested this substance?
9. Have you previously testified as an expert in any court of law?
   (a) How many times?
   (b) In what courts?
10. Why do you consider yourself an expert in the field of forensic chemistry?

The following court-related questions are provided for informational purposes only. They are intended to provide you with a glimpse of an actual trial setting.

5.14 Questions

1. Describe the differences between natural drugs and synthetic drugs.
2. Discuss the difference between an analog and a precursor.
3. Could a precursor be an analog of a controlled substance?
4. Define an isomer and list three types.
5. List four physical effects of drugs use.
6. Define physiological effects of drugs use.
7. What is the difference between psychological dependence and physical dependence?
8. What differentiates synthetic drugs from natural drugs?
9. What is the difference between a synthetic drug and a designer drug?
10. List two routes of administration of drugs.
11. Describe the difference between potency and physical dependence.
12. Describe the three stages of substance identification.
13. Name three schedule III drugs.
14. What is the usable quantity of morphine?
15. List three characteristics of analogs.

Suggested Reading

Abadinsky, H. *Drug Use and Abuse: A Comprehensive Introduction*, 6th ed.; Brooks/Cole: New York, 2008; chaps. 1 & 7.
California Law: California Health and Safety Code. http://www.leginfo.ca.gov/.html/hsc_table_of_contents.html (accessed April 2009).
Doweiko, H.E. *Concepts of Chemical Dependency*, 7th ed.; Brooks/Cole Publishing: New York, 2009; chap. 4.
Faupel, C.; Horowitz, A.; Weaver, G. *The Sociology of American Drug Use*, 2nd ed.; Oxford Press: New York, 2009; chaps. 1 & 3.
U.S. Drug Enforcement Agency. http://www.usdoj.gov/dea/pubs/csa.html (accessed April 2009).

6 Forensic Documentation

6.1 Introduction

Documentation is the foundation of science as well as the hallmark of a sound legal system. It is often said, "If it is not written down it does not exist." It is absolutely essential to properly document each step of any examination process.

Investigators and forensic personnel are required to submit evidence that appropriate tests were performed and that the results presented in court are (in fact) the test results observed either in the laboratory or at the crime scene. Memory is no longer considered either reliable or trustworthy, and proof through documentation has become the standard for acceptance.

Scientific results must be reproducible before they are recognized by the scientific community. Documentation provides a mechanism for peer review, replication, and research advancement. Examination procedures performed on physical evidence must be properly recorded, along with results. The results must support and justify the conclusions presented in the final report. More importantly, an expert in the same field should agree with the report's conclusion. The reviewer may have a different opinion, but he must agree that the records support the conclusion.

Documentation procedures used in forensic laboratories generally have three common components, each with its own requirements. They are chain of custody, case notes, and the case report.

6.2 Chain of Custody

The chain of custody is a document (or series of documents) that tracks the location of evidence from collection (crime scene) to final disposition (court). The chain of custody maintains the integrity of evidence through accountability. It requires the examiner to document when, and from whom, the evidence was initially received and when, and to whom, the evidence was transferred. It is generally accepted that the evidence is in the examiner's sole care and custody during this period. The documentation procedure usually requires an official form with supplemental notations in the examiner's working notes.

The details of examination procedures are rarely addressed in the chain of custody. Typically, they include the date and time of an examination, the identity of the person performing the examination, and the date and time the evidence was transferred, returned, or released.

The law does not stipulate categories of transfers. Therefore, the transfer of evidence within a laboratory is documented in the same manner as a transfer of evidence between laboratories.

The early use of mass spectrometry illustrates the importance of chain-of-custody documentation. The mass spectrometer is a highly specialized analytical instrument. The analyst had to rely on the operator to perform this specific examination and often had no direct knowledge of how the evidence was handled once it was transferred. Although a forensic chemist could interpret the resulting data and knew the theoretical basis of the instrument's operation, they could not provide direct knowledge that every procedure was followed. As a result, chain-of-custody documentation was required.

J.I. Khan et al., *Basic Principles of Forensic Chemistry*, DOI 10.1007/978-1-59745-437-7_6,
© Springer Science+Business Media, LLC 2012

6.3 Case Notes

6.3.1 Types

Methods of documenting case-related information are contained in the laboratories' standard-operating-procedures manual. However, these manuals often provide only a list of requirements and rarely address the details of recording information. Case notes may be either very simple or highly detailed, depending on the individual practice of the analyst. Regardless, they must be accurate and comprehensive. Incomplete case notes reflect negatively on the analyst and can cause complications at trial. There is a notable difference between incomplete notes and messy notes; incomplete notes lack accuracy, while messy notes are simply difficult to read.

When documenting evidence, remember the old saying, "A picture is worth a thousand words." It is a good practice to take pictures of unusual evidence (Fig. 6.1). If an item contains a large number of specimens, each piece must be individually documented (Fig. 6.2). It is often helpful to use abbreviations in case notes for simplicity and clarity, but do not overuse them (Fig. 6.3). Be sure abbreviations, if used, are from an approved list, and never use those that are self-created or ambiguous.

If an error is discovered in your notes, simply cross out and initial the error, and date and initial the correction (Fig. 6.4). Never obliterate errors or use whiteout; this may cause suspicion (Fig. 6.5).

Fig. 6.1 Pictures always assist in understanding a crime scene. (**a**) A single picture is often better than pages of descriptive notes. (**b**) Samples of cocaine pricks. (**c**) An imprint on a cocaine brick.

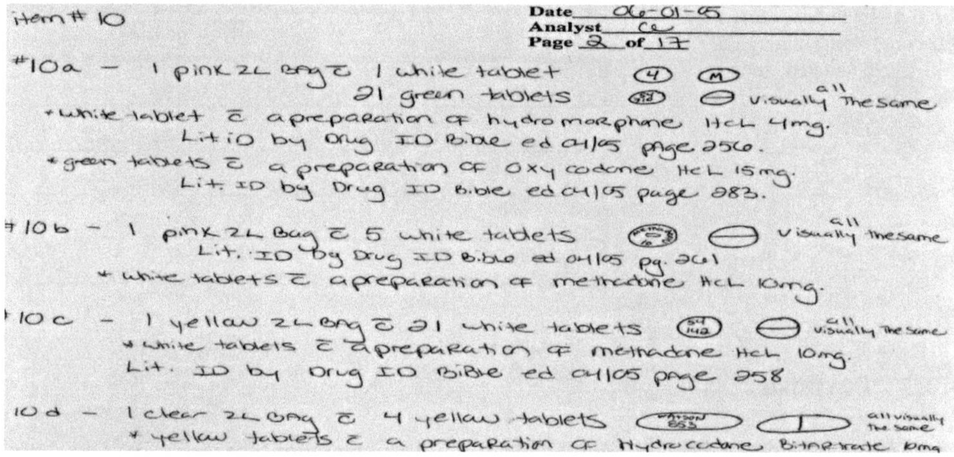

Fig. 6.2 Note how useful it is when each item is described separately. This minimizes possible ambiguity.

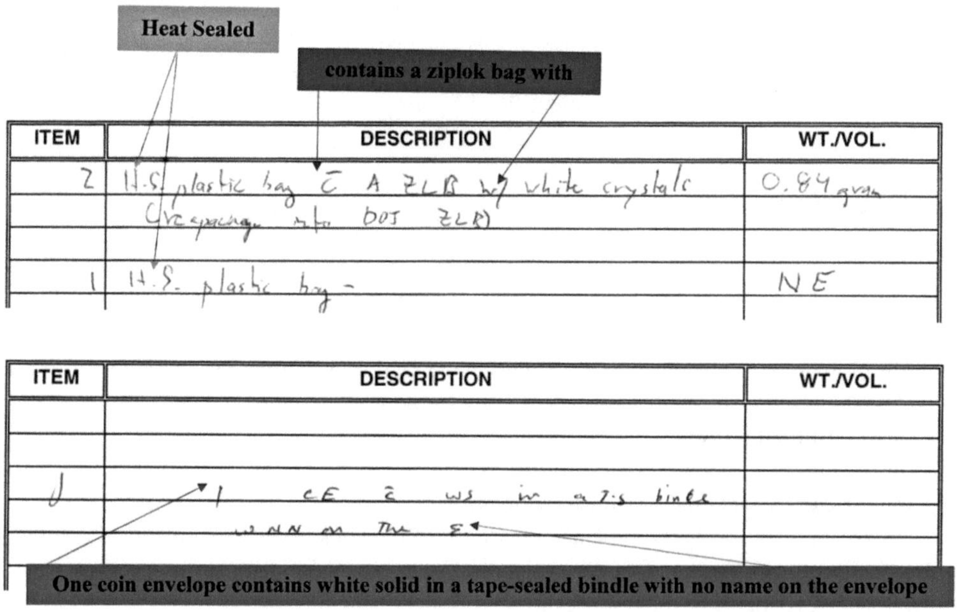

Fig. 6.3 Examples of notes with abbreviations.

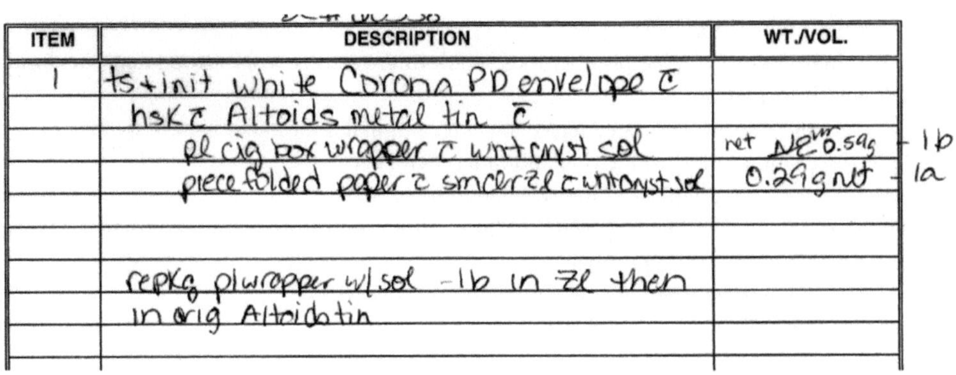

Fig. 6.4 Properly made corrections are initialed and dated to maintain the integrity of the documentation process.

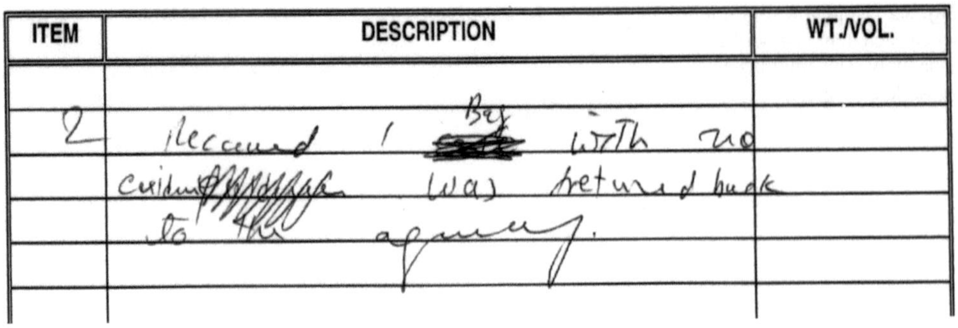

Fig. 6.5 Example of unacceptable corrections: material is obliterated, undated, and not initialed.

6.3.2 Purpose

The case notes are the second component in the documentation process. They consist of handwritten notes, worksheets, examination results, analytical data, and administrative paperwork. The case notes serve two functions; first, they document all aspects of the examination process, and second, they are used as references by the examiner for report writing and preparation of testimony.

Case notes are a complete record of all case activity in the criminal-justice system. Although the final report will provide a summary and conclusion of the entire process, the history of the case is contained within the pages of the case notes.

In recognition of the fact that examinations produce more information than that contained in laboratory reports, regulatory and accreditation agencies require that case notes be maintained. Also, peer review requires the availability of information (case notes) used to formulate scientific opinions.

6.3.3 Content

The American Society of Crime Laboratory Directors (ASCLD) criteria states that a case file must be generated for each case and that each file must be uniquely identified. The International Organization for Standardization (ISO-17025:2005) criteria are slightly more specific; it details the type of information that must be included in technical records. Surprisingly, accreditation agencies, such as ASCLD and the ISO, do not regulate the contents of the case files.

The following list contains items commonly found in a case file.

- Copy of all final reports
- Copy of all evidence-submission forms
- Copy of all relevant chain-of-custody documents
- Itemized description of the type and condition of the packaging when received (*sealed* vs. *unsealed*)
- Itemized description of evidence received
- Detailed description of the items examined
- Description of the examination(s) performed
- Handwritten notes on examinations performed with observations and evaluations, to include but not limited to:
  - Original sample weight
  - Results of wet chemical tests
  - Results of instrumental examinations
  - Sample-preparation techniques
- All original charts, graphs, photographs, photomicrographs worksheets, analytical data or any other type of laboratory generated information
  - Photocopies are an acceptable replacement for original information that is not in a form conducive to storage in the case file
- Copies of written reports related to submitted evidence
- Correspondence or telephone notes related to the case

Each page of the case file is required to have additional information as a quality-assurance measure. These items are:

- The agency's unique case number
- The forensic chemist's handwritten signature or initials
- The date each page was generated
- The page number and total number of pages
  - Machine-generated dates, record numbers, and pagination are acceptable.

6.3.4 Format

Accreditation agencies do not regulate the format of case files. Laboratory-approved forms, official police reports, and instrumental data sheets are usually provided. The format of handwritten notes (case notes) may, or may not, be specified in the laboratories standard-operating-procedures manual.

Handwritten notes have two common formats; one is simply a series of notations on a blank sheet of paper, and the other consists of preprinted worksheets designed to streamline the documentation of repetitive testing procedures. Regardless of format, each page of notes should contain the examiner's initials, the date of the examination, the case number, exhibit number, page number, and the total number of pages (including pages of instrumental data).

Legibility is a key component of handwritten notes. If shorthand is used, it must be defined and clarified in order for the peer-review process to be effective. Ambiguity may lead to misinterpretation.

6.3.5 Dissemination

Whether or not case notes are admissible as evidence or are subject to discovery is an ongoing debate. Some laboratories feel that case notes are the protected, personal property of the forensic chemist. They will go to great lengths to prevent their discovery, citing the case notes are summarized in the final report. Others will openly release case notes upon request as a supplement to the final report. The case-note dissemination policy is ultimately regulated by local statutes, case law, and individual laboratory policy.

6.4 Case Report

6.4.1 Purpose

The case report is the final component in the documentation process. This report contains a summary of examination procedures, the case notes, analytical data, and the professional opinion of the examiner. This report should be a stand-alone document requiring little, if any, clarification from the examiner.

6.4.2 Format and Content

Generally, case-report formats are determined by individual laboratories; however, most formats are based on the criteria set forth by the American Society of Testing Materials (ASTM), the ISO 17025, or ASCLD. Regardless of format, each case report includes:

- Name of the laboratory performing the examination
- Case file number
- Name of the individual requesting the examination(s);
- Examiner's name
- A list and description of the evidence submitted for examination
- Description of the examination(s) performed
- Results of the examination
- Chain of custody of documentation

As previously stated, the case report should be a stand-alone document. While this is true for some sections of the report, it does not necessarily apply to the entire document. For example, courts often require additional testimony to clarify the examination description and result sections, while the administrative segment may be accepted as written in the report. Some formats separate all the sections, while others do not; for example, a description of the testing process could be included in the results narrative. In either case, the reader should be able to confirm the identity of the controlled substance along with the testing procedures used to make that determination.

6.5 Examples

Below are two examples of different formats used to report examination results.

6.5.1 Example One

Items
1. White powder
2. Plant material

Exam

Drugs

Results
1. Contained cocaine, a narcotic drug. Substance mass 1.32 g. A usable quantity.
2. Contained marijuana. Substance mass 6.29 g. A usable quantity.

6.5.2 Example Two

Items
1. Item 1: A paper packet containing a white powder.
2. Item 2: A plastic bag containing a green leafy plant material.

Results
1. Examination of Item 1 using wet chemical tests, microcrystal tests, gas chromatography, and infrared spectroscopy concludes that Item 1 contained a usable quantity of cocaine. The total substance mass was 1.32 g, which is considered a usable quantity. Cocaine is defined as a narcotic drug under ARS 13-3401.20.
2. Examination of Item 2 using microscopic and wet chemical techniques concludes that Item 2 contained marijuana. The total substance mass was 6.29 g, which is considered a usable quantity. Marijuana is defined as a narcotic drug under ARS 13-3401.20.

Example One provides the information in a basic format. The reader can quickly identify each exhibit, the quantity of substance, its classification under governing statutes, and a case-law opinion on the amount of substance seized. However, information justifying the examiner's conclusions is not included. This omission may lead to an unnecessary and time-consuming court appearance.

Example Two represents a more complete report. It not only contains the same information as example One but also includes a description of the tests used to justify the examiner's conclusions. This addition may not require a court appearance by the examiner.

6.6 Questions

1. Describe the process of chain of custody.
2. What are the advantages of chain of custody?
3. Please explain to the jury the procedure of evidence receipt.
4. Name two organizations that regulate case-documentation procedures.
5. List seven items often included in a case report.
6. What information is commonly included in case notes?
7. What do you do with the case notes after completion of the analysis?

8. Describe the difference between incomplete notes and messy notes.
9. Is it possible to alter case notes? Explain.
10. Describe the process of correcting errors found in case notes.
11. What activity initiates evidence analysis?
12. What did you do with the evidence after the analysis is complete?

Suggested Reading

The American Society of Crime Laboratory Directors. http://www.ascld.org/ (accessed May 2009).
International Organization for Standardization (ISO). http://www.iso.org/iso/home.htm (accessed May 2009).

7 Chemical Screening

7.1 Introduction

Chemical-screening methods are presumptive tests commonly used to initiate the process of substance identification. These simple reactions cannot identify the substance without uncertainty; however, they do provide preliminary confirmation of the presence of either a particular functional group or a generic molecular structure. Chemical-screening tests produce a distinct color when the reagents are mixed with compounds containing a specific functional group. Although not highly specific, these preliminary tests will determine which subsequent method is best suited to identify the substance.

Unfortunately, it is not uncommon for two people to describe the same color in different terms. The interpretation and reporting of colors can be influenced by the concentration of the sample, the presence of diluents and adulterants, the age of the reagent, and test reaction times. Also, color transitions and instabilities may result in the formation of multicolored complexes. The subjective and inconsistent nature of color formation and reporting may call into question the use of color screening as a viable method, especially in the area of controlled-substance examinations. This fact should be recognized and allowances made, and the forensic chemist must be prepared to justify reported conclusions.

7.2 Chemistry of Color Formation

Visible light (white light) contains a mixture of wavelengths in the electromagnetic spectrum, ranging from approximately 350 nm (violet) to 750 nm (red). If some of these wavelengths are removed from white light, it is no longer observed as white. All matter is composed of atoms and/or molecules. If an object is colored, the atoms or molecules absorb a portion of white light, thus removing it from the visible range. The observed color of the object is not the light that is absorbed; it is the remaining wavelengths that are reflected (not absorbed). For example, if red light (~750 nm) is absorbed by an object, it is removed from white light and the object appears green (a mixture of the reflected wavelengths). The ability (or capacity) of a substance to absorb (or not absorb) light depends on its chemical properties, i.e.; molecular structure, bond energies, electron arrangement, etc. A change in chemical properties can result in a change in color. A chemical reaction is any process that results in a chemical change. Therefore, a chemical reaction will produce different colored products if it changes the light-absorbing capacities of the reagents (starting material). Color-screening tests produce distinct colors by changing the light-absorption properties of controlled substances. These changes are often directly related to a small change in either the orientation or the location of electrons in the structure. For example, primary amines will react differently than secondary amines when mixed with specific reagents. Tertiary amines may or may not react with reagents that produce a change in primary and/or secondary amines.

The location of electrons in the three-dimensional structure of a molecule is one of the factors that determine color. Color-screening reagents utilize four general mechanisms to produce a characteristic color change; all are based on changing either the location or the orientation of electrons. A characteristic color is produced when:

J.I. Khan et al., *Basic Principles of Forensic Chemistry*, DOI 10.1007/978-1-59745-437-7_7,
© Springer Science+Business Media, LLC 2012

1. A screening reagent(s) removes an electron directly from the test compound.

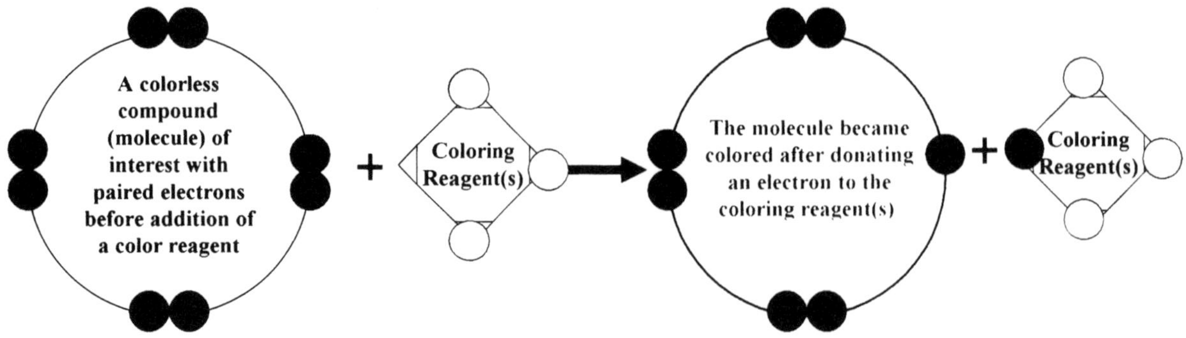

Diagram 7.1

2. A screening reagent(s) adds an unpaired electron directly to the test compound.

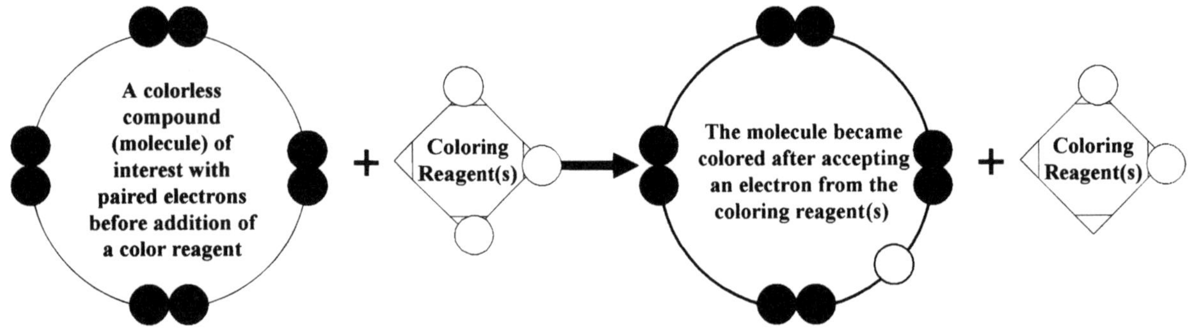

Diagram 7.2

3. A screening reagent(s) complexes directly with the test compound, resulting in the addition of an unpaired electron.

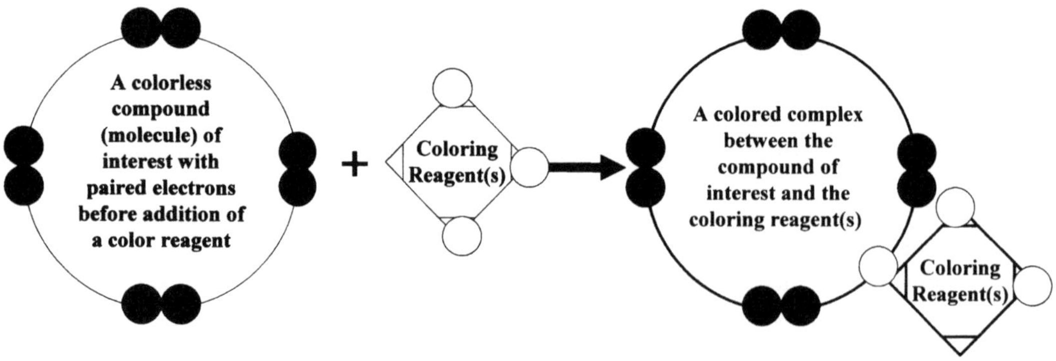

Diagram 7.3

4. A screening reagent(s) bridges two or more test compounds.

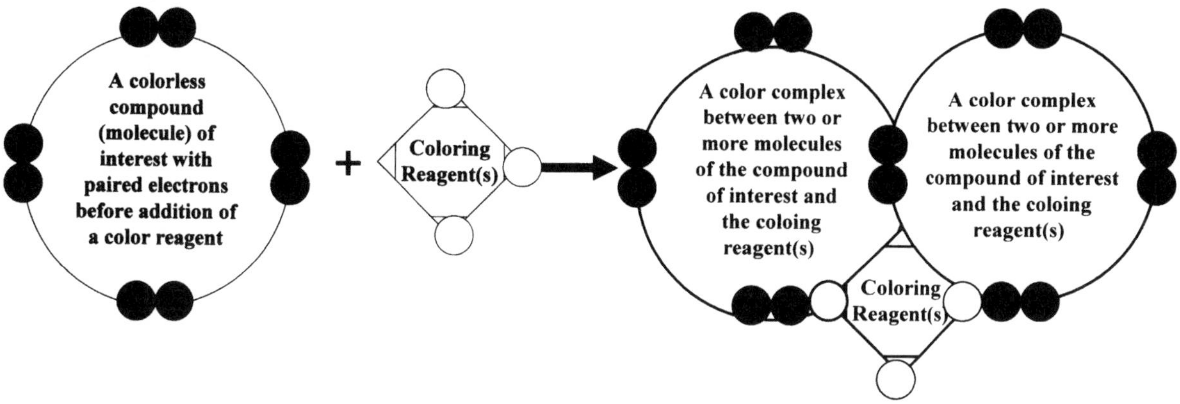

Diagram 7.4

The color and intensity (shade) of the products produced in chemical-screening tests may be affected by the acidity of the test solution. The Duquenois–Levine test, the recognized standard in the identification of cannabis resin, relies heavily on the pH of the solution to produce a color change. Addition of the Duquenois reagent to a sample of suspected cannabis resin produces a color change only after the pH is adjusted with hydrochloric acid.

The Chen's test is an example of how intensity is affected by pH. In the Chen's test (like the Duquenois–Levine test), the color is not observed until the pH of the test solution is adjusted. However, the intensity (shade) of the color is dependent on *how* the pH is adjusted. The use of a weak base (bicarbonate solution) to adjust pH will produce a shade that is noticeably different from the shade produced using a strong base (hydroxide solution).

The addition sequence of reagents in multistep color tests is very important because an error may produce a different result (color). For example, the use of the Marquis' reagent with 3,4-methylenedioxyamphetamine (MDA) or 3,4-methylenedioxymethamphetamine (MDMA) should produce a green-to-black color transition. However, switching the sequence will produce a purple-to-black transition.

7.3 Limitations of Chemical Color Tests

Color-screening tests are nonspecific. They are only used to confirm the presence of either a functional group or a characteristic structure. They cannot be used to positively identify any substance; however, they can indicate the presence of a specific class of compound.

Unfortunately, not all compounds respond to chemical color tests and, in some instances, the screening tests are significantly more complicated than the confirmatory methods. For example, some tryptamines have no known chemical color test and the color test for GHB (γ-hydroxybutyrate) is more complicated than the confirmatory examination.

These limitations should not preclude the use of chemical color tests. They are an excellent method to effectively differentiate specific classes of compounds.

7.4 Chemical Color-Test Methods

Chemical color tests are generally performed by transferring a small amount of the substance under investigation to either the well of a spot plate or a test tube (Fig. 7.1). Next, the test reagent is added to the substance. Some tests may be conducted in a sequential manner using multiple reagents. In these cases, the results of each step in the sequence are observed and noted. Positive and negative controls should be run on a regular basis to ensure the reliability of the testing reagents. The following is a basic procedure for performing a chemical color test:

- Place a small amount of a test sample in either a well of a clean spot plate or an unused culture tube.
- Add a few drops of the chemical color reagent and record the *immediate* color change.
  - Continue if the test requires more than one reagent.
- Add a few drops of the subsequent reagent(s) and record the *immediate* color change.
  - Continue adding test reagents as required.

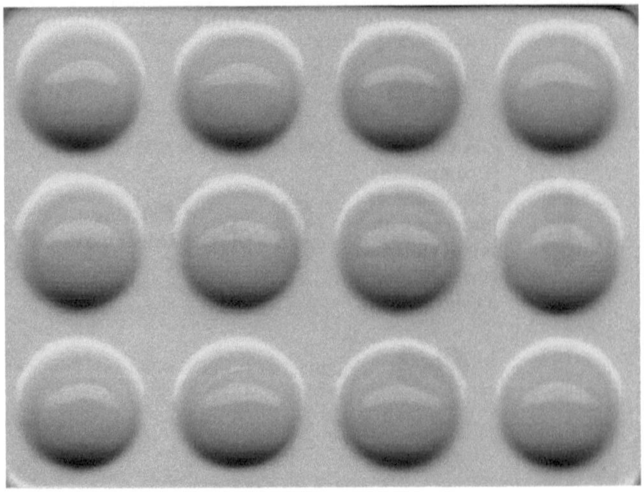

Fig. 7.1 A spot plate typically used in color-screening tests.

7.5 Documentation

Comprehensive documentation of chemical color tests would include (at minimum): a complete description of reagents (expiration dates, color, physical properties, photographs, etc.), a complete description of test substance (color, physical properties, irregularities, notable markings, identifying characteristics, photographs, etc.), observations during test performance (testing conditions, testing equipment, glassware, spot plate, transition colors from initial mixing to end of test, photographs, etc.), complete description of results (final color, positive or negative, comparison to published results, deviations, supporting evidence for observed results, photographs, etc.). Simply recording a positive (+) or negative (−) result should be avoided because it does not provide adequate information for subsequent peer or technical review.

Note: Photographing a chemical color test may or may not provide adequate documentation. A photograph cannot illustrate short-lived color transitions that may have been observed during the examination; however, they can accurately prove selected results.

7.6 Chemical Color Tests

We have selected some of the most common and most reliable color-screening tests for discussion. It should be noted that this list is not intended to represent a comprehensive collection of color-screening tests.

7.6.1 Chen's Test

Reagent 1:

- 1% (m/v) copper(II) sulfate ($CuSO_4$) in water

Reagent 2:

- 8 g sodium hydroxide in 100-ml water (2 M NaOH)

Place either 1–2 mg or 1–2 drops of sample in a spot plate, add two drops of reagent 1, then two drops of reagent 2 and note the color.

Results:

- Produces purple color with ephedrine/pseudoephedrine, phenylpropanolamine, and lidocaine.

Note: This test requires a blank because the reagents are light blue.

The formation of a copper complex produces the color. Copper(II) acts as a chelating agent connecting two of the target molecules. Ephedrine is shown as an example.

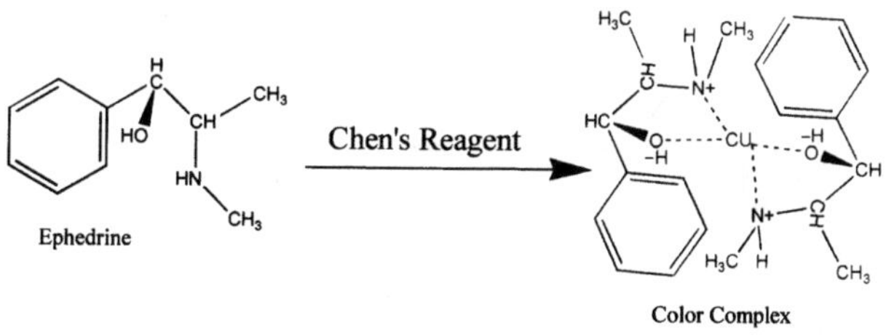

Diagram 7.5

7.6.2 Dille–Koppanyi's Test

Reagent 1:

- 0.1 g cobalt(II) acetate or, 0.1 g cobalt(II) acetate tetrahydrate
- 0.2 ml glacial acetic acid
- 100 ml methanol (absolute)

Reagent 2:

- 5 ml isopropylamine
- 95 ml methanol (absolute)

Place either 1–2 mg or 1–2 drops of sample in a spot plate, add three drops of reagent 1, then three drops of reagent 2 and note the color.

Results:

- Purple: glutethimide, theophylline, chlorzoxazone, all barbiturates (except thiobarbs)
- Blue/purple: dilantin

The colored complex contains cobalt(II) and two target molecules stabilized by two molecules of isopropylamine. Barbiturates are shown as an example.

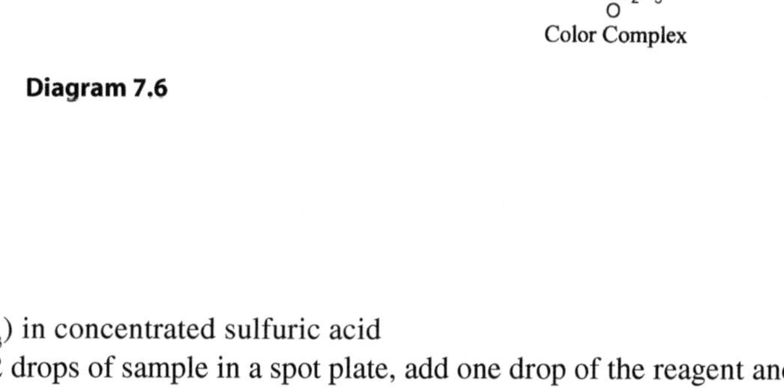

Diagram 7.6

7.6.3 Mecke's Test

Reagent:

- 1% selenous acid (H_2SeO_3) in concentrated sulfuric acid

Place either 1–2 mg or 1–2 drops of sample in a spot plate, add one drop of the reagent and note the color.

Results:

- Purple: codeine, diazepam, methcathinone, flunitrazepam, phenylacetone, and oxycodone
- Green: opiate alkaloids, i.e., morphine and heroin

The reagent oxidizes opiates to a green-colored ortho-quinone of apomorphine. Heroin is shown as an example.

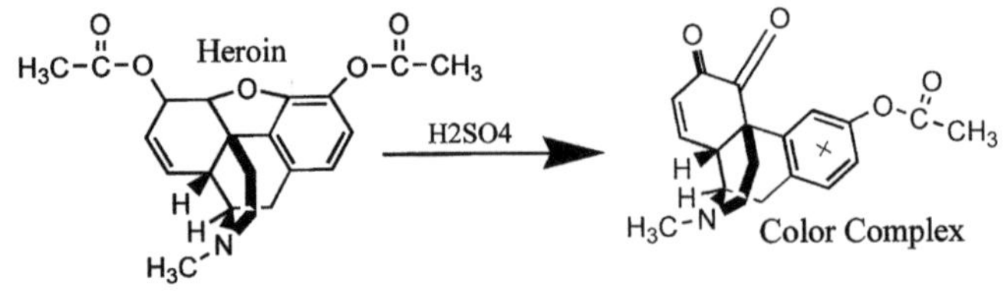

Diagram 7.7

7.6.4 Marquis' Test

Reagent 1:

- Concentrated sulfuric acid

Reagent 2:

- Eight to ten drops of 37% formaldehyde in 10 ml of glacial acetic acid

Place either 1–2 mg or 1–2 drops of sample in a spot plate, add one drop of reagent 1, then one drop of reagent 2 and note the color.

Results:

- Orange-to-brown: *N*,*N*-dimethylamphetamine, amphetamine, methamphetamine, mescaline, and the pethidine group
- Purple: opiate alkaloids, morphine, heroin, and codeine
- Brown-red-purple: opium
- Pink-to-violet: methadone
- Green-to-black: MDA and MDMA (when reagent 2 is added first)
- Purple-to-black: MDA and MDMA (when reagent 1 is added first)

Color formation from opiate alkaloids is most likely from a complex containing two molecules of the opiate and two molecules of formaldehyde. Its formation is promoted by the presence of a strong acid (e.g., sulfuric acid). The colored complex of heroin is shown.

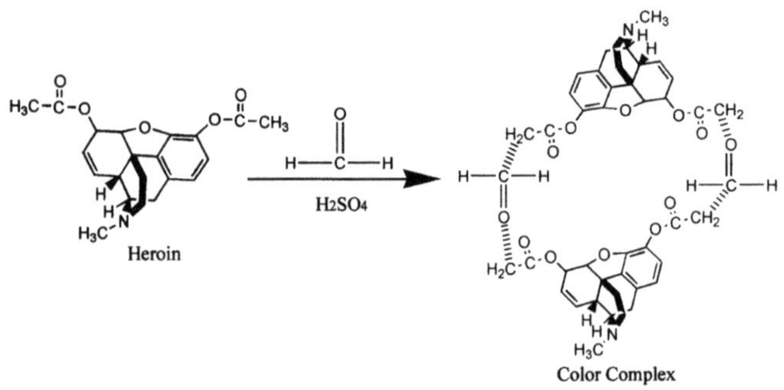

Diagram 7.8

A bimolecular orange-to-brown-colored carbenium ion complex is formed between two amphetamine or methamphetamine molecules in the presence of the Marquis' reagent. Amphetamine is shown as an example.

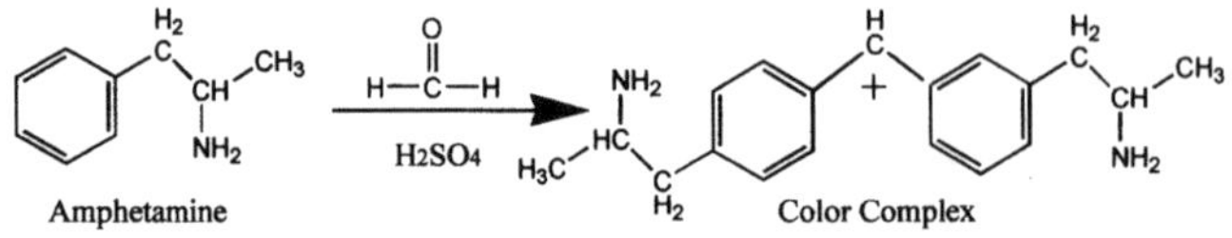

Diagram 7.9

A bimolecular-molecular green-to-black-colored complex containing two carbenium ions is formed when either MDMA or MDA reacts with formaldehyde in the presence of sulfuric acid. This will occur when reagent 2 is added first.

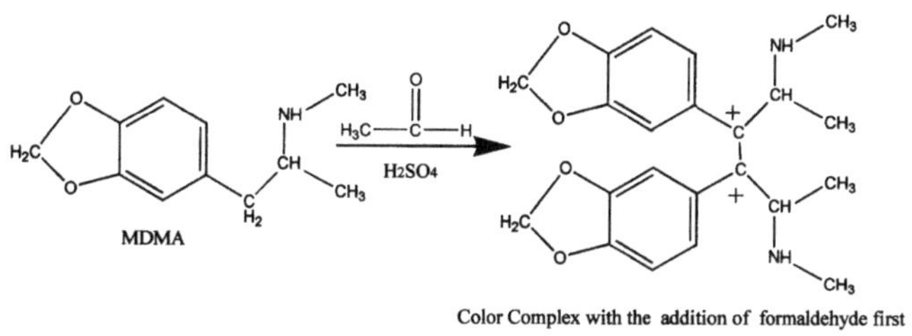

Diagram 7.10

A bimolecular-molecular purple-to-black-colored complex containing two carbenium ions is formed when either MDMA or MDA reacts with sulfuric acid in the presence of formaldehyde. This will occur when reagent 1 is added first.

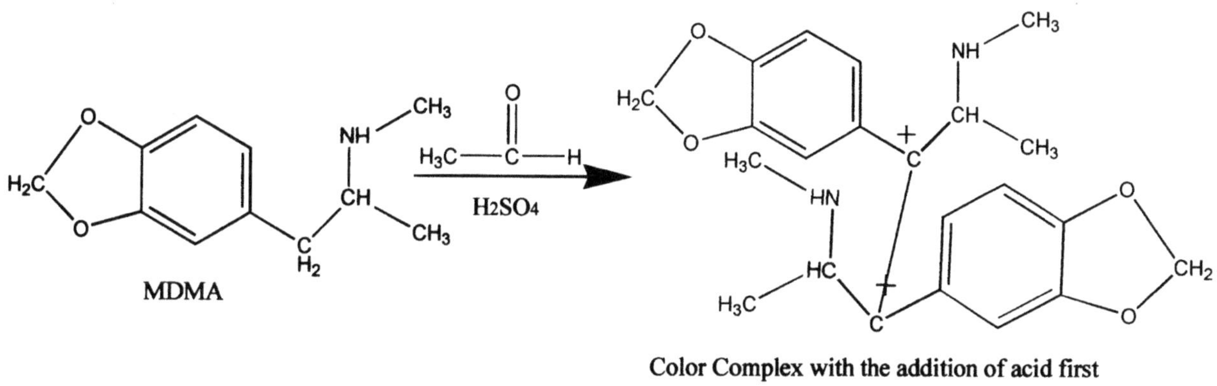

Diagram 7.11

7.6.5 Nitric Acid Test

Reagent:

- Concentrated nitric acid

  Place either 1–2 mg or 1–2 drops of sample in a spot plate, add one drop of the reagent and note the color.

Results:

- Orange-to-red: morphine
- Orange: codeine
- Yellow: heroin

  The aromatic portion of heroin, codeine, and morphine (benzene ring) is nitrated at the ortho position. The highly polar nitro group (NO_2) generates the colored complex through intramolecular ring closure via hydrogen bonding. Heroin is shown as an example.

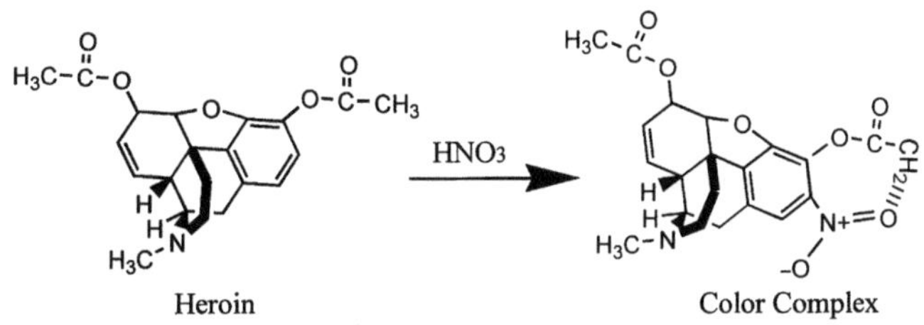

Diagram 7.12

7.6.6 Primary Amine Test

Reagent 1:
- 1 g sodium nitroprusside (nitroferricyanide)
- 10 ml acetone
- 90 ml water

Reagent 2:
- 2% sodium carbonate in water

Place either 1–2 mg or 1–2 drops of sample in a spot plate, add one drop of reagent 1, then one drop of reagent 2 and note the color.

Results:
- Blue is positive for the presence of primary amines.

7.6.7 Secondary Amine Test

Reagent 1:
- 1 g sodium nitroprusside (nitroferricyanide)
- 10 ml acetaldehyde
- 90 ml of water

Reagent 2:
- 2% sodium carbonate in water

Place either 1–2 mg or 1–2 drops of sample in a spot plate, add one drop of reagent 1 then one drop of reagent 2 and note the color.

Results:
- Blue color is positive with some secondary amines, such as MDMA and methamphetamine. This test cannot be used to screen the secondary amines pseudoephedrine, ephedrine, and ketamine.

7.6.8 Tertiary Amine Test

Neutral reagent:
- 2% cobalt(II) thiocyanate in water solution

Acidified reagent:
- 2% cobalt(II) thiocyanate in water solution
- Add a few drops of concentrated HCl.

Place either 1–2 mg or 1–2 drops of sample in a spot plate, add one drop of the reagent and note color.

Results on neutral test:
- Blue: cocaine HCl, ketamine, pethidine, methadone, methylphenidate, and methaqualone

Results on acidified test:
- Blue: cocaine base, phencyclidine (PCP), pethidine, methadone, methylphenidate, and methaqualone

7.6.9 Van-Urk's Test

Reagent:
- 1 g *para*-dimethylaminobenzaldehyde (p-DMBA)
- 10 ml concentrated HCl
- 90 ml ethanol
- Preparation: Dissolve p-DMBA in ethanol, then add HCl.

Place either 1–2 mg or 1–2 drops of sample in a spot plate, add one drop of the reagent and note the color.

Results:
- Purple: LSD (lysergic acid diethylamide)
- Blue: indoles, pyrroles, and tryptophan
- Yellow: procaine and benzocaine

The purple-colored trimolecular complex formed with LSD involves two molecules of LSD and one modified reagent molecule.

7.6.10 Duquenois–Levine Test

Reagent 1:
- Petroleum ether

Reagent 2:
- 97.5 ml of 2% vanillin solution in methanol (absolute)
- 2.5 ml of acetaldehyde

Reagent 3:
- Concentrated hydrochloric acid

Reagent 4:
- Chloroform

Procedure:
- Wash plant material with petroleum ether.
- Place the petroleum-ether extract in a spot-plate well and allow the ether to evaporate.
- Add a few drops of reagent 2.
- Add a few drops of reagent 3 and note the color.
- Add a few drops of reagent 4 and note the color in the chloroform layer.

Results:
- A positive test for cannabis resin requires two observations: (1) a purple color must form after the addition of reagent 3, and (2) the color must transfer into the chloroform layer after the addition of reagent 4.

7.6.11 Froehde's Test

Reagent:
- 0.5% sodium molybdate (Na_2MoO_4) solution in concentrated sulfuric acid

Place either 1–2 mg or 1–2 drops of sample in a spot plate, add one drop of the reagent and note the color.

Results:
- Purple: opiate alkaloids

7.6.12 Janovsky Test

Reagent 1:
- 0.2% (m/v) *meta*-dinitrobenzene in 2-propanol

Reagent 2:
- 10% (m/v) potassium hydroxide in methanol (absolute)

Place either 1–2 mg or 1–2 drops of sample in a spot plate, add one drop of reagent 1, then one drop of reagent 2 and note the color.

Results:
- Purple: diazepam, methcathinone, flunitrazepam, phenylacetone, oxycodone

7.6.13 Weber Test

Reagent 1:

- Freshly prepared 0.1% (m/v) Fast Blue B or Diazo Blue B (O-dianisidine, tetrazotized) solution in water

Reagent 2:

- Concentrated hydrochloric acid

Place either 1–2 mg or 1–2 drops of sample in a spot plate, add one drop of reagent 1, then one drop of reagent 2 and note the color.

Results:

- Red color after the addition of reagent 1, followed by a blue-green color after the addition of reagent 2 indicates the presence of psilocin or psilocybin.

7.7 Summary of Chemical Color Tests

The flowchart below is a summary of the color-screening methods discussed in this chapter (Fig. 7.2). The chart illustrates the progressive nature of chemical screening and the correlation that exists among and between different tests. Although the chart could be used to isolate atypical controlled substances (using different combinations), the results shown are representative of those most frequently encountered. Remember, color-screening methods are presumptive tests and should never be used as definitive proof in the identification of a substance.

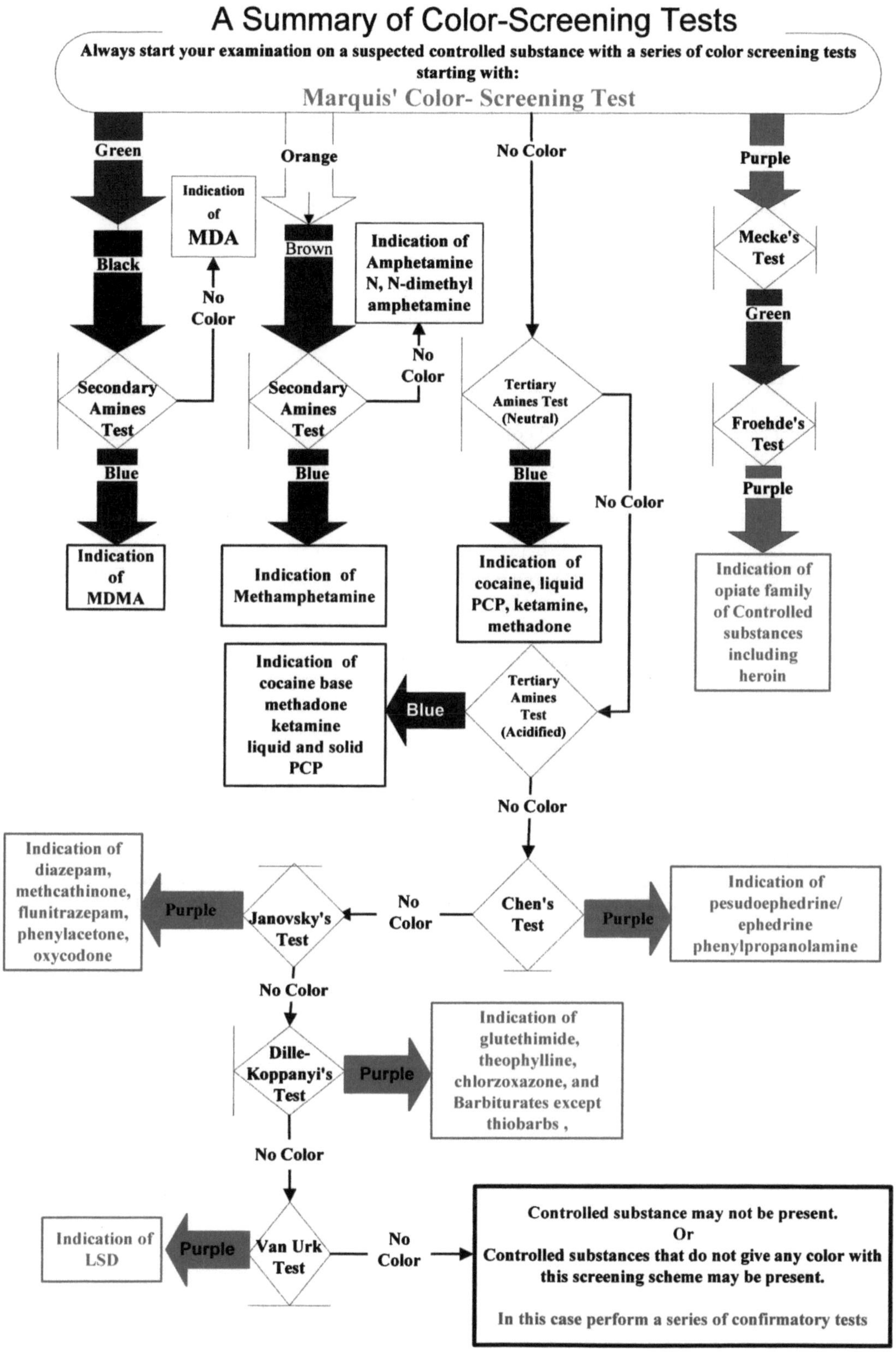

Fig. 7.2 Results of color-screening tests on frequently encountered controlled substances.

7.8 Questions

1. Describe the use of color-screening methods in forensic chemistry.
2. Describe the chemistry of color formation.
3. List three ways colors are produced through chemical reactions.
4. List two limitations of chemical color tests.
5. Outline the documentation process for color tests.
6. Please explain to members of the jury why the Marquis' test used in this case to test MDA produced two different results.
7. Why is it necessary to run a blank control when using Chen's test?
8. Please explain to members of the jury what test you used to determine the sample was cannabis resin. Explain the procedure and results.
9. Describe two tests used to screen for heroin.
10. What test is commonly used to screen for LSD?
11. Why are blank controls always used in chemical color tests?
12. Identify the substance indicated from the following results: Marquis' test, no color; tertiary amine test (neutral), no color; tertiary amine test (acidified), blue.
13. Identify the substance indicated from the following results: Marquis' test, orange-to-brown; secondary amine test, blue.
14. Identify the substance indicated from the following results: Marquis' test, no color; tertiary amine test (neutral), no color; tertiary amine test (acidified), no color; Chen's test, no color; Janovsky's test, no color; Dille–Koppanyi's test, no color; Van Urk's test, purple.

Suggested Reading

Christian, D. R. Jr. Analysis of Controlled Substances. In *Forensic Science: An Introduction to Scientific and Investigative Techniques*, 3rd ed.; James, S. H.; Nordby, J. J., Eds.; CRC Press: Boca Raton, FL, 2009.

Cole, M. D. *The Analysis of Controlled Substances*; John Wiley & Sons: New York, 2003; Appendix 1. Presumptive Color Tests.

Jeffery, W. Chapter 19. In *Clarke's Analysis of Drugs and Poisons 2004*; Moffat, A. C.; Osselton, M. D.; Widdop, B., Eds.; Pharmaceutical Press: London, 2004.

Jones, H. S.; Wist, A. A.; Najam, A. R. Spot Tests: A Color Chart Reference for Forensic Chemists. *J. Forensic Sci.* **1979**, 24, 631–649.

Jones, L.; Atkins, P. *Chemisty: Molecules, Matter, and Change*, 4th ed.; W.H. Freeman and Company: New York, 2000; p 267 & chapter 21.

King, L.; McDermott, S. Chapter 2. In *Clarke's Analysis of Drugs and Poisons 2004*; Moffat, A. C.; Osselton, M. D.; Widdop, B., Eds.; Pharmaceutical Press: London, 2004.

United Nations. *Methods for Testing Barbiturate Derivatives Under International Control. A Manual for Use By National Narcotics Laboratories*; ST/NAR/18; United Nations: New York, 1989.

United Nations. *Rapid Testing Methods of Drugs of Abuse. A Manual For Use by National Law Enforcement and Narcotics Laboratory Personnel*; ST/NAR/13/REV.1; United Nations Publication: New York, 1994.

United Nations. *Recommended Methods for The Identification and Analysis of Amphetamine, Methamphetamine and Their Ring-Substituted Analogs in Seized Materials. Manual for Use by National Drug Testing Laboratories*; ST/NAR/34; United Nations Publication: New York, 2006.

Microcrystal Techniques

8

8.1 Introduction

Microcrystal test techniques are based on highly developed chemical-precipitation reactions in which a polarized microscope is used to observe and distinguish the different types of crystals formed. Most of these tests were developed in the late nineteenth century for the identification of alkaloids. Over the years, they have been modified and are currently used in the identification of a majority of controlled substances. Despite the fact that they were developed more than 100 years ago, microcrystal tests still have a role in modern forensic chemistry.

Microcrystal tests are confirmatory techniques often used to verify the results of preliminary screening methods. They are fast, easy to perform, and can be highly specific. However, there is considerable debate on whether they are specific enough to be used as a confirmatory test. The forensic community is divided on this issue into three main groups.

Traditionalists use microcrystal techniques as a confirmatory test in the forensic examination of controlled substances. This older generation of chemists has used wet chemical techniques to identify compounds since the 1960s and 1970s. Although not fully understood at the time, the chemistry of color formation is different from the chemistry of crystal formation. Traditionalists view microcrystal tests as independent tests and perform them in conjunction with color-screening methods. Positive results obtained from two independent tests would represent definitive proof that the sample under investigation is a controlled substance.

The modern, younger generations of forensic chemists use microcrystal tests as a preliminary screening tool. They believe the true chemistry behind microcrystal tests is unknown and that analytical examination (i.e., gas chromatography mass spectrometry (GCMS) or Fourier transform infrared (FTIR) spectroscopy) is required before a positive identification can be made.

The clinical group does not use microcrystal tests and prefers more sophisticated instrumental analysis. This decision appears to be based more on economic, rather than scientific, reasons. This group believes the automation of GCMS and the chemical reliability of FTIR represent a more efficient and economical method of examination. They consider microcrystal tests to be laborious techniques that require extensive training and skill. To address caseload requirements, the clinical group prefers a single trained technician (to operate either the GCMS or FTIR instruments) over a group of scientists performing wet techniques.

8.2 Advantages of Microcrystal Techniques

Microcrystal tests are a low-cost alternative to GCMS and FTIR that are recognized by the scientific community. The Scientific Working Group for the Analysis of Seized Drugs (SWGDRUG) has established criteria for their use in the identification of controlled substances. Additionally, the American Society of Testing and Materials (ASTM) have established methods for identifying cocaine and methamphetamine using microcrystal tests.

In general, microcrystal tests are safe and environmentally friendly. In most cases, the entire analysis can be performed with as little as a single drop of reagent. This is in contrast to the volumes of chemicals required to prepare samples for

J.I. Khan et al., *Basic Principles of Forensic Chemistry*, DOI 10.1007/978-1-59745-437-7_8,
© Springer Science+Business Media, LLC 2012

instrumental analysis. For example, a mere 2 ml of organic solvent used in sample preparation for automated FTIR analysis is roughly 40 times the volume required for a typical microcrystal test.

Purification of the controlled substance is not required for microcrystal testing. Diluents and impurities generally do not interfere with the crystal features used in the identification process. Characteristic crystals can be observed at sample concentrations approaching 2% (by mass). This sensitivity can detect microgram quantities of controlled substance in a particular sample.

A clear advantage of microcrystal tests is their use in the detection of optical isomers (see Chap. 4; Chirality). The identification of enantiomers (optical isomers) is complicated by the fact that, except for optical rotation, they have the same chemical and physical properties. Microcrystal tests can quickly and accurately differentiate optical isomers by forming different crystal structures with each enantiomer. This can play an important role in the analysis of controlled substances when one isomer is controlled and the other is not, or when one isomer can establish a particular method of drug production.

It should be noted that microcrystal formation does not affect the chemical and physical properties of the substance; thus, it can be recovered for subsequent testing (i.e., instrumental analysis). This is clearly an advantage if sample quantity is limited.

8.3 Disadvantages of Microcrystal Techniques

The principal disadvantage of microcrystal testing is that it is not applicable to all substances. Crystal-testing procedures do not exist for a number of commonly encountered controlled substances and, given the current debate over their use, this is not likely to change in the near future.

Microcrystal testing can produce more than one type of arrangement. The presence of additional crystal-forming agents may interfere with the precipitation of the target compound. This interference may cause either distortions or variations in the expected crystal form (polymorphism). This may complicate the identification process. In such cases, a purification procedure, such as thin-layer chromatography or extraction, is recommended before microcrystal analysis.

The formation of a solid crystal in solution begins when individual molecules or ions cluster together. This process of nucleation continues until a visible particle appears. The speed of the nucleation process can influence the shape of the crystal. Crystals with definitive features are formed very slowly; while those formed rapidly can "mechanically trap" undesired particles (i.e., solvent, impurities, dilutants, etc.). The trapped particles can distort the overall structure of the crystal and complicate identification. Highly concentrated samples and reagents will develop crystals rapidly, resulting in polymorphism. Therefore, reagents and samples may require dilution to produce standard crystal forms for comparison and identification. Also, reference samples should be run with all reagents to verify reagent activity.

Microcrystal testing is a manual technique. The individual handling of samples and reagents requires care and consistency. The analyst's results must be reproducible for definitive identification. This aspect is a barrier to automation and may preclude its use in forensic laboratories with a high-volume caseload.

Microcrystal techniques lack the versatility of chromatographic methods offering single-step identification, quantization, and documentation. Consequently, many would argue that they are not well suited for the production-oriented clinical environment of the modern forensic laboratory.

Identifying compounds using standard crystal features is not inherently subjective. Nonetheless, a degree of subjectivity is always present in the interpretation of crystal test results. Characterizing shapes and structural features is influenced by the experience and training of the analyst, concentration of sample and reagents, presence of interfering compounds, reagent age, and crystal polymorphism. The effective use of microcrystal tests requires training and experience. The analyst must develop recognition of unpredictable reagent behavior through practice. Unfortunately, this requires countless hours behind a microscope.

8.4 Documentation

The results of microcrystal testing should be documented completely. Table 8.1 provides a list of terms and diagrams commonly used to describe crystals. In addition, comprehensive documentation would include: a complete description of reagents (expiration dates, color, physical properties, photographs, etc.), a complete description of test substance (color, physical properties, irregularities, notable markings, identifying characteristics, photographs, etc.), observations during test performance (testing conditions, testing equipment, glassware, spot plate, colors, photographs, etc.), complete description

Table 8.1 Microcrystal descriptions

Crystal	Shape	Description
Blade		Broad needle
Bunch/Bundle		Cluster with the majority of the crystals lying in one direction
Burr/Hedgehog		Rosette, which is so dense that only the tops of the needles show
Cluster		Loose complex of crystals
Cross		Single cruciform crystal
Dendrites		Multibrachiate branching crystals
Grains		Small lenticular crystals
Needles		Long thin crystals with pointed ends
Plates		Crystals with the length and width that are of the same magnitude
Prisms		Thick tablet
Rod		Long thin crystals with square cut ends
Rosette	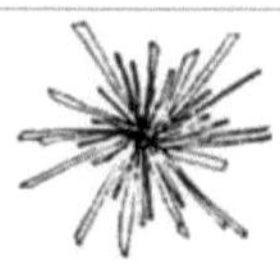	Collection of crystals radiating from a single point
Sheaf		Double tuff
Splinters		Small irregular rods and needles
Star		Rosette with 4 or 6 components
Tablet		Plates with appreciable thickness
Tuff/Fan		Sector of a rosette

of results (crystal features, sketch, comparison to standard features, deviations, supporting evidence for conclusion, photographs, etc.).

Supporting documentation may or may not be required when microcrystal tests are used as a screening method. The documentation requirements are flexible on presumptive tests because the final opinion does not necessarily rely on the results. Regardless, a physical description and diagram of the crystals must be documented for peer review.

Microcrystal tests used as a confirmatory method require documentation. It is recommended that a photomicrograph (photograph) be taken of the crystals used for identification. Microcrystal results can easily be challenged as evidence, if the examiner fails to provide documentation of performance and results.

8.5 Microcrystal Test Techniques

In microcrystal tests, the test sample is dissolved in a solution. A test reagent either is added to the solution or is already present in the solvent. A reaction between the compound of interest and the test reagent forms a solid compound that is not soluble in the test drop. The solid forms uniquely shaped crystals that can be observed with a microscope. Microcrystal-test techniques are divided into two broad categories: aqueous techniques or volatility techniques.

8.5.1 Aqueous Test Technique

Despite considerable variation in testing reagents, the technique for aqueous testing remains unchanged. A reference standard is required and must be run concurrently:

- A small sample is placed on a microscope slide and dissolved in one drop of water or one drop of diluted acetic acid (Fig. 8.1).
- Place one drop of reagent next to the sample on the slide or place one drop of reagent directly into the test drop (Fig. 8.2).
- Mix the two drops (if side by side) (Fig. 8.3).
- Place the slide under a microscope and observe crystal formation.
  - A slide cover is not required.

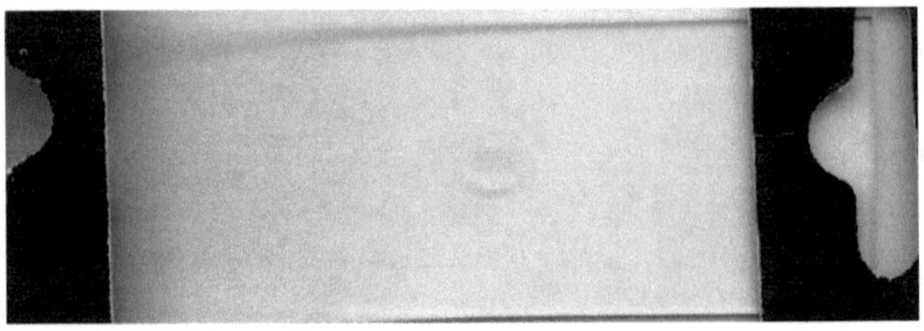

Fig. 8.1 Small sample placed on a microscope slide and dissolved in one drop of water or one drop of diluted acetic acid.

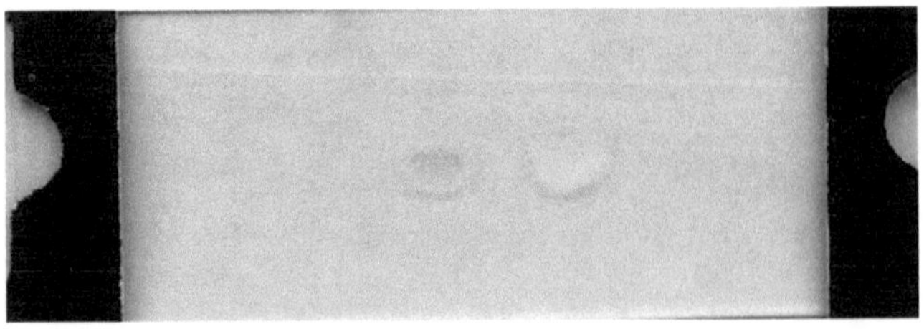

Fig. 8.2 Place one drop of reagent next to the sample on the slide or place one drop of reagent directly into the test drop.

Fig. 8.3 Mix the two drops (if side by side).

8.5.2 Volatility Test Technique

The volatility technique is used when the test substance is volatile (easily vaporized) or when a solvent is chosen that causes the substance to be volatile. The sample vapors rise and react with a drop of reagent suspended on a slide over the substance. Crystals form in the solution on the cover slide. A reference standard is required and must be run concurrently.

- The test sample is placed into the depression of either a clean spot plate or a volatility chamber.
- A drop of crystal reagent is placed onto a microscope slide.
  - The reagent may contain a viscous material to aid in suspension.
- The microscope slide is inverted and placed over the depression containing the test sample. Align the reagent drop over the test sample.
- The test sample is vaporized.
  - It may require placing the spot plate onto a warm hot plate.
- Allow reagent drop and sample vapors to react.
- Place the slide under a microscope and observe crystals.

This technique is particularly useful in the detection of volatile poisons containing the aldehyde and ketone functional groups. Controlled substances containing primary and secondary amines have been isolated using this technique. Also, it may have possible applications in the identification of γ-hydroxybutyric acid (GHB) and in the detection of explosive residues.

8.5.3 Acid and Anionic Test Technique

This technique combines portions of both the aqueous and volatility techniques. A few crystals of sample are placed in the cavity of a cavity slide, and one drop of alcohol solution (methanol or ethanol) is added. One drop of reagent is added immediately after the alcohol, and a cover slide is placed over the cavity to prevent evaporation. Crystal formation is observed by placing the cavity slide (with cover) directly under a microscope. This method is often used to crystallize steroid hormones and barbiturates, including phenobarbital.

8.6 Aqueous Test Reagents

The protocols for microcrystalline testing, including recommended procedures for the preparation of a vast number of reagents and solvents, are readily available in a variety of scientific publications. However, only a few are used in forensic investigation. The following tests are representative of those commonly used in most forensic laboratories, with only minor variations to reagent preparation.

8.6.1 Gold Chloride Test

Reagent: 3-g gold(III) chloride ($AuCl_3$) + 100-ml water + 0.25-ml concentrated HCl

Target: Cocaine	Crystals: Combs and rosettes of needles

8.6.2 Gold Chloride in Phosphoric Acid Test

Reagent 1: 5-g gold(III) chloride ($AuCl_3$) in 100-ml water
Reagent 2: Concentrated phosphoric acid (H_3PO_4)
Mix two drops Reagent 1 with one drop Reagent 2.

Target: Methamphetamine	Crystals: Long needles and long barbs
Target: D-Amphetamine	Crystals: Long yellow rods and blades

8.6.3 Platinum Chloride Test

Reagent: 5-g platinum(IV) chloride ($PtCl_4$) in 100 ml of 1 M hydrochloric acid (HCl)

Target: Cocaine	Crystals: Combs of needles

8.6.4 Mercuric Iodide Test

Reagent: 10% hydrochloric acid (HCl) saturated with mercury(II) iodide (HgI_2)

Target: Heroin	Crystals: Rosettes of dendrites

Note: The test is extremely sensitive, even with highly contaminated samples. Use caution when interpreting results; product crystals are colorless and may be difficult to differentiate from undissolved reagent particles.

8.6.5 Mercuric Chloride Test

Reagent: 5-g mercury(II) chloride ($HgCl_2$) in 100-ml water

Target: Methadone	Crystals: Small rosettes of rods

8.6.6 Potassium Permanganate Test

Reagent: 2-g potassium permanganate ($KMnO_4$) + 100-ml water + 0.25-ml phosphoric acid (H_3PO_4)

Target: Phencyclidine (PCP)	Crystals: purple H-shaped plates

Note: The test is extremely sensitive, and the best results are obtained using very dilute samples. It is recommended to use just enough sample to produce a light amorphous precipitate when reagent is added.

8.6.7 Sodium Acetate Test

Reagent: 10-g sodium acetate in 100-ml water

Target: Heroin	Crystals: Hexagonal plates
Target: Quinine	Crystals: Irregular logs

8.7 Critical Considerations

The interpretation of microcrystalline test results requires a great deal of care and attention. The subtle details of structural features are often either lost or hidden by factors influencing crystal formation. The following list contains suggestions designed to minimize the complexities associated with microcrystalline examinations.

- The best crystals form very slowly.
- Crystals get bigger with time, and larger crystals are more easily interpreted.
- Diluted test samples produce better crystals.
- Test solutions should never evaporate to dryness.
  - The solid should separate almost immediately. Extended periods of time promote evaporation and increase the probability that undesired crystalline compounds will form, complicating the results.
- Crystals may be affected by changes in ambient temperature and humidity.
- Always run reference controls concurrently with samples.
- Reagent age can affect crystal formation.

The process of microcrystal formation in solution is not fully understood. Nonetheless, there appears to be no shortage of theories that attempt to describe the procedure. Most have narrow applications and have trouble addressing even the simplest of arguments. The dynamics of crystal formation is really quite simple; crystals will form in any solution when the limit of solute solubility has been reached. But what factors determine a solvent's capacity to dissolve solute? Alas, this is the real question and, at present, that question is unanswered. There will be no theory that is universally applied to precipitate formation in aqueous-solution chemistry. The complexities of crystal formation will ensure that.

8.8 Questions

1. Cite two reasons why microcrystal techniques would be used as a confirmatory test.
2. Describe the complete documentation of a microcrystalline-test result.
3. Cite two advantages of microcrystalline techniques.
4. Can you please explain to the jury how a basic microcrystal test is performed?
5. Please explain to the jury how you determined the substance was heroin using the sodium acetate test.
6. How would you test for amphetamine?
7. In your opinion, is this technique in question #6 a confirmatory or a screening examination? Explain.
8. Is it possible for two different substances to produce identical crystals? Explain.
9. Cite three disadvantages of microcrystalline techniques.
10. Discuss two critical factors that need to be considered when evaluating microcrystalline-test results.
11. What group considers microcrystalline tests obsolete?
12. Describe the crystals formed from cocaine and methadone.
13. Discuss an instance when complete documentation of testing results would not be required.
14. Compare and contrast the aqueous technique and the volatility technique.

Suggested Reading

California Department of Justice. *Technical Procedures Manual for Controlled Substance Analysis*; Sacramento, CA., 2006.

Chamot, E. M.; Mason, C. W. *Handbook of Chemical Microscopy, Volume II: Chemical Methods and Inorganic Qualitative Analysis*; McCrone Research Institute: Chicago, 1989, chaps. 11–13.

De Forest, P. R.; Gaensslen, R. E.; Lee, H. C. *Forensic Science: An Introduction to Criminalistics*; McGraw-Hill: New York, 1983, chap. 5.

Evans, H. K. Drug and Microcrystal Tests for Forensic Drug identification. *Microscope*. 1999, 47, p.147.

Fulton, C. *Modern Microcrystal Tests for Drugs*; John Wiley & Sons. New York, 1969.

Julian, E. A. Microcrystalline Identification of Drugs of Abuse: The Psychedelic Amphetamines. *J. Forensic Sci.* **1990**, 35, pp. 821–830.

Julian, E. A. Microcrystalline Identification of Drugs of Abuse: The White Cross Suite. *J. Forensic Sci.* **1981**, 26, pp. 358–367.

Moorehead, W. A Brief Background and Justification for the Continued Use of Microcrystal Tests. *CAC News*, **2000**, 3rd quarter, pp 11–15.

9 Chemical Extractions and Sample Preparation

9.1 Introduction

Solutions are homogeneous mixtures (see Chap. 1) containing two or more substances. The component of a solution present in the greatest amount is called the *solvent* and the dissolved substances are the *solutes*. It is possible to have more than one solute in a particular solution; however, it is not possible to have more than one solvent. Solvents have a varying capacity to dissolve particular solutes; for example; sodium chloride (NaCl) will readily dissolve in water, while silver chloride (AgCl) will not. Solubility refers to the maximum amount of solute particles that can be dissolved in a specified volume of solvent at a given temperature. Temperature affects the solubility properties of a solvent and, in general, solubility increases with increasing temperature. A common example of this is illustrated using a simple cup of coffee. Have you ever wondered about the dark sediment that mysteriously appears at the bottom of a cool cup of coffee? Stop blaming your dishwasher and the coffee filters you purchased at the discount store; they are not the culprits. When hot water is added to solid coffee, only specific components are dissolved (extracted) from the coffee grinds into water, that is, flavor, odor, caffeine, etc. The concentration of each component in the resulting solution is dependent on the temperature of water. If the water is very hot, a "strong" cup of coffee results because the water can dissolve a higher concentration of the components (increased solubility). As the coffee cools, the solubility decreases and the components precipitate out as dark sediment in the bottom of the cup. The coffee/water system is an example of solid/liquid extraction because soluble components are transferred from a solid phase (coffee) into a liquid phase (water).

Extraction is a general term used to describe a number of chemical techniques that separate the components of a mixture using the solubility properties of various solute/solvent systems. In the context of extraction, solubility often refers only to a solvent's *ability* to dissolve a particular solute and not necessarily a quantitative maximum amount. Most extraction techniques are slight variations of three general procedures: solid–liquid extraction, liquid–liquid extraction, or acid–base extraction.

9.2 Techniques

9.2.1 Solid–Liquid Extraction

Solid–liquid extraction is most often used to extract a natural component from a solid natural source, such as a dried plant. This technique is found in the isolation procedures of morphine from the opium poppy and cocaine from coca plants. Although the basic concept is used in forensic analysis, a slight modification adds versatility to this technique. This method relies heavily on the selective extraction (transfer) of soluble components from a solid phase into solution. Ideally, a carefully chosen solvent will dissolve only the target compound and no other components. Isolation is accomplished by filtering out the insoluble contaminants and recovering the target compound from the filtrate (solution that passes through filter). This method generally requires a single extraction, and an outline of the general procedure is below.

- Identify the components in the solid sample.
- Identify the solubility properties of each component.
- Select a suitable solvent, ideally one with a *high* solubility for the target compound and a *low* solubility for the remaining components.

J.I. Khan et al., *Basic Principles of Forensic Chemistry*, DOI 10.1007/978-1-59745-437-7_9,
© Springer Science+Business Media, LLC 2012

- Add the solvent and agitate.
- Allow the insoluble components to settle.
- Filter the solution to remove insoluble contaminants.
- Analyze the filtrate.

The Modification: Alternatively, the solvent can be used to dissolve all components in the sample except the target compound. Isolation is again accomplished using filtration and the target compound is recovered as crystals on the filter paper. This method may require multiple extractions with different solvents, depending on the characteristics of the contaminants. The general procedure is a slight variation of the one described earlier.

- Identify the components in the solid sample.
- Identify the solubility properties of each component.
- Select a suitable solvent, ideally one with a *low* solubility for the target compound and a *high* solubility for the remaining components.
- Add the solvent and agitate.
- Allow the insoluble component to settle.
- Filter the solution to remove target compound.
  - Repeat as required to remove other components.
  - May require different solvents.
- Analyze the crystals on the filter paper.

9.2.2 Liquid–Liquid Extraction

Solubility is a term used to describe solutions containing a solid dissolved in a liquid. It is rarely associated with solutions containing only liquids. If two liquids dissolve in one another, they are said to be *miscible* and the resulting solution forms a single, continuous layer (i.e., ethanol and water). If two liquids do not dissolve in one another, they are said to be *immiscible*, and the resulting solution forms two distinct layers. We refer to each layer of an immiscible mixture as a *phase*. A number of organic solvents are immiscible in water (i.e., oil and water) (Fig. 9.1). Liquid–liquid extractions use carefully chosen immiscible solvent pairs to isolate a target compound in one layer and the impurities in the other. Preferred solvents will readily dissolve the sample, have low boiling points, not react with solutes, and not be either flammable or toxic. Water is commonly one solvent, and any of a number of organic solvents, such as diethyl ether, methylene chloride, or chloroform, is often chosen as the second solvent. The effectiveness of extraction depends on the relative affinity (attraction or preference) of the solute for each solvent. Ideally, one solvent has a high affinity for the target compound and a low affinity for the impurities, while the other's affinity is low for the target and high for impurities. It is worth noting that affinity is not solubility. A solvent can have a low affinity for a particular solute and still dissolve appreciable amounts (high solubility). Affinity merely refers to a solvent's ability to retain solute in the presence of another solvent. When performing liquid–liquid extractions, the solid sample is dissolved in one solvent and then mixed with the other. The target compound is extracted (transferred) into the added solvent and

Fig. 9.1 Liquid–liquid extraction utilizes immiscible solvent pairs to isolate target compounds in a single layer of the mixture.

isolated when the two layers are separated. This process can be repeated to maximize recovery of the target compound. A general procedure is described below.

- Identify the components in the solid sample.
- Identify the solubility properties of each component.
- Select a suitable immiscible solvent pair; ideally, one solvent's affinity is *low* for target and *high* for impurities; the other's is *high* for target, and *low* for impurities.
- Dissolve the solid in one solvent, generally the one with *low* affinity for target and *high* affinity for impurities.
- Add the other solvent (*high* affinity for target, *low* affinity for impurities) and mix.
- Let the solution stand undisturbed and allow the phases (layers) to form in the immiscible solution.
- Separate the layers to isolate the target compound.
  - Repeat as needed to maximize recovery.
- Remove solvent to recover target crystals.

9.2.3 Acid–Base Extraction

Acid–base extractions rely on the fact that the relative solubility of some compounds is affected by pH. This method uses simple acid–base reactions to isolate strong organic acids, weak organic acids, neutral organic compounds, and basic organic substances. When any acid is mixed with any base, the subsequent neutralization reaction (by definition) produces an ionic salt and water. Most ionic salts are soluble in water and insoluble in organic solvents. For example, benzoic acid is an organic acid that is slightly soluble in water, but soluble in a variety of organic solvents. To isolate benzoic acid, it is first dissolved in an organic solvent (i.e., ether) and then mixed with an aqueous weak base (i.e., sodium bicarbonate in water). Water and ether are immiscible. The weak base will selectively react with the benzoic acid in solution. The resulting neutralization reaction will convert benzoic acid into sodium benzoate, a salt that is soluble in water but insoluble in organic solvents. The sodium benzoate will be extracted into the aqueous phase and isolated. Addition of acid to the aqueous phase converts sodium benzoate (soluble salt) back into benzoic acid (insoluble), which can be isolated either by direct precipitation from the aqueous solution or by extraction into the organic layer. Barbiturates and their associated sodium salts are examples of acidic drugs isolated using this method. The following is a general procedure for isolating acidic compounds from basic and neutral compounds.

- Dissolve the sample in a suitable organic solvent.
  - Solvent selection is critical.
  - Ethers, ketones, chloroform, or methylene chloride are common choices.
- Add a suitable aqueous base.
  - If extracting a strong acid, use a weak base. If extracting a weak acid, use a strong base. In either case, carefully monitor the amount of base added because any excess will require neutralization.
  - Agitate and allow the mixture to stand undisturbed.
  - The added base will convert the acid (target compound) in the organic layer into a soluble ionic salt that will be extracted into the aqueous layer.
- Remove the aqueous layer (usually the bottom layer) and save.
  - Repeat base addition to the organic layer two more times to maximize yield.
  - Combine all aqueous portions.
  - At this point, you have isolated the soluble salt, not the target compound.
- Decrease the pH of the combined aqueous portions to <6.
  - 1.0 M sulfuric acid solution works well.
  - Adding an acid converts the salt back into the original target compound, which is soluble in organic solvents.
- Add a suitable organic solvent and agitate.
  - The target compound is extracted into the organic layer.
- Remove the organic layer (usually the top layer) and save.
  - Repeat organic solvent addition to the aqueous phase two more times to maximize the yield.
  - Combine all organic portions.
- Prepare the organic layer for analysis.
  - May be used in current form.
  - May require solvent evaporation using a steady stream of air (most organic solvents are volatile).

Many skilled and highly educated scientists have faced the embarrassment of discarding the wrong layer from an extraction. You would be well advised to never discard any extraction layers until the entire procedure is complete. If, however, layers must be discarded, run this simple verification test first. Remove a drop of either layer using a glass pipette (dropper). Add the drop to a small beaker containing distilled water. If the drop dissolves and mixes, you have the aqueous layer; if it forms an observable droplet that does not dissolve, you have the organic layer.

The procedure used to extract basic organic substances is very similar to that used for acids. For example, aniline is an organic base that is soluble in both water and organic solvents. The aniline (weak base) is dissolved in an organic solvent (i.e., ether), and then mixed with dilute hydrochloric acid (strong acid). Water and ether are immiscible. The acid will selectively react with aniline (base) in solution. The resulting neutralization reaction will convert aniline into the ionic anilinum chloride (highly soluble in water) which will be extracted into the aqueous phase and isolated. The addition of base to the aqueous phase converts the anilinum chloride back to aniline. The aniline can be extracted using an organic solvent because it has a higher affinity for organic solvents than it does for water. Most naturally occurring alkaloids are isolated using this method. The following is a general procedure for isolating basic compounds from acidic and neutral compounds.

- Dissolve the sample in a suitable organic solvent.
  - Solvent selection is critical.
  - Ethers, ketones, chloroform, or methylene chloride are common choices.
- Add a suitable aqueous acid.
  - If extracting a strong base, use a weak acid. If extracting a weak base, use a strong acid. In either case, carefully monitor the amount of acid added, as any excess will require neutralization.
  - Agitate and allow the mixture to stand undisturbed.
  - The added acid will convert the base (target compound) in the organic layer into a soluble ionic salt that will be extracted into the aqueous layer.
- Remove the aqueous layer (usually bottom) and save.
  - Repeat the acid addition to the organic layer two more times to maximize yield.
  - Combine all aqueous portions.
  - At this point, you have isolated the soluble salt, not the target compound.
- Increase the pH of the combined aqueous portions to >8.
  - Concentrated ammonium hydroxide or 2.0 M sodium hydroxide works well.
  - Adding a base converts the salt back into the original target compound, which is soluble in organic solvents.
- Add a suitable organic solvent and agitate.
  - The target compound is extracted into the organic layer.
- Remove the organic layer (usually top) and save.
  - Repeat organic solvent addition to the aqueous phase two more times to maximize yield.
  - Combine all organic portions.
- Prepare the organic layer for analysis.
  - May be used in current form.
  - May require solvent evaporation using a steady stream of air (most organic solvents are volatile).

9.2.4 Neutral Compound Extraction

Controlled substances can be categorized as acidic, basic, or neutral. We have outlined general procedures for the isolation of acidic and basic forms, and the success of these methods is directly linked to the formation of soluble ionic salts. Unfortunately, neutral drugs have no transitional ionic forms. Do not despair; neutral drugs do have a preferential affinity for organic solvents and can be isolated by removing acidic and basic contaminants. Benzodiazepines are an example of controlled substances that are neutral compounds, and the following is a general isolation procedure.

- Dissolve the sample in a suitable aqueous acid.
  - 0.1 M H_2SO_4 is suggested.
  - Any basic contaminants will be converted to soluble ionic salts that will remain in the aqueous layer.
- Add a suitable organic solvent and agitate.
  - Ethers, ketones, chloroform, or methylene chloride are common choices.
  - Basic contaminants remain in the aqueous layer as soluble salts.
  - Acidic contaminants and the neutral target compound will be extracted into the organic layer.

- Remove the organic layer (usually top) and save.
  - Repeat organic solvent addition to the aqueous layer two more times to maximize yield.
  - Combine the three organic layers
  - Test the aqueous layer and discard.
- Add a suitable aqueous base to the combined organic layers.
  - 0.2 M NaOH is suggested.
  - The added base will convert acidic contaminants to soluble salts that will be extracted into the aqueous layer.
  - The neutral target compound is isolated in the organic layer.
- Remove the aqueous layer (usually bottom) and discard (test first!).
  - Repeat aqueous-base addition to combined organic layers two more times to maximize yield.
  - Discard the aqueous layers.
  - Save the organic layer.
- Prepare the organic layer for analysis.
  - May be used in current form.
  - May require solvent evaporation using a steady stream of air (most organic solvents are volatile).

9.3 Sample Preparation

Extraction techniques are used to isolate target compounds for further analysis. In some instances, the sample is ready for immediate use, while in others, it may require further preparation. The level of sample preparation is largely dependent on the chosen method of confirmation.

9.4 Gas Chromatography/Gas Chromatography Mass Spectrometry

Gas chromatography separates the components of a gaseous mixture and introduces them sequentially into a detector. The following standard techniques are often used to prepare samples for gas chromatography (Fig. 9.2).

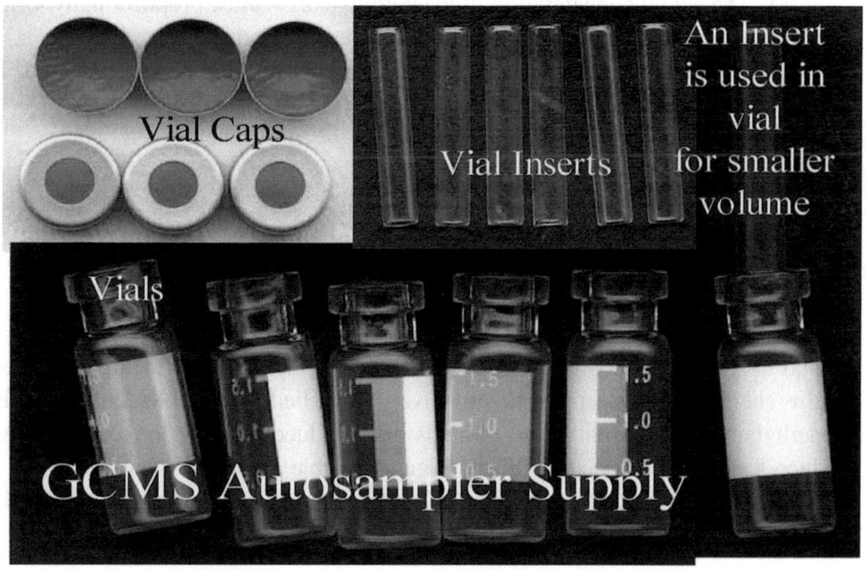

Fig. 9.2 Sample preparation for gas-chromatography mass-spectrometry (GCMS) analysis.

9.5 Dry-Extraction Gas-Chromatography Modification

This method is used for raw samples and samples purified using thin-layer chromatography.

- Approximately 1 mg of sample is placed into a sample container using:
  - 2-ml autosampler vial or
  - 3 × 50-mm culture tube
- Approximately 1 ml of organic solvent is added to the sample container.
  - The organic solvent used will depend on the solubility properties of the target compound.
  - A 1-mg/ml sample concentration is an acceptable analytical standard.
    (a) The actual concentration of components will be proportionally <1 mg/ml.
  - Proper technique requires the use of an internal standard.
    (a) An internal premix standard of 1 mg/ml concentration is recommended.
- The sample is agitated, and undissolved solids are allowed to settle.
  - A filtration step may be required at this point.
- A sample of the solution is drawn from the vial and analyzed.

9.6 Acid–Base-Extraction Gas-Chromatography Modification

The method is a slight modification of an acid/base extraction. It is typically used on impure samples and minimizes the introduction of contaminates that could either adversely affect results or block the injection port of the gas chromatograph.

- Approximately 1 mg of sample is placed into a sample container using:
  - 2-ml autosampler vial or
  - 3 × 50-mm culture tube
- Approximately 1 ml of saturated sodium bicarbonate solution is added to the sample vial.
  - Saturated sodium bicarbonate solutions have a pH = 8, which will create an environment conducive to the extraction of acidic, neutral, and amphoteric (can be acidic or basic) compounds.
- Approximately 1 ml of organic solvent is added to the sample container.
  - The organic solvent used will depend on the solubility properties of the target compounds.
    (a) Chloroform extracts a wide range of compounds.
    (b) Hexane is more selective.
  - A 1-mg/ml sample concentration is an acceptable analytical standard.
    (a) The actual concentration of the components will be proportionally <1 mg/ml.
  - Proper technique requires the use of an internal standard.
    (a) An internal premix standard of 1 mg/ml concentration is recommended.
- The sample is agitated and the phases are allowed to separate.
- A sample of the organic liquid is drawn from the vial and analyzed.

9.7 Infrared Spectroscopy

Infrared (IR) spectroscopy requires a highly purified sample for analysis. The slightest contamination can result in unexplained absorption bands or intensity shifts that complicate the identification process. Therefore, some compounds may require chemical processing to be effectively identified using IR spectroscopy. For example, the free-base forms of many phenethylamines are volatile, oily liquids that produce nondescript spectra. However, the salts of these same compounds produce well-defined spectra with sharp, reproducible absorption bands.

The following methods are slight modifications of acid–base extraction and are required for IR-spectroscopy sample preparation.

9.8 Acid–Base-Extraction Infrared-Modification-I

Acid–base-extraction IR-modification-I is used when the target compound is a solid at room temperature.

- Perform a complete acid–base extraction procedure for isolating an acid, base, or neutral compound depending on the properties of the target compound.
- Evaporate the organic solvent to dryness.
- Analyze using IR spectroscopy after:
  - Preparing a sample pellet using an IR-transparent material (dried potassium bromide, KBr). Grind 1 mg of sample with 80 mg of dried KBr using a mortar and pestle, or
  - Dissolve and recrystallize a thin film of sample on IR-transparent material, or
  - Place a drop of solution on IR-transparent material, typically salt plates.

9.9 Acid–Base-Extraction Infrared-Modification-II

Acid–base-extraction IR-modification-II is used when the target compound is either a basic liquid or a volatile oil at room temperature.

- Perform the complete acid–base extraction procedure for isolating a base.
- Bubble hydrogen chloride gas into the isolated organic layer containing the target compound.
  - The HCl gas reacts with the free-base compound to produce a salt that is insoluble in organic solvents.
  - HCl gas can be obtained by transferring the headspace air from a bottle of hydrochloric acid into the organic liquid containing the free-base compound.
  - HCl gas can also be generated through the reaction of sulfuric acid and sodium chloride.
- Evaporate the organic solvent to dryness.
- Analyze via IR spectroscopy after:
  - Preparing a sample pellet using an IR-transparent material (dried potassium bromide, KBr). Grind 1 mg of sample with 80 mg of dried KBr using a mortar and pestle, or
  - Dissolve and recrystallize a thin film of sample on IR-transparent window material, or
  - Place a drop of solution on IR-transparent material, typically salt plates.

9.10 Methanol Extraction

Methanol is an excellent solvent for use in gas-chromatography mass-spectrometry (GCMS) examinations; however, it is seldom used because it also dissolves a variety of impurities. Methanol extraction is preferred for the examination of residue only because there is a high risk of sample loss during extraction. In residue analysis, the deposits are washed with approximately 2 ml of methanol, and a sample of the resulting solution is run on the GCMS.

Most samples submitted to forensic laboratories for analysis require chemical processing to isolate the controlled substance. In some cases, separation can be accomplished using extraction techniques, while, in others, analytical instrumentation is the only viable alternative. Preliminary screening methods are often used to establish the direction of subsequent analyses. Although extraction techniques are not technically classified as screening methods, the isolation of a compound using extraction indicates the presence of fundamental properties that determine the appropriate confirmatory test.

It is important to recognize that no single extraction technique has universal applications in all instances. There is considerable variation among procedures, and the choice of particular technique will be based on the chemical and physical properties of the target compound. Notwithstanding these variations, product recovery is always greater when using multiple extractions with smaller volumes as opposed to a single extraction with a larger volume. This is because the solubility limit of the extracting solvent is often reached during the procedure. Also, the order of extraction steps may be changed with little, if any, affect on the overall success of the technique. For example, is extraction more effective if you dissolve the sample and extract the target, or dissolve the sample and extract impurities? The answer to this question will establish the order of extraction steps and is usually determined by availability of resources, time constraints on the analysis, and desired yield (quantity) of the target compound. Regardless, extraction techniques represent some of the most reliable and cost-effective methods used to isolate controlled substances.

9.11 Questions

1. Please describe chemical extraction to the jury.
2. Discuss the difference between solubility and affinity.
3. Describe the basic principles of liquid–liquid extraction.
4. What factors would be considered in the choice of liquid–liquid extraction over solid–liquid extraction?
5. Please explain to the jury the basic principles of acid–base extraction and cite any procedural differences when using this technique to isolate an acid and a base.
6. Discuss a procedure used to isolate a neutral compound.
7. Briefly explain why IR spectroscopy requires highly purified samples.
8. Define the terms miscible and immiscible.
9. Provide two examples of immiscible solvent pairs.
10. Discuss why immiscible solvent pairs are required for extraction procedures.
11. Outline a common test used to verify phase identity in extraction procedures.
12. Discuss why multiple extractions maximize product yield.
13. In your opinion, would it be more efficient to perform a single extraction using a larger volume, or multiple extractions using smaller volumes. Justify your choice.

Selected Reading

Bell, C. E. Jr.; Taber, D. F.; Clark, A. K. *Organic Chemistry Laboratory with Qualitative Analysis*, 3rd ed.; Harcourt College Publishers: New York, 2001; pp. 33–37.

Fieser, L. E.; Williamson, K. L. *Organic Experiments*, 8th ed.; Houghton Mifflin Company: Boston, 1998; pp. 104–110.

Pavia, D. L.; Lampman, G. M.; Kriz, G. S.; Engel, R. G. *Introduction to Organic Laboratory Techniques*, 3rd ed.; Saunders College Publishing: New York, 1990; pp. 595–616.

10 Chromatography and Mass Spectrometry

10.1 Introduction

A vast majority of samples submitted to forensic laboratories for identification are complex mixtures. Suspected drugs or controlled substances recovered from crime scenes are often contaminated with various solvents or by-products from manufacturing and distribution. The identification process begins with the separation of the mixture into individual components. The purified components can then be identified using various chemical or analytical techniques.

Chromatography (from Latin for *color writing*) is a general term used to describe any physical method of separation in which the components to be separated are distributed between a mobile and stationary phase. Although several types of chromatography exist, the underlying chemical principles are the same. The most common systems consist of a mobile phase that is either a liquid or gas and a stationary phase that is a solid. The components of the analyte (mixture) will interact with both phases with varying relative affinities. The physical process of separation will depend on these interactions as the mobile phase pushes the components through the system. Generally, components with a higher affinity for the mobile phase will move faster, while those with a higher affinity for the stationary phase will move slower. A common river provides a good analogy to most chromatographic techniques. The rocks (analyte) in the riverbed are pushed down the river by the running water (mobile phase). The actual separation depends on how the rocks interact (affinity) with the riverbed (stationary phase) and the running water. Normally, the smaller rocks move downstream faster, while the larger rocks move slower.

There are several different types of chromatography in common use and considerable variations exist. However, some are more frequently used in forensic analysis than others. We have selectively chosen representative methods for discussion.

10.2 Chromatographic Techniques

10.2.1 Paper Chromatography

Paper chromatography is the simplest, and perhaps oldest, of all chromatographic techniques with references dating back to the late 1800s. The stationary phase is a special paper that may or may not be pretreated. The inert (nonreactive) solid support medium is cellulose (polymer of glucose), a major component in paper manufacturing. A small amount of water is adsorbed from air at the hydroxyl groups of cellulose forming the stationary phase. Adsorption is different from absorption. *Adsorption* refers to water molecules loosely bound at the surface of a molecule, while *absorption* refers to water molecules integrated into the interior. A simple sponge illustrates these subtle differences quite effectively. Water molecules are absorbed into a sponge when it is placed in water. Sand particles are adsorbed to the surface of a wet sponge once it is thrown into a sand box.

Paper chromatography is performed in a clear container called a *developing chamber*. Initially, the developing chamber contains only the mobile phase, a suitable solvent often called the *running* or *developing solvent*. Covering the chamber allows an equilibrium to be established between solvent molecules in the liquid phase and those evaporating from the surface. Thus, the environment in the developing chamber is saturated with solvent molecules.

J.I. Khan et al., *Basic Principles of Forensic Chemistry*, DOI 10.1007/978-1-59745-437-7_10,
© Springer Science+Business Media, LLC 2012

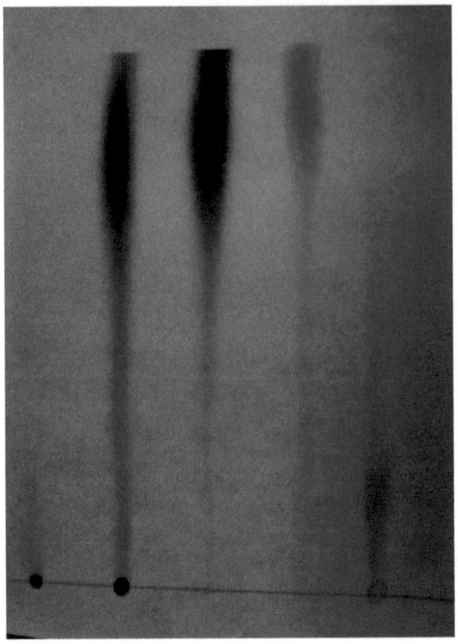

Fig. 10.1 Separation of samples using paper chromatography with a polar running solvent. Left-to-Right: Column 1 – a nonpolar component. Column 2 – reference. Column 3 – mixture of polar components. Column 4 – a polar component. Column 5 – a component of intermediate polarity. Note beginning reference line and solvent front (not drawn).

Small drops of sample and reference solutions are placed at regular intervals (spotted) at one end of the paper on a horizontal line drawn in pencil. Samples are identified using numbers, labels, or codes commonly written directly on the paper (in pencil) under each sample. The samples must be high enough to ensure that they will not be immersed in the running solvent once the paper is placed in the developing chamber. The samples are allowed to dry and the paper is placed in the developing chamber with the samples toward (but not immersed in) the solvent.

Capillary action draws the mobile phase up the paper and the eluting solvent contacts each sample positioned on the stationary phase. The components in each sample will interact with the mobile and stationary phases with varying affinities and separation is accomplished using differential migration (different migration rates). In general, this is directly related to differences in polarity; for example, if a polar running solvent is used, polar components will travel more rapidly in the mobile phase, while nonpolar components will move slowly or not all. A quick note, polarity can be equated to a bar magnet. The characteristics and behavior of polar molecules are very similar to those of a magnet. When you encounter terms such as polarity, dipole, polar, etc., simply think of a magnet to clarify the concept.

The procedure is complete when the running solvent stops eluting (moving) or when it approaches the end of the paper. The running solvent should never elute to the end of the paper. The paper is removed from the developing chamber and a line is immediately drawn in pencil across the solvent front. This must be done quickly because volatile solvents evaporate quickly and the solvent front may soon be indistinguishable. The paper is allowed to dry and the separated components are observed as spots. They are readily observed if they happen to be colored; if they are colorless, visual enhancement techniques will be required, i.e., oblique lighting with short- and long-wavelength UV (a black light) or iodine fuming. Iodine fuming involves placing the paper in a covered beaker containing a few crystals of iodine. Iodine will readily sublime (convert from solid to gas) and the gaseous molecules adhere to the components coloring them light purple-violet (Fig. 10.1). The color intensity of each component can be used to compare relative concentrations. More intensely colored components (darker spots) are present in higher concentrations. In addition, an assessment of purity can be made using the results. Samples containing a single spot are highly purified, while those containing many spots are contaminated. The paper, or a clear sketch of the results, including visual enhancement method, should always be documented in the case file notes.

Paper chromatography produces both qualitative and quantitative results. Qualitative results are obtained by simply observing the spots on the paper, i.e., purity, number of components, relative concentrations, degree of separation, characteristics of each component based on running solvent properties, etc. Quantitative results are obtained using *retention factors* (R_f-factor). R_f-factors can be calculated for each component by dividing component travel by solvent travel. Component travel is the measured distance from the line where the sample was initially spotted to the final location of the component on the paper. Solvent travel is the measured distance from the same beginning line to the solvent front. R_f-factors can be used to identify components in different samples separated under the same chromatographic conditions.

There are several factors to consider when performing paper chromatography:

- The best resolution (observed separation) is accomplished using small, highly concentrated drops when spotting. The drops will diffuse radially outward when they are applied to the paper. If care is not taken, the drops may contaminate one another.
- Never use ink when drawing reference lines on the paper. Ink will often separate and migrate in a fashion similar to the samples.
- Never allow the running solvent to elute to the end of paper. This affects overall resolution by clustering components at the top of the paper.
- Never immerse the components in the running solvent. Components with characteristics similar to the solvent will simply dissolve directly off the paper.

Paper chromatography is classified as a type of *solid–liquid chromatography* because separation is achieved using a liquid mobile phase and a solid stationary phase.

10.2.2 Thin-Layer Chromatography

The procedure used to perform thin-layer chromatography (TLC) mirrors that used for paper chromatography with one notable exception. The stationary phase is a thin layer of adsorbent material coated on a small sheet of glass or plastic. TLC plates can be cut in any size or shape and most often resemble a coated microscope slide.

There are many types of adsorbents available but alumina (Al_2O_3) and silica (SiO_2) are by far the most common. Anhydrous alumina (no water adsorbed) adsorbs substances more strongly and is often used to separate nonpolar components such as hydrocarbons and alkyl halides, as well as molecules containing ether, aldehyde, and ketone functional groups. Silica is less active and has common applications separating polar components containing alcohol, carboxylic acid, and amine functional groups.

Thin-layer chromatography is a sensitive, fast, and simple analytical technique; however, it gained significant popularity with the development of cost-effective commercially available plates. The automated industrial application of adsorbent material to plates creates a solid layer of uniform size and thickness. This produced a more stabile, even flow of solvent and a smooth path for component travel. The consistency of separation using commercially available TLC plates was a dramatic improvement over paper chromatography (Fig. 10.2).

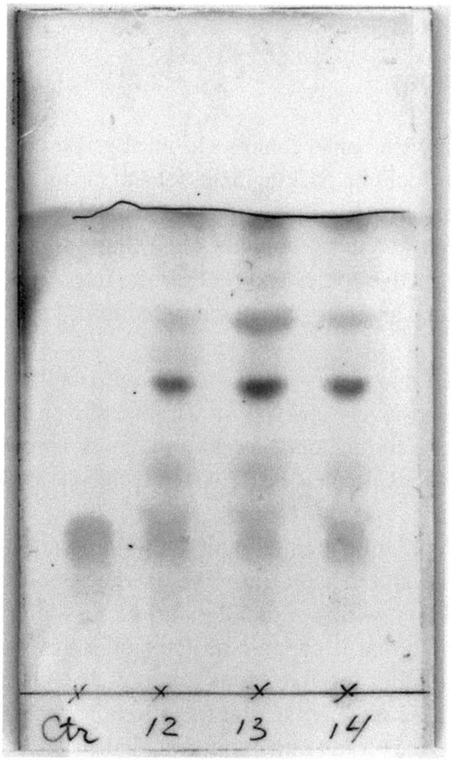

Fig. 10.2 Component separation using a commercially available TLC plate. CTR is control, a reference sample containing the purified drug. All samples test positive for the presence of the drug. Note the well-defined columns of separation, sample labels, and drawn reference lines. Color intensity can be used to determine relative concentration. Comparison of spots midway through the each sample column indicates sample 13 has the highest concentration of these components (*darkest spots*).

10.2.3 Column Chromatography

Column chromatography is a macroscopic version of TLC often used to separate large samples. In this method, the "column" is a specially designed glass or plastic tube open at one end and fitted with a polyethylene frit and stopcock at the other. The frit holds the stationary phase in the column while allowing the mobile phase to pass. The stopcock is used to control the flow of the mobile phase.

The vertical column is filled with adsorbent material (stationary phase) in a process called *packing*. Like TLC, alumina and silica are most often used, but hydroxapetite and cellulose are also quite common. Uniform packing of the column is critical to the success of column chromatography. For this reason, the most efficient columns are rarely packed with powdered adsorbent, although this is an option. Instead, a slurry is prepared by mixing the adsorbent with the running solvent (mobile phase) and poured into the column. The stopcock is opened and the slurry uniformly packs the column as gravity pulls the solvent downward. The draining process continues until the solvent level is just above the top of the packed adsorbent. Care must be taken to ensure that the last portion of slurry forms a flat surface at the top of the column. Under no circumstances should the level of the solvent fall below the packing, and, if this occurs, the procedure must be repeated. The size of the column can vary depending on the desired resolution (separation). In general, the height of a packed column should be at least ten times the column's diameter and the amount of adsorbent used should weigh at least 30 times the amount of sample.

Highly concentrated samples work best; therefore, it is recommended that samples be dissolved in a minimum amount of solvent. The sample is *carefully* added to the column without disturbing the top. The stopcock is opened and the solvent is drained until the sample solution level is just above the packing. The sample must be added uniformly to the column in a small, discrete band.

The first running solvent is added to the top of the column and the stopcock is opened. Gravity pulls the running solvent (mobile phase) through the column (stationary phase) and soluble components are separated as they migrate (travel) down the column at different rates (differential migration). Individual components arrive at the lower end of the column at different times and are collected as they drain through the stopcock. The process is repeated with solvents of different polarity to isolate the remaining components.

10.2.4 Ion-Exchange Chromatography

Ion-exchange chromatography (IEC) is very similar to column chromatography in that both use packed columns to achieve separation. The two procedures differ in the column packing material and the method used to isolate individual components. Ion-exchange chromatography relies on electrostatic forces of attraction (opposites attract) to bind charged atoms or polar molecules to an oppositely charged resin packed in the column. The column is subsequently treated with an eluting solvent of varying concentration to disrupt the attractive forces and recover the fixed molecules. Ion-exchange chromatography is typically used in forensic analysis to isolated anions from aqueous solutions, i.e., chloride (Cl^-), sulfate (SO_4^{2-}), and nitrate (NO_3^-).

Ion-exchange chromatography is divided into two broad classes differing only in the composition of the stationary phase. Cation-exchange chromatography isolates positively charged ions from a sample mixture using a negatively charge resin to selectively bind the cations. Anion-exchange chromatography does the polar opposite (no pun intended!), using a positively charged resin to isolate negatively charged ions. Regardless of which technique is used, ions are fixed to oppositely charged resins as the sample passes down the column.

Once the components are bound, the column is treated with a low-ionic-strength solution (low salt concentration) to establish an equilibrium between the resin and the solution. The fixed components are removed from the column using a *salt gradient* applied to the running solvent, which gradually increases its ionic strength. The changing salt concentration will eventually reach an ionic strength that is capable of displacing the ions from the resin. Alternatively, the pH of the running solvent can be changed to alter the fixed ions' charge. The change in charge will disrupt the ion-resin attractive forces and free the bound ions. In either case, the fixed ions are released from the resin and collected at the bottom of the column (Fig. 10.3). Sometimes it may be more practical to select conditions that bind contaminants and allow the target ions to flow through. In these cases, ion-exchange chromatography is virtually identical to column chromatography.

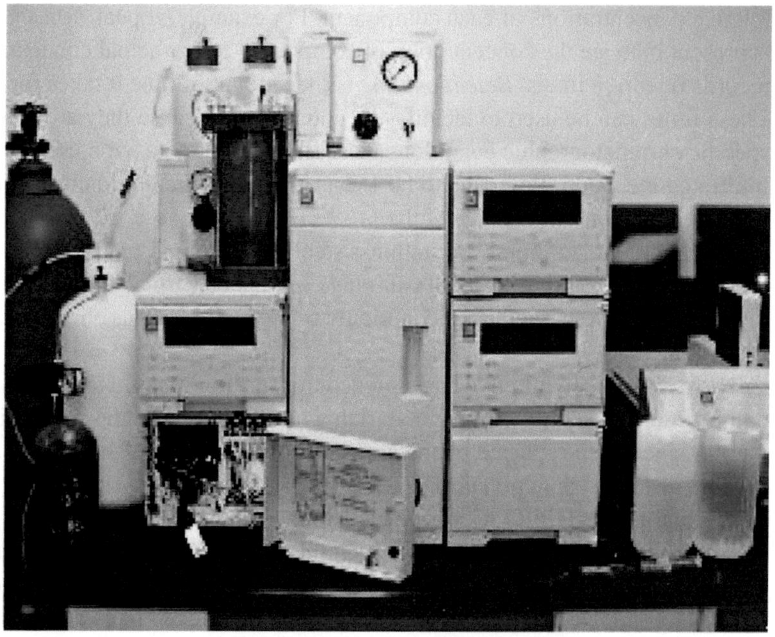

Fig. 10.3 A typical instrument used to perform ion-exchange chromatography. The automated computer workstation is capable of performing single-step qualitative and quantitative analysis using several variations of ion exchange.

Fig. 10.4 Integrated HPLC workstations have dramatically advanced the practical applications of column chromatography. HPLC is a highly sensitive technique that is capable of differentiating structural isomers. It is commonly used in forensic analysis to separate drugs and controlled substances from complex mixtures with expanding applications in the isolation of designer drugs.

10.2.5 High-Performance Liquid Chromatography

High-performance liquid chromatography (HPLC) is a more controlled variant of column chromatography. Measurements are performed using an automated instrument called a *liquid chromatograph*, which houses, among other things, a closed system of pumps and the column (Fig. 10.4). In HPLC analysis, high pressures (up to 400 atmospheres!) force the liquid phase (mobile phase) through a column containing densely packed, small-diameter microspheres (stationary phase). The nature of the packing material dramatically increases the surface area for column adsorption, a significant improvement over column chromatography. Separation is achieved in the same manner as column chromatography. The mobile phase pushes the sample through the stationary phase and components travel at different rates, depending on their selective interactions with both phases. The capacity to precisely regulate pressure provides direct control over solvent velocity, which enhances resolution (degree of separation) and decreases the level of component diffusion onto the column. Several types of detectors are used to differentiate components as they emerge from the column.

The detector records and displays component information on a *chromatogram*, a series of peaks on a graph of relative abundance (y-axis) versus time (x-axis). The area under each peak is integrated by the instrument and is used to compare the relative concentrations of each component. For example, if peak 1 has an integrated area twice that of peak 2, component 1 is present in twice the concentration of component 2. The actual concentrations however are not known. The instrument also records retention times. *Retention time* is the amount of time it takes for each component to pass through the entire column. These times can be used to identify the component because they are well defined for a variety of substances isolated under specific chromatographic conditions. A subtle point of note – the use of high pressures in HPLC has somehow (erroneously) infiltrated the name. Referring to HPLC as high-*pressure* liquid chromatography is incorrect and its use can sometimes call into question your knowledge of the process.

Typical HPLC columns are stainless steel with a length of 10–30 cm and an internal diameter of 4–10 mm. The particle size of column packing material can range in diameter from 1 to 10 μm (μm is micrometer, 1 $\mu m = 10^{-6}$ m; 1 millionth of a meter). The stationary phase is commonly treated with various chemical agents, depending on the nature of the molecules to be separated.

HPLC is divided into two basic procedures: *normal*-phase HPLC and *reverse*-phase HPLC. The main difference is in the composition of column packing material, which necessitates the use of different solvents. Normal phase was developed first, but has since been replaced by the almost exclusive use of reverse phase. Therefore, any discussion of normal-phase HPLC is more historical than practical.

Normal-phase HPLC uses a polar stationary phase and a nonpolar mobile phase to achieve separation. Columns are packed with tiny particles of silica containing polar amino, diol, or cyano groups. The sample is dissolved in a nonpolar solvent such as hexane (mobile phase) and pushed through the column. Retention times for polar components are longer than those for nonpolar components in yet another example of "likes dissolve likes." Essentially, the polar components adhere to the modified silica particles (adsorbed) and are retained longer on the column. Nonpolar components proceed through the column at different rates, depending on their relative interaction with the mobile and stationary phase.

Reverse-phase HPLC uses a nonpolar stationary phase and a polar mobile phase to achieve separation (the reverse of normal phase). The column is packed with silica containing a bound hydrocarbon component. Alkyl groups (the hydrocarbon component) containing 8 or 18 carbons are preferentially used to make the silica particles nonpolar. A solution of water and alcohol (methanol) is commonly used to create a moderately polar mobile phase. The sample is dissolved in the mobile phase and pushed through the column. The nonpolar components are adsorbed on the surface of the modified silica particles and are retained longer on the column (likes dissolve likes). The polar components travel through the column at different rates and are separated. Retention times can easily be modified using reverse-phase HPLC. In general, adding more water to the mobile phase will increase retention times of nonpolar components and decrease those of polar. Adding an organic solvent will decrease nonpolar retention times and increase those of polar components.

10.2.6 Gas Chromatography

Gas chromatography (GC), also called vapor-phase chromatography (VPC) and liquid-gas chromatography (LGC), is the most popular method used to separate the individual components of a volatile mixture. It is, without question, the most frequently used chromatographic technique in modern forensic laboratories. Measurements are performed using a complex, highly integrated instrument called a *gas chromatograph* (Fig. 10.5). The vital components of a gas chromatograph are, among other things, the injection port, the column, detectors, and an oven. All components are integrated using a computerized workstation with highly programmable functions. The injection port is a heated chamber typically located on the top of instrument. The column is contained in an insulated oven at the front of the instrument, which is easily accessible for maintenance, repair, or replacement. Detectors are generally located at the back and can be individually accessed through the workstation. Injection port and column temperatures are independently regulated and are capable of reaching temperatures in excess of 400°C.

In GC analysis, the mobile phase is an inert gas such as helium, argon, or nitrogen called the *carrier gas*. The stationary phase is a high boiling liquid coated on the interior wall of the column (capillary GC) or on the surface of a solid support medium (packed-column GC). The sample is vaporized and carried through the column by the mobile phase (carrier gas). A detector at the end of the column records and displays component information on a *chromatogram*. GC chromatograms closely resemble the appearance of those produced using HPLC. They provide the same information and are interpreted in a similar manner.

The sample is introduced into the gas chromatograph using a glass microliter (μL) syringe. Typically, a 2–4-μL sample is sufficient, but actual sample size may vary from 1 to 25 μL. The needle is pushed through a rubber septum at the injection port. Care must be taken to avoid bending the syringe needle during placement. Also, the injection port is usually very hot and

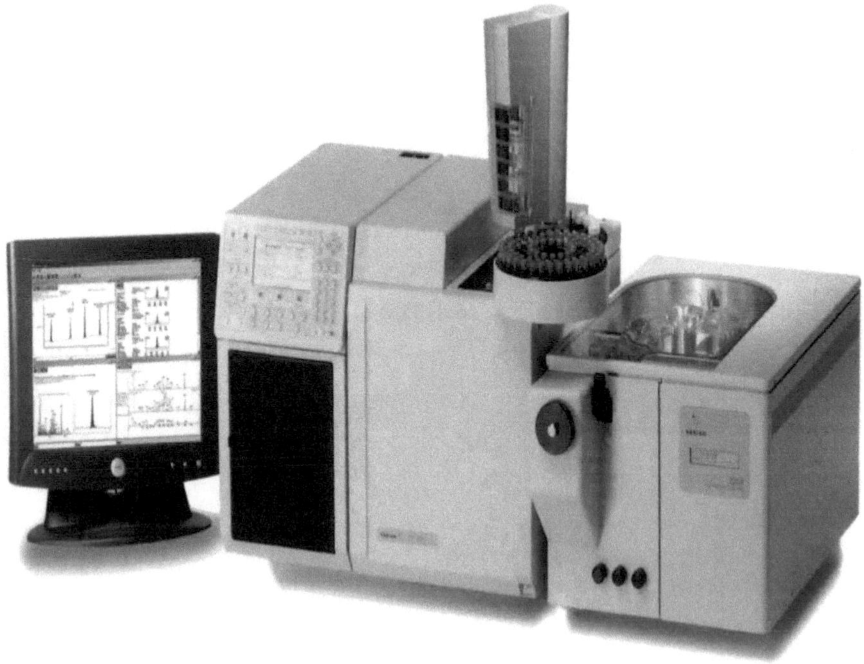

Fig. 10.5 Automated gas chromatographs are commonly used in forensic analysis to separate volatile drugs and controlled substances. The above instrument is equipped with an autosampler located on the top of the instrument. This configuration permits the sequential examination of several mixtures with minimal supervision.

direct contact should be avoided. The sample is vaporized in the injection port and mixed with a continuous flow of carrier gas. Components with low boiling points will vaporize first, followed by those with higher boiling points. The sample–carrier gas mixture runs through the column at a preset flow rate. The individual components are differentially adsorbed on the surface of the stationary phase and an equilibrium is established between molecules in the vapor and liquid phases. Easily adsorbed components travel slowly through the column (high retention times), while those not readily adsorbed travel more rapidly (low retention times). Flow rates must be carefully regulated to ensure good resolution. If the rate is too high, an equilibrium will not be established between molecules in the vapor and liquid phase. If it is too slow, the molecules will emerge from the column over an extended period of time producing band (peak) broadening. Both conditions result in poor resolution (separation).

There are two types of columns used in GC analysis. The stationary phase is the same in both, but the column dimensions and the method used to affix the stationary phase are different. The type of column is incorporated into the name of the technique and differentiates the two procedures.

Packed-column GC uses a column constructed of stainless steel or glass tubing typically 1.5–10 m (meters) long with an internal diameter of 2–4 mm (millimeters). The stationary phase is a thin layer of high boiling liquid coated on the surface of a solid support medium. The composition of the stationary phase can vary depending on the characteristics of the molecules to be separated, hydrocarbon greases, silicone-based oils, and carbowaxes (derivatives of polyethylene glycols) are most common. The solid support media is typically crushed firebrick, glass or nylon beads, silica, or alumina.

Capillary GC uses a wall-coated open-tubular (WCOT) column made of small-bore flexible tubing (often copper). WCOT columns are generally 10–60 m long with an internal diameter of 0.2–0.5 mm. The tubing is wound around a solid support to accommodate space limitations in the gas chromatograph (Fig. 10.6). The inner wall of the tube is coated with a stationary phase of similar composition to that used in packed-column GC. Capillary GC produces more effective separation than techniques using larger packed columns because the small diameter and extreme length of WCOT columns increase the interactions between sample components and the stationary phase.

One of the clear advantages of GC analysis is the ability of gas chromatographs to accommodate a variety of detectors in a single instrument. Switching detectors or using detectors in series is a relatively simple process, often requiring only a sequence of keystrokes at the workstation. Although the type of detector used will depend heavily on the specific application, *flame-ionization detectors* (FIDs) and *thermal conductivity detectors* (TCDs) are very common. Flame-ionization detection is a destructive technique because the components are burned (ionized) as they elute off the column. The ions create (induce) an electrical current that is measured and recorded as a peak on the chromatogram. Thermal conductivity detection is a nondestructive technique and would be preferred when sample quantities are limited. In this method, two separate filaments are heated to a constant temperature using an electrical current. Both filaments are exposed to a continuous flow of carrier gas,

Fig. 10.6 A typical wall-coated open-tubular (WCOT) capillary GC column. Notice the ends of the column on the right and left of the center support. The small diameter and extreme length of WCOT columns offer significant advantages over most chromatographic techniques.

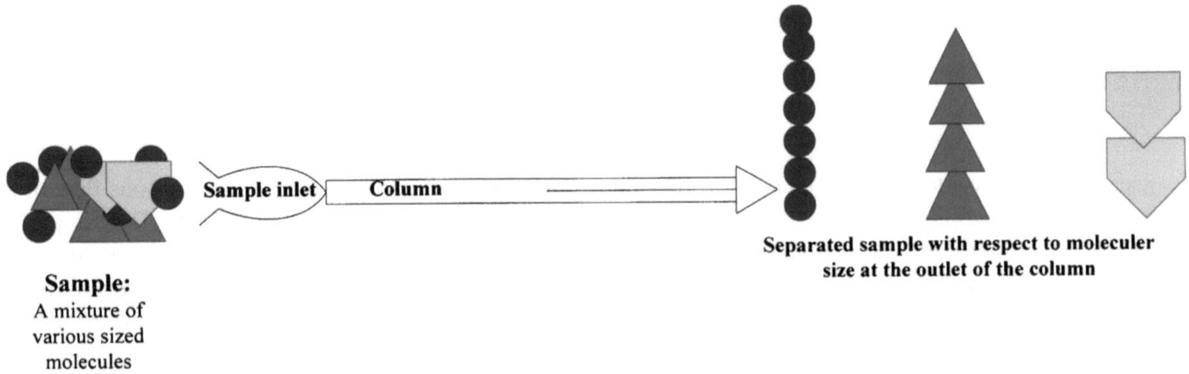

Fig. 10.7 Gas chromatography (GC) can be compared to the operation of coin-separating machines. This useful analogy can be used to clarify GC analysis to members of a jury.

which cools the filaments. The final temperature of both filaments will be the same and represents the base line (starting point). The eluting components of the mixture are directed over one filament and cause a temperature change specific to that component. The current required to bring the filament back to base-line temperature is recorded as a peak on the chromatogram.

Flame ionization and thermal conductivity are versatile, highly sensitive techniques and both are commonly used to detect a wide range of organic compounds. However, in cases requiring a higher degree of specificity, a more selective detector may be desired. For example, electron capture detectors (ECDs) are often used exclusively to detect halogens (Group VII), organometallic compounds (contain metals), nitriles (contain CN group), and nitro-compounds (contain NO_2). Also, nitrogen–phosphorus detectors (NPDs) are frequently used to distinguish compounds containing nitrogen or phosphorus. Without question, the most versatile and well-known GC detector is the mass selective detector (MSD). This detector is actually a mass spectrometer connected directly to the gas chromatograph. This analytical instrument will be discussed in more detail below.

10.2.6.1 Gas Chromatography: A Simple Analogy

In many cases, the forensic chemist is called upon to describe complex chemical procedures to individuals that have a limited understanding of scientific principles. Courtroom testimony is carefully prepared using common terminology and the presentation must be in a clear, simple manner that avoids confusion and misinterpretation. A gas chromatograph is a highly integrated instrument used to separate a gaseous mixture into individual components. The complex details of the separation process may easily confuse most people on a jury. Therefore, a description of how a gas chromatograph functions may contain a reference to coin-separating machines frequently found in local grocery stores. These machines separate the mixture of coins based on size, and totals each pile based on weights. This analogy would illustrate how a gas chromatograph functions and may help members of a jury be more comfortable with testimony about this complex instrument (Fig. 10.7).

10.2.7 Chromatography: Limitations

Chromatography is a purification technique that is generally considered a nonspecific form of identification, that is, it does not necessarily provide definitive proof in the identification of a substance. The separated components often require further analysis for identification. Although retention times (R_t) are characteristic for a given compound under a specific set of analytical conditions, you must have a reference for comparison and identification. In addition, the area under a chromatographic peak can be used to establish the relative abundance of a component, but not its exact concentration. It is important to realize that taller peaks do not necessarily have a greater area. It is not uncommon for a peak with half the vertical height of another to have twice the abundance. This is because the compound represented by the shorter peak eluted over a longer period of time producing a broader base, thus increasing its area. Modern instruments avoid these complications by integrating peak areas and recording the results directly on the chromatogram. A more rudimentary method would be to cut out the peaks and measure their individual mass using an analytical balance.

10.2.8 Interpretation of GC Chromatograms

A typical GC chromatogram is represented in Fig. 10.8. The x-axis is analysis run time and the y-axis is relative abundance. It is possible to determine general characteristics of the separated components by simply examining the chromatogram. For example, there are seven measurable peaks in Fig. 10.8; therefore, there are seven different components in the sample. The first peak appears to have the greatest area and is therefore present in the greatest relative amount. It is also, most likely, the smallest molecule because it was the first to elute off the column (shortest retention time). We cannot determine the actual concentration or identity of each component using only the chromatogram; however, valuable information can be obtained through relative comparisons of all components present. Some additional qualitative observations are shown in the figure.

Retention times (R_t) are generally calculated by the instrument and recorded on the chromatogram. They can also be determined by drawing a vertical line from the apex of a peak down to the x-axis. The retention time is the measured distance on the x-axis from the y-axis (origin) to the apex of the peak (drawn vertical line). This value is generally very stable, but fluctuations can be observed if there are variations in the analytical method. The units of retention time are those specified for the x-axis, typically minutes or seconds.

Calculating a relative retention time (RR_t) is a more accurate representation of the actual time a component takes to pass through the column. It normalizes the retention time of each component so that accurate comparisons can be made. The RR_t is a ratio of the R_t of a particular component to the R_t of an internal standard run under the same conditions. It is represented in the following equation where x is a chosen peak on the chromatogram. Simply measure the retention times of the component and internal standard as outlined above. Then divide the component R_t by the standard R_t to obtain the relative retention time ($RR_t x$).

$$RR_t x = R_t x \,/\, R_{t\,\text{Internal Standard}}$$

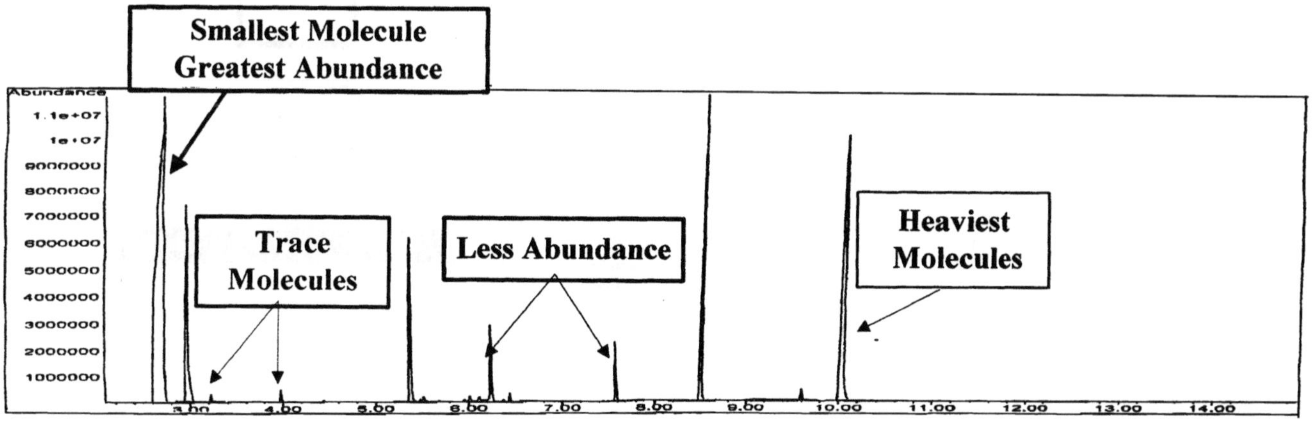

Fig. 10.8 Chromatogram peaks can provide valuable qualitative information. Each peak represents an individual component separated using gas chromatography. The figure contains a few general observations obtained through relative comparison.

The use of an internal standard to establish a compound's RR_t is a reliable method to normalize retention times, but the internal standard also demonstrates that the instrument is operating properly. The GC retention times of commonly used internal standards are well known and any deviation indicates a variation in the instrument's operating parameters. This is not necessarily bad because small discrepancies in operating parameters will affect the internal standard and each component equally.

The *y*-axis represents relative abundance and the value assigned to a particular peak is used to calculate the area under the peak. The area under a chosen peak can be normalized to the area under the largest peak. In this method, the largest peak is assigned a value depending on the range of the *y*-axis, typically 100, 1,000, or 1,000,000 units. The normalized area of the remaining peaks is a ratio of each peak area to that of the largest peak. The equation below illustrates this relationship.

$$\text{Normalized Area}_{\text{Peak X}} = \left(\text{Area}_{\text{Peak X}} / \text{Area}_{\text{Maximum Peak}}\right)$$

Once again, normalized areas are a more accurate representation of actual values and can be used to make precise relative comparisons.

10.3 Mass Spectrometry

Mass spectrometry (MS) is an analytical method with a surprising number of applications. In the field of forensic investigation, it is often used to confirm the identity of controlled substances. Measurements are performed using an analytical instrument called a *mass spectrometer*. Mass spectrometry quantitatively measures the relative abundance of positively charged ion fragments created from the destruction of the sample. A detector records the data on a *mass spectrum*, a graph illustrating each fragment's mass-to-charge ratio (*m*/*z*) versus intensity (abundance). The mass spectrometer is an extremely versatile instrument capable of supporting a variety of ionization sources and detectors. Early models were very large, cumbersome, and quite expensive, which virtually precluded their use in many laboratories. Today, advancements in technology have made these instruments an affordable asset found in many forensic laboratories.

An interesting combination of chromatography and mass spectrometry produced arguably the most versatile and most recognizable instrument in quantitative analysis, the GCMS (gas-chromatography mass spectrometry) (Fig. 10.9). Although mass spectrometry is an independent analytical technique, it is typically used in tandem with gas chromatography because reliable MS measurements require highly purified samples. In this configuration, the mass spectrometer functions as a detector for gas chromatography. Gas-chromatography mass spectrometry (GCMS) is the single most frequently used method in forensic analysis to confirm the identity of drugs and controlled substances. It is extremely sensitive over a broad range of compounds and provides consistent, definitive results.

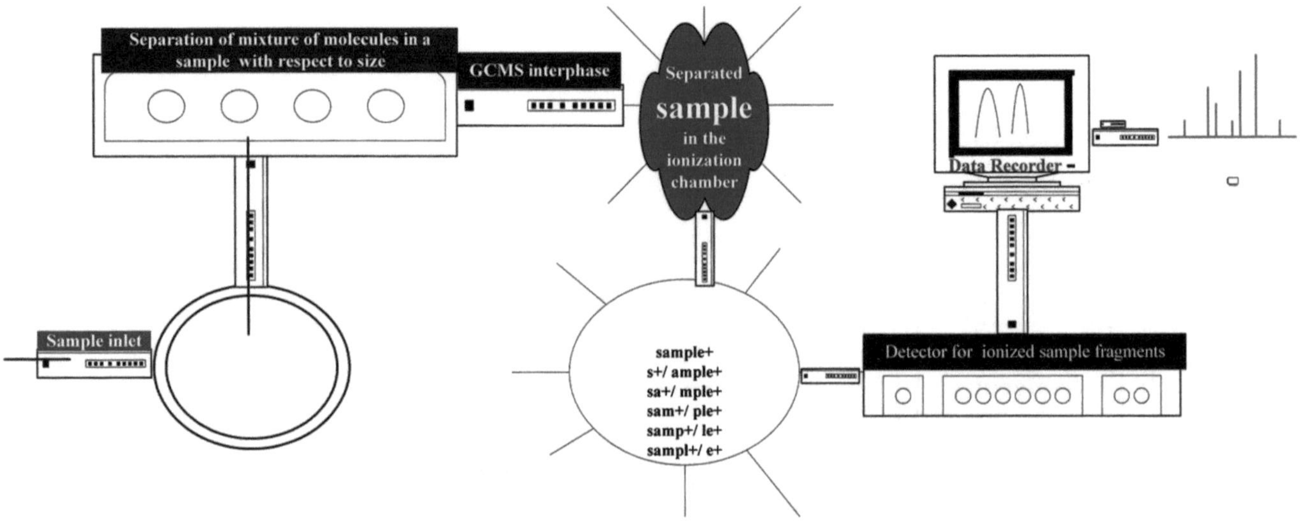

Fig. 10.9 Summary of gas-chromatography mass spectrometry (GCMS). A sample mixture is vaporized and separated in gas chromatography. The individual components are passed into a mass spectrometer where ionized fragments are produced and recorded.

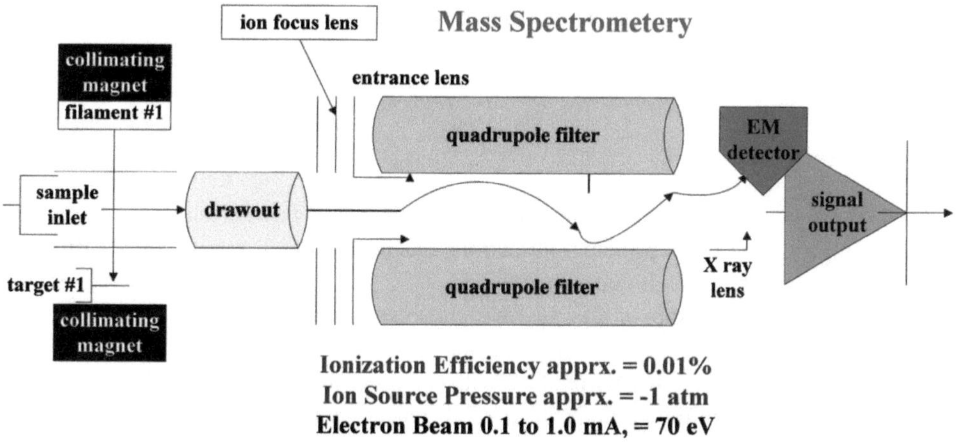

Fig. 10.10 Mass spectrometry is an analytical technique commonly used to identify drugs and controlled substances. Distinctive ion fragments are produced from the destruction of the sample and then focused on the detector. The above example contains a quadrupole mass analyzer.

In GCMS, a vaporized sample is separated into pure components using gas chromatography (GC). As each component elutes off the GC column, it is analyzed separately in the mass spectrometer, which is directly connected to the gas chromatograph. The sample (component) enters a heated vacuum chamber called the *ionization chamber* where it collides at right angles with a beam of high-energy electrons. The electron beam shatters the molecule into a series of positively charged ion fragments. Although the fragments have different masses, a vast majority carry a charge of $^{+}1$. Therefore, each fragment has a unique *mass-to-charge ratio* which is commonly represented as m/z (m is mass, z is charge). Since we are usually dealing with ions that have a $^{+}1$ charge ($z = {}^{+}1$), mass-to-charge ratios are often interpreted as simply mass. Multiple ionization does occur, but it is extremely rare and will not be considered. The mass-to-charge ratio of each ion fragment is characteristic and will vary depending on the structure of the molecule under analysis. The ion fragments are filtered by the *mass analyzer* using their mass-to-charge ratio and directed toward the *detector*. The detector records the mass and relative abundance of each fragment and displays the information as a band (peak) on the *mass spectrum*. The location of a peak on the x-axis is used to determine the mass of the ion (more specifically, charge-to-mass) and the peak height determines its relative abundance. A particularly useful ion is formed when the electron beam knocks a single electron off the molecule. This produces a structurally intact, positively charged form of the molecule called the *molecular ion* (M^+). The location of the M^+ peak on the x-axis gives the mass of the molecule. Also, the highest peak on the spectrum is called the *base peak* and represents the most stable ion fragment formed. Identity of the component is achieved using the molecular ion peak and supported by ion fragments which are pieced together like a jigsaw puzzle to reconstruct the original molecule. Mass spectrometry creates a unique fragmentation pattern for each molecule that can be used for definitive identification (Fig. 10.10).

10.3.1 Ionization

A compound must be broken into ionized particles before it can be introduced into the mass analyzer. The ionization can be achieved by "*hard*" (*electron ionization*) or "*soft*" (*chemical ionization*) techniques.

10.3.2 Electron Impact

Electron-impact ionization (EI) is the most frequently used ionization technique. In a vacuum of 10^{-5}-torr pressure (10^{-5} atmosphere), an electron beam emitted from a tungsten or rhenium filament is accelerated toward a plate with a variable voltage of 5–100 V. The ionizing beam is controlled either by the filament dimensions or by regulating the current reaching the trap plate. To increase the electron mean free path, a magnetic field along the principal axis is applied. The sample is crossed with this beam of electrons and the beam impacts a cross section of molecules such that maximum ionization occurs in a narrow energy window, typically 50–100 electron volts. Therefore, in a normal electron source, the electron energy is fixed at 70 eV.

Ions formed in the ionization volume are extracted through the ion exit slit by a small potential applied to the ion repeller plate. In magnetic sector instruments, an accelerating voltage is applied through the ion exit slit.

10.3.3 Chemical Ionization

Chemical ionization (CI) mass spectrometry is a technique in which the characteristic ionization is produced by ionic reactions rather than electron impact. One of its principal attractions has been its ability to produce information on molecular mass in cases where electron impact mass techniques have failed.

Chemical ionization requires a high pressure of reagent gas in the ion source, usually about 1 torr. A torr is a unit of pressure equal to the amount of pressure required to support 1 mm of mercury at 0°C. Electron-impact ionization of the reagent gas produces ions that are either nonreactive or slightly reactive with the reagent gas itself, but which readily react to ionize the sample. The sample itself, as in electron impact ionization, is introduced at a much lower pressure. The partial pressure of the gaseous sample is rarely greater than 0.01% of the reagent gas pressure and usually much less.

One of the advantages of CI is that the characteristics of a CI mass spectrum are highly dependent on the nature of the reagent gas used. As a consequence, it is possible to control the structural information observed by varying the nature of the reagent gas used.

Any ionization process produces both negatively and positively charged ions. Normally, the negative ions remain undetected because ion source and focusing potentials allow only the extraction of positive ions from the source. Other changes are also necessary because magnetic analyzers do not transmit negative ions unless the magnet current is reversed. Also, a standard electron multiplier does not usually detect negative ions without modification.

10.3.4 Mass Spectral Fragmentation

The fragments created during ionization are assembled to reconstruct the original molecule. The two most significant ions are the *base peak* and the *molecular ion peak*. The remaining ion peaks are considered "*grass*" and provide the detail for specific identification (Fig. 10.11).

The tallest peak in a mass spectrum is the *base peak*. It represents the ion that is most easily formed (greatest stability). Base peaks are not unique to specific compounds; however, it is not unusual for members of a particular class to share a

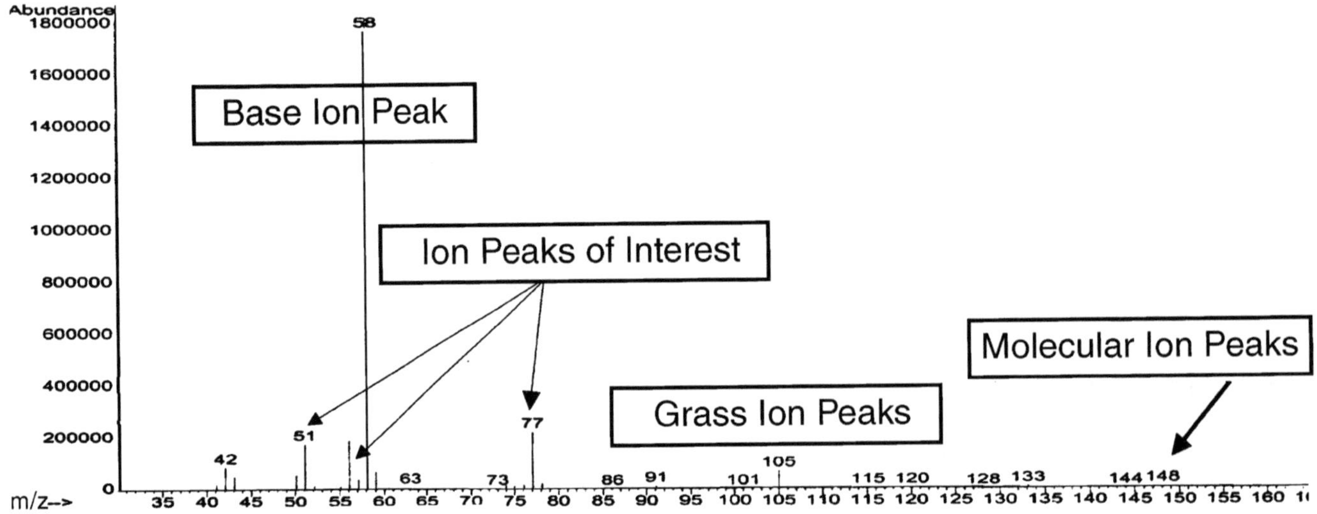

Fig. 10.11 A typical mass spectrum. The numbers above each peak represent the mass of the ion and the height is used to determine relative abundance. Peak labels are provided for ion peaks most often used in the identification process.

common base peak. For example, a base peak of 44 is often characteristic of primary amines such as amphetamine and phenylpropanolamine. These generalities do exist, but exercise great caution because they are not observed in all cases. For example, a base peak of 58 is typically associated with secondary amines such as methamphetamine, but the primary amine phentermine also has a base peak of 58.

The molecular ion (M^+) is produced when a single electron is lost from the molecule during ionization. The molecular ion is actually a radical cation, a positively charged ion containing an unpaired electron. Radicals are typically short-lived and extremely unstable. Therefore, the molecular ion may, or may not, exist long enough to reach the detector. If it reaches the detector, the location of molecular ion peak on the mass spectrum indicates the mass of the molecule, thus providing useful analytical information. If the ion does not reach the detector, it is degraded in a stability-driven process that produces a positive fragment of the original molecule and a hydrogen radical. In this instance, the molecular ion peak reflects the loss of a single hydrogen and appears on the spectrum at a position one unit less than the actual mass. Also, many elements such as carbon, oxygen, and the halogens (Group VII) have isotopes, forms of the element that have different masses. The presence of isotopes in a small number of molecules in the sample can appear on the spectrum as peaks located at one (M+1), two (M+2), or even three (M+3) units above the M^+ peak. Lastly, some molecules are extremely fragile and do not form molecular ions, or, if one is formed, the concentration is extremely low and may not be detected. These factors must be recognized when evaluating spectra (Fig. 10.12a–d).

The remaining peaks in the spectrum are a result of fragmentation of the main molecule or the fragments themselves reacting with the ionization source (fragments of fragments). The height of each peak and its size in relation to other peaks provide the details used to distinguish compounds.

Figure 10.12c, d are the normalized spectra of methamphetamine and phentermine, compounds that differ only in the location of a methyl ($-CH_3$) group. These compounds have the same base peak (58) and the remaining peaks are difficult to distinguish on both spectra. In forensic analysis, the identification and differentiation of methamphetamine from phentermine require a much broader approach, one that is less reliant on the M^+ peak. First, the molecular ion peaks are slightly different on each spectrum and could be used to differentiate the two. However, notice the location of the molecular ion peak is not consistent with the molecular weight in both cases. Justification of this fact will require a detailed explanation of radical-induced rearrangement as well as a discussion on isotopic contribution to molecular mass. Although this is certainly an option, bear in mind that this type of technical testimony could easily confuse jury members. A less complicated approach involves a careful examination of the *grass*, which reveals that the spectrum for phentermine lacks peak 56 (a minor peak), and peaks 42 and 134 are more prominent with phentermine than methamphetamine. In addition, the observed retention time of each compound in GC analysis (not shown) is different. Moreover, methamphetamine would test positive using the Marquis' and secondary amine chemical tests, while phentermine would test negative. This example illustrates the importance of using collective information in the identification process and not relying solely on mass spectrometry (Table 10.1). Table 10.2 lists mass spectral data for drugs commonly encountered in forensic analysis.

10.3.5 Mass Analyzers (Filters)

Mass analyzers filter ion fragments and direct them to the detector. The three methods most often used are quadrupole, magnetic sector, and ion trap. All techniques use mass-to-charge (m/z) ratio in the filtration process.

10.3.6 Quadrupole Mass Analyzers

The quadrupole mass analyzer uses combined DC and RF potentials applied to a set of four rods (quadrupole). The potentials can be adjusted so that only a selected m/z ratio is allowed to pass. Ions that do not have a stable trajectory through the quadrupole configuration will collide with the rods (are filtered) and never reach the detector (Fig. 10.13).

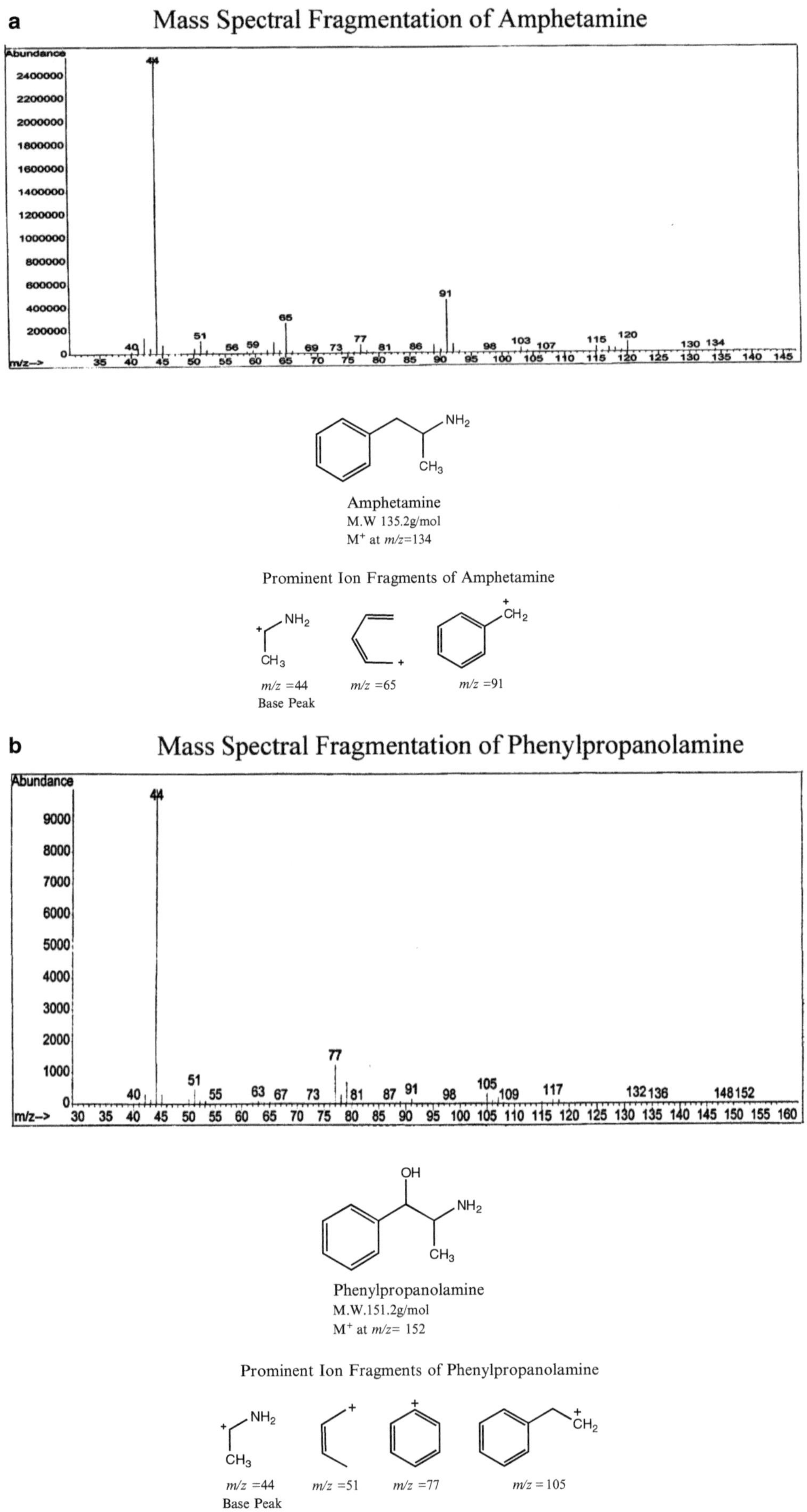

Fig. 10.12 Mass spectra of (**a**) amphetamine, (**b**) phenylpropanolamine, (**c**) methamphetamine, and (**d**) phentermine.

c

Mass Spectral Fragmentation of Methamphetamine

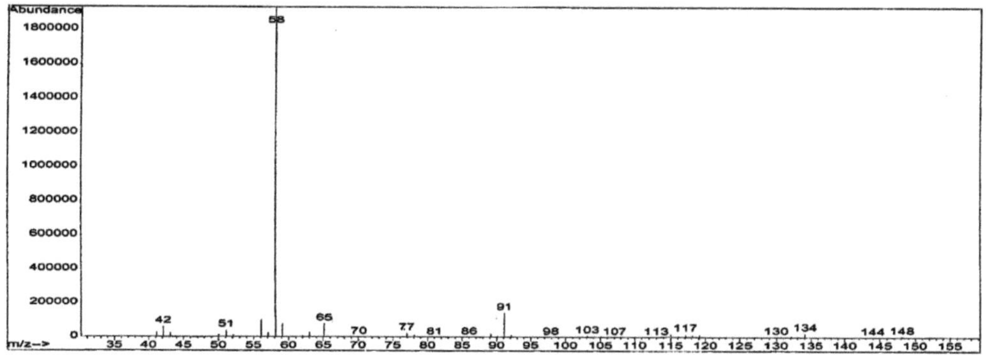

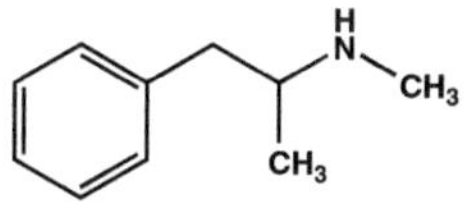

Methamphetamine
M.W.149.2g/mol
M^+ at m/z =148

Prominent Ion Fragments of Methamphetamine

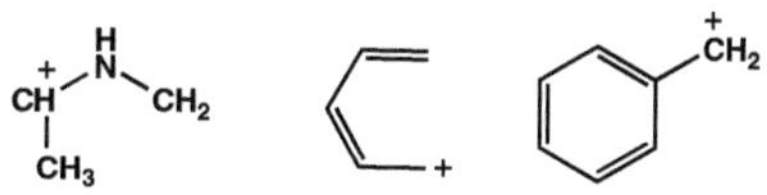

m/z = 58 Base Peak | m/z = 65 | m/z = 91

d

Mass Spectral Fragmentation of Phentermine

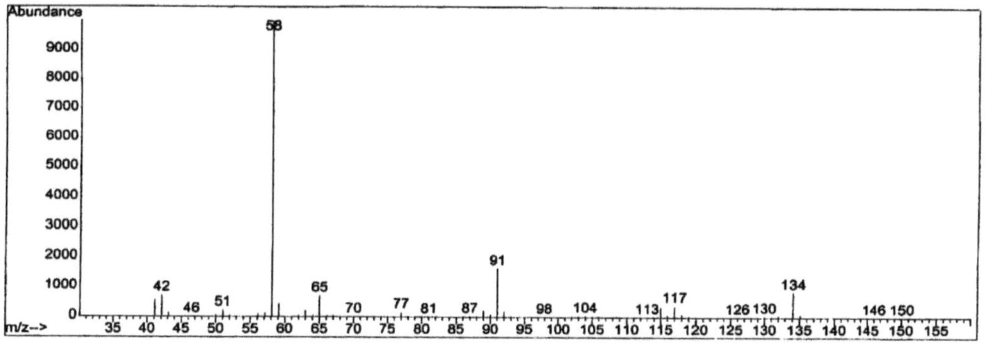

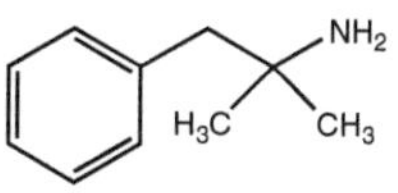

Phentermine
M.W.149.2 g/mol
M^+ at m/z = 150

Prominent Ion Fragments of Phentermine

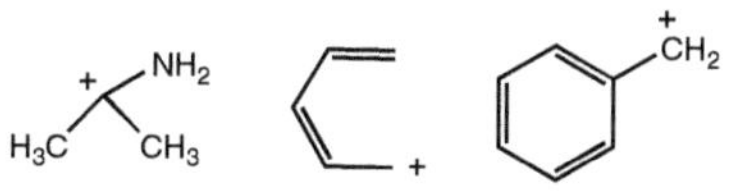

m/z = 58 Base Peak | m/z = 65 | m/z = 91

Fig. 10.12 (continued)

Table 10.1 Examination of phentermine and methamphetamine

Examination type	Phentermine	Methamphetamine
MS spectra/base peak	Same	Same
MS spectra/prominent peaks	Same	Same
GC retention time	Different	Different
MS spectra/grass	Different	Different
Marquis' test	Weak yellow	Orange-brown
Secondary amines test	Nonresponsive	Blue

MS mass spectrometry, *GC* gas chromatography

Table 10.2 Mass spectral peaks of frequently encountered drugs

Drug name	Molecular mass	Base peak	Prominent peaks				
Amphetamine	135	44	65	77	91	92	120
Butabarbital	212	141	41	98	155	156	157
Butalbital	224	168	97	124	141	153	167
Cannabinol	310	295	165	223	238	295	296
Tetrahydrocannabinol (THC)	314	231	91	174	243	258	271
Cathinone	149	44	50	51	77	78	105
Cocaine	303	82	77	83	94	105	
Codeine	299	299	42	124	162	214	229
Diazepam	284	256	77	151	165	221	
Ephedrine/pseudoephedrine	165	58	51	56	77	79	105
Fentanyl	336	245	77	91	105	146	189
GHB/GBL	86	42	41	56	85	55	45
Heroin	369	327	43	204	215	268	310
Hydrocodone	299	299	96	115	185	214	228
Ketamine	237	180	102	115	138	152	182
LSD	323	221	72	167	181	196	207
MDA	179	44	51	81	77	135	136
MDMA	193	59	51	89	105	135	
Mescaline	211	182	136	148	151	167	
Methadone	309	72	57	91	115	165	178
Methamphetamine	149	58	56	65	91	115	119
PCP	281	200	85	91	104	117	129
Psilocin/psilocybin	204	58	42	77	117	146	159

GHB gamma-hydroxybutyric acid, *GBL* gamma-hydroxybutyrolactone, *LSD* lysergic acid diethylamide, *MDA* 3,4-methylenedioxyamphetamine, *MDMA* 3,4-methylenedioxymethamphetamine, *PCP* phencyclidine

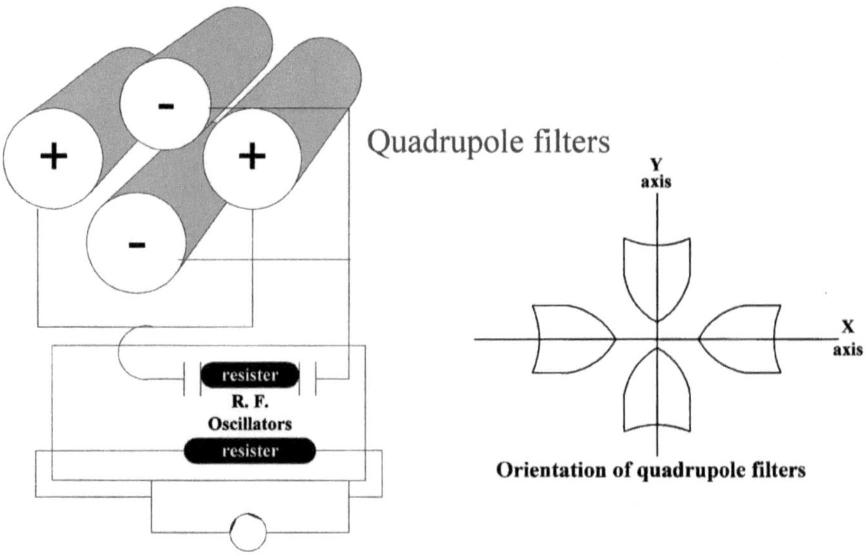

Fig. 10.13 Quadrupole mass analyzers use a central cavity created by four rods to filter ion fragments. The potentials at each rod can be independently regulated to direct ions of a specific mass-to-charge (mass) toward the detector.

10.3.6.1 Benefits

- Good reproducibility
- Relatively small, cost-effective systems

10.3.6.2 Limitations

- Limited resolution
- Peak heights vary as a function of mass (mass discrimination) and require tuning.
- Not well suited for pulsed ionization methods

10.3.6.3 Applications

- Majority of bench-top GC/MS and LC/MS systems (LC, liquid chromatography)

10.3.7 Magnetic Sector Mass Analyzers

In a magnetic sector analysis, the ion fragments are accelerated out of the chamber by a strong electric field. Generally, light fragments reach higher speeds than heavier ones but, ultimately, relative speed is determined by mass-to-charge ratio. The accelerated ions pass through a magnetic field applied at right angles, which curves their path. The extent of deviation (curvature) depends on how fast the ion is traveling and is therefore directly related to its mass. The magnetic field strength is varied until a force is produced that bends the path toward the detector. Ions traveling at different speeds require a distinctive magnetic field strength in order to reach the detector. The mass of the ion is calculated from the accelerating voltage and magnetic field strength and plotted as a peak on the mass spectrum.

10.3.7.1 Benefits

- Very high reproducibility
- High quantitative performance
- High resolution
- High sensitivity
- High dynamic range

10.3.7.2 Limitations

- Not well suited for pulsed ionization methods (e.g., matrix-assisted laser desorption/ionization, MALDI)
- Usually larger and more expensive than other mass analyzers

10.3.7.3 Applications

- All organic MS methods
- Accurate mass measurements
- Quantitation
- Isotope ratio measurements

10.3.8 Ion Trap Mass Analyzers

In ion trap analysis, ions are effectively trapped in a three-dimensional quadrupole storage device using an applied RF and DC potential. A common design employs a "bath gas" to help contain the ions. The RF and DC potentials are varied resulting in the selective ejection of ions with specific *m*/*z* ratios. The released ions are directed into the detector in a process called *mass-selective ejection*.

10.3.8.1 Benefits

- High sensitivity
- Multistage mass spectrometry
- Compact mass analyzer

10.3.8.2 Limitations

- Poor quantization
- Very poor dynamic range (can sometimes be compensated using automatic gain control)
- Collision energy not well defined
- Several parameters in the analytical sequence affect the quality of the mass spectrum (excitation, trapping, and detection conditions)

10.3.8.3 Applications

- Bench-top GC/MS and LC/MS
- Target compound screening
- Ion chemistry

10.4 Advantages of Gas Chromatography Mass Spectrometry

The clear advantage of GCMS lies in its ability to identify the components of a volatile mixture with minimal sample preparation. Consequently, the large volumes of aqueous and organic solvents required to extract and purify compounds using other techniques are no longer needed.

The sensitivity of GCMS can easily detect milligram quantities (or less) of a controlled substance compared to the gram-size (or larger) samples required for solvent extraction techniques. Also, with detection capabilities in the nanogram range and below, GCMS an excellent tool for a toxicologist searching for drugs of abuse or poisons in blood and urine samples.

The programmable nature of integrated GCMS can accommodate automated sampling methods. This capability allows the instrument to run overnight which can dramatically increase the laboratory's case output.

10.5 Disadvantages of Gas Chromatography Mass Spectrometry

The biggest disadvantage of GCMS is that it cannot easily distinguish most isomers, i.e., geometric, positional, and enantiomers. Retention times are the same and the mass spectra are identical. Compounds such as ephedrine and pseudoephedrine cannot be distinguished without modification prior to GCMS analysis (Fig. 10.14). Also, highly volatile compounds with low molecular weights are not easily analyzed. The weight may be outside the scanning range of the MS or the compound may elute from the GC column either before or with the solvent. This is not prohibitive because an experienced operator can correct these issues and identify the compounds.

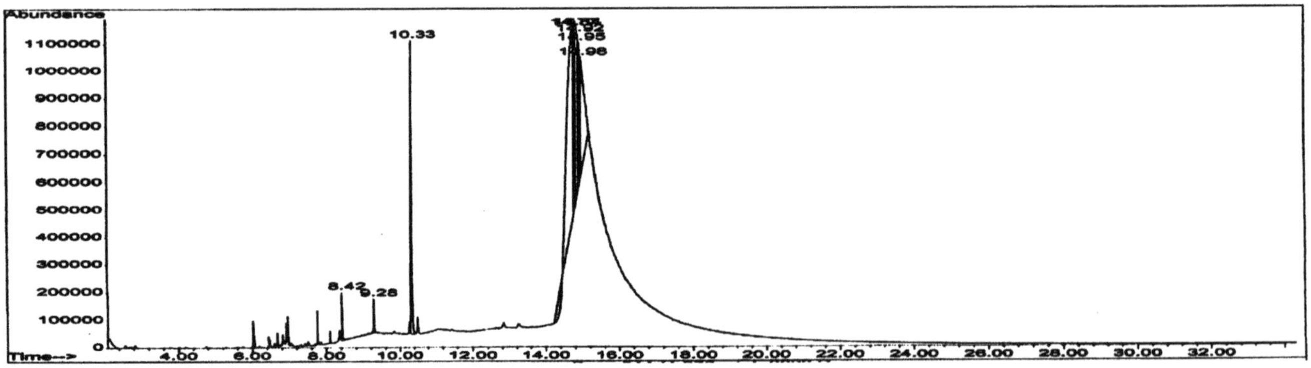

Fig. 10.14 A clear disadvantage of GCMS is its failure to distinguish isomers. Isomers can have similar retention times and identical fragmentation patterns. Note the unresolved peaks in the GC chromatogram above.

10.6 Questions

1. Define chromatography.
2. Define mass spectrometry.
3. What qualitative observations can you make from a TLC plate?
4. What is an R_f-value and how is it calculated?
5. Describe the procedure used to pack a column used in liquid chromatography.
6. Define the molecular ion peak and base peak.
7. How is mass spectrometry used to definitively identify a drug?
8. Please describe to the jury how a typical GCMS functions.
9. Define retention time.
10. What information is gained from the peaks on a chromatogram?
11. Is it possible for two compounds to have identical retention times? If so, describe how you would differentiate them.
12. Describe the difference between paper chromatography and TLC.
13. Cite two advantages of GCMS.
14. Describe the operation of a magnetic sector mass analyzer.
15. How is relative abundance determined using a chromatogram peak?
16. Describe the process of electron impact ionization.
17. How do you calculate a relative retention time?
18. Describe how a flame-ionization detector functions.
19. Using Table 10.2, sketch a mass spectrum for cocaine.
20. Refer to Fig. 10.13 and draw the ion fragments of methamphetamine produced during MS.
21. How can you differentiate phentermine from methamphetamine?
22. Cite two disadvantages of GCMS.
23. Describe reverse-phase HPLC. In this context, what does reverse mean?
24. Describe a case when a thermal conductivity detector would be preferred over a flame-ionization detector.
25. Provide a common analogy for the gas chromatography.

Suggested Reading

Audier, E.; Millet, A.; Sozzi, G. Mass Spectrum of Amphetamine and Related Compounds. *Org. Mass Spectrum.* **1984**, 19, 522 – 523.

Baldwin, M. A. Mass Spectral Analysis. *J. Org. Chem.* **1979**, 14, 601.

Bax, A. K.; Summers, M. F. Spectrometric Identification of Compounds. *J. Am. Chem. Soc.* **1986**, 108, 2093–2094.

Crain, P. F. Mass Spectrum Review. *Mass Spectrum.* **1990**, 9, 505–554.

De Forest, P. R.; Gaensslen, R. E.; Lee, H. C. *Forensic Science: An Introduction to Criminalistics*; McGraw-Hill: New York, 1983, pp. 65–68.

Dommrose, F.; Gritzmacher, H. F. Ion Processes. *Int. J. Mass Spectrum.* **1987**, 76, 95.

Eadon, G. Mass Spectrum of Organic Compounds. *Mass Spectrum.* **1977**, 12, 671.

Ion Exchange Chromatography. http://www.proteinchemist.com/tutorial/iec.html (accessed July 2009).

Jones, L.; Atkins, P. *Chemistry: Molecules, Matter, and Change*, 4th ed.; W. H. Freeman and Company: New York, 2000; pp. 10, 530–531.

Jones, M. Jr. *Organic Chemistry*; W. W. Norton & Company: New York, 1997; pp.686–690.

Pavia, D. L.; Lampman, G. M.; Kriz, G. S.; Engel, R. G. *Introduction to Organic Laboratory Techniques*, 3rd ed.; Saunders College Publishing: New York, 1990; pp. 710–716.

Silverstein, R. M.; Bassler, G. C.; Morill, T C. *Spectroscopic Identification of Organic Compounds*, 4th ed.; John Wiley & Sons: New York, 1981; pp. 4–5.

10.6 Questions

11 Infrared Spectroscopy

11.1 Introduction

Spectroscopy is the study of the interaction of matter with electromagnetic radiation. All forms of electromagnetic radiation are *transverse waves*. This type of wave has a crest (peak) and a trough (valley) and closely resembles those commonly seen on the ocean. The *wavelength* is the measured distance between two adjacent peaks or, alternatively, two adjacent troughs.

Transverse waves possess energy and anyone who has spent time in the ocean has experienced this. The amount of energy is inversely proportional to wavelength, that is, shorter wavelengths of radiation have greater energy, while those with longer wavelengths have less. The electromagnetic spectrum is divided into regions based on wavelength and therefore energy. The visible region is radiation with wavelengths between 350 and 750 nm (nanometers). The energy associated with this region differentially excites specific areas of the retina and sends a pulse down the optic nerve interpreted as color. For example, a wavelength of 700 nm (red) has a different energy than one of 400 nm (blue). Each wavelength excites a different region of the retina and sends a characteristic pulse through the optic nerve. Since each pulse originated from a different region, it is interpreted as a different color.

The ultraviolet (UV) and infrared (IR) regions are adjacent to the visible section of the electromagnetic spectrum. These wavelengths are not visible because they possess energy outside the operational range of the retina. UV radiation has a shorter wavelength (higher energy) than visible light and is damaging to biological tissue – a characteristic that most are familiar with and the reason for applying UV protection (SPF lotion) during prolonged exposure to the sun. Conversely, IR radiation is *heat*, a low-energy form of radiation with a somewhat longer wavelength (lower energy) than visible light.

Infrared (IR) spectroscopy has been a long-established method of confirming the identity of a controlled substance. Traditionally, the sample was subjected to a series of screening tests to establish the identity of the suspected compound and any adulterants and diluents. The controlled substance was then extracted and purified. Finally, an IR spectrum was obtained. Modern technology has introduced instrumentation that can obtain an IR spectrum from a single particle or from a peak in gas chromatography (GC), thus eliminating the need for complicated extraction procedures.

IR spectroscopy requires highly purified samples, and advances in technology have drastically reduced sample preparation and analysis times. The Fourier transform IR (FTIR) spectrophotometer can obtain an IR spectrum of an individual peak eluting off a GC column, and a micro-FTIR can isolate and obtain IR spectra of individual particles within a mixture.

11.2 Theory of Infrared Spectroscopy

IR spectroscopy relies on a compound's ability to absorb IR radiation as a means of identification. All of the atoms in a molecule are in continuous vibration with respect to one another. A molecule absorbs IR radiation when the frequency of a specific vibration is equal to the frequency of the IR radiation directed on the molecule. Absorbed frequencies are recorded and displayed as *bands* on an *IR spectrum.*

J.I. Khan et al., *Basic Principles of Forensic Chemistry*, DOI 10.1007/978-1-59745-437-7_11,
© Springer Science+Business Media, LLC 2012

An IR spectrum is typically a graph of percent transmittance (y-axis) versus wave numbers (x-axis). Absorbance (Abs) is also commonly graphed on the y-axis and may be unitless or expressed in percent. The units of wave numbers are reciprocal (or inverse) centimeters expressed as cm^{-1} (1/cm). This unusual x-axis label is an obvious point of confusion and requires clarification. First, it is a matter of practical convenience. Wave numbers are directly proportional to vibrational energy and most modern instruments are linear in the cm^{-1} scale. Wave numbers and frequency are terms that are often used interchangeably and, although the terms are related, this practice is not technically correct. Frequency is defined as the number of cycles (wavelengths) that pass a given point per unit of time, usually expressed in units of cycles per second (s^{-1}). Wave numbers expressed in units of cm^{-1} are the number of wavelengths in one centimeter (wavelengths per centimeter). The concepts are clearly related, but not identical. Both frequency and wave number are inversely proportional to wavelength; therefore, a transverse wave of short wavelength will have a high frequency and a high wave number. Therefore, absorption frequencies measured in IR spectroscopy can be accurately represented using wave numbers because the term is closely related to frequency.

The total number of observed absorption bands is generally quite different from the total number of fundamental vibrations. It is reduced because some modes are not IR active or a single frequency may cause more than one mode of motion to occur. Additional overtone bands can be generated by combinations of fundamental frequencies, differences in fundamental frequencies, and coupling interactions of fundamental frequencies. The observed intensities of overtone bands are generally less than those of the fundamental bands. The combination and blending of all the factors create a unique IR spectrum for each compound (Fig. 11.1).

IR radiation contains frequency ranges between 13,000 and 30 cm^{-1} that lie between visible light and microwave radiation in the electromagnetic spectrum (Fig. 11.2). The area most commonly used in forensic examination lies in the middle IR region between 4,000 and 400 cm^{-1}. The region on the spectrum between 2,000 and 400 cm^{-1} is commonly referred to as the fingerprint region.

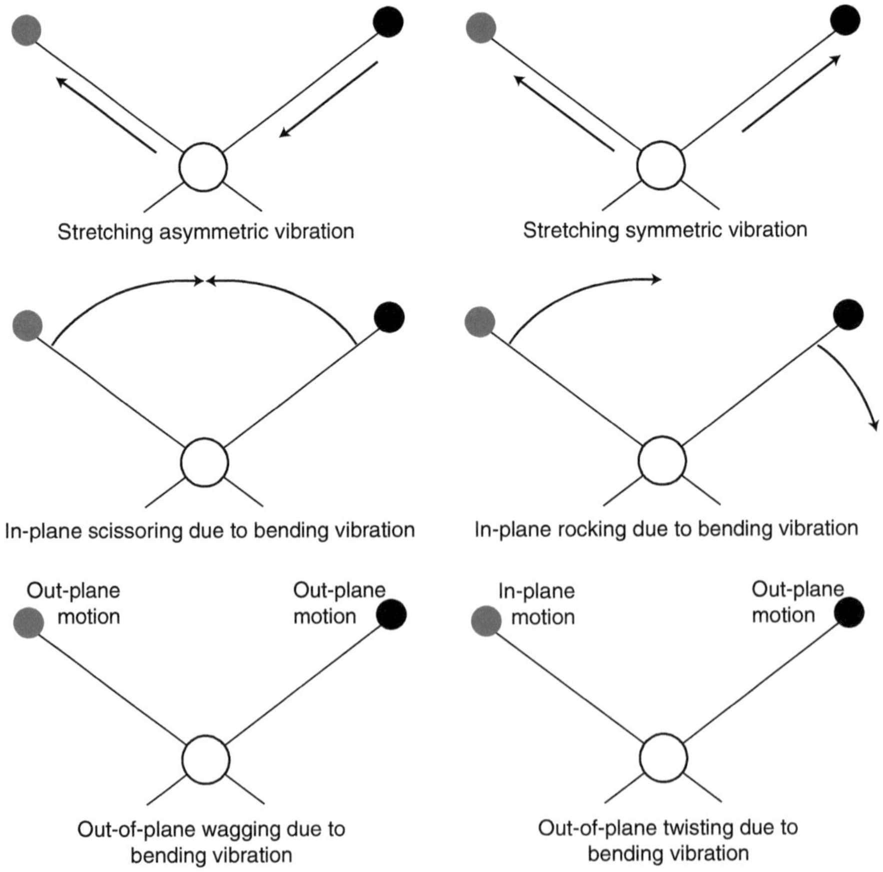

Fig. 11.1 Common examples of bond activity in a molecule. The vibrations occur at characteristic frequencies that are detected using IR spectroscopy. Overtones can be created by various combinations of vibrational frequencies.

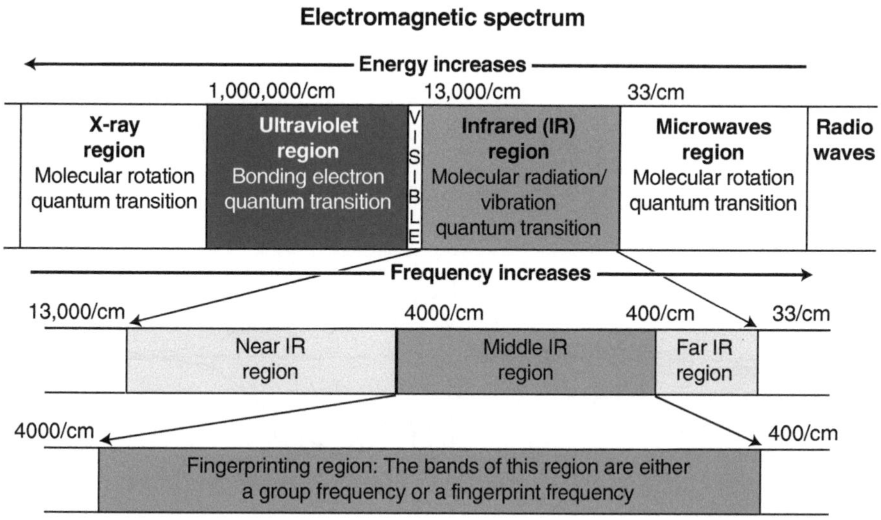

Fig. 11.2 Infrared (IR) spectroscopy relies on the absorption of low-energy IR radiation to distinguish bond vibrational frequencies. This analytical technique is commonly used in forensic investigation to confirm the presence of functional groups in a wide variety of drugs and controlled substances.

11.3 Infrared Spectrum

The IR spectrum is a measure of the amount of IR radiation that is absorbed by the sample. A traditional format of linear percent transmittance (%T) versus linear wave numbers, usually from 4,000 to 400 cm^{-1}, is used in most atlases. A 2× scale expansion of the fingerprint region below 2,000 cm^{-1} is used to enhance the spectral detail of the fingerprint region.

The transmittance spectra provide better contrast between intensities of strong and weak bands because transmittance ranges linearly from 0% to 100% *T*, whereas absorbance ranges nonlinearly (exponentially) from infinity to zero. The analyst should be aware that the same sample would give quite different profiles on an IR spectrum which is linear in wave number, compared to an IR plot which is linear in wavelength. It will appear as if some IR bands have been contracted or expanded.

The use of absorbance is not common in IR spectral atlases but is quite common in journal articles and quantitative work. Digital databases use Abs and therefore most searches are based on Abs spectra; hence, it is becoming more widely used.

The choice of whether to display the *y*-axis as %*T* or Abs depends on the purpose at hand. It is much easier to mentally perform the vertical mirror-image transformation of Abs to %*T* than it is to do the *x*-axis transformation of microns to wave numbers. Quantization and search matching require Abs; however, observation of features just above the baseline seems better in %T.

The conversion of Abs to transmittance is a simple mathematical process ($A_\lambda = \log(1/T_\lambda)$). Figure 11.3 demonstrates the effect of this conversion on the spectrum.

11.4 Instrumentation

The IR spectrophotometer is the instrument used to acquire and display the IR spectrum (Fig. 11.4). As with all modern scientific instrumentation, the ability to obtain IR spectra has constantly improved. Despite significant advancements in technology, there are only two basic instrument formats in common use: the dispersive instrument and the Fourier transform instrument.

11.4.1 Dispersive Infrared Spectrometer

Since the introduction of dispersive spectrometers in the mid-1940s, they have been widely used. Their solid design and robust nature can withstand the repeated application of this technique.

11.4.2 Spectrometer Components

An IR spectrometer consists of three basic components: radiation source, monochromator, and detector. A schematic diagram of a typical dispersive spectrometer is shown in Fig. 11.5.

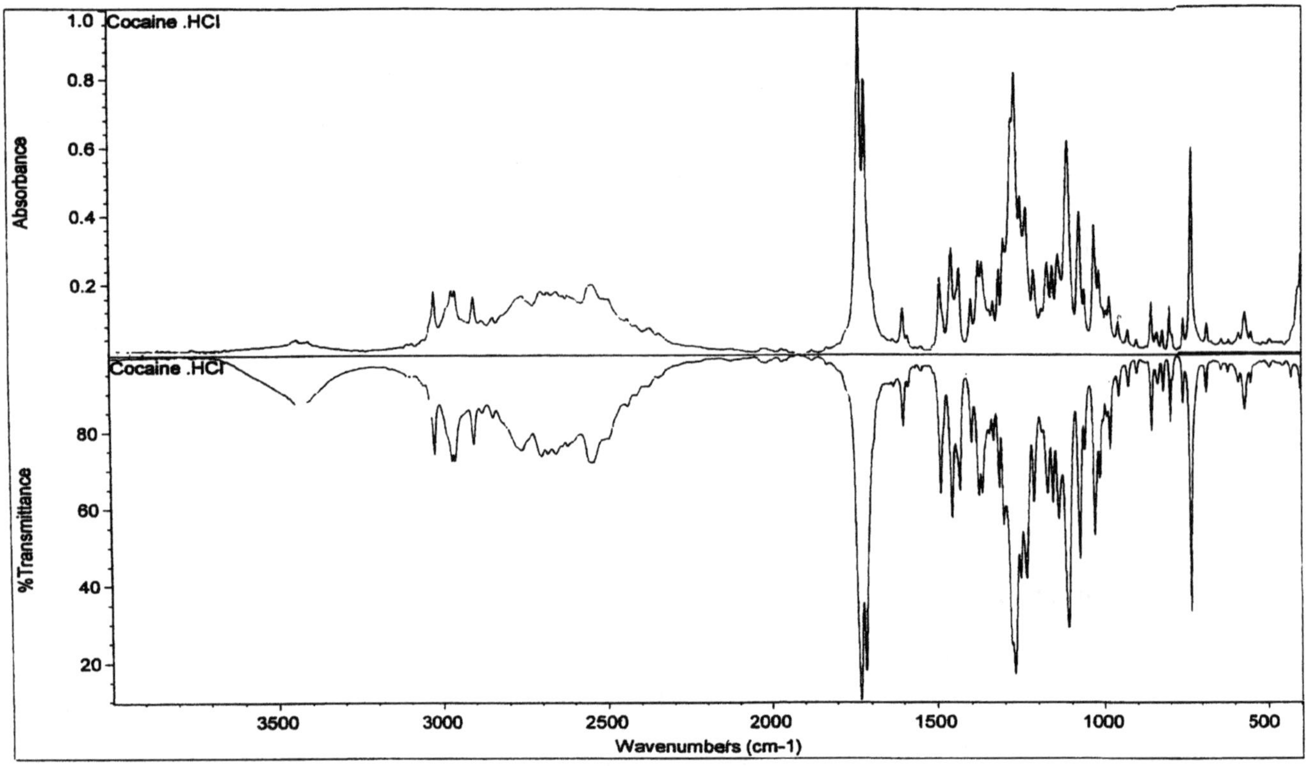

Fig. 11.3 IR spectra of cocaine illustrating Absorbance and % transmittance formats. The spectra are essentially vertically flipped mirror images. Note that high absorbance bands are reflected as low transmittance.

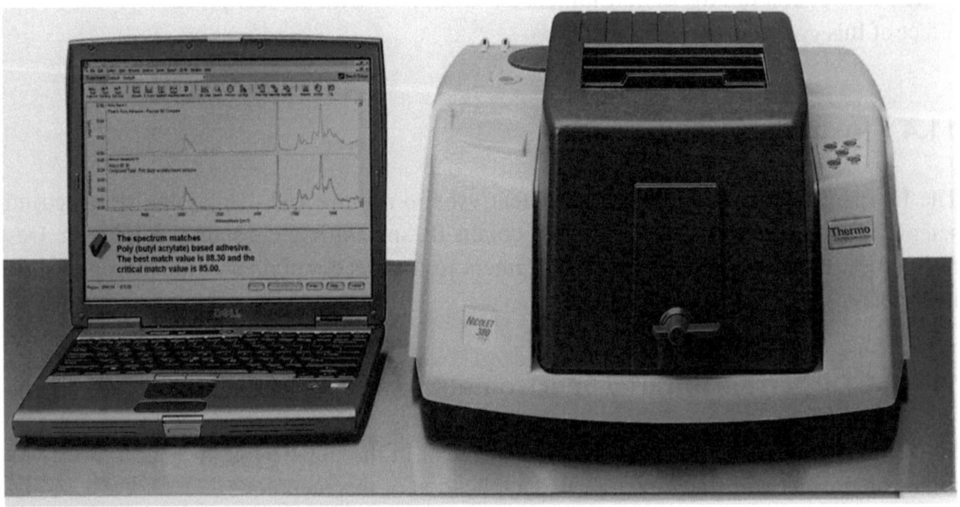

Fig. 11.4 A typical IR spectrophotometer. Integrated computer workstations are used to access the highly programmable functions of this portable instrument.

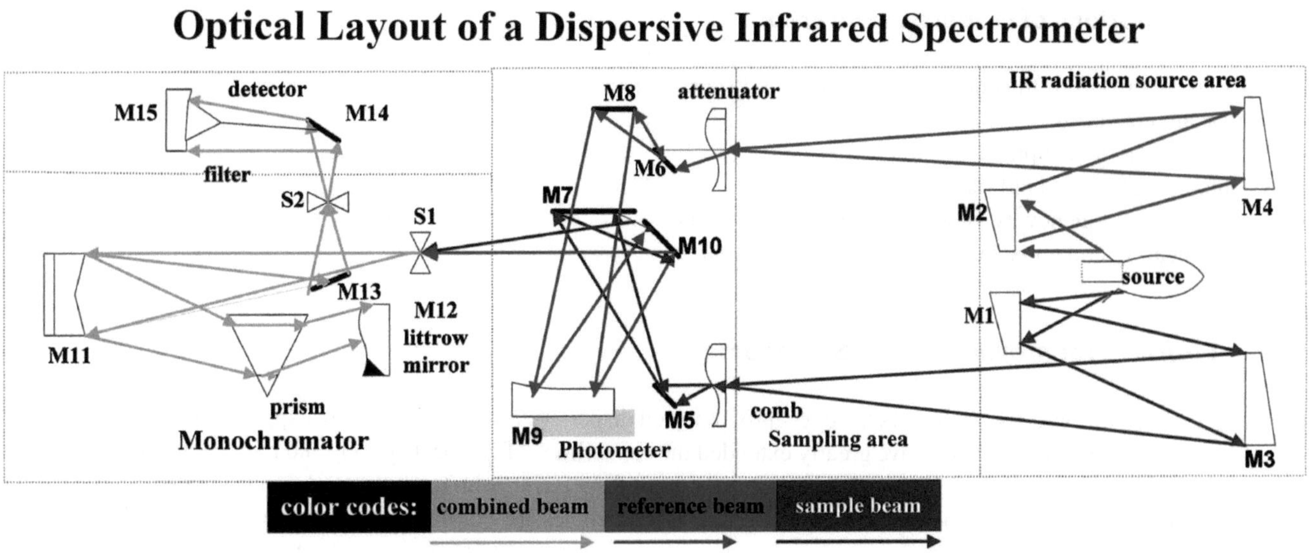

Fig. 11.5 The complicated path of IR radiation through a dispersive IR spectrometer. A series of precisely aligned mirrors are used to direct the beam through the various chambers of the instrument.

The IR radiation source is an inert solid, electrically heated to 1,000–1,800°C. Three of the most common types of sources are the Nernst glower (constructed of rare-earth oxides), the Globar (constructed of silicon carbide), and the Nichrome coil. Each source produces continuous radiation with different energy profiles.

The monochromator is a device used to disperse a broad spectrum of radiation and provides a continuous calibrated series of electromagnetic energy bands of known wavelength or frequency range. Prisms or gratings are used to separate the radiation emanating from the source into a spectrum of wavelengths. A variable-slit mechanism, mirrors, and filters are used to introduce bands of radiation to the sample.

The size of the slits is used to establish a balance between resolution and sensitivity during analysis. Narrow slits enable the instrument to better distinguish more closely spaced frequencies of radiation, resulting in better resolution. Smaller slits allow less light to reach the detector, thus reducing the sensitivity, while wider slits allow extra light to reach the detector, providing enhanced system sensitivity. However, wider slits also expose the sample to a broader range of wavelengths, which reduces the resolving power. Thus, a degree of compromise must be exercised when setting the desired slit size.

Most detectors used in dispersive IR spectrometers can be categorized into two classes: thermal detectors and photon detectors. Thermal detectors measure the heating effect produced by IR radiation and include thermocouples, thermistors, and pneumatic devices (Golay detectors). Photon detectors rely on the interaction of IR radiation with a semiconductive material. Nonconducting electrons are excited to a conducting state that generates a small current, or voltage.

There are trade-offs with detectors as there are when selecting slit size. Thermal detectors provide a linear response over a wide range of frequencies but exhibit slower response times and lower sensitivities. Photon detectors are more sensitive and provide faster response times; however, their linear response over a range of IR frequencies is narrower than that found in thermal detectors.

11.4.3 Spectrometer Design

In a typical dispersive IR spectrometer, radiation from a broadband source passes through the sample and is dispersed by a monochromator into component frequencies (Fig. 11.5). The beams then fall on the detector, which generates an electrical signal that is recorded and displayed.

Most dispersive spectrometers have a double-beam design. Two equivalent beams from the same source pass through the sample and reference chambers, respectively. Using a sector mirror, the reference and sample beams are alternately focused on the detector. The change of IR radiation intensity caused by sample absorption is detected as an off-null signal that is translated into a recorder response through the actions of synchronous motors.

11.4.4 Limitations of Dispersive Infrared

Limitations as a result of the complex mechanics of dispersive IR are shown below:

- Slow scanning speed
- Low sensitivity to sample
- No internal reference calibration
- Insensitive to frequency by viewing one element at a time
- Sample heating and emissions from the heated sample

11.5 Fourier Transform Infrared Spectrometer

Fourier transform infrared spectrometers have recently replaced dispersive instruments for most applications because of their superior speed and sensitivity. They have greatly extended the capabilities of IR spectroscopy and have been applied to many areas that are very difficult or nearly impossible to analyze by dispersive techniques. Instead of viewing each component frequency sequentially, as in a dispersive IR spectrometer, all frequencies are examined simultaneously in an FTIR spectrometer (Fig. 11.6).

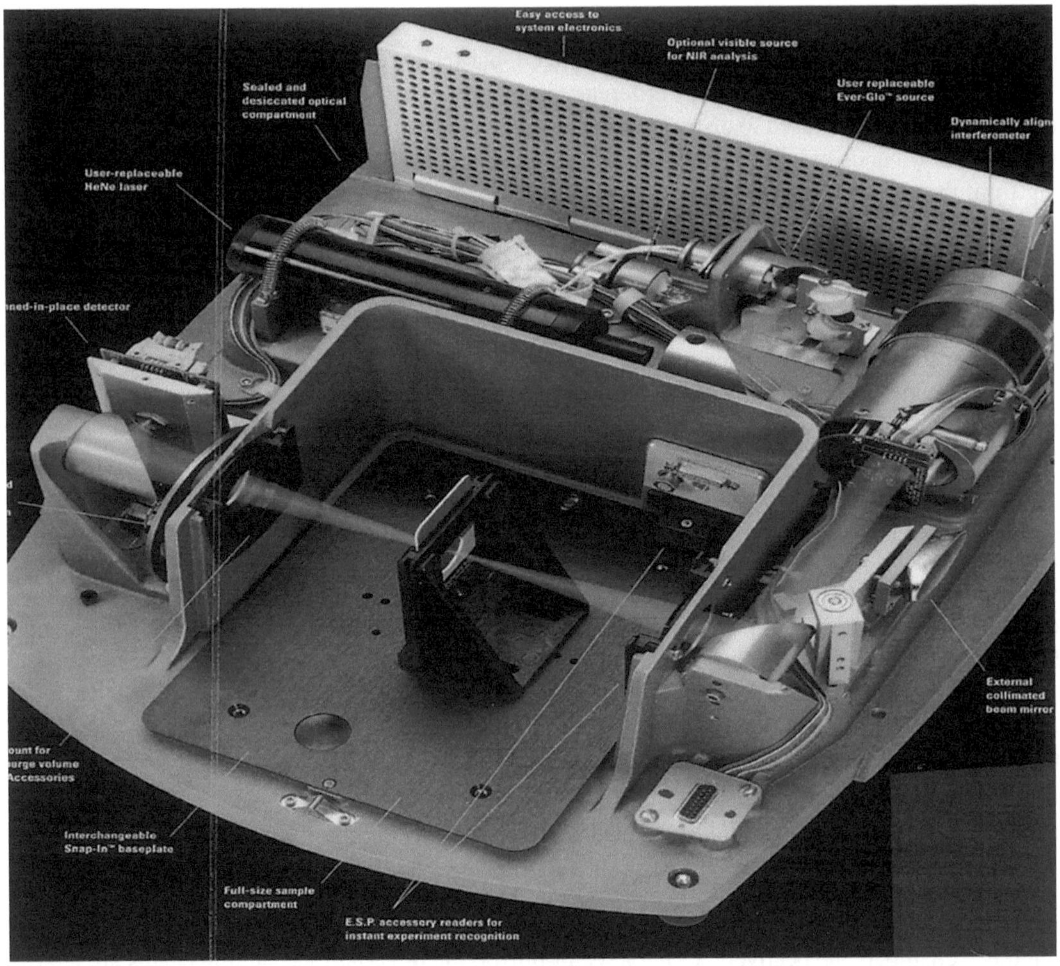

Fig. 11.6 Fourier transform infrared (FTIR) spectroscopy is superior to most IR analytical techniques. The modern design of the spectrometer has dramatically increased sensitivity, reliability, and analysis speed.

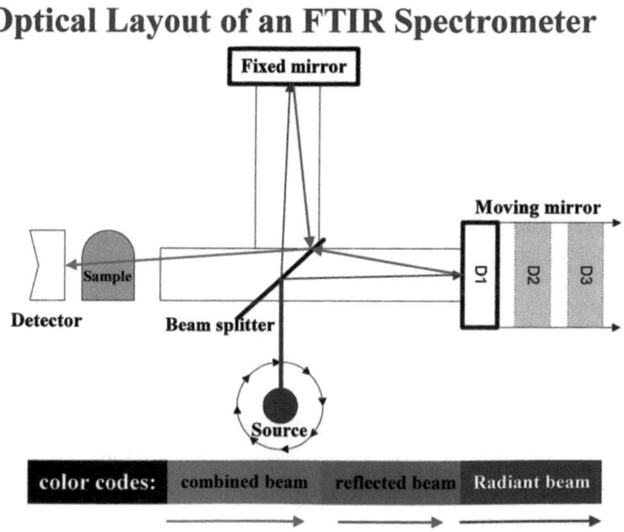

Fig. 11.7 The simplified beam path of IR radiation through a Fourier transform infrared spectrometer translates to significant reductions in maintenance and operational costs compared to dispersive techniques.

11.5.1 Spectrometer Components

The three basic FTIR components are the radiation source, interferometer, and detector. Figure 11.7 shows a simplified optical layout of a typical FTIR spectrometer.

Dispersive and FTIR spectrometers use the same types of radiation sources. However, FTIR instrument sources are often water cooled to enhance power output and stability.

FTIR spectrometers use a completely different approach to differentiate and measure the absorption at component frequencies. The monochromator is replaced by an interferometer. The interferometer divides *radiant beams*, generates an optical path difference between the *reflected radiant beams*, and then recombines the beams. The recombination produces repetitive interference signals measured as a function of optical path difference by the detector. The resulting interference signals contain IR spectral information generated after passing through a sample.

The most frequently used interferometer is the Michelson, which consists of three active components: a moving mirror, a fixed mirror, and a beamsplitter. The fixed and moving mirrors are perpendicular to each other. The beamsplitter is a semi-reflecting device often containing a thin film of germanium applied to a flat piece of potassium bromide (KBr). Radiation from the broadband IR source is collimated and directed into the interferometer, and impinges on the beamsplitter. At the beamsplitter, half the IR beam is transmitted to the fixed mirror and the remaining half is reflected toward the moving mirror. After the divided beams are reflected from their respective mirrors, they recombine at the beamsplitter. An interference pattern is generated as a result of changes in the relative position of the moving mirror to the fixed mirror. The resulting beam then passes through the sample and is eventually focused on the detector.

The detector signal is sampled at small, precise intervals during the mirror scan. The sampling rate is controlled by an internal, independent reference, commonly a modulated monochromatic beam from a helium neon (HeNe) laser focused on a separate detector.

The two most popular detectors used in FTIR spectrometers are deuterated triglycine sulfate (DTGS) and mercury cadmium telluride (MCT). The response times of many detectors used in dispersive IR instruments are too slow for the rapid scan times of the interferometer. The DTGS is a pyroelectric detector that delivers rapid responses because it measures changes in temperature rather than actual temperature values. The MCT is a photon (or quantum) detector that depends on the quantum nature of radiation and also exhibits very fast responses. Whereas DTGS detectors operate at room temperature, MCT detectors must be maintained at liquid nitrogen temperature (77 K) to be effective. In general, the MCT detector is faster and more sensitive than the DTGS detector.

11.5.2 Spectrometer Design

The basic instrument design is quite simple. The IR radiation from a broadband source is first directed into an interferometer where it is divided and then recombined after the split beams travel different optical paths to generate constructive and destructive interference. Next, the resulting beam passes through the sample compartment and on to the detector.

Most bench-top FTIR spectrometers are single-beam instruments. Unlike double-beam grating spectrometers, single-beam FTIR does not obtain transmittance or Abs IR spectra in real time. A typical operating procedure is described as follows:

1. A background spectrum is first obtained by collecting an interferogram (raw data), followed by data processing using the Fourier transform conversion.
2. Next, a single-beam sample spectrum is collected. It contains absorption bands from the sample and the background (air or solvent).
3. The ratio of the single-beam sample spectrum against the single-beam background spectrum results in a "double-beam" spectrum.

11.5.3 Advantages of Fourier Transform Infrared Spectrometers

FTIR instruments have distinct advantages over dispersive spectrometers:

- Better speed and sensitivity (Felgett advantage).
- A complete spectrum can be obtained during a single scan of the moving mirror while the detector observes all frequencies simultaneously.
- Increased optical throughput (Jaquinot advantage).
- Energy-wasting slits are not required in the interferometer because dispersion or filtering is not needed. Instead, a circular optical aperture is usually used in FTIR systems. The beam area of an FT instrument is usually 75–100 times larger than the slit width commonly used in dispersive techniques. Thus, more radiation energy is made available. This constitutes a major advantage for many samples or sampling techniques that are energy limited.
- Internal laser reference (Connes advantage). The use of a helium neon laser as the internal reference in many FTIR systems provides an automatic calibration with an accuracy of more than 0.01 cm^{-1}. This eliminates the need for external calibrations.
- Simpler mechanical design. There is only one moving part, the moving mirror, resulting in less wear and better reliability.
- Elimination of stray light and emission contributions. The interferometer in FTIR modulates all the frequencies. The unmodulated stray light and sample emissions (if any) are not detected.
- Modern FTIR spectrometers are usually equipped with a powerful, computerized data system.

11.5.4 Fourier Transform Infrared Sample Preparation Techniques

11.5.4.1 Liquid/Vapor Phase FTIR

FTIR spectra can be obtained from liquid and gas phase samples; however, these are not usually the phases of choice. The resulting spectra have broader absorption bands which reduce the selective power of the spectra. However, vapor-phase spectra have applications in GC/FTIR.

To obtain a gas- or liquid-phase FTIR spectrum, a background spectrum of the gas or liquid substrate is acquired. Next, a sample spectrum is collected which contains absorption bands from the sample and the background (air or solvent). The ratio of the sample spectrum against the background spectrum results in the spectrum of the compound of interest.

11.5.4.2 Solid-Sample FTIR

Solid samples are the preferred form for IR analysis because they place the molecules into a solid crystal lattice (structure). The fixed structure limits the ability of a functional group to vibrate, rotate, or bend. Limiting this type of movement, in turn, limits the wavelengths of IR radiation that can be absorbed. This specificity of absorbable wavelengths leads to sharp, narrow peaks in the IR spectra.

The crystal-lattice effect can be demonstrated through the comparison of solid-sample spectra to vapor-phase spectra. Vapor-phase spectra have broader bands when compared to the solid-sample counterpart. This is attributed to the ability of the various functional groups to more freely vibrate, rotate, or bend in the vapor phase. Figure 11.8 shows a comparison of free-base cocaine in a solid sample (pressed KBr pellet preparation) and in the vapor phase as one would see in a GC/FTIR spectra.

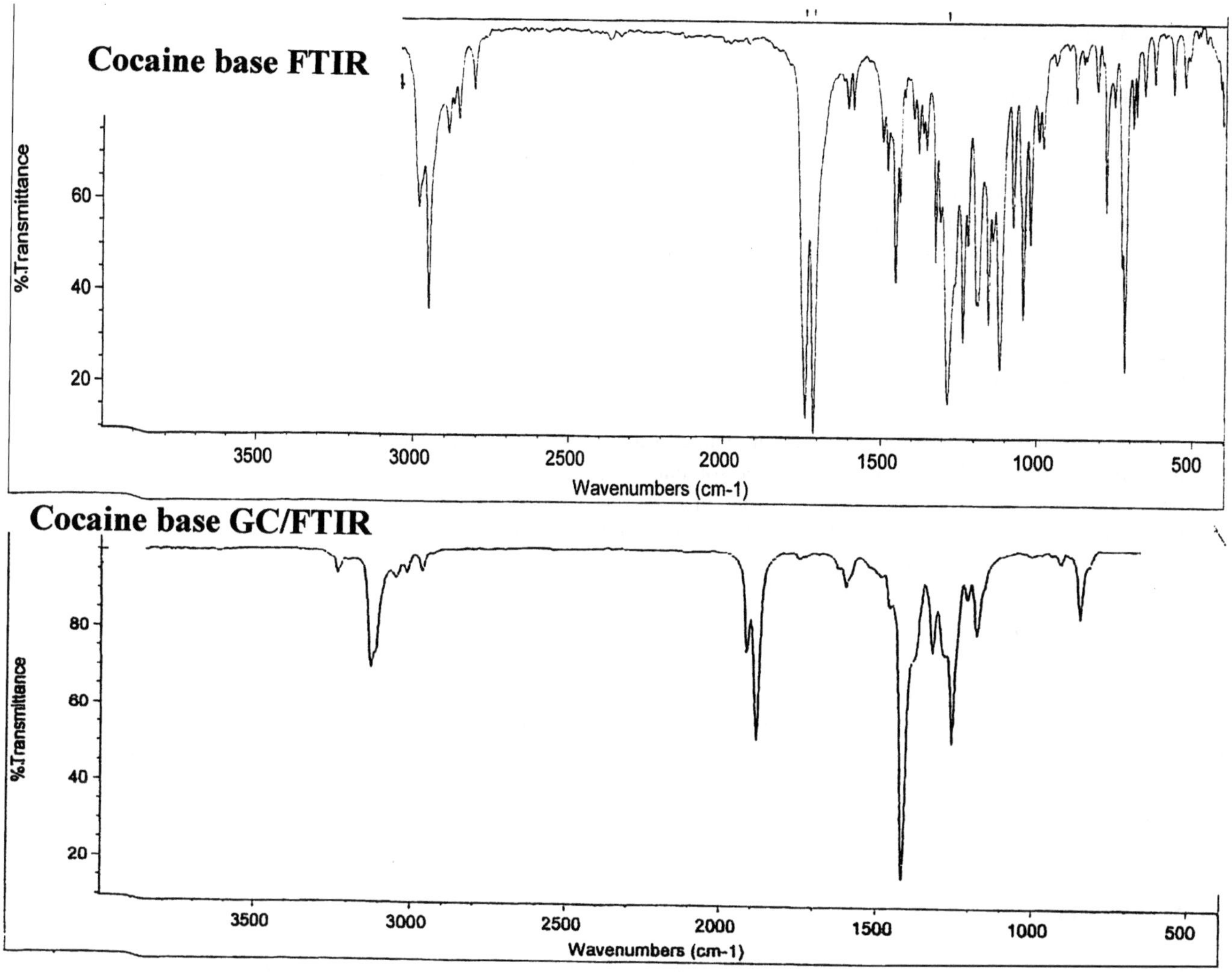

Fig. 11.8 Comparison of cocaine samples run in solid (*top*) and vapor (*bottom*) phase. Bond vibrational modes are restricted in the solid phase producing sharp, distinct absorption bands.

IR spectroscopy analyzes the vibrations of different parts of a molecule exposed to IR light. Changing the sample preparation method may affect the way different parts of the molecule vibrate. This will cause shifts in the peak intensities in the resulting IR spectra. The way the compound is crystallized (or not crystallized) within the sample matrix will affect the resulting IR spectrum. The IR spectra of a sample crystallized within a KBr matrix (pellet) can differ from that of the same compound directly recrystallized on a salt plate.

Polymorphism can affect a compound's IR spectrum. A single compound may have more than one crystalline form, along with an amorphous form (no definite structure). The way a compound crystallizes influences the vibrations within the molecule, which in turn affects the resulting IR spectrum. These variations can occur within the same sampling technique and will result in the production of a slightly different IR spectrum.

Variations in IR spectra are somewhat influenced by sampling techniques. Therefore, a library of known spectra from traceable sources should be maintained for compound identification purposes. The various IR spectra libraries that are available should only be used as a screening tool, not as a reference for identification. As with mass spectrometry (MS), final identification is only accomplished by comparing the IR spectrum of the unknown to the IR spectrum of a traceable reference standard. The spectra should be produced on the same instrument, under the same conditions.

11.6 Sampling Techniques

11.6.1 Nujol Mull

- One to two milligrams of sample is placed into an agate mortar and ground into a fine powder. The finer the particles, the less light scattering effects.
- Nujol (mineral oil) is added to the sample and ground into a homogeneous mixture.
- A film of the nujol/sample mixture is placed onto an IR transparent window material.
- The window material is placed into the sample chamber and analyzed.

11.6.2 Cast Film A

- One to two milligrams of sample is placed onto an IR transparent window material.
- The sample is dissolved with a drop of volatile solvent.
- The sample creates a crystalline film as the solvent evaporates.
- The window material is placed into the sample chamber and analyzed.

11.6.3 Cast Film B

- One to two milligrams of sample is placed onto an IR transparent window material.
- The window material with the sample is placed on a hot plate, heat is added, and the sample melted.
- The sample creates a crystalline film as the window material cools to ambient temperature after it is removed from the hot plate.
- The window material is placed into the sample chamber and analyzed.

11.6.4 Pellets

- One to two milligrams of sample is placed into an agate mortar and ground into a fine powder.
- An IR transparent matrix powder (usually dried KBr) is added to the sample and ground into a homogeneous mixture.
- The mixture is placed into a pellet mold and compressed into a transparent window (Fig. 11.9). The resulting pellet is commonly called a KBr press.
- The pellet is placed into the sample chamber and analyzed.

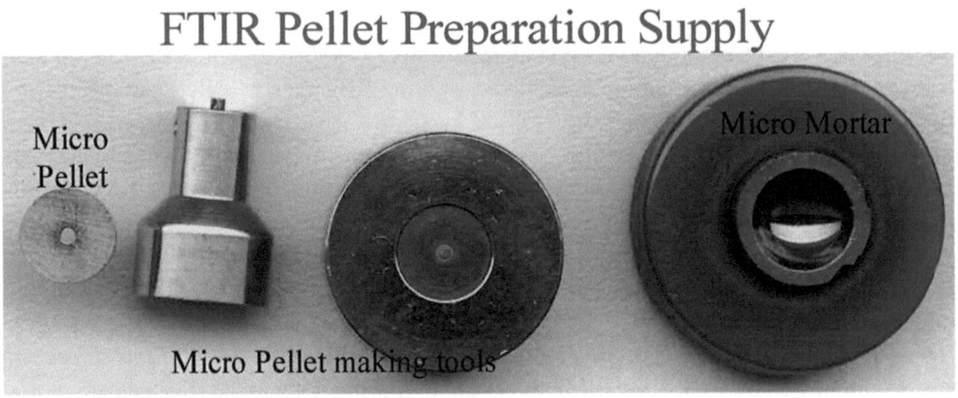

Fig. 11.9 Solid samples are prepared for FTIR measurements using a pellet mold. The sample is mixed with a solid matrix (usually KBr) that is transparent to IR radiation and ground to a fine powder. A sample of the powder is placed between two anvils and compressed using a hand press or minipress. Although the resulting micropellet appears opaque, the matrix is actually transparent to IR radiation while the sample is not.

11.6.5 Synthetic Membrane Sample Cards

- A reference spectrum of a blank sample card is taken.
- A liquid sample containing the compound of interest is deposited on the sample card's membrane and allowed to evaporate.
- The sample card is placed into the sample chamber and analyzed.
  - Additional sample can be added if the signal is too weak.

11.7 Reflectance

A number of reflectance techniques can be used to acquire IR spectra. A few of the most common are listed below.

- Attenuated total reflectance (ATR) resembles the conventional IR spectrum, but with some notable differences: the absorption band positions are identical in the two spectra, but the relative intensities of corresponding bands are different. ATR spectra can be obtained using either dispersive or FT instruments.
- Specular reflectance provides a nondestructive method for measuring thin coatings on selective, smooth substrates without sample preparation. It uses mirror-like reflections to analyze reflective materials, or a reflection–absorption for measurements on surface film placed on a reflective surface.
- Diffuse reflectance technique is mainly used for acquiring IR spectra of powders and rough surface solids such as coal, paper, and cloth. It can be used as an alternative to pressed-pellet or mull techniques.

11.8 Fourier Transform Infrared Spectroscopy

FTIR spectroscopic determination is based on the presence and orientation of functional groups in compounds. Each functional group uniquely absorbs a characteristic frequency of IR radiation. The same functional group absorbs a different band of IR radiation if oriented differently in a molecule.

Figure 11.10 shows the IR spectra of ephedrine and pseudoephedrine. The molecular formula of both compounds is the same and the only difference is the orientation of the hydroxyl group on carbon-one in the molecule. These molecules generate identical GCMS spectra; however, a noticeable difference in the FTIR spectra is observed at 2,500 cm^{-1} and the region between 500 and 1,500 cm^{-1} (variable region). This sensitivity illustrates the selective capabilities of FTIR commonly used to distinguish isomers.

11.9 Advantages of Fourier Transform Infrared Spectroscopy

1. Inexpensive with no hazardous waste.
2. Can be used on gas, liquid, and solid organic samples.
3. Excellent technique to differentiate isomers.
4. Can detect very small quantities of sample.
5. Very fast analysis times.

11.10 Disadvantages of Fourier Transform Infrared Spectroscopy

1. Manual operation, lack of automated capabilities requires the presence of an analyst.
2. Highly purified samples are required. Very small quantities of impurity cause interference which complicates spectra.

11.11 Instrument Selection for Forensic Identification

IR spectroscopy has a longstanding history in the identification of organic compounds. However, in forensic analysis, this method is generally considered a nonspecific form of identification because it is often used to indicate only the presence or orientation of a particular functional group. It provides no detail on the specific structural arrangement of the molecule. For

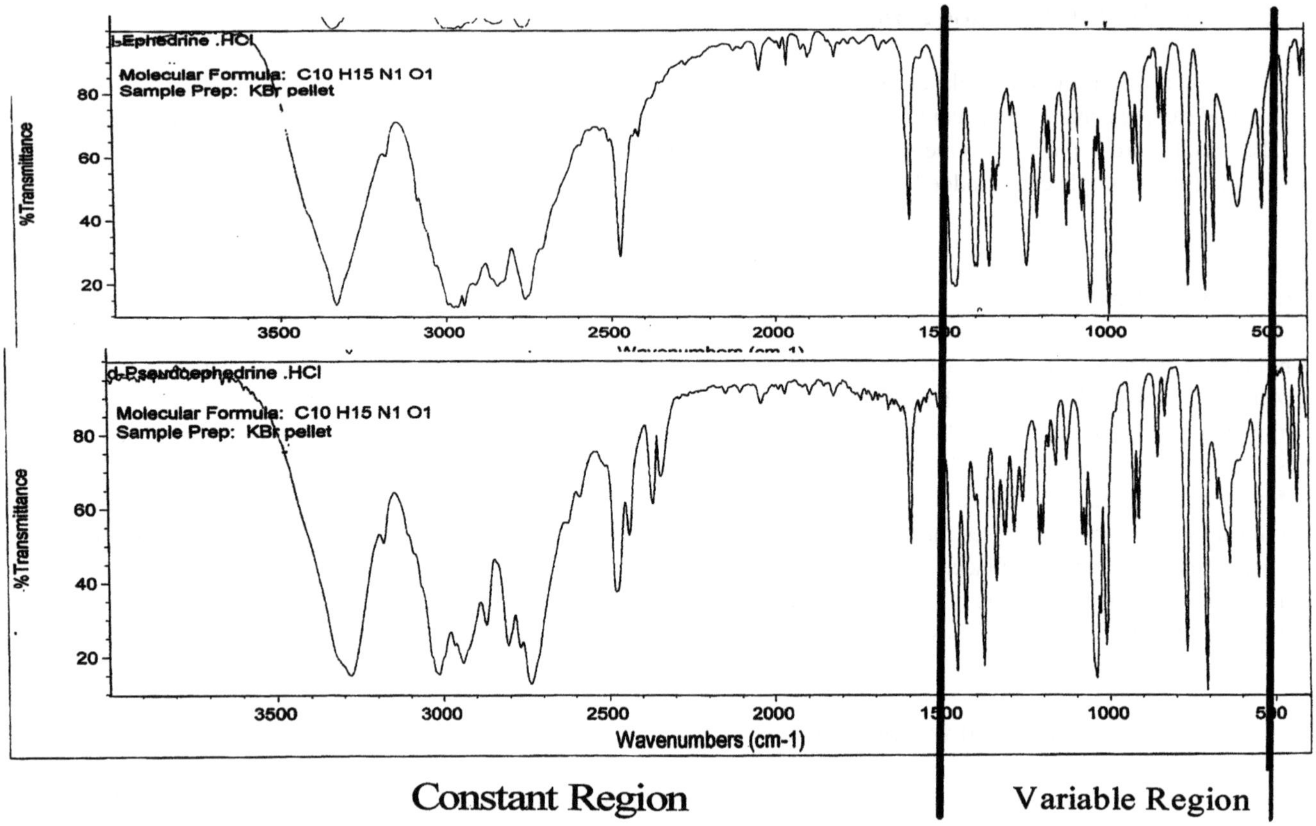

Fig. 11.10 Variable regions of FTIR spectra are useful in differentiating isomers. The extreme sensitivity of FTIR spectroscopy is a clear advantage over most IR analytical methods. The positional isomers ephedrine (*top*) and pseudoephedrine (*bottom*) are shown.

example, a sharp absorption band in the 1,700-cm^{-1} region of a spectrum indicates the presence of a carbonyl group (carbon–oxygen double bond), but many classes of drugs contain carbonyl groups, i.e.; aldehydes, ketones, esters, carboxylic acids, etc. If a broadband centered around 3,000 cm^{-1} is also present, it is highly likely that the compound is a carboxylic acid. However, the specific structure of the carboxylic acid cannot be determined from the IR spectrum alone. For this reason, IR spectroscopy is often used to provide supporting information to more selective methods (i.e., GCMS). Still, this is not universally applied because, in some cases, FTIR may be the only analytical method that can definitively identify a particular controlled substance (i.e., isomers). Ultimately, the choice of confirmatory method is directly dependent on the nature of the target compound.

11.12 Inorganic Analysis

The limitations described above do not necessarily apply to the IR analysis of *inorganic* compounds where FTIR is commonly used as a stand-alone technique for identification.

Two factors are considered in the use of IR spectroscopy to identify inorganic compounds: sample preparation and peak identification. The broad Abs bands present in the IR spectra of inorganic compounds make it difficult to identify peak maximums. Therefore, the sample concentration should be diluted until a definite peak observed in the primary Abs band is at a 5% transmittance level (95% abs). This will allow the examiner to identify the wavelength at maximum Abs for each peak in the spectrum. This is critical because a shift of only 10 cm^{-1} can represent the difference between the sodium (Na) and potassium (K) salts of a particular compound. These minor shifts may seem insignificant; however, if they are reproducible in properly prepared samples, they provide the specificity required for identification purposes.

Table 11.1 Infrared absorbance frequencies for selected anions

Anion	Absorbance (cm^{-1})
BO_2^-	1,300–1,360
$B_4O_7^-$	990–1,000, 1,340–1,375
CO_3^{-2}	1,425–1,455
HCO_3^-	690–710, 830–840, 990–1,010
SCN^-	2,020–2,100
SiO_3^{-2}	950–1,010
NO_2^-	1,225–1,250
NO_3^-	1,340–1,390
NH_4^+	1,390–1,440
PO_4^{-3}	1,000–1,040
HPO_4^{-2}	1,010–1,060, 840–900
$H_2PO_4^-$	1,025–1,090, 900–950
SO_3^{-2}	920–980
SO_4^{-2}	1,075–1,130
HSO_4^-	845–855, 1,030–1,070, 1,160–1,175
$S_2O_3^{-2}$	950–1,000, 1,100–1,120
$S_2O_5^{-2}$	975–990, 1,175–1,190
$S_2O_8^{-2}$	700–710, 1,050–1,075, 1,275–1,300
ClO_3^-	940–970
ClO_4^-	1,050–1,120
BrO_3^-	795–805
IO_3^-	725–750
$Cr_2O_7^{-2}$	795–860, 860–875
$Cr_2O_7^{-2}$	725–775, 845–890
WO_4^{-2}	790–825
MnO_4^-	890–915
CN^-	2,077–2,100
OCN^-	2,100–2,180

Many anions have well-defined, characteristic Abs bands that can be used as a screening tool to classify the type of inorganic compound. Table 11.1 lists anions and their corresponding IR absorption wave numbers (cm^{-1}).

In some instances, IR analysis cannot distinguish the salt forms of a given compound. In these cases, the results of other testing procedures can be used for identification. For example, the IR spectra of sodium and potassium cyanide are almost identical, but sodium and potassium are easily distinguished using microcrystal techniques.

11.13 Organic Analysis

Organic salts are usually differentiated from their free-base counterpart using hydrogen chloride (HCl) attached to the salt's name or chemical formula. Free-base forms are generally more pure but typically exhibit limited stability. Conversion into the salt form is surprisingly simple in most cases and increases the overall stability of the compound. Figure 11.11 differentiates free-base cocaine (crack) and cocaine hydrochloride (salt). Figure 11.12 contains the IR spectra of HCl and free-base forms of methamphetamine. The most obvious difference is demonstrated in the spectra's front portion (4,000–2,000 cm^{-1}). The identification of specific forms of illegal drugs can be used to establish particular manufacturing processes.

The highly purified samples necessary for effective IR analysis require extensive purification. For this reason, many view IR spectroscopy as a tedious and time-consuming analytical technique. The advent of FTIR spectroscopy has reduced analysis times from minutes to seconds and has allowed FTIR detectors to be used in conjunction with GC. This technique allows the forensic chemist to identify the components of a mixture using IR without lengthy purification steps.

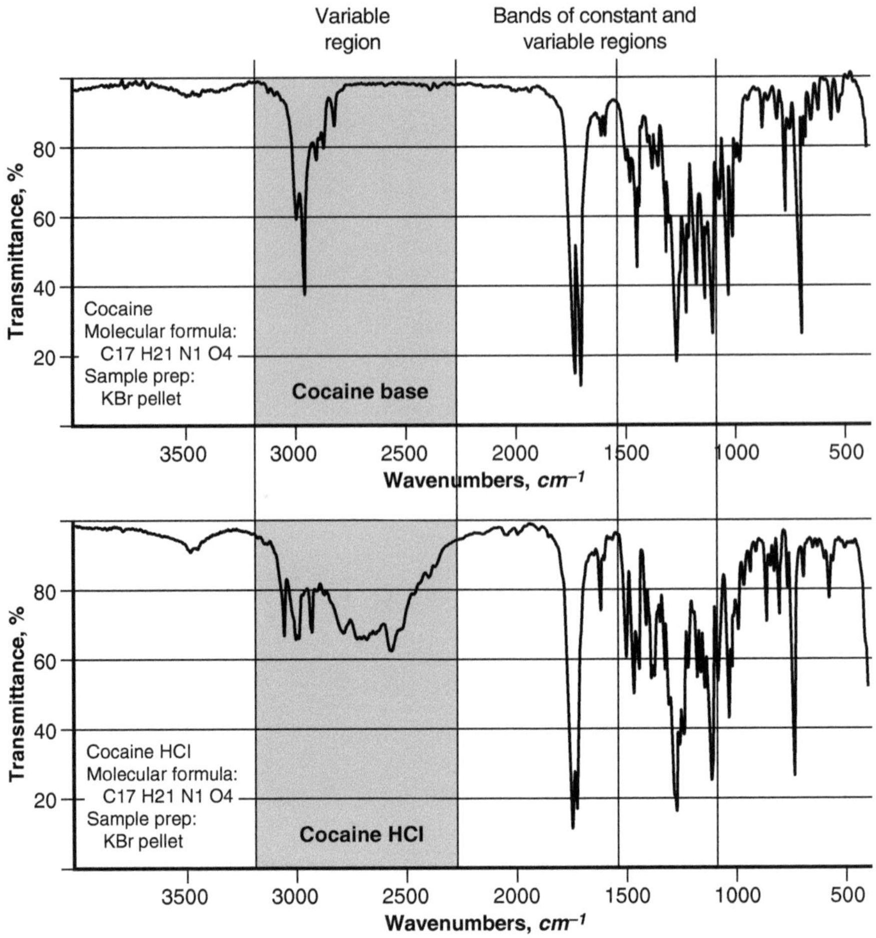

Fig. 11.11 IR spectra of free-base (*top*) and salt (*bottom*) forms of cocaine. Free-base cocaine acquired the popular street name "crack or crack cocaine" from the characteristic "cracking" sound produced while smoking this form.

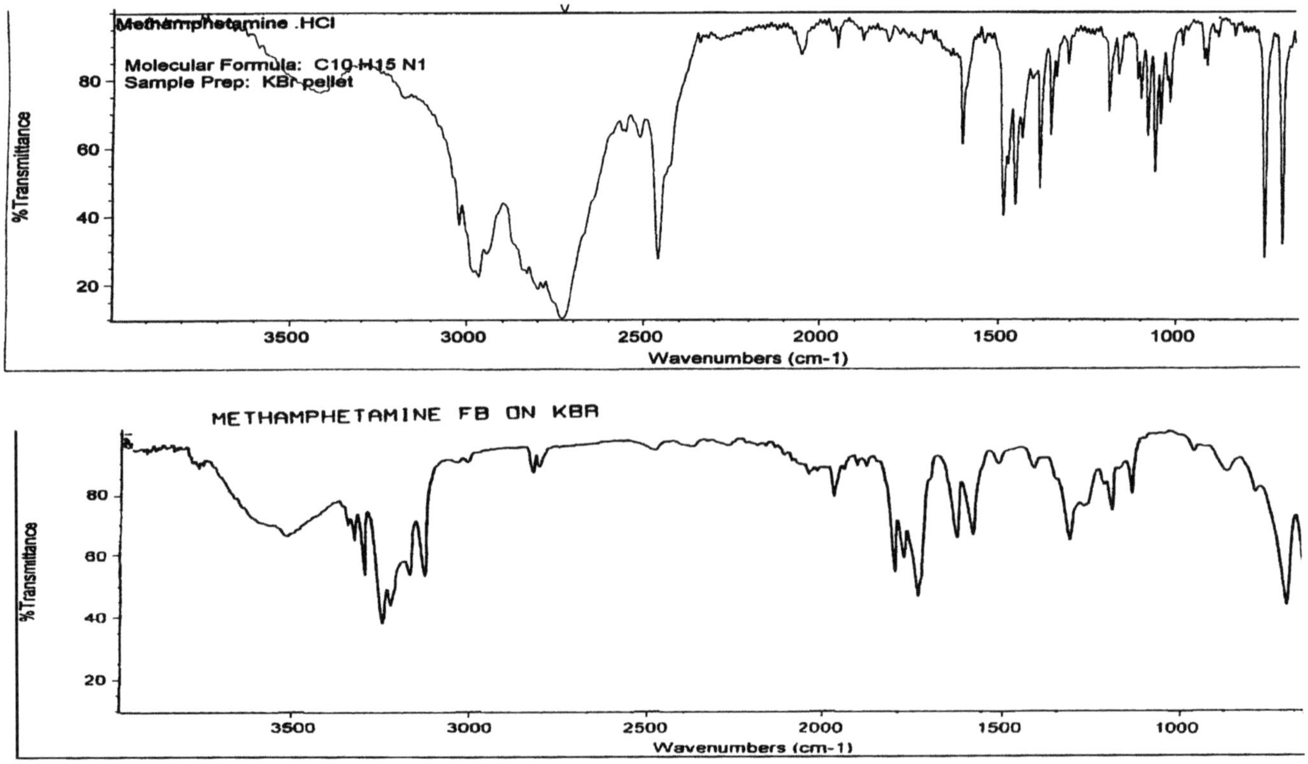

Fig. 11.12 IR spectra illustrating the differences between the salt (*top*) and free-base (*bottom*) forms of methamphetamine. Free-base forms are generally produced first in the manufacturing process and subsequently converted to salts using relatively simple techniques.

11.14 Questions

1. Define spectroscopy.
2. Define wavelength for a transverse wave and discuss its relationship to energy and frequency.
3. List and describe three modes of bond vibration.
4. Please define the term wave numbers in units of cm^{-1} to members of the jury.
5. Please explain to the jury how wave numbers differ from frequency.
6. Define IR radiation and explain how it differs from UV and visible radiation.
7. Give two advantages of FTIR over dispersive techniques.
8. Explain why IR analysis is a form of nonspecific identification.
9. What advantages does FTIR have over the GCMS in the examination of cocaine?
10. Why are solid sampling methods preferred in IR analysis?
11. Describe the preparation of a KBr pellet for FTIR analysis.
12. Describe the importance of variable regions in spectral interpretation.
13. How are free-base and salt forms differentiated.
14. Cite three advantages of FTIR.
15. Compare and contrast IR analysis of organic and inorganic compounds.

Suggested Reading

Allen. A.C.; et al. Cocaine Diastereomers. *J. Forensic Sci.* **1981**, 26, 12–26.

Bellamy, J. *Infrared Spectra of Complex Molecules,* Vol. I.; Chapman and Hall: New York, 1975.

Christian, D. R. Jr. Analysis of Controlled Substances. In *Forensic Science: An Introduction to Scientific and Investigative Techniques*, 3rd ed.; James, S. H.; Nordby, J. J., Eds.; CRC Press: Boca Raton, FL, 2009.

Cole, M. D.; Caddy, B. *The Analysis of Drugs of Abuse: An Introduction Manual*; Taylor & Francis: New York, 1994.

Daley, H. L.; Wiberly, S. E. *Introduction to Infrared and Raman Spectra*; Academic Press: New York, 1975.

Griffiths, P. R.; De Haseth, J. A. *Fourier Transform Infrared Spectroscopy*; Wiley & Sons: New York, 1986.

Hsu, C. P. S. Infrared Spectroscopy. http://www.prenhall.com/settle/chapters/ch15.pdf (accessed August 2009).

Infrared Spectroscopy. http://www.umsl.edu/~orglab/documents/IR/IR2.html (accessed August 2009).

Infrared Spectroscopy. http://www.spectroscopynow.com/coi/cda/detail.cda?id=18427&type=EducationFeature&chId=2&page=1 (accessed August 2009).

Palenik, S. *Particle Atlas of Illicit Drugs*; Walter McCrone Associates: Chicago, 1974.

Rao, A. B.; Mathews, C. W. *Molecular Spectroscopy: Modern Research*; Academic Press: New York, 1972.

Razumovski, L. Infrared Spectroscopy. http://www.medlabs.com/file.aspx?FileID=58 (accessed August 2009).

Sverdlov, M., Kovner, M. A., Krainov, E. P., *Vibrational Spectra of Polyatomic Molecules*; John Wiley & Sons: New York, 1974.

Wielbo, D.; Tebbett, I. R. The Use of Microcrystal Tests with Fourier Transform Infrared Spectroscopy for the Rapid Identification of Elicit Cocaine. *Microgram.* **1990**, 23, 258–261.

Part III

Examination of Drugs/Narcotics

12 Cannabis

12.1 Introduction

Marijuana is not a scientific classification; it is a term typically used to describe the dried leaves of cannabis plants and flowering portions of the female cannabis. Cannabis contains the psychoactive drug tetrahydrocannabinol which acts on the central nervous system producing both physical and psychological effects. The *trans*-Δ^9-isomer is the main active form of tetrahydrocannabinol (THC) (Fig. 12.1). The delta-nine symbol (Δ^9) indicates the presence of a carbon–carbon double bond (Δ) located between carbons 9 and 10 (exponent is the first carbon in the double bond). This terminology is used quite often in organic chemistry and biochemistry where it is frequently encountered in the abbreviated forms of fatty acids. The *trans*-Δ^9-isomer is classified in the controlled substance act and is the form most often referred to when using the acronym THC. It can be extracted from the herbal form of cannabis using a variety of techniques. The chosen method of extraction determines the overall concentration of THC in the final product as well as its physical appearance. Marijuana is described above and typically contains 7–25% THC. Hemp is a form grown for industrial (nondrug) purposes and the concentration of THC (less than 1%) is typically too low to produce euphoric effects. Hashish (hash) is a THC resin extracted from the female flowers and is somewhat more concentrated than marijuana. Hash oil is a more concentrated form of hashish and can easily approach 50% THC content. Kief is a powder form commonly called (incorrectly) crystal or pollen. It contains a THC content comparable to that of hash; in fact, a type of hash is produced from highly compressed kief. Resin is a thick tar by-product of smoking cannabis and contains trace amounts of THC. Smoking resin vapors can cause irritation to the throat and lungs.

Historically, THC has been the most frequently analyzed controlled substance in forensic laboratories where it can exceed 50% of the caseload. Programs implemented by law-enforcement agencies allow trained personnel to conduct preliminary tests to identify cannabis. This practice has dramatically reduced the workload in forensic laboratories.

The visual examination of suspected cannabis, especially marijuana, requires great care. In this instance, it is identity of plants and plant material that is in question, not a specific drug. The forensic chemist is educated and trained in areas of chemistry, not biology or botany. This fact must be recognized when performing and documenting visual inspections. Most jurisdictions, however, recognize the informed opinion of an analyst trained in the identification of specific plants, despite a lack of formal education in this area.

The preliminary examination of plants or plant material requires techniques that are inherently subjective and most cannot be documented in a manner that can be objectively reviewed. Therefore, the peer-review process relies solely on the working notes for evaluation, unless some form of photography is used to document the procedure (always a good idea). Consequently, the results of visual examination should be recorded in great detail and include as much information as possible.

12.2 History

Marijuana has a long-standing, documented history of use as a euphoric drug. It is referenced in Chinese medical compendiums (abstracts) dating back to 2700 BC. Its use spread from China to India and on to North Africa and Europe as early as 500 AD. A major crop in colonial North America, marijuana (hemp) was grown as a source of fiber. It was extensively cultivated in the United States during World War II when Asian sources of hemp were cut off.

J.I. Khan et al., *Basic Principles of Forensic Chemistry*, DOI 10.1007/978-1-59745-437-7_12,
© Springer Science+Business Media, LLC 2012

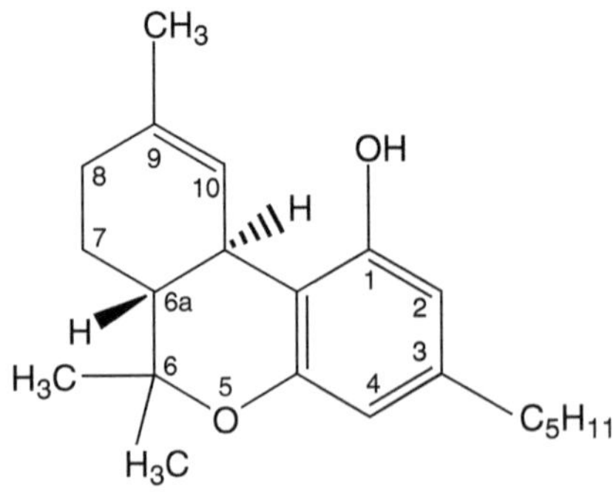

Fig. 12.1 The structure of *trans*-Δ^9-tetrahydrocannabinol. This active isomer of THC produces a variety of physical and psychological effects. It is classified as a Schedule I hallucinogen in the Controlled Substance Act.

Table 12.1 Regional names of marijuana

Country	Commonly known as
India	Bhang, Ganja
Algeria, Morocco	Kif
Tunis	Takrouri
Turkey	Kabak
Syria/Lebanon	Hashish El Keif
Africa (Central)	Djamba, Liamba, Riamba
Africa (South)	Dagga
America (South)	Marihuana
Mozambique	Suruma
Madagascar	Rongony
Brazil	Maconha
United States	Mary Jane, Grass, Pot, Weed

Marijuana was listed in the United States Pharmacopoeia (USP) from 1850 to 1942 when it was prescribed for various conditions, including labor pains, nausea, and rheumatism. Its use as an intoxicant has also been documented from the 1850s to the 1930s. A rigorous campaign conducted in the 1930s by the U.S. Federal Bureau of Narcotics (now the Bureau of Narcotics and Dangerous Drugs) portrayed marijuana as a powerful, addictive substance that would eventually lead to narcotics addiction. To date, this opinion is still held by some who consider marijuana a "gateway" drug. In the 1950s, it symbolized the "beat generation," and, in the 1960s, it became a dubious (no pun intended!) symbol of rebellion against authority and was closely associated with college students and "hippies."

The Controlled Substances Act of 1970 classified marijuana with heroin and LSD as a schedule I drug (highest abuse potential and no accepted medical use). Most marijuana at that time came from Mexico, but, in 1975, the Mexican government agreed to eradicate the crop by spraying it with paraquat (herbicide). Fears concerning the toxic side effects of the herbicide served as a deterrent to potential users. As a result, Colombia became the primary source of marijuana. The "zero-tolerance" policy of U.S. President's Reagan and Bush (1981–1993) produced strict laws and mandatory sentences for possession of marijuana. This had a direct impact on smuggling at the southern borders. The "war on drugs" prompted a shift in U.S. reliance on imported drugs to one on domestic cultivation. In 1982, the Drug Enforcement Administration (DEA) began targeting marijuana farms in the United States. As a result, indoor cultivation became widespread. Cross-pollination created new and innovative breeding techniques that altered genetic structure and produced small plants with elevated levels of THC. These plants were easily cultivated and concealed. After more than a decade of decline, marijuana use in the mid-1990s began to increase, especially among teenagers. The new millennium brought a slight decrease in use and current levels appear to be stabile.

A rose by any other name is still a rose. Table 12.1 lists common regional names of marijuana. The physical form may change, the region of the world may change, and the method of cultivation may change, but the plant remains the same: *Cannabis sativa* is still *Cannabis sativa*.

Fig. 12.2 Decayed marijuana plants stored in plastic bags. Levels of THC in cannabis are drastically reduced by the decomposition of plant matter.

12.3 Packaging for Forensic Examination

Paper bags or envelopes should always be used for packaging and storing marijuana plants. Marijuana – *especially fresh marijuana plants* – should not be stored in plastic bags because deterioration or fungal infection may deplete THC levels (Fig. 12.2).

12.4 Forms of Cannabis

Cannabis is submitted to forensic laboratories in many forms. The two most common are marijuana and hashish. Marijuana is the herbal form and may be leaves or flowering tops from cannabis plants. Hashish is an oily resin isolated from cannabis. Both contain the psychoactive drug THC, which is the target compound in forensic analysis.

Forensic laboratories receive marijuana in all conceivable forms for examination. Plants range from seedlings to mature stalks with flowering tops and quantities range from hand-rolled cigarettes (commonly called joint or doobie) to multi-kilogram bales.

Hashish is encountered less often than marijuana and is usually submitted as either a solid or oil. The solid form is smoked through some type of pipe, while oil forms are usually applied to the surface of plant material such as marijuana, tobacco, or mint, and smoked. Figure 12.3 shows some forms of cannabis submitted to forensic laboratories for analysis.

12.5 Psychoactive Ingredient

Cannabinoids are a class of compounds derived from terpenes and phenol. Terpenes are hydrocarbon derivatives of turpentine that show considerable variation in chain length and branching. Phenol is a derivative of benzene containing a hydroxyl group (OH) bound to the aromatic ring. In pure form, it is a toxic, white crystalline substance. Cannabinoids can also be defined as any compound sharing the basic structural features of THC. A large number of cannabinoids have been isolated from the herbal form of cannabis and not all are psychoactive. Cannabinol, cannabidiol, and THC receive the most attention because of their ubiquitous nature (Fig. 12.4).

Two of the most common psychoactive forms of THC are *trans*-Δ^9-tetrahydrocannabinol and *trans*-Δ^8-tetrahydrocannabinol, with the Δ^9-isomer generally present in higher concentrations. The two structures differ only in the location of a carbon–carbon double bond; however, on this particular point, there is considerable debate. Unfortunately, two numbering systems are commonly used to locate the double bond. If the method used to number fused-ring systems is applied, the major form is called Δ^9-THC and the minor form is called Δ^8-THC. If the method commonly applied to terpenes is used, the major form is Δ^1-THC and the minor is Δ^6-THC. The fused-ring application is much more common, most likely due to its extensive use in areas of organic and biochemistry. Conversely, the terpene system is generally considered an "older" method (Fig. 12.5).

Fig. 12.3 Various forms of cannabis commonly submitted to forensic laboratories for analysis. Clockwise from top left corner: harvested mature plants, intact mature plants just before harvesting, indoor cultivation, and compressed bricks.

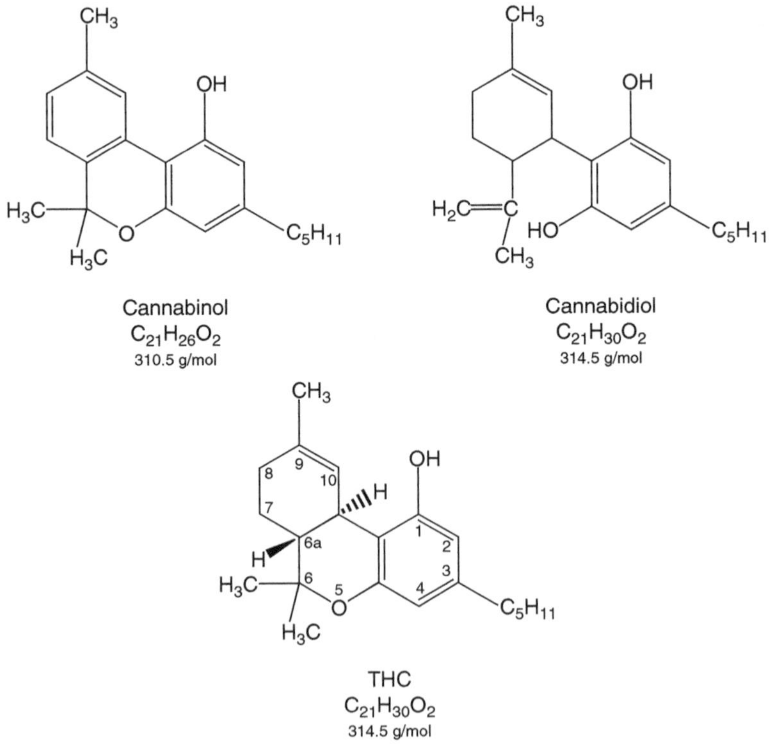

Fig. 12.4 Cannabinoids are a class of compounds that are structurally related to THC. Unlike THC, cannabinol (hemp) and cannabidiol are not considered psychoactive drugs. Note the structural similarities.

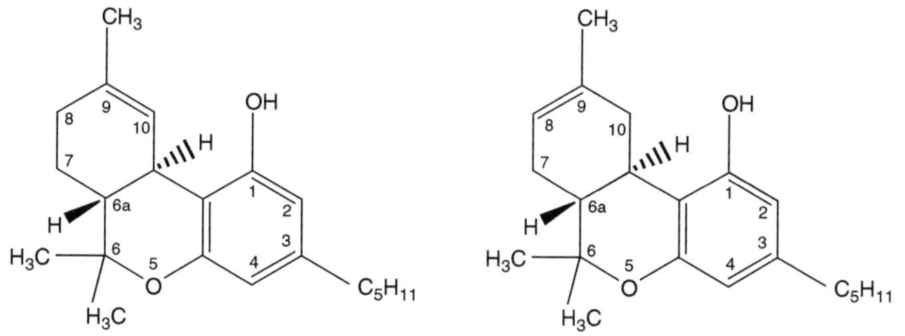

Fig. 12.5 The *trans*-Δ^9 (*left*) and *trans*-Δ^8 (*right*) isomers are the psychoactive forms of THC isolated from cannabis. Although the *trans*-Δ^9 isomer is normally present in higher concentrations, both produce euphoria and alterations in visual, auditory, and olfactory senses. Note the subtle differences in structures.

Table 12.2 Scientific classification of marijuana

Kingdom	Plant
Subkingdom	Vascular plants
Super division	Seed plants
Division	Flowering plants
Class	Dicotyledons
Subclass	Hamamelidae
Order	Urticales
Family	Cannabaceae
Genera	Cannibis
Species	*C. sativa*

12.6 Forensic Identification of Marijuana

The procedure used by forensic laboratories to identify cannabis is one of the oldest internationally accepted methods in forensic science. In 1938, the United States Treasury Department published a pamphlet that outlined the steps used in the botanical identification of cannabis. In 1950, the League of Nation's Subcommittee on Cannabis adopted the original Duquenois reaction as a preferential test. In 1960, The United Nation's Committee on Narcotics acknowledged this test with a Levine modification. Today, the Duquenois–Levine color test is universally accepted as a specific method for testing marijuana.

A combination of a botanical examination and chemical testing is used to identify cannabis and is commonly accepted by many jurisdictions. Microscopic examination of raw plant material followed by the Duquenois–Levine color test is the internationally recognized procedure. In addition, analytical methods may be used to provide definitive confirmation. If botanical examinations are not possible (i.e., hashish), other procedures such as the chromatography or instrumental analysis are required for identification.

12.6.1 Botanical Identification

Marijuana is the common name for the plant *Cannabis sativa*. Although many different types of marijuana exist (i.e., indica, rhutamalus, and Americana), these are simply variations of the sativa species. To avoid confusion and misinterpretation, some jurisdictions have opted to control all varieties by defining "cannabis" as a controlled substance. Table 12.2 identifies the scientific classification of marijuana. Figure 12.6 illustrates the physical transformation of marijuana at different growth phases.

12.6.2 Macroscopic Properties

Marijuana is an annual plant with separate male and female types (dioecious). The stem is fluted, and the plant has a primary root system. Leaves are simple, palmate, with an odd number of foliolates (leaflets), usually five or seven. Each foliolate has pinnate veination with a saw-toothed (dentated) edge. Most leaves cluster around a central axis (inflorescence) toward the top of the stalk (Fig. 12.6).

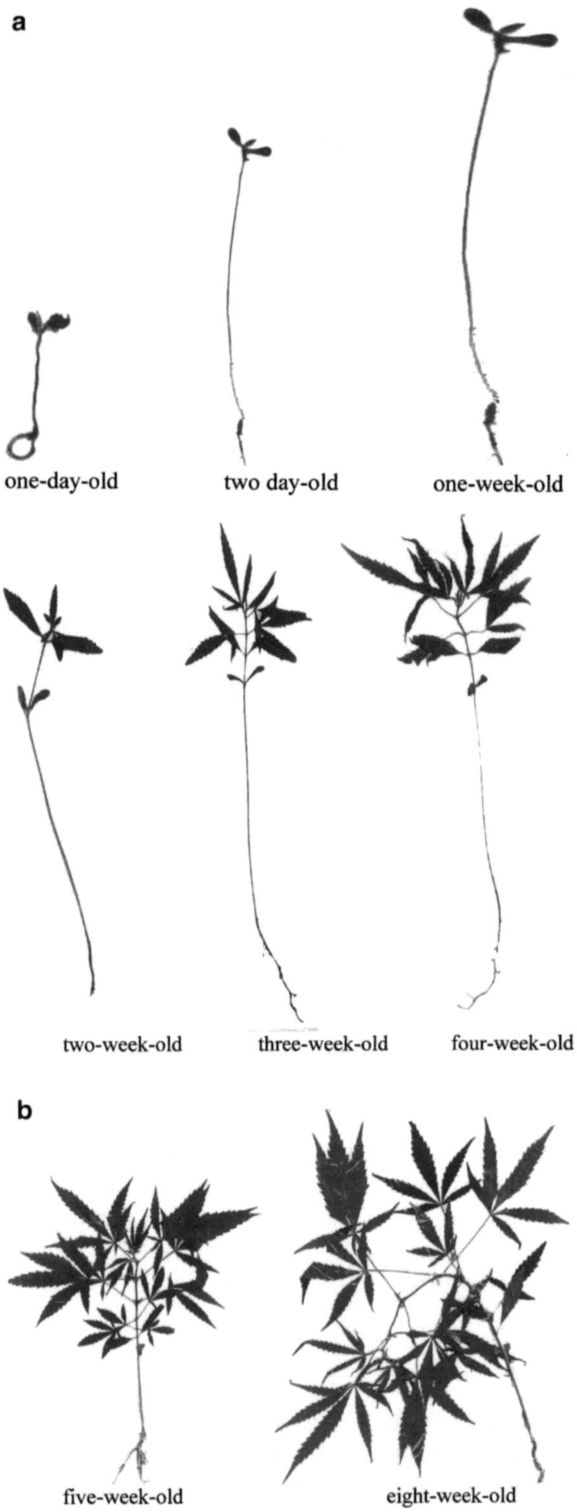

Fig. 12.6 The aging process produces distinct physical changes in marijuana plants. Detectable levels of THC are found in all stages.

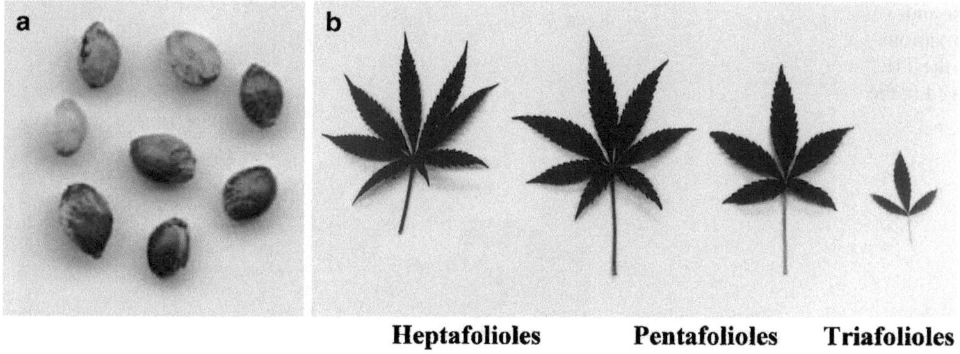

Fig. 12.7 Examples of marijuana seeds and leaves. Striated edges and inflorescence (clustering) are characteristic of cannabis leaf material. Note the odd number of leafs in each example, including the immature specimen (*on right*). The exterior protective coating of seeds is extremely variable in both color and texture.

The stem of the male is straight, small, and slender compared to female specimens. The flowers are grouped in loose panicles composed of five sepals and five episepal stamens with an introrse anther. The female plant is somewhat shorter and generally has thicker foliage. The flowers are topped by two long, stigmas that are pink in color. The seeds are generally oblong in shape (Fig. 12.7) and have a characteristic lace-like exterior. A particularly potent form of cannabis is *sinsemilla*. This variant is produced by removing the male plants from the local environment of females before they have a chance to pollinate. The females produce very little, if any, seeds. As a result, the plant's resources are focused on the production of psychoactive compounds and not on reproduction.

12.6.3 Microscopic Identification

Cannabis has a unique surface texture that is readily observed under low-power magnification, typically 10–40 times. The top surface exhibits fine hairs, while the underside contains glandular and cystolith hairs. Cystolith hairs are unicellular appendages containing calcium carbonate that closely resemble a bear-claw shape. The mushroom-shaped glandular hairs are multicellular units that secrete cannabis resin (Fig. 12.8).

12.6.4 Chemical Identification (Duquenois–Levine Test)

Chemical analysis of cannabis resin is the second component in the identification process of marijuana. With hashish, two separate chemical tests are required to confirm the presence of cannabis resin. The Duquenois–Levine test is one of the most widely used and accepted chemical tests for marijuana.

12.6.4.1 Proposed Reaction Mechanism

In acidic solutions, protinated aldehydes (at carbonyl oxygen) are strongly electrophilic (electron loving) at the (now) positively charged oxygen. The hydroxyl group of phenols and phenol derivatives is a strong *ortho*/*para* director (carbons 2 and 4 respectively on Δ^9-THC). Aromatic π-electrons from the benzene ring can attack protinated aldehydes at the carbonyl carbon or the protinated carbonyl oxygen. It seems likely that the oxygen is targeted more often because of its positive charge. Substitution at the *ortho* and *para* positions would be expected with the product possibly undergoing further condensation to yield a resinous material of considerable complexity. Oxidation of this product could lead to quinone structures that produce an intensely colored solution.

Independent testing suggests that an aldehyde–phenol reaction leading to resin formation by *ortho*- and *para*-electrophilic aromatic substitution is the likely mechanism involved in the Duquenois reaction. Although this mechanism is reasonable, and is consistent with experimental observations, it has yet to be proven.

A modification of the Duquenois test incorporates extraction of the blue-colored aqueous solution into a purple-colored product in chloroform (organic layer). The extraction is repeated until the blue color is extracted entirely into chloroform. The chloroform layers are combined and evaporated to dryness under mild heat. Upon drying, the color turns back to blue, indicating that the color of the organic layer is somehow influenced by solubility and pH.

Fig. 12.8 Glandular and cystolith hairs under various magnification. Note the THC resin droplets observed in the bottom examples.

12.6.4.2 Test Reagents

The Duquenois–Levine chemical color test requires three reagents. The test can be conducted directly on the suspected plant material. Although not required, specificity can be increased, and potential sources of interference eliminated, if the resin is extracted before treatment.

The Duquenois–Levine reagents:

Reagent A: Petroleum ether
Reagent B: 97.5 ml of 2% vanillin solution in methanol (absolute)
2.5 ml of acetaldehyde
Reagent C: Concentrated hydrochloric acid
Reagent D: Chloroform

12.6.4.3 Test Technique

- After microscopic examination, a small amount of the suspected marijuana plant material is placed into a culture tube with a small amount of reagent A and agitated (Fig. 12.9-1a).
  - Cannabinoids are selectively soluble in hydrocarbon solvents such as petroleum ether.
- The petroleum ether is transferred to a clean culture tube or spot plate and allowed to evaporate to dryness.
- Two to four drops of reagent B are added to the test sample and observations are documented (Fig. 12.9-1b).
- Two to four drops of reagent C are added to the mixture and observations are documented.
  - If cannabis resin is present, a transition of colors will occur, culminating in a shade of purple (Fig. 12.9-2).
  - The exact shade of purple will vary depending on the relative concentration of cannabinoids in the sample.
- Two to four drops of reagent D are added to the mixture and observations are documented (Fig. 12.9-3, 4).
  - If cannabis resin is present, a purple color will extract into the chloroform (bottom) layer.
  - The exact shade of purple will vary depending on the relative concentration of cannabinoids in the sample.

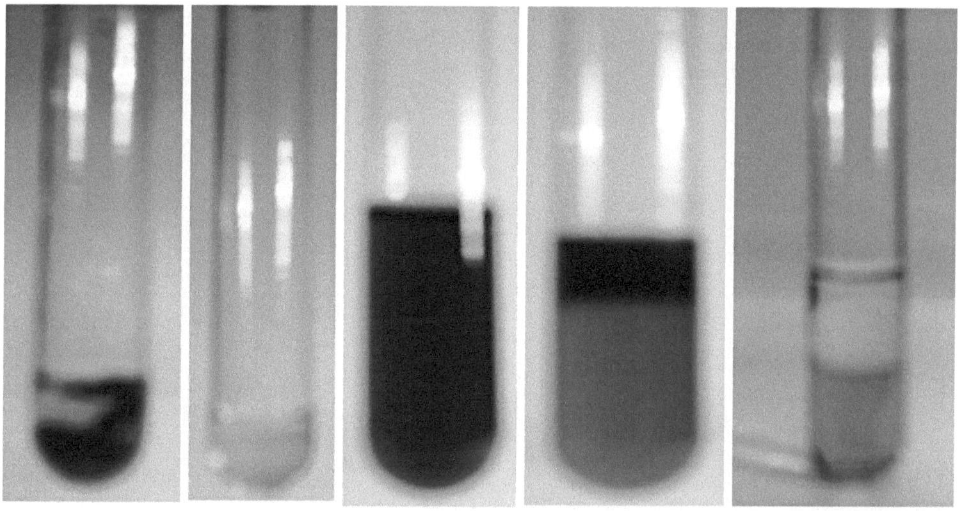

Fig. 12.9 The Duquenois–Levine chemical color test. (1a) Extraction of THC from plant material. (1b) Vanillin and acetaldehyde are added to the extract. (2) Addition of acid produces observable color. (3) The first chloroform extraction produces an intensely colored organic layer (*bottom*). (4) Subsequent extractions produce lightly colored organic layers.

12.6.5 Thin-Layer Chromatography

Thin-layer chromatography (TLC) is a wet chemical technique used to separate and identify the various components in cannabis resin. TLC combines chromatography and chemical color tests in the identification process. Capillary action, solubility of the sample, and the tortuosity of the path through the TLC plate all contribute to the technique's ability to resolve (separate) the components. Coloring reagents are used to locate the individual components on the TLC plate.

12.6.5.1 Reagents

A number of solvent pairs can be used for cannabis resin identification. Three common solvent systems include:

- 8% diethylamine in toluene
- Hexane/chloroform (9:1)
- Cyclohexane/diethylamine (5:1)

12.6.5.2 Test Technique

TLC uses internal standards and retention factors (R_f) to identify separated components. The following is a sequence of steps frequently used in TLC analysis.

Plate preparation:

- A 5×20 cm commercially prepared TLC plate is used.
- A reference line is drawn in pencil approximately 1 cm up from the bottom edge of the 5 cm side.
- A minimum of two marks are placed on the reference line and labeled.
  - One mark is for the suspected sample.
  - One sample mark is for a THC reference standard.

Sample preparation:

- A sample of unknown plant material or resin material is placed into a culture tube with an appropriate amount of petroleum ether and agitated.
- A 1–5-μl micropipette is used to draw a sample of mixture.
- Small micro drops of sample are applied to the suspected sample mark on the TLC plate.
- A separate 1–5-μl micropipette is used to draw a sample of the THC reference standard.
- Small micro drops are applied to the THC standard mark on the TLC plate.
- The samples are allowed to dry.

Separation:

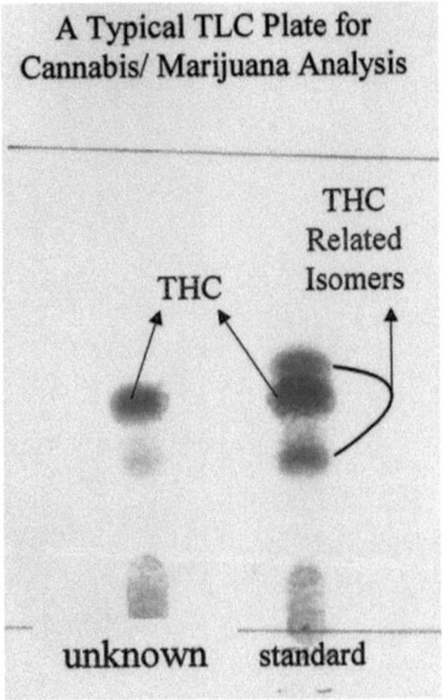

Fig. 12.10 TLC analysis of cannabis. Components are identified by comparing colors and R_f-values against known standards. Note the number of components (*spots*) present in each sample.

- The mobile phase is placed into the developing chamber.
  - The solvent level should be about 0.75 cm from the bottom of the chamber.
- The TLC plate is placed into the chamber with the sample side down and the chamber is covered.
- The solvent travels up the plate for a fixed period of time or until the solvent front reaches a predetermined point on the plate.
- The plate is removed from the chamber and a line is drawn across the solvent front in pencil.
- The solvent is allowed to evaporate.

12.6.5.3 Visualization

Common coloring agents for TLC identification of cannabis are a 0.5% solution of Fast Blue B Salt (tetrazotized o-dianisidine zinc chloride salt) or Fast Blue BB Salt [4-benzoylamino-2,5-diethoxy-benzenediazonium chloride hemi (zinc chloride) salt] in water. The order of the components may vary depending on the mobile phase selected; however, the color of a particular component is constant. The characteristic color of the three most common cannabinoids is given below:

- Cannabidiol: orange
- Δ^9-Tetrahydrocannabinol: red
- Cannabinol: purple (Fig. 12.10)

12.6.5.4 Interpretation of TLC Results

- The dried TLC plate is observed under long- and short-wave ultraviolet light.
- The positions of any observed spots are marked on the plate with pencil, and colors are documented in case notes.
- The plate is lightly sprayed with Fast Blue B or Fast Blue BB reagent.
- The position and color of the unknown spots are compared with those of the reference sample.
- R_f-values are calculated for each component.
  - R_f=measured distance from reference line to center of spot divided by measured distance from reference line to solvent front.

12.6.6 Gas Chromatography Mass Spectrometry

Petroleum ether is used in GCMS to extract cannabis resin from suspected plant material. An internal standard containing a known concentration of THC is analyzed under the same conditions as the suspected sample (Fig. 12.11).

12.7 Documentation

In general, case notes are the only form of documentation produced during visual inspection. Although they do not independently demonstrate that an examination occurred, they should include details of visual observations that contributed to the conclusions. Drawings or photographs of the observed structural characteristic can support such conclusions. Detailed descriptions of the color changes observed during Duquenois–Levine testing are equally important because definitive identification requires positive results in both tests.

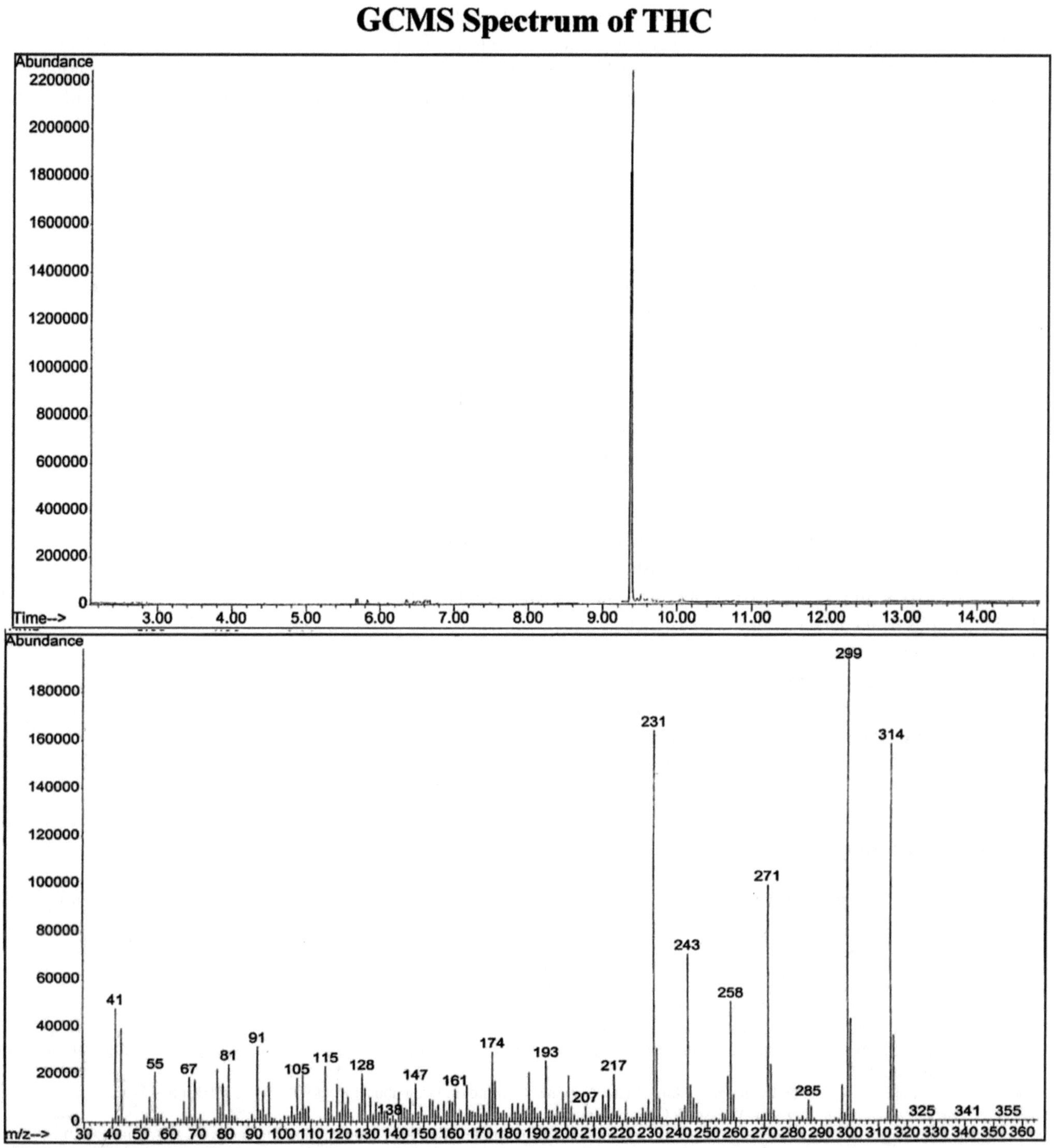

Fig. 12.11 A GC chromatogram (*top*) and MS spectrum (*bottom*) of THC from GCMS analysis. Note the sharp, clean peak on the chromatogram. This illustrates the resolving (separation) capabilities of GC and gives a precise retention time. The peak at 314 on the mass spectrum is the molecular ion peak (M^+) for THC. Note the complicated fragmentation pattern with a base peak at 299.

Essentially, it is difficult to verify that TLC was performed unless photographs of the developed plates are taken. A sketch will aid in documenting the test results, but sketches cannot be precisely interpreted by a case reviewer or independent examiner. GC and GCMS analyses are much more reliable because testing conditions, parameters, and results are recorded in great detail by the instrument. Therefore, all documents related to GC and GCMS analyses should be clearly marked and labeled with appropriate case information.

12.8 Questions

1. What is the scientific name of marijuana?
2. Please explain to the jury how it is possible for a forensic chemist to render an opinion on the structural features of marijuana, a plant.
3. Please define THC to the jury.
4. Define the term psychoactive drug.
5. Name the two psychoactive forms of THC and describe the structural differences.
6. What are the three common forms of THC in cannabis? Which are active and which are not.
7. How is cross-pollination used in the cultivation of marijuana?
8. Define hemp.
9. Why is sinsemilla a potent form of marijuana?
10. Please describe to the jury why suspected marijuana is never stored in plastic bags.
11. Compare and contrast marijuana and hashish.
12. Please clarify to the jury the two numbering systems used to name THC.
13. Cite two physical characteristics of marijuana.
14. Where is cannabis resin produced in marijuana plants?
15. Outline the procedure for performing the Duquenois–Levine color test.
16. Explain a positive result for THC using the Duquenois–Levine test.
17. Sketch a TLC plate testing positive for THC.
18. What is the MS molecular ion (M^+) and base peak for THC?

Suggested Reading

Athanaselis, S. S. *et al.* Cannabis: Methods of Forensic Analysis. In *Handbook of Forensic Analysis*; Smith, F. P. Ed.; Elsevier Academic Press: St. Louis, MO, 2005.

Christian, D. R. Jr. Analysis of Controlled Substances. In *Forensic Science: An Introduction to Scientific and Investigative Techniques*, 3rd ed.; James, S. H.; Nordby, J. J., Eds.; CRC Press: Boca Raton, FL, 2009.

Cole, M. D.; Caddy, B. *The Analysis of Drugs of Abuse: An Introduction Manual.* Taylor & Francis: New York, 1994.

Core, L. *Plant Taxonomy*. Prentice-Hall: Englewood Cliff, NJ, 1955.

Lewis, W. H.; Elvin-Lewis, M. P. F. Medical Botany. John Wiley & Sons: New York, 1977.

National Highway Safety Administration. Cannabis/ Marijuana. http://www.nhtsa.dot.gov/people/injury/research/job185drugs/cannabis.htm (accessed August 2009).

National Institute on Drug Abuse. Hallucinogens: An Update. National Institute on Drug abuse: Rockville, MD, 1994; pp. 43–67.

Palenik, S. *Particle Atlas of Illicit Drugs*; Walter McCrone Associates: Chicago, 1974.

Prater, A. M. The origins of Marijuana. http://war-on-drugs.suite101.com/article.cfm/the_origins_of_marijuana (accessed August 2009).

Szara, S.; Lin, G. C.; Glennon, R. A. *Are Hallucinogens Psychoheuristic?* National Institute on Drug abuse: Rockville, MD, 1994; pp. 33–51.

United Nations. *Recommended Methods for Testing Cannabis. Manual for Use by National Narcotics Laboratories*; ST/NAR/8; United Nations Publication: New York, 1987.

13 Phenethylamines

13.1 Introduction

Phenethylamines are a broad and diverse class of compounds that include neurotransmitters, hormones, stimulants, hallucinogens, entactogens, anorectics (appetite loss), bronchodilators, and antidepressants. Natural forms are commonly produced from the amino acid phenylalanine using enzyme-catalyzed decarboxylation, while synthetic forms are alkaloids derived from 1-amino-2-phenylethane (alternatively, 2-phenylethylamine or β-phenylethylamine). The nature of the substituted group and its location has an effect on the overall activity of the resulting compound. Figure 13.1 illustrates the various positions commonly substituted to produce members of the phenethylamine class. The carbons in ethane are labeled alpha (α) and beta (β) to avoid confusion with those in the aromatic ring. For simplicity, only a single hydrogen and corresponding R-group is shown attached to the alpha and beta carbons. In reality, each carbon contains two hydrogens that are equivalent, but both are rarely simultaneously substituted.

Table 13.1 contains a list of phenethylamine derivatives along with their characteristic groups and the location of each group on 1-amino-2-phenylethane. Although hundreds of synthetic phenethylamines are known, the derivatives most often targeted in forensic analysis contain methyl ($-CH_3$), hydroxyl (–OH), ketones (=O), methylene dioxy ($-O-CH_2-O-$), and methoxy groups ($-O-CH_3$).

A large number of substituted phenethylamines are biologically active because of their similarity to monoamine neurotransmitters. Representative examples include bronchodilators; stimulants such as ephedrine and cathinone; the anorectics phentermine, fenfluramine, and amphetamine; most natural and synthetic hallucinogens (i.e., mescaline); and the empathogen–entactogens 3,4-methylenedioxymethamphetamine (MDMA) a.k.a ecstacy and 3,4-methylenedioxyamphetamine (MDA).

Substitutions at either the α- or the β-position in 1-amino-2-phenylethane produce a chiral molecule. The number of stereoisomers can vary between two and four, depending on the number of carbons substituted. For example, a single methyl group substituted at the α-carbon produces amphetamine, which has two stereoisomers. Note the substitution is alpha carbon methyl on phenethylamine or amphetamine. Adding a hydroxyl group to the β-carbon of amphetamine produces phenylpropanolamine, a chiral molecule with four stereoisomers. If the two groups are in opposite planes, cathine, one of the optical isomers of phenylpropanolamine, is produced.

Phenethylamines can be administered using a variety of innovative techniques. Often, the drug is snorted, swallowed, injected, or inhaled by users. Typically, a single method is preferred, but it not uncommon for more than one method to be used either simultaneously or in combination.

13.2 Methyl Derivatives

The addition of a single methyl group ($-CH_3$) to the phenethylamine skeleton converts a naturally occurring neuromodulator into various drugs with legitimate anorectic properties. However, recreational intake of 10× the therapeutic dosage has become widely used to produce very different effects.

J.I. Khan et al., *Basic Principles of Forensic Chemistry*, DOI 10.1007/978-1-59745-437-7_13,
© Springer Science+Business Media, LLC 2012

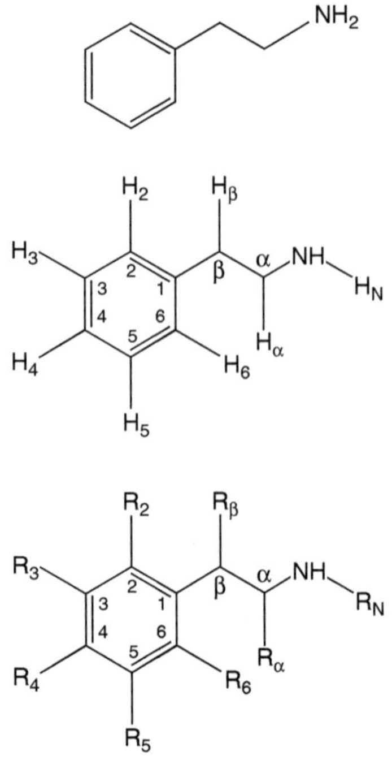

Fig. 13.1 *Top-to-bottom*. Structure of 1-amino-2-phenylethane, also called 2-phenylethylamine or β-phenylethylamine. The second structure contains *numbers* and *symbols* to identify the location of carbons and hydrogens within the structure. The third structure illustrates the various positions available for substitution. The labels correspond to groups shown in Table 13.1.

Table 13.1 Phenethylamine substitutions

Common name	R_α	R_β	R_2	R_3	R_4	R_5	R_N
Tramline					OH		
Dopamine				OH	OH		
Epinephrine (adrenaline)		OH		OH	OH		CH_3
Norepinephrine (noradrenaline)		OH		OH	OH		
Amphetamine	CH_3						
Methamphetamine	CH_3						CH_3
Levmetamfetamine	CH_3						CH_3
Ephedrine, pseudoephedrine	CH_3	OH					CH_3
Cathine	CH_3	OH					
Cathinone	CH_3	=O					
Methcathinone	CH_3	=O					CH_3
Phentermine	CH_3, CH_3						
Mescaline				OCH_3	OCH_3	OCH_3	
MDA	CH_3			$-O-CH_2-O-$			
MDMA	CH_3			$-O-CH_2-O-$			CH_3
DOM	CH_3		OCH_3		CH_3	OCH_3	

MDA 3,4-methylenedioxyamphetamine, *MDMA* 3,4-methylenedioxymethamphetamine, *DOM* 2,5-dimethoxy-4-methylamphetamine

13.2.1 Amphetamine

13.2.1.1 Introduction and History

Amphetamine was originally synthesized for medical purposes. It was first used in the 1920s as a decongestant and also to treat obesity and depression. During World War II and the Korean and Vietnam Wars, soldiers were commonly given amphetamines to stay awake for extended periods of time without food.

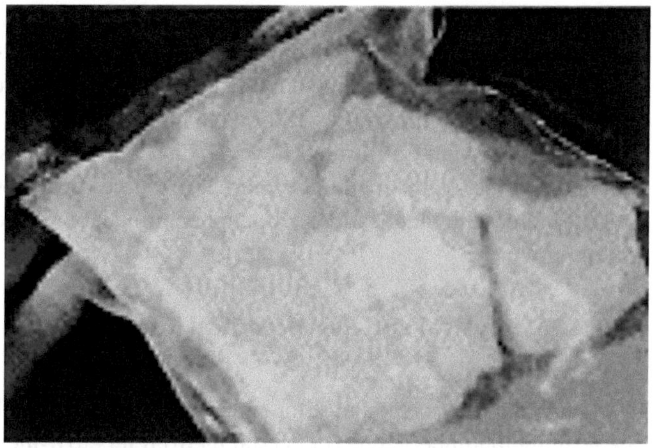

Fig. 13.2 The solid form of amphetamine can be administered using a variety of methods. When snorted, it deteriorates the thin membranes lining the nose.

The α-carbon in amphetamine is chiral and produces two optical isomers (*d* and *l*) with different pharmacological effects. The *d*-isomer (2S) is commonly prescribed under the brand names Dexamphetamine, Duromine, and Ritalin for attention deficit hyperactivity disorder. The *l*-isomer (2R) is often found in inhalers prescribed for asthma and congestion.

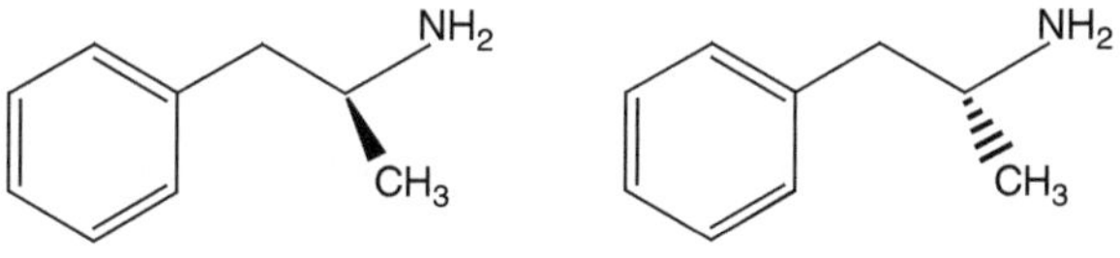

Structure 13.1

13.2.1.2 Physical and Psychological Effects

The short-term physical effects of amphetamine abuse include decreased appetite, increased stamina and physical energy, increased sexual drive/response, involuntary body movements, increased perspiration, hyperactivity, nausea, blotchy or greasy skin, increased or irregular heart rate, increased blood pressure, and headaches. Fatigue is a common side effect that follows the period of effectiveness.

Long-term (or overdose) effects can include tremors, restlessness, changes in sleep patterns, poor skin condition, tachypnea, gastrointestinal narrowing, and immune-system depression. The initial stages of exhilaration are often followed by periods of fatigue and depression. In addition, erectile dysfunction, heart problems, stroke, and liver, kidney, and lung damage can result from prolonged use. Amphetamine can cause a deterioration of the nostril lining when snorted (Fig. 13.2).

Short-term psychological effects include alertness, euphoria, increased concentration, rapid talking, increased confidence, increased social responsiveness, nystagmus, hallucinations, and loss of sleep.

Long-term psychological effects include insomnia, schizophrenia, aggressiveness (not associated with schizophrenia), irritability, confusion, panic, and addiction or dependence, including symptoms of withdrawal. Chronic use can lead to amphetamine psychosis which causes delusions and paranoia. These effects are uncommon when the drug is taken under the supervision of a physician.

Amphetamine is highly addictive and tolerance develops very quickly. Withdrawal is usually an extremely unpleasant experience, but it is not generally life threatening. Typical symptoms include paranoia, depression, difficulty breathing, dysphoria, gastric fluctuations or pain, and lethargy. Unfortunately, a large number of chronic users relapse.

Amphetamine exists in many different forms and is identified using various street names, such as amp, speed, crank, dolls, crystal, black birds, leapers, pixies, uppers, and whites.

13.2.2 Methamphetamine

13.2.2.1 Introduction and History

Methamphetamine was developed by the Japanese chemist Akira Ogata in 1919 using the reduction of ephedrine with red phosphorus and iodine. During World War II, it was used by the Japanese to help soldiers stay alert and to energize factory workers. A massive supply of methamphetamine was stockpiled by the Japanese military after World War II. This supply was later made available to the civilian population and addiction skyrocketed.

In the 1950s and 1960s, methamphetamine was widely prescribed as a medication for depression and obesity, reaching a peak in 1967 of 31 million prescriptions in the United States alone. During the late 1980s, illicit use and manufacturing was primarily centered in California. However, distribution and use have become widespread because the raw materials are easily obtained and the sophistication of Internet resources provides unlimited access to information on synthetic procedures. The greatest increase in clandestine methamphetamine laboratories was observed in the Midwestern states at the start of the new millennium.

13.2.2.2 Physical and Psychological Effects

Methamphetamine is a potent stimulant, even in small doses. Like amphetamine, the α-carbon in methamphetamine is chiral. The term methamphetamine (or crystal meth) refers to *d*-methamphetamine (2S), a powerful central nervous stimulant. The *l*-isomer (2R) is most often found in inhalers to treat nasal congestion and has no central nervous system activity or addictive properties.

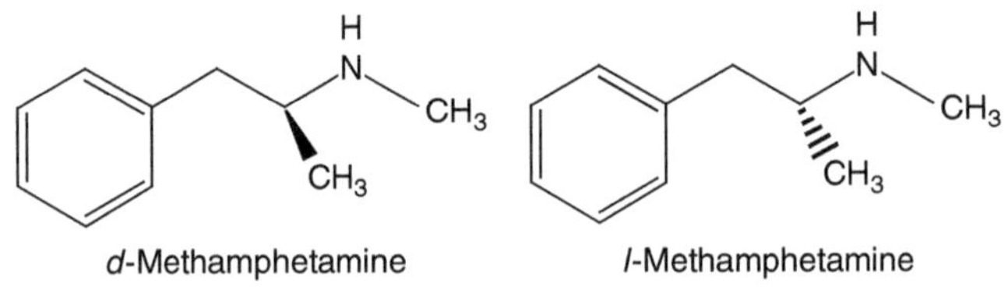

Structure 13.2

Methamphetamine increases alertness and physical activity, while decreasing appetite. Those who either smoke or inject methamphetamine have reported a short, intense sensation termed "a rush." Oral ingestion or snorting produces a much longer-lasting high, which may continue for half a day (Fig. 13.3). Methamphetamine is believed to cause the release of high levels of the neurotransmitter dopamine into areas of the brain that regulate feelings of pleasure. These elevated levels can damage nerve terminals in the brain and have been implicated in the overall toxic effects of the drug.

Short-term effects include increased attention and activity, decreased fatigue and appetite, euphoria, rush, elevated respiratory rates, and hyperthermia. High doses can immediately elevate body temperature to dangerous, sometimes lethal, levels, as well as cause life-threatening convulsions.

Long-term abuse can result in addiction and brain damage, which is manifested as violent behavior, anxiety, confusion, and insomnia. Addicts can also display a number of psychotic features including paranoia, auditory hallucinations, mood disturbances, and delusions accompanied by repetitive motor activity, weight loss, and increased risk of stroke. These symptoms can result in homicidal behavior and thoughts of suicide.

Tolerance does develop with long-term use. In an effort to maintain the desired effects, users increase dosage, increase frequency of use, or change the method of administration. In some cases, abusers go without food and sleep for extended periods, while indulging in a form of chronic use known as a "run." During extreme episodes, addicts inject as much as a gram

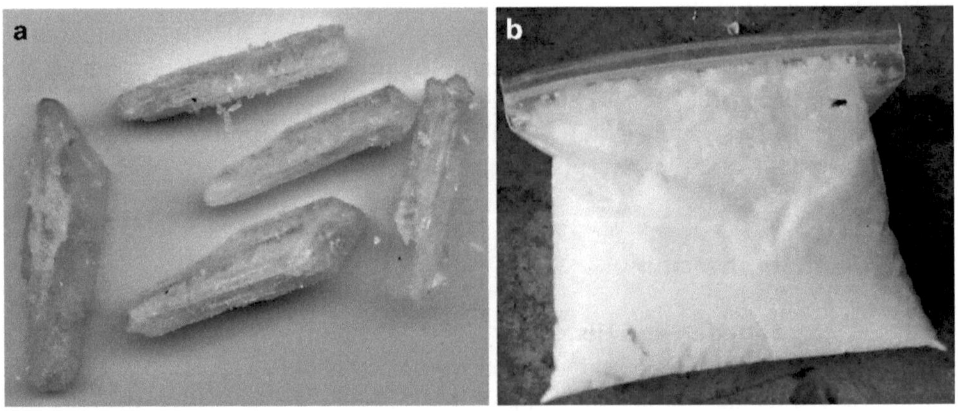

Fig. 13.3 Methamphetamine crystals (**a**) and powder (**b**). This drug is a powerful stimulant that affects the central nervous system.

Fig. 13.4 Phentermine is a rare example of α,α-simultaneous substitution of 1-amino-2-phenylethane. The salt form (*above*) is commonly prescribed to treat obesity in patients with high blood pressure and diabetes.

of the drug every 2–3 h over several days. This continues until either the supply is depleted or the user is too disoriented to continue. The user typically displays elevated physical and psychological symptoms during this time including intense paranoia, visual and auditory hallucinations, and out-of-control rages with extremely violent behavior.

Although there are no physical symptoms of a withdrawal syndrome, depression, anxiety, fatigue, paranoia, aggression, and an intense craving for the drug do occur.

Street names for methamphetamine are speed, crank, meth, crystal, crystal meth, base, L.A. ice, ice, shabu, ox blood, chalk, glass, tina, and white cross.

13.2.3 Phentermine

13.2.3.1 Introduction and History

Phentermine first received approval from the Food and Drug Administration (FDA) in 1959 as an appetite suppressant for the short-term treatment of obesity. The resin form became available in 1959, and the hydrogen-chloride-salt form appeared in the early 1970s (Fig. 13.4).

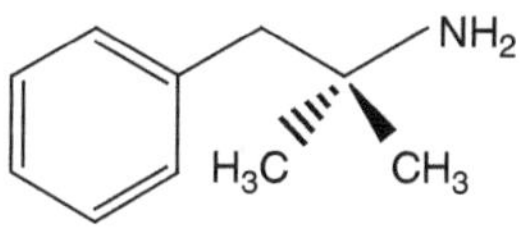

Phentermine

Structure 13.3

Phentermine gained notoriety as the most frequently prescribed drug for appetite suppression, most likely because it is significantly cheaper than the other major FDA-approved diet drugs, Meridia and Xenical.

In the early to mid-1990s, the diet-pill cocktail called phen-fen, a combination of phentermine and fenfluramine (a substituted phenethylamine), was introduced. Another phen cocktail, dexfen-phen, containing dexfenfluramine (a substituted phenethylamine), was developed shortly after dexfenfluramine received FDA approval as an appetite suppressant in 1996.

Fen-phen became an overnight sensation in dietary medicine. In 1996, 6.6 million prescriptions were written in the U.S. alone. Dexfen-phen became equally popular, despite the fact that both were never thoroughly tested for safety.

By the summer of 1997, the Mayo Clinic had reported 24 cases of heart-valve disease related to the use of fen-phen. In July 1997, the FDA issued a Public Health Advisory containing a report of these findings, which were later published in the New England Journal of Medicine.

Further evaluation of patients using either fenfluramine or dexfenfluramine revealed that approximately 30% had some indication of heart-valve abnormalities. This figure was much higher than expected and suggests that fenfluramine and dexfenfluramine are implicated as the cause of primary pulmonary hypertension and valvular heart disease.

The FDA responded promptly in September 1997 by requesting that drug manufactures voluntarily withdraw fenfluramine and dexfenfluramine from the market. At the same time, the FDA highly recommended that patients using either drug discontinue use immediately. Surprisingly, the FDA did not request the withdrawal of phentermine until a year later.

13.2.3.2 Physical and Psychological Effects

Phentermine is well known for its role as an anorectic. It stimulates the adrenal glands located on top of the kidneys to produce the catecholamines epinephrine (aka adrenaline) and norepinephrine (noradrenaline). These chemical messengers trigger the fight-or-flight response and also suppress appetite. Epinephrine produces weight loss through direct action on fat cells, triggering a break down in fats. Phentermine also acts on several regions of the body producing a variety of hormones and neurotransmitters. In high doses, it stimulates various regions of the brain, producing the neurotransmitters dopamine and serotonin. Dopamine is a precursor to epinephrine and norepinephrine that regulates movement, emotional response, and perception of pain and pleasure. Serotonin regulates sleep, mood, attention, appetite, muscle contraction, memory, and learning.

13.2.3.3 Side Effects

Phentermine mimics the actions of the sympathetic nervous system, in particular, the fight-or-flight response. Excessive use can produce unwanted side effects including hypertension (high blood pressure) and tachycardia (increased heart rate), but the incidence and intensity are typically less than those related to amphetamines. It can also cause heart palpitations, loss of sleep, and restlessness. Long-term use can result in both physical and psychological addiction.

13.3 Hydroxyl Derivatives

Substituted phenethylamines containing hydroxyl groups have diverse functions. Epinephrine (adrenaline), norepinephrine (noradrenaline), tramline, and dopamine are examples of naturally occurring variations found in the body. Phenylpropanolamine, ephedrine, and pseudoephedrine are hydroxyl- and methyl-substituted phenethylamines that are frequently found in over-the-counter pharmaceuticals.

13.3.1 Phenylpropanolamine

13.3.1.1 Introduction and History

Phenylpropanolamine is produced by substituting a hydroxyl group at the β-carbon of amphetamine. This simple addition alters the pharmacological effects of the drug. Although phenylpropanolamine retains some of the stimulant and anorectic characteristics of amphetamine, the overall effect is drastically reduced.

The α and β carbons in phenylpropanolamine are chiral, resulting in four possible stereoisomers: the *d*- and *l*-optical isomers (enantiomers) of norephedrine and the *d*- and *l*-optical isomers (enantiomers) of norpseudoephedrine. The prefix "nor" is often used to indicate the replacement of a methyl group on the parent molecule with a hydrogen atom. Therefore, norephedrine would be ephedrine missing the methyl group attached to nitrogen which is replaced with a hydrogen atom. Differentiating these isomers can be somewhat simplified by referring to Table 13.1. Notice the only substituted phenethylamine containing a methyl (CH_3) group on the α carbon *and* hydroxyl group (OH) on the β carbon is cathine. When groups are substituted on both carbons, they can have two different orientations with respect to one another because of the planar nature of the molecule. They can both be either on the same side of the molecule or on opposite sides. The methyl and hydroxyl groups are on the same side in the two enantiomers of norephedrine. The orientation at the α and β carbons producing the *l*- and *d*-optical isomers is either 1R, 2S or 1S, 2R. The 1R, 2S isomer is the one often referred to as phenylpropanolamine. The groups are on opposite sides in two enantiomers of norpseudoephedrine, and the orientation at the α and β carbons is either 1R, 2R or 1S, 2S. Cathine is *d*-norpseudoephedrine (1S, 2S) and is the isomer of forensic interest. It is a stimulant isolated from the *Catha edulis* (khat) plant.

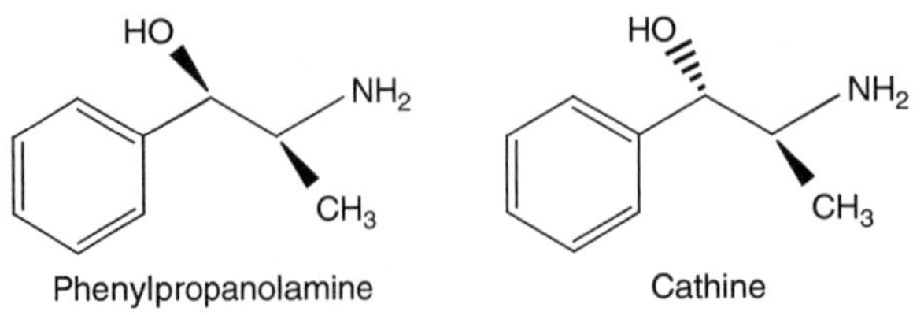

Structure 13.4

Phenylpropanolamine is a common precursor used in clandestine drug manufacturing. It is easily reduced to amphetamine under the same conditions used to reduce ephedrine and pseudoephedrine to methamphetamine. Also, a relatively simple oxidation reaction converts phenylpropanolamine into cathinone, another psychoactive stimulant.

13.3.1.2 Physical and Psychological Effects

Phenylpropanolamine is used to treat nasal congestion associated with the common cold, allergies, hay fever, and other respiratory conditions (e.g., rhinitis and sinusitis). It has also been used as a diet aid for weight loss. Side effects of use include dizziness, headache, loss of appetite, nausea, dry mouth, and restlessness.

13.3.2 Ephedrine/Pseudoephedrine

13.3.2.1 Introduction and History

Ephedrine and pseudoephedrine contain a hydroxyl group substituted at the β carbon of methamphetamine. Again, the configuration at the α and β carbons is important and is used to differentiate these two stereoisomers.

The methyl and hydroxyl groups are on the same side in the two enantiomers of ephedrine, and the orientation at the α and β carbons is either 1R 2S (*l*-isomer) or 1S, 2R (*d*-isomer). The term ephedrine often refers to *l*-ephedrine, the isomer commonly found in over-the-counter medications and is the one of forensic interest. The groups are on opposite sides in two enantiomers of pseudoephedrine, and the configuration is either 1R, 2R (*l*-isomer) or 1S, 2S (*d*-isomer). Although both isomers target the central nervous system, *d*-pseudoephedrine is significantly more active.

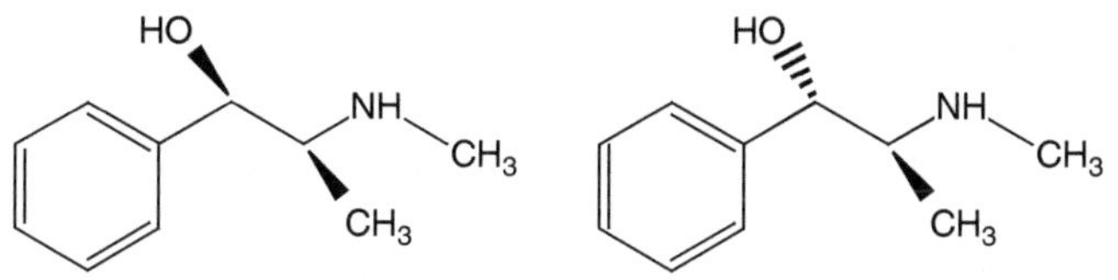

Ephedrine Pseudoephedrine

Structure 13.5

Ephedrine is an alkaloid found in the stem of plants in the genus *Ephedra*. It is the primary active component in many dietary supplements taken for either weight loss or energy enhancement. Recently, these supplements have become the target of intense scrutiny because a number of cardiovascular and central-nervous-system disorders have been associated with their use. Also, the use of ephedrine in the illicit production of methamphetamine has resulted in restrictions on its distribution and use.

Ephedrine and other alkaloids (i.e., norephedrine, pseudoephedrine, and norpseudoephedrine) are naturally produced through decarboxylation of the amino acids phenylalanine and tyrosine. Norpseudoephedrine is also found in khat (*Catha edulis*), a plant native to the Arabian Peninsula. In Eurasian *Ephedra* plants, the two most prevalent alkaloids are ephedrine and pseudoephedrine (Fig. 13.5).

The vast majority of pseudoephedrine submitted to forensic laboratories for analysis is produced synthetically. It is derived from the fermentation of dextrose in the presence of benzaldehyde. In this process, genetically designed strains of yeast are added to large vats containing water, dextrose, and the enzyme pyruvate decarboxylase. Benzaldehyde is added producing *l*-phenylacetylcarbinol (L-PAC), which is subsequently converted into pseudoephedrine through reductive amination.

13.3.2.2 Physical and Psychological Effects

Ephedrine is a stimulant that acts on the central nervous system. It is commonly used to treat respiratory conditions (bronchodilator), nasal congestion (decongestant), low blood pressure (orthostatic hypotension), and myasthenia gravis. It also has general applications in the treatment of certain sleep disorders (narcolepsy), menstrual-cycle abnormalities (dysmenorrhea), and urinary-control problems (incontinence or enuresis).

A majority of the adverse side effects associated with ephedrine use are cardiovascular in nature and include hypertension, palpitations, arrhythmia, myocardial infarction, cardiac arrest, stroke, transient ischemic attack, and seizures. Less serious effects are related to the central nervous system and include tremors, anxiety, nervousness, hyperactivity, and insomnia.

Fig. 13.5 Examples of ephedra plants. A wide range of ephedrine alkaloid derivatives are naturally produced in the stems of indigenous species of ephedra plants.

Pseudoephedrine is commonly used to treat nasal and sinus congestion caused by either the common cold or allergies. Common side effects include central-nervous-system stimulation, nervousness, excitability, dizziness, and insomnia. Tachycardia and/or palpitations are infrequent, but do occur. In rare instances, pseudoephedrine has been associated with hallucinations, arrhythmias, hypertension, seizures, and ischemic colitis.

13.3.3 Ephedra Plant: Introduction and History

The use of ephedra plants in clandestine manufacturing is becoming more frequent. This is mostly likely the result of strict policies regulating the distribution of ephedrine and pseudoephedrine. The most commonly encountered plant, known as *ma huang* (*Ephedra sinica*), is a member of the *Ephedraceae* family. It has been used in China for more than 4,000 years to treat symptoms of asthma and upper respiratory infections. Varieties are also found in Europe, India, Australia, and Afghanistan. American ephedra is native to the Southwest and is commonly used to treat headaches, fevers, colds, and hay fever. Early settlers used the plant to make tea called "Mormon tea" or "Squaw tea." Today, compounds derived from this herb are found in many over-the-counter cold and allergy medications.

Ephedra suppresses the appetite and increases metabolism through thyroid-gland stimulation. Recently, ma huang has been the subject of scientific research for obesity because of its thermogenic fat-burning effects. Ephedra can cause peripheral vasoconstriction, elevation of blood pressure, and cardiac stimulation. As a result, it is often combined with other tonic herbs to help counteract these effects.

13.4 Ketone Derivatives

Hydroxyl groups (–OH) are easily oxidized to ketones (C=O) under relatively mild conditions. A good working definition of oxidation is any process that results in the production of additional bonds to oxygen. Compounds containing hydroxyl groups have one carbon–oxygen bond (C–OH) for each OH present. Oxidation produces a carbonyl group containing a carbon–oxygen double bond (C=O) at each OH position, thus increasing the number of bonds to oxygen. A subtle point of note: the term oxidation is often loosely applied to entire molecules when, in fact, only specific positions actually undergo oxidation, that is, the carbons containing OH groups. Phenethylamine derivatives containing hydroxyl groups can be converted into ketones, creating a new class of phenethylamines. Cathinone, a primary amine, and methcathinone, a secondary amine, are examples. It should be noted that the definition of oxidation presented above is broad and somewhat incomplete, but it is more than adequate for applications in forensic analysis.

13.4.1 Cathinone

Cathinone is a naturally occurring stimulant found in khat (*Catha edulis*). It is a schedule I controlled substance that is illegal under any circumstances in the U.S. Cathinone is structurally related to cathine, a less potent stimulant. The two differ only in the substitution at the β-position; cathinone contains a carbonyl group, while cathine contains an OH group.

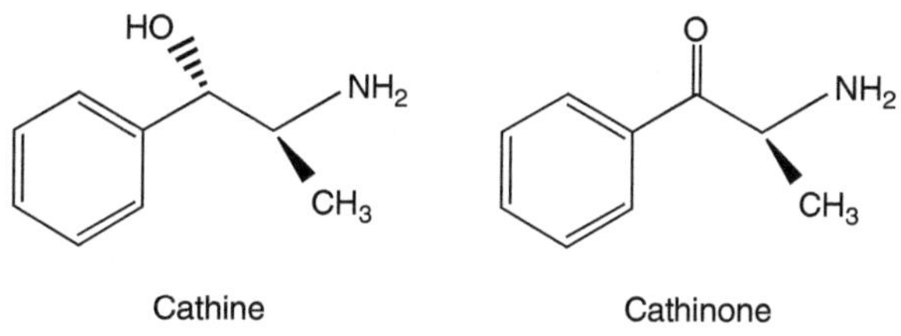

Structure 13.6

Cathinone is commonly isolated from leaves of khat or synthetically produced from propiophenone. It is not typically produced from the oxidation of cathine, as one might expect because the oxidation process produces low yields contaminated with a variety of toxic by-products. Cathinone is most active when isolated from fresh khat. The reduction (reverse of oxidation) of cathinone to cathine occurs over time, decreasing the concentration of cathinone in aging (dried) leaves. Also, isolated cathinone tends to lose potency after 48 h. Although the exact mechanism is unknown, it is highly likely that some form of reduction is involved.

The activity of cathinone is similar to amphetamines. The primary effects are caused by stimulating the release of the neurotransmitter dopamine. Short- and long-term adverse side effects are consistent with those caused by amphetamine abuse.

13.4.2 Methcathinone

Methcathinone is a naturally occurring alkaloid stimulant also found in khat. It is structurally similar to cathinone and is often synthetically produced from the oxidation of ephedrine. It was extensively used in the 1930s and 1940s as an antidepressant but has since been used only for recreational purposes. It is classified as a schedule I controlled substance.

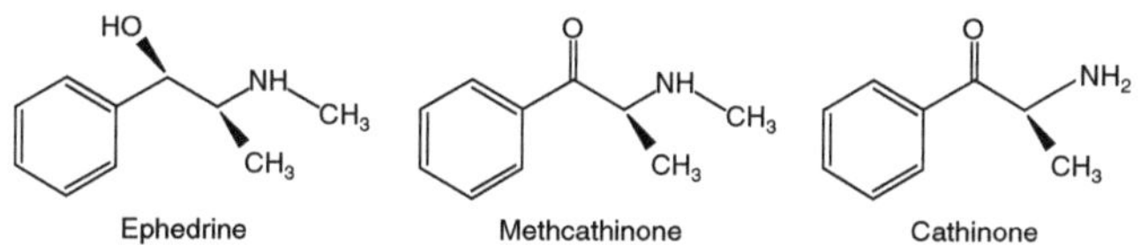

Structure 13.7

The effects of methcathinone are similar to those produced from other alkaloid stimulates (i.e., amphetamine, methamphetamine, cathinone, and cathine) and include euphoria, rapid breathing, increased heart rate and alertness, and dilated pupils. Methcathinone stimulates the release of high levels of norepinephrine and dopamine in the brain. The elevated concentrations amplify the activity of these neurotransmitters, producing adverse side effects, such as anxiety, convulsions, hallucinations, insomnia, paranoia, irregular heart rate, restlessness, tremors, headaches, and convulsions.

13.4.3 Khat

Today, khat (*Catha edulis*) is recognized as a well-known source of naturally occurring alkaloid stimulants (Fig. 13.6; Table 13.2).

Khat consumption produces mild euphoria and excitement. Individuals become very talkative under the influence and may appear unrealistic and emotionally unstable. Khat can induce manic behavior and hyperactivity. It is an effective anorectic that can also produce constipation. Symptoms of withdrawal include lethargy, mild depression, nightmares, and slight tremors.

The use of khat is accepted in Somali, Ethiopian, and Yemeni cultures. In these countries, khat is not regulated or controlled and is openly sold at public markets. Unfortunately, emigrants from these countries often continue its use in the United States, and khat is often readily available in ethnic restaurants, bars, grocery stores, and smoke shops. Also, Muslims commonly use khat during the religious month of Ramadan.

Fig. 13.6 Khat contains a variety of naturally occurring alkaloid stimulants. The native form (*right*) is harvested from its natural environment and packaged (*left*) for transport. Forensic analysis is often performed immediately on leaf material because some stimulants are unstable and degrade or lose potency over time.

Table 13.2 Classification of khat

Kingdom	Plantae
Division	Magnoliophyta
Class	Magnoliopsida
Order	Celastrales
Family	Celastraceae
Genus	*Catha*
Species	*C. edulis*

Unlike marijuana, khat is not directly regulated in the U.S., but some of its alkaloids are. Section 812 of Title 21 United States Code (21 USC sec. 812) defines cathine as a schedule IV controlled substance and cathinone as a schedule I. This type of detailed regulation requires the isolation and specific identification of each substance, unlike the generic botanical identification used in cannabis examinations.

Other alkaloids in khat are celastrin, edulin, chroline, ratine, tannis, and ascorbic acid. Common street names of khat are Cat, Abyssinian Tea, African Tea, African Salad, Catha, Chat, Mirra, Qat, Quat, Tohai, and Tschat.

13.5 Methylenedioxy Derivatives

The methylenedioxy (–O–CH_2–O–) substituted phenethylamines that are most often encountered in forensic analysis bridge the R3 and R4 ring positions of amphetamine and methamphetamine. The selective involvement of these positions in the formation of the dioxy-5-membered ring is a stability-driven process. Rings are generally considered fixed structures. Bond angles are usually well defined, and rotation around the bonds in the ring is restricted. In most cases, normal bond angles are altered to accommodate the angles required in the ring geometry. The extent of deviation from the ideal bond angles affects the overall energy of the resulting ring. Torsional strain is a measure of the ring-imposed resistance to the twisting action of bonds in the ring as the distance between adjacent ring-substituted positions is maximized. Ring-closure reactions that produce five- and six-membered rings are favored because these structures possess geometries that minimize the torsional and angle strain introduced by closure.

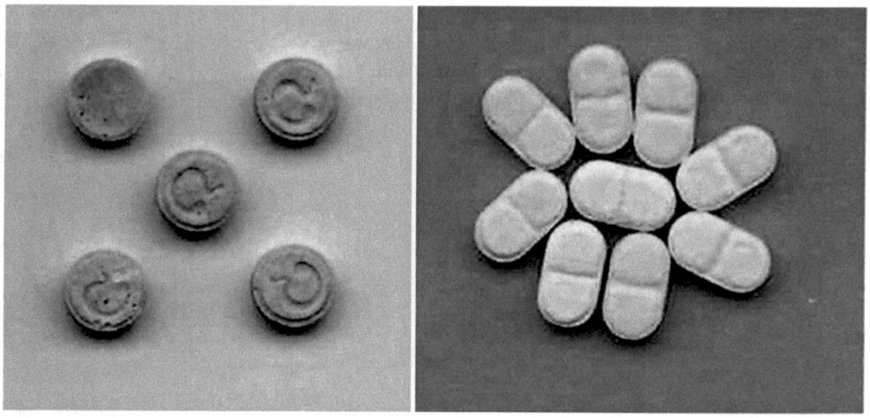

Fig. 13.7 Tablets of 3,4-methylenedioxyamphetamine (MDA). Although capable of producing effects ranging from a potent stimulant to a powerful hallucinogen, this psychedelic stimulant is best known for its soothing effects associated with a general state of well-being.

13.5.1 3,4-Methylenedioxyamphetamine

3,4-Methylenedioxyamphetamine (MDA) is produced by bridging the R3 and R4 positions on amphetamine with the methylenedioxy group. The characteristic two-fused ring system is easily recognized and distinguishes the structure of MDA from most substituted phenethylamines.

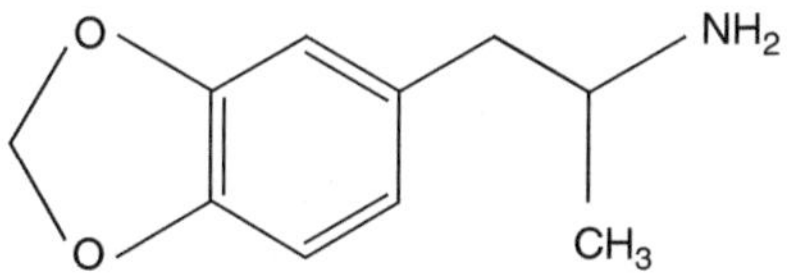

3,4-Methylenedioxyamphetamine (MDA)

Structure 13.8

MDA is a psychedelic stimulant and an empathogen–entactogen. It is classified as a schedule I controlled substance, with no approved medical use. It is strictly a recreational drug that, while technically classified a stimulant, is best known for its calming affects.

MDA is administered orally in the form of either a capsule or pill (Fig. 13.7). The effects become apparent in 20–60 min and may persist for 10–12 h. Users perceive the onset of effects quite differently; some experience initial nausea, while others feel a warm glow spreading throughout their body. Most experience of a sense of physical and mental well-being that intensifies gradually and steadily.

MDA commonly induces a state of profound relaxation and patience, with no anxiety, aggression, or thoughts of violence. Habitual users of tobacco feel no need to smoke; nail biters leave their fingers alone; compulsive talkers become quiet; and compulsive eaters do not think about food. Moreover, this condition feels normal and natural because MDA does not significantly affect either the senses or perception. An intense aura of peace and calm is experienced with rare instances of hallucinations, illusions, or paranoia.

13.5.2 3,4-Methylenedioxymethamphetamine

3,4-Methylenedioxymethamphetamine (MDMA) is the methylated "cousin" of MDA. It is produced by bridging the R3 and R4 positions on methamphetamine with the methylenedioxy group.

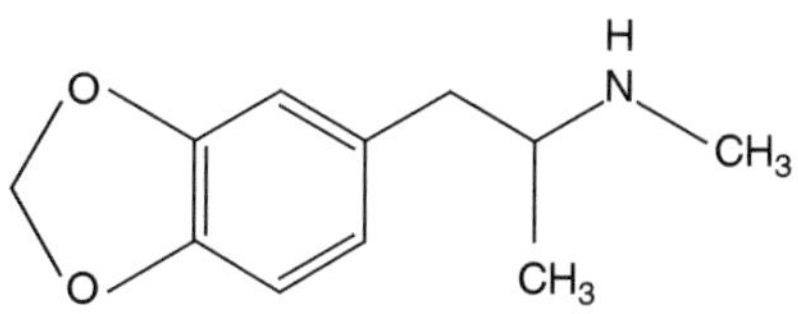

3,4-Methylenedioxymethamphetamine (MDMA)

Structure 13.9

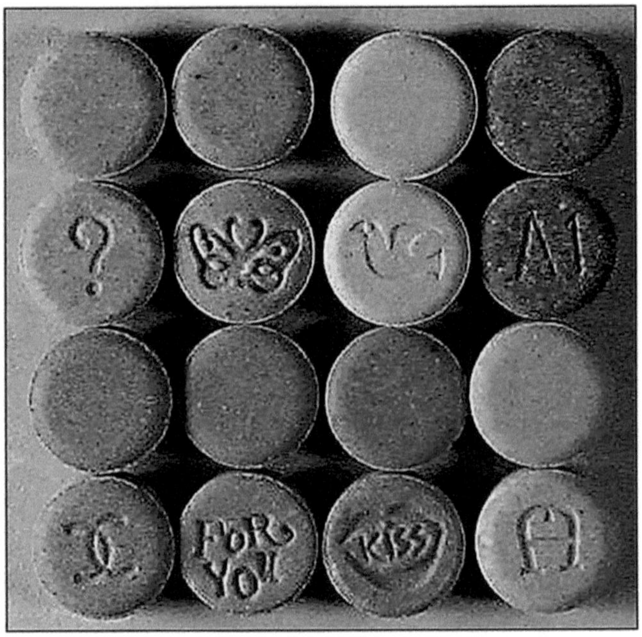

Fig. 13.8 Tablets of 3,4-methylenedioxymethamphetamine (MDMA). This psychedelic stimulant is commonly known by its street name "ecstasy." MDMA, cocaine, heroin, and marijuana are the four most widely used illegal drugs in the US.

Like MDA, it is a synthetic psychedelic stimulant classified as a schedule I controlled substance. MDMA is commonly known by its street name "ecstasy" (or "XTC"), and its use is more widespread than the less famous MDA. It gained notoriety in the mid- to late 1980s as a "party drug," where it was often found in nightclubs, "raves" and "techno-parties." MDMA produces as general state of relaxation and well-being that is comparable to MDA, but it is generally considered more dangerous and unpredictable. It is used solely for recreational purposes and appears to be most popular with adolescents and young adults. Chronic use can cause permanent brain damage to serotonin nerve terminals and less severe effects including confusion, depression, paranoia, blurred vision, and an inability to control body temperature leading to liver, kidney, and cardiovascular failure.

MDMA is usually self-ingested in tablet form, but other common methods of administration include snorting and injection (Fig. 13.8). Chronic users take MDMA by "staking" (taking multiple tablets at once) or "piggy backing" (taking multiple tablets over a short time). The practice of "candy flipping" (taking MDMA with LSD or lysergic acid diethylamide) and "hippie flipping" (taking MDMA with hallucinogenic mushrooms) is also common. The effects of MDMA become apparent in 30–45 min and persist for 4–6 h.

13.6 Methoxy Derivatives

Methoxy ($-OCH_3$) substitution on the aromatic ring of phenethylamine produces a compound with a variety of hallucinogenic effects. The number of methoxy groups substituted and their location determine the pharmacological effects of the resulting compound.

13.6.1 Mescaline

Mescaline (3,4,5-trimethoxyphenethylamine) is one of the more prominent methoxy-substituted phenethylamines.

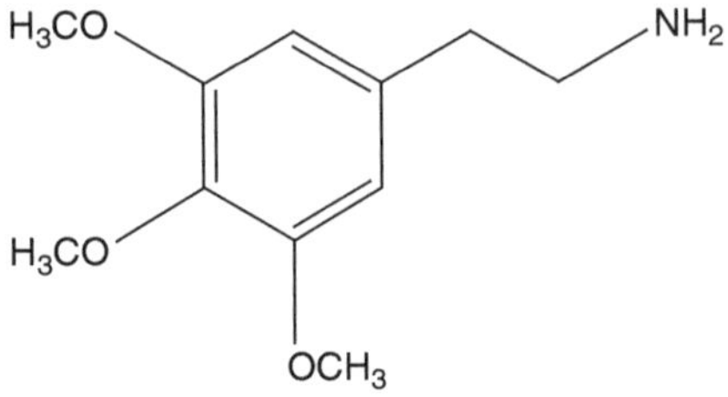

Mescaline

Structure 13.10

Fig. 13.9 Confiscated peyote plants. The specimens above are about the size of a golf ball. All species of peyote have extremely slow growth rates, often requiring 30 years to reach flowering age. Slow growth and overharvesting have placed peyote in danger of extinction.

Table 13.3 Classification of peyote

Kingdom	Plantae
Division	Magnoliophyta
Class	Magnoliopsida
Order	Caryophyllales
Family	Cactaceae
Genus	*Lophophora*
Species	*williamsii* and *diffusa*

It is naturally occurring in peyote and can be extracted as organic mescaline. It can also be synthetically produced as mescaline sulfate, a white solid observed as needle-point crystals. Both peyote and mescaline are listed in the Controlled Substances Act as schedule I hallucinogens.

Peyote is a small, spineless cactus whose principal active ingredient is the hallucinogen mescaline (Fig. 13.9). The top of the cactus is called the *crown* and consists of disk-shaped buttons that are cut from the roots and dried. The buttons are generally chewed or soaked in water to produce an intoxicating liquid or tea. The tea is bitter and, in most cases, the user experiences some degree of nausea before the onset of the psychedelic effects. A hallucinogenic dose is typically 300–800 mg of mescaline and lasts about 12 h. Peyote produces rich, visual hallucinations that are considered important in many Native-American cultures.

Mescaline is an atypical hallucinogen in that true hallucinations do not actually exist. The user is aware that illusions and fantasies are of their own creation. Past experiences are not only recalled, but may be relived, in the mind. While perceptions of sight, sound, taste, touch, and smell exist in amusing chaos, intellect and judgment are clear and functional. Visual perception is altered to such a degree that the most common things seem miraculous. Ordinary objects shift and change in color and shape, creating a unique reality. The mood, expectation, and personality of the user affect the overall experience, as do the physical surroundings. There is a fine line between a mystical, dream-like sensory experience and a schizophrenic episode characterized by extreme mood changes and unprovoked outbursts of emotion, anxiety, confusion, and depression. Nausea, anorexia, and insomnia are usually present, regardless of which "trip" is experienced. Other undesirable side effects include pupil dilation, dizziness, vomiting, tachycardia, diarrhea, headaches, and a fear of not returning to normal consciousness. In rare instances, lung and heart disease and diseases of the blood vessels have been implicated in mescaline use. Table 13.3 contains information on the classification of peyote.

13.7 Analytical Methods

13.7.1 Visual Inspection

Figure 13.10

13.7.2 Chemical Screening

The comprehensive flowchart in Fig. 13.11 illustrates the results of chemical color tests (Chap. 7) commonly performed on controlled substances in the phenethylamine class.

13.7.3 Microcrystal Tests

Although the use of microcrystalline testing as either a screening or confirmatory method is still debated, there are standardized methods in microscopy that are accepted as confirmatory techniques. The American Society of Testing Materials (ASTM) standard E1969-01 (Standard Guide for Microcrystal Testing in the Forensic Analysis of Methamphetamine and Amphetamine) describes the use of either gold chloride ($HAuCl_4$) or platinum chloride (H_2PtC1_6) in the identification of amphetamine and methamphetamine. The ASTM also describes procedures using these reagents to identify optical isomers (Fig. 13.12).

Fig. 13.10 Amphetamine (*top*) and methamphetamine (*bottom*). Visual inspection is an important part of the identification process. Although not conclusive in most cases, the results can determine which confirmatory method will be used for definitive identification.

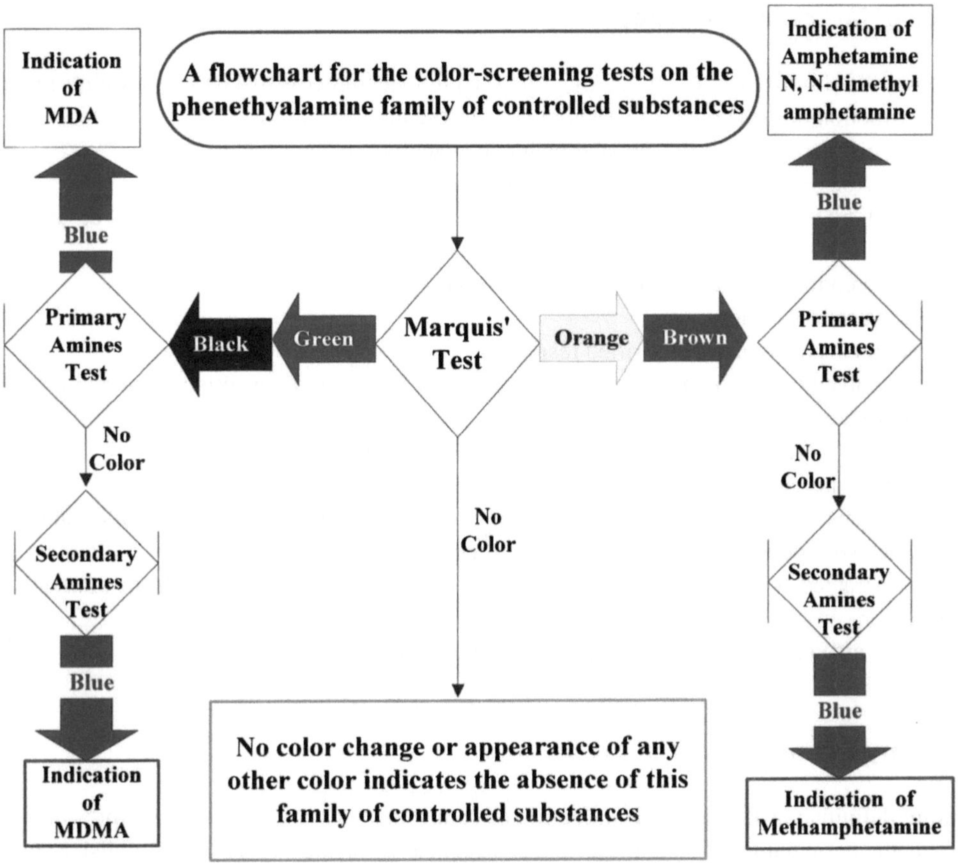

Fig. 13.11 Flowchart for the color screening of phenethylamines. These presumptive tests will determine subsequent steps in the identification process.

Fig. 13.12 Differentiation of the optical isomers of methamphetamine using gold chloride. The *d*-isomer (*right*) is a powerful central nervous system (CNS) stimulant. The *l*-isomer (*left*) has no CNS activity or addictive properties.

13.7.4 Extraction Techniques

In general, most amines are bases due to the presence of nitrogen. Phenethylamines will react with acids to form water-soluble salts. Therefore, acid/base extractions as well as dry extractions are effective in isolating these compounds for analysis. Dry extractions using methanol can be used to isolate cathinone and methcathinone from khat. Mescaline extraction from peyote is somewhat more complicated.

13.7.5 Extraction of Mescaline from Peyote

The extraction of a controlled substance from plant material is a multistep process that usually requires more than one extraction technique. The extraction of mescaline from peyote is a representative example.

- Up to 2–5 g of dried peyote is ground to a powder using an agate mortar.
- A dry extraction is performed on the powdered plant material.
  - Approximately 10 ml of alcohol (usually methanol) is added to the powder, agitated, and allowed to stand undisturbed.
  - The solution is filtered to separate it from the plant material.
- The solution is evaporated to dryness.
- An acid/base extraction is performed on the dried residue:
  - The residue is dissolved in approximately 10 ml of 0.1 M HCl. The acid converts the mescaline (base) into a water-soluble ionic salt (aqueous solution).
  - The aqueous solution is washed three times with ethyl ether to extract impurities.
  - The organic layers (ether) are discarded. Test the layers!
  - The pH of the aqueous solution is increased to >8, using a base, such as concentrated ammonium hydroxide or 2.0 M sodium hydroxide. The added base converts the mescaline salt back to mescaline.
  - The solution is extracted three times with ethyl ether. The mescaline is extracted from the aqueous layer into the organic layer (ether).
  - The organic layers are combined and evaporated to isolate mescaline.

13.7.6 Confirmatory Examination

Legal statutes require definitive proof in the identification of controlled substances. Visual and presumptive screening tests are used to determine the basic nature of a controlled substance, and the results are used to establish a suitable confirmatory method. Gas-chromatography mass spectrometry (GCMS) and Fourier transform infrared spectroscopy (FTIR) are reliable methods that are universally accepted and recognized by most jurisdictions.

13.7.6.1 Gas-Chromatography Mass Spectrometry

The structure of controlled substances in the phenethylamine class can be quite similar. Therefore, great care must be exercised during analytical testing and subsequent data interpretation.

Baseline chromatographic resolution is required to distinguish phenethylamines that differ only in the position of a methyl group. Carrier-gas flow rates and vaporization temperatures must be carefully regulated to produce conditions that will separate individual phenethylamines.

The mass spectra of amphetamine and MDA are almost identical when normalized to their respective base peaks because they have the same base peak and the balance of ions in the fragmentation pattern typically exhibit <10% relative abundance. The differences are easily observed when the spectra are normalized to the second highest peak (Fig. 13.13).

13.7.6.2 Fourier Transform Infrared Spectroscopy

The use of gas chromatography to separate phenethylamines can be complicated by the low molecular weight of some derivatives. FTIR analysis requires highly purified samples, and ineffective separation can create issues with its use. Low-molecular-weight phenethylamines are typically volatile oils at room temperature and often produce amorphous IR spectra containing broad, poorly resolved absorption bands.

The use of the acid/base-extraction infrared modifications (Chap. 9) eliminates this problem by converting the phenethylamine into its hydrogen-chloride salt. The salts produce spectra with sharp, well-resolved absorption bands.

GCMS is a powerful and extremely versatile analytical method, but it cannot differentiate diastereomers. Infrared spectroscopy is the preferred method used to achieve this (Fig. 13.14).

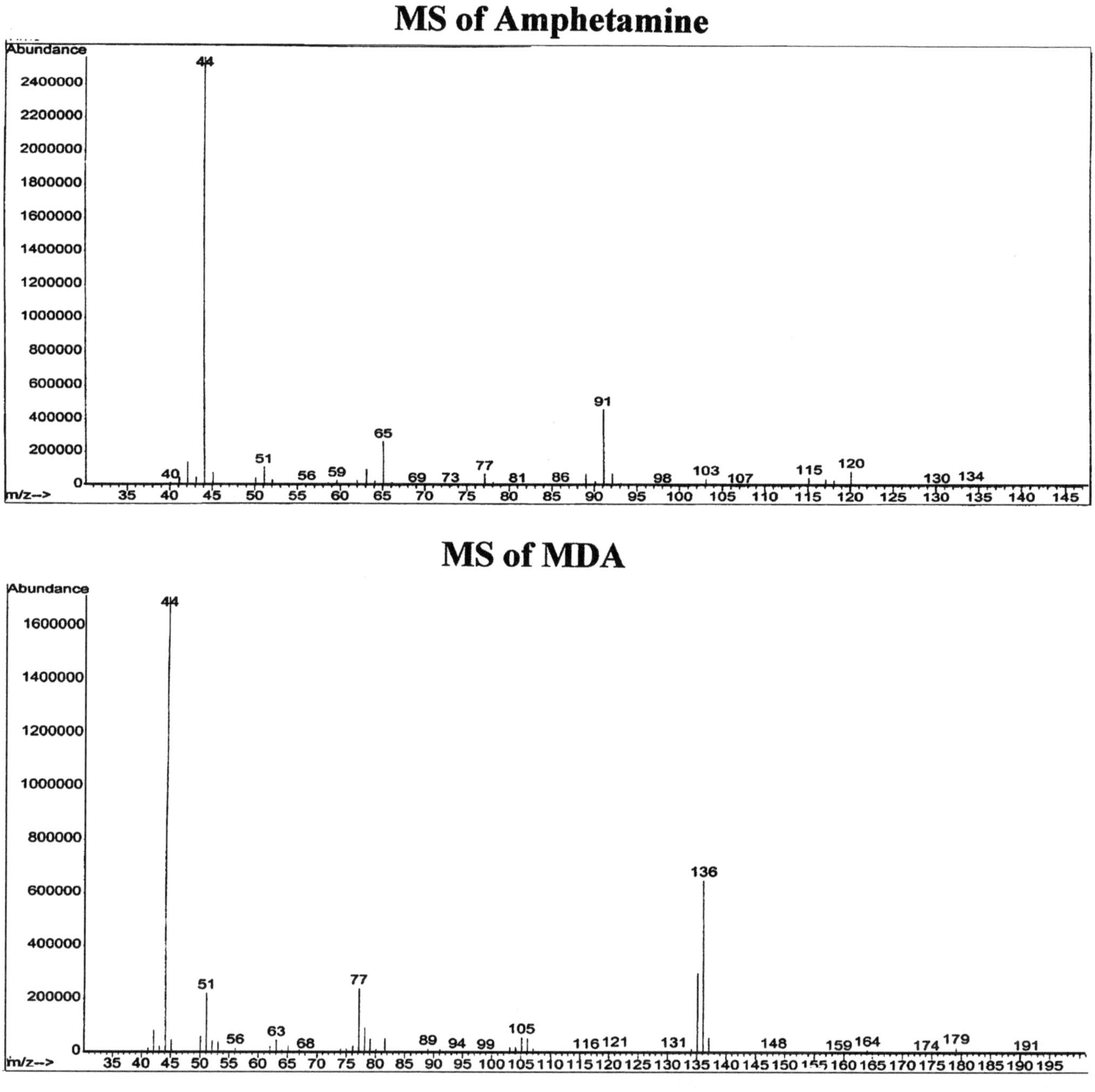

Fig. 13.13 Mass-spectra of 3,4-methylenedioxyamphetamine (MDA) and amphetamine. These two controlled substances are difficult to differentiate unless the spectra are enhanced. The above spectra were normalized to the second most abundant peak in each ion-fragment pattern. The differences are clearly illustrated at m/z peaks 65 and 91 in amphetamine and 77 and 136 in MDA.

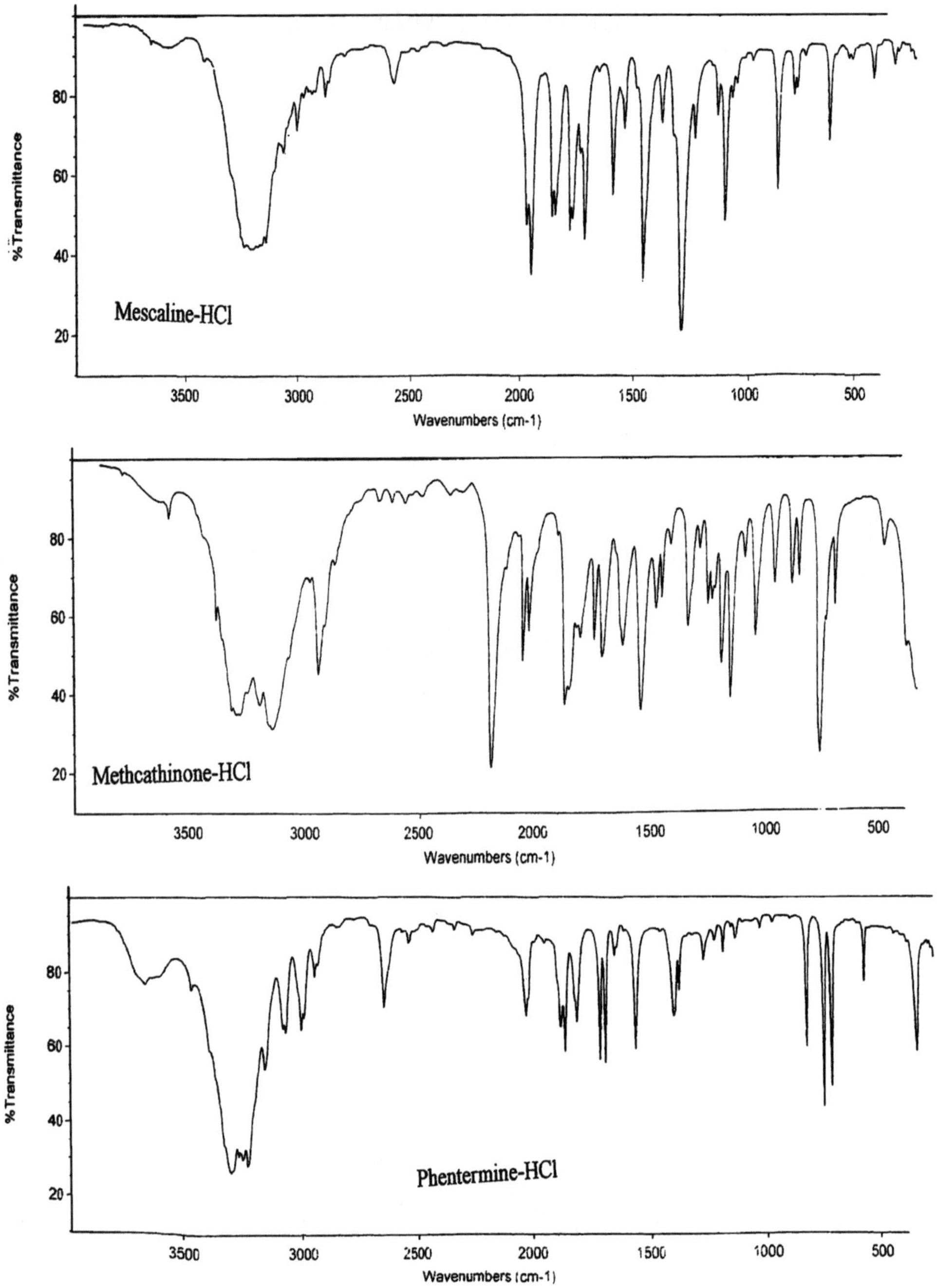

Fig. 13.14 The IR spectra of mescaline, methcathinone, and phentermine salts. The primary regions used to differentiate these controlled substances are 3,000–3,500 cm^{-1} and 1,500–2,200 cm^{-1}.

13.8 Questions

1. Draw the structure of 1-amino-2-phenylethane.
2. Draw the following molecules. Refer to Table 13.1.
   (a) Dopamine
   (b) Amphetamine
   (c) Methamphetamine
   (d) Ephedrine
   (e) Pseudoephedrine
   (f) 3,4-Methylenedioxymethamphetamine (MDMA)
3. Briefly describe the term optical isomer to members of the jury.
4. Draw the two optical isomers of methamphetamine and briefly discuss each.
5. List three long- and short-term effects of methamphetamine abuse.
6. Draw the structure of phentermine.
7. Name two infamous diet pills containing phentermine.
8. List the classification of each of the following controlled substance, according to the Controlled Substance Act.
   (a) MDA
   (b) Cathinone
   (c) Mescaline
   (d) MDMA
9. What specific neurotransmitter is commonly released by phenethylamines?
10. Name two hydroxyl substituted phenethylamines and draw their structure.
11. Name two controlled substances isolated from khat.
12. Briefly describe to the jury why analysis of khat must be performed immediately.
13. What is the common street name of 3,4-methylenedioxy-methamphetamine?
14. Please describe to members of the jury, the difference between MDA and MDMA?
15. Classify mescaline.
16. What are the common forms of mescaline?
17. Discuss why mescaline is considered an atypical hallucinogen?
18. Why is peyote in danger of extinction?
19. Name the color-screening tests used to indicate methamphetamine and include test results.
20. What analytical method is particularly effective in differentiating diastereomers?
21. Explain how IR modifications to acid/base extractions are used in IR-spectroscopy.

Suggested Reading

Allen, A. C.; Kiser, W. O. Methamphetamine from Ephedrine: Chloroephedrine and Aziridine. *J. Forensic Sci.* **1987**, 32, 953–962.

Auerbach, L. Microcrystalline Identification Test of Some Amphetamines and hydrochlorothiazide: A Collaborative Study. *JAOAC*, **1978**, 61, 1435–1440.

Chamakura, R. P. The MDMA Tablets. *Microgram* **1994**, 27, 316–329.

Christian, D. R. Jr. Analysis of Controlled Substances. In *Forensic Science: An Introduction to Scientific and Investigative Techniques*, 3rd ed.; James, S. H.; Nordby, J. J., Eds.; CRC Press: Boca Raton, FL, 2009.

Dal Cason, T. A. A Re-examination of the Mono-Methoxy Positional Ring Isomers of Amphetamine, Methamphetamine, and Phenul-2-Propanone. *Forensic Sci. Int.* **2001**, 119, 168–194.

Dal Cason, T. A. The Characterization of Some 3,4-Methylenedioxyphenylisopropylamine (MDA) Analogs. *J. Forensic Sci.* **1989**, 34, 928–961.

Davenport, T. W.; Allen, A. C.; Cantrell, T. S. Synthetic Reductions in Clandestine Amphetamine and Methamphetamine Laboratories. A Review. *Forensic Sci. Int.* **1989**, 42, 183–199.

Emboden, W. A. *Narcotic Plants*. McMillan Publishing: New York, 1972.

Erickson, Carl. Dopamine-A Sample Neurotransmitter. http://www.utexas.edu/research/asrec/dopamine.html (accessed August 2009).

Esnu, K. *Plant Anatomy*. John Wiley & Sons: New York, 1965.

Glaston, T. R.; Rasmussen, G. T. Identification of 3,4-methylenedioxyamphetamine. *Microgram*. **1972**, 5, 60–63.

Glennon, R. A. Stimulus Properties of Hallucinogenic Phenylalkylamines and Related Designer Drugs: Formulation of Structure-Activity Relationships. In *Classical Hallicinogens: An introductory Overview*; Lin, G. C.; Glennon, R. A., Eds.; National Institute on Drug Abuse: Rockville, MD, 1994, pp. 4–32.

Gouzoulis-Mayfrank, E.; Hemley, L. Are "Entactogens" a Distinct Psychoactive Class? The Contribution of Human Experimental Studies to the Classification of MDMA and Other Chemically Related Methylenedioxyamphetamine Derivatives. *Heffter Rev. Psychedelic Res.* 1988, 1, 46–51.

Henderson, G. et al. Designer Drugs: The California Experience. In *Clandestinely Produced Drugs, Analogs, and Precursors.* U. S. Department of Justice, Drug Enforcement Administration: Washington, D. C., 1989.

Kovar, K. A. Chemistry and Pharmacology of Hallucinogens, Entactogens and Stimulants. *Pharmacopsychiatry.* 1998. 31 (suppl.), 69–72.

Kram, T. C. Approaches to Drug Sample Differentiation. *J. Forensic Sci.* **1979**, 24, 596–599.

Mescaline. http://www.drugtext.org/library/books/recreationaldrugs/mescaline.htm (accessed August 2009).

Methcathione. http://www.a1b2c3.com/drugs/meth1.htm (accessed August 2009).

National Institute on Drug Abuse. Methamphetamine. http://www.drugabuse.gov/drugpages/methamphetamine.html (accessed August 2009).

Office of National Drug Control Policy (ONDCP). Street Terms. http://www.whitehousedrugpolicy.gov/streetterms/default.asp (accessed August 2009).

Shulgin, A. T. Basic Pharmacology and Effects. In Hallucinogens: A Forensic Handbook; Laing, R.; Siegel, J. A., Eds.; Academic Press: New York, 2003, pp 67–137.

The Partnership for a Drug Free America. Methcatinone. http://www.drugfree.org/Portal/Drug_Guide/Methcathinone (accessed August 2009).

United Nations. *Recommended Methods for The Identification and Analysis of Amphetamine, Methamphetamine and Their Ring-Substituted Analogs in Seized Materials. Manual for Use by National Drug Testing Laboratories*; ST/NAR/34; United Nations Publication: New York, 2006.

U. S. Department of Justice Drug Enforcement Administration. Drugs and Chemicals of Concern: 3,4-Methylenedioxymethamphetamine. http://www.deadiversion.usdoj.gov/drugs_concern/mdma/mdma.htm (accessed August 2009).

Wielbo, D.; Tebbett, I. R. The Use of microcrystal Tests in Conjuction with Fourier Transform Infrared Spectroscopy for the Rapid Identification of Street Drugs. *J. Forensic Sci.* 1992, 37, 1134–1148.

14 Tertiary Amines

14.1 Introduction

Amines are organic derivatives of ammonia (NH_3) produced by replacing the hydrogen atoms with various alkyl groups. Tertiary amines contain nitrogen bound to three alkyl groups. There are no limitations on the alkyl substitutions: they may be identical or different, straight chains, branched chains, or cyclic, including aromatic rings. Amines represent some of the most important and vital compounds in biochemistry. However, they are also some of the most destructive and addictive substances known. In this chapter, we examine cocaine, a few common natural and synthetic opiate alkaloids, and phencyclidine (PCP). All are members of the tertiary amine class of controlled substances. In most cases, the highly complex structures are often differentiated by only subtle differences in substituted groups. The tertiary amine group, however, is often readily observed.

14.2 Natural Tertiary Amines

14.2.1 Cocaine

Cocaine is a potent stimulant that is highly addictive. In pure form, it is either a powdered hydrogen-chloride salt or a rock-crystal free base (crack).

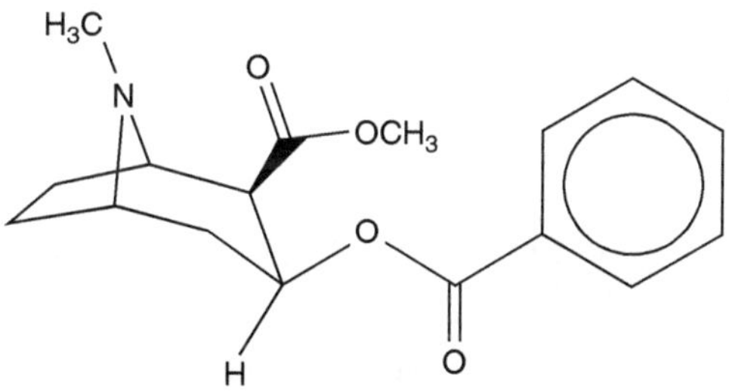

Cocaine

Structure 14.1

J.I. Khan et al., *Basic Principles of Forensic Chemistry*, DOI 10.1007/978-1-59745-437-7_14,
© Springer Science+Business Media, LLC 2012

The acid form (salt) is typically either snorted or dissolved in water and injected, while the base form (free base) is usually vaporized and inhaled. The term "crack cocaine" originates from the characteristic "crackling" sound that is heard when the free-base form is heated. Cocaine is a Schedule II controlled substance that has a high potential for addiction. It also has legitimate applications in medicine that are primarily related to its anesthetic or numbing effects. Anyone who has ever had either a tooth pulled or a root canal has experienced the effects of the "caine" family of drugs. Novocaine (procaine) and xylocaine (lidocaine HCl) are commonly used as local anesthetics in dentistry. Cocaine, however, is a highly active member of the "caine" family, producing more pronounced and severe effects.

The effects of cocaine appear almost immediately after ingestion and can last from minutes to hours, depending on the method of administration. The user commonly feels euphoric and energetic while experiencing elevated states of mental awareness, especially to the sensations of sight, sound, and touch. These effects are often accompanied by a reduced need for sleep and food. The duration and intensity of the "high" is largely dependent on the rate of absorption. Cocaine that is rapidly absorbed into the bloodstream is delivered to the brain sooner producing more intense, but shorter periods of effectiveness. Snorting may last 15–30 min, while smoking is comparable to injection and may last only 5–10 min. For this reason, users often require repeated doses to sustain their high.

Cocaine is known by the street names blow, nose candy, snowball, tornado, and wicky sticks.

The short- and long-term effects of cocaine use are usually related to the inhibition of dopamine reabsorption by nerve cells in the brain. Elevated levels of dopamine produce a variety of physical and psychological effects including increased body temperature, heart rate and blood pressure, constriction of blood vessels, dilated pupils, tremors, muscle twitches, vertigo, paranoia, and anxiety. Excessive use can cause irritability, erratic or violent behavior, periods of paranoid psychosis including auditory hallucinations, seizures resulting in respiratory failure, and death from cardiac arrest. Some effects are directly related to the method of administration. For example, excessive snorting can cause nosebleeds, hoarseness, chronic runny nose, problems with swallowing, and damage to nasal membranes resulting in a loss of the sense of smell.

Cocaine is arguably the oldest known drug of abuse, with a documented history dating back thousands of years. It is extracted from the coca plant, a member of the order Geraniales and the family *Erythroxylaceae*. There are four genera with an estimated 200 species in *Erythroxylaceae*.

Evidence shows that coca-leaf chewing was prevalent before the rise of the Incan Empire around 3000 BC. South-American Incas, who lived in present-day Peru, Bolivia, Ecuador, and parts of Chile and Colombia, regarded coca leaves as a gift from their gods. Ingesting coca enabled the Incas to work at high altitudes because it suppressed appetite and increased energy.

Early botanists believed that all coca plants belonged to a single species. However, subsequent research revealed two species of domesticated coca: *Erythroxylum coca Lam.* and *Erythroxylum novogranatense*. The two species have two varieties each: *Erythroxylum coca Lam.* var. *coca* and var. *Lpuda Plowman* and *Erythroxylum novogranatense* var. *novogranatense*, and var. *truxillense Plowman*.

Coca is grown in South America, Africa, Ceylon, Taiwan, and Indonesia. But it is most commonly associated with its center of origin, the South-American Montana zone of the eastern Andes below 2,000 m (Fig. 14.1).

Fig. 14.1 Cocaine is a naturally occurring stimulant found in the leaves of the coca plant (*left*). It is extracted as white crystalline powder, which is compressed into a variety of shapes for transport and distribution (*right*).

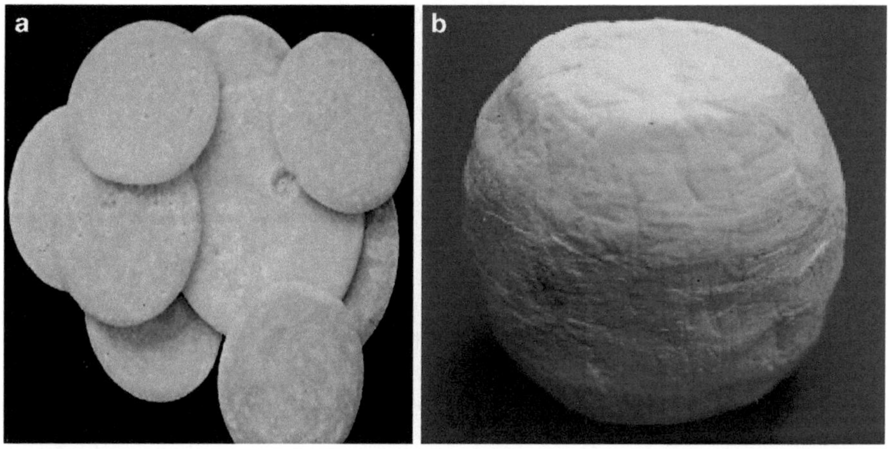

Fig. 14.2 Chewing coca counters the symptoms of "mountain sickness" and oxygen deprivation and is often part of the daily intake of people living in high-altitude regions. The daily dose is approximately 200 mg (**a**) sliced from bulk reserves (**b**).

Huánuco, or Bolivian coca (*E. coca* var. *coca*), is an ancestral variety that grows in the moist tropical forests of the eastern Andes of Peru and Bolivia. This variety has the distinction of being the only type that grows in the wild.

Amazon Coca (*E. coca* var. *Ipuda*) is cultivated in the lowland areas of the Amazon. It is most likely a lowland derivative of Bolivian coca but is not indigenous to the region.

Trujillo coca (*E. novogranatense* var. *truxillense*) is a hardy, drought-resistant variety found in the river valleys of coastal Peru and surrounding regions. It displays characteristics of both *E. coca* var. *coca* and *E. novogranatense* var. *novogranatense* and may represent an evolutionary stage between the two species.

Colombian coca (*E. novogranatense* var. *novogranatense*) is cultivated in the moist, dry areas of the Colombian mountains. It is also drought resistant and is not native to this region. Evidence suggests that this variety may be the most evolved species.

Natural selection favors the biological synthesis of cocaine because it acts as a natural plant pesticide. In humans, cocaine is a potent inhibitor of neuronal reabsorption of the "reward neurotransmitter" dopamine. Insects possess a neurotransmitter called octopamine, which is analogous to dopamine: for example, it is released in honeybees in anticipation of a flower offering high levels of nectar. Cocaine is a more powerful inhibitor of octopamine reabsorption and insects that feed on coca can "overdose" on their own octopamine.

Coca typically thrives in warm, moist, frost-free valleys between 1,500 and 6,000 m above sea level. The plant can grow to heights exceeding 8 ft. with leaves rich in vitamins, proteins, calcium, iron, and fiber. The cocaine content in leaves range from 0.1 to 0.9% and (like its users) tends to get higher with altitude. Diurnal fluctuations of cocaine within the leaf occur in 24-h cycles (Fig. 14.2).

14.2.2 Opiates

Opiates refer to any of the naturally occurring narcotic alkaloids found in the latex of opium poppy plants as well as synthetic derivatives of the natural alkaloids. Latex is a milky, sap-like emulsion containing proteins, alkaloids, sugars, oils, and resins that are commonly used in non-vulcanized rubber manufacturing (i.e., latex gloves). Morphine and codeine are naturally occurring alkaloids of forensic interest, as is the synthetic alkaloid heroin.

14.2.2.1 Morphine

Morphine is the most prevalent alkaloid in latex, often representing 10–15% of the total mass. It crystallizes as morphine sulfate, a salt that is usually either a white powder or white silky crystals.

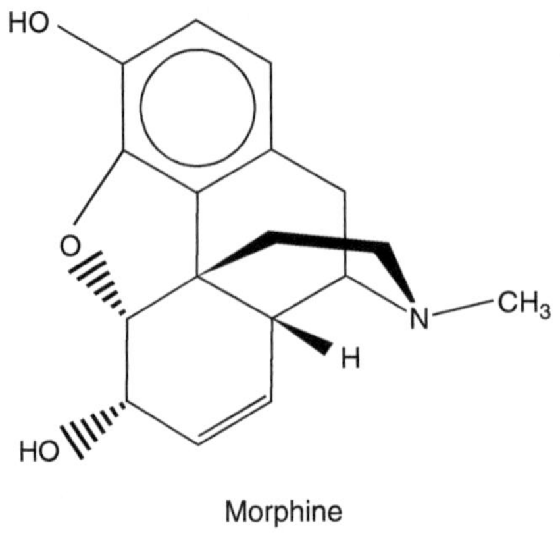

Structure 14.2

It is a potent narcotic that directly affects the central nervous system and gastrointestinal tract. It is primarily used as an analgesic and sedative to relieve pain. Side effects include euphoria, alteration in mood, impairment of mental and physical performance, reduced fear and anxiety, decreased hunger, renal failure, and depression of respiratory functions.

Morphine can be administered orally, rectally, subcutaneously, intravenously, or epidurally. Abusers usually inhale morphine in a process termed "chasing the dragon." The solid is heated forming a thick, smoking liquid. The user "chases" the liquid while inhaling the vapors through a tube. Morphine is highly addictive and can cause physical and psychological dependence as well as tolerance and withdrawal. Withdrawal symptoms typically develop just before the next scheduled dose and include watery eyes, insomnia, runny nose, sweating, severe headaches, irritability, severe abdominal pain, and depression. In extreme cases, heart attack, stroke, and blood clotting can occur.

14.2.2.2 Codeine

Codeine is structurally similar to morphine and close examination of the two reveals a single difference in the aromatic substituted group: the H in the hydroxyl group on morphine is replaced with a methyl group (CH_3).

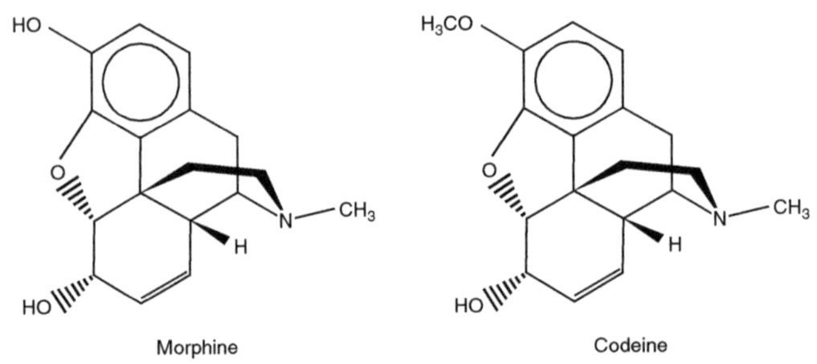

Structure 14.3

Although codeine can be isolated as a natural product from latex, it is usually synthetically produced using O-methylation (adding a methyl group to oxygen) of morphine. Codeine is a Schedule II controlled substance.

Codeine is typically used as an analgesic, cough suppressant, or antidiarrheal agent. It is often prescribed with acetaminophen (Tylenol-3), aspirin (acetylsalicylic acid), or ibuprofen because of the synergistic pain-relieving effect of these combinations. Although codeine is significantly less active than morphine, administration must be closely supervised because roughly 10% will be metabolized to morphine in the liver. Like all alkaloids, excessive use of codeine can result in physical dependence and psychological addiction. The short- and long-term side effects, including symptoms of withdrawal, are very similar to those associated with morphine but are less severe in nature.

14.2.2.3 Heroin

Heroin (diacetylmorphine) is synthetically produced from morphine using a chemical process called acetylation in which the hydroxyl groups on morphine are replaced with acetyl groups.

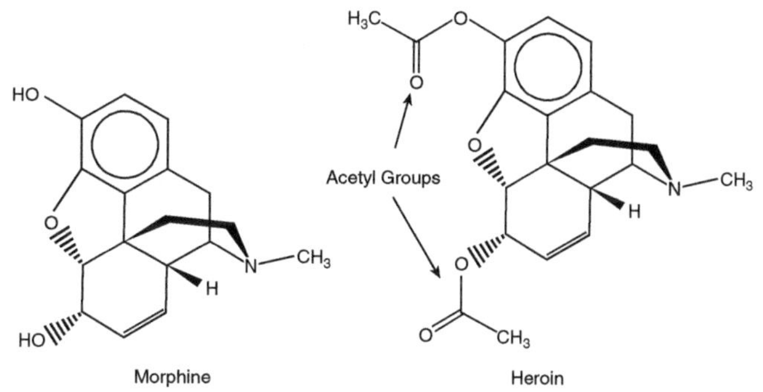

Structure 14.4

Heroin is typically a white crystalline solid in both the hydrogen-chloride salt and free-base forms. It can be administered orally, by inhalation ("chasing the dragon"), by snorting (rarely), or by intravenous injection. Injecting or "shooting up," a solution of heroin is the preferred method of illicit use. The solubility properties of the hydrogen-chloride salt and free-base forms are very different. The salt readily dissolves in water when heated, but the free base does not unless a weak acid (commonly citric acid or lemon juice) is added. Abusers will initially inject easily accessible veins in the arm, but these will eventually collapse from damage caused by the acid. As a result, users will seek alternative veins for injection. Heroin is a Schedule I controlled substance.

The onset of the effects of heroin depends on the method of administration, and the acetyl groups play a significant role. Inhalation and injection are felt almost immediately because the presence of the acetyl groups increase the lipid solubility of heroin, resulting in direct crossing of the blood–brain barrier. When taken orally, the acetyl groups are effectively removed through metabolism (deacetylation), and heroin is essentially delivered to target systems as morphine.

Heroin's potential for addiction is much greater than most narcotics. The short- and long-term side effects of heroin use, including symptoms of withdrawal, are very similar to those associated with morphine. However, they are often complicated by the fact that heroin is often used in combination with other drugs. Cocaine can be fatal when used in combination with heroin. Speedballs (when injected) and moonrocks (when smoked) are popular combinations, but mixing stimulants and depressants can have unpredictable effects that are often fatal.

14.2.2.4 Poppy

Poppy (*Papaver somniferum L.*), or opium poppy, is a plant native to southeastern Europe and western Asia. The species is cultivated extensively in many countries, including Iran, Turkey, Holland, Poland, Romania, Czechoslovakia, Yugoslavia, India, Canada, and many regions on the Asian and South-American continents. The erect plant can have white, pink, red, or purple flowers and can grow to heights exceeding 4 ft. Seeds range in color from white to purple-gray and grow best at temperatures between 7 and 23°C in soil with a pH of 4.5–8.3, receiving an annual rainfall of 1–5 ft. (Fig. 14.3).

Fig. 14.3 An artist's rendering of a flowering poppy (*left*) illustrating the flowers (*center*) and seed capsules (*right*). The plant grows best in rich, moist soil and tends to be frost sensitive.

Fig. 14.4 Poppy seed capsules prior to alkaloid extraction (*left*). Sliced seed capsules oozing latex (*middle*). Harvesting of seed capsules from a poppy plant field (*right*). The total yield of alkaloids is dependent on light, temperature, plant species, and the time of harvest.

Latex is obtained from immature seed capsules 1–3 weeks after flowering. Incisions are made in the walls of the green seed pods, and the milky, sap-like liquid is collected and dried. Opium and the alkaloids morphine, codeine, noscapine, papaverine, and thebaine are subsequently isolated from the dried material. The poppy seeds themselves, along with fixed oil components in the seeds, are not narcotic because they develop after the capsule has lost its opium-producing capabilities (Fig. 14.4).

Poppy seeds are generally considered safe for human consumption and are commonly used in baked goods and pastries because of their "nutty" taste and odor. Poppy oil is widely used as cooking oil and also has applications in the manufacturing of paints, varnishes, and soaps. Surprisingly, ornamental poppies are viewed in many cultures as an important part of prestigious flower gardens.

The use of poppy for medicinal purposes is well documented and historically, dry opium is considered an astringent, antispasmodic, aphrodisiac, diaphoretic, expectorant, hypnotic, narcotic, and sedative. It is commonly used to treat toothaches and coughs, but opium, and its derivatives, are highly addictive and can have toxicological effects.

Morphine is clearly the most dangerous component found in latex. Aside from its toxic effects, it is also the preferred precursor used in the production of several harmful drugs. Currently, there is significant interest in the genetic development of a poppy rich in thebaine and low in morphine. This variant could be used to produce codeine and other legal pharmaceutical drugs with less morphine available for the manufacturing of illegal drugs.

Corn poppy (*Papaver rhoeus L.*) is an annual plant native to Europe and Asia containing extracts used in medicine and beverages. *Papaver orientale L.*, formerly *Papaver bracteatum Lindl.*, is a morphine-free alkaloid source commonly used for medicinal purposes. Mexican poppy (*Argemone mexicana L.*) has toxicological properties with no significant medicinal uses.

All opiates except heroin and opium are prescribed and regulated in tablet, caplet, or liquid form (for injection). Therefore, they are easily distinguished from heroin, which is often submitted to crime laboratories as a white powder.

14.3 Synthetic Tertiary Amines

14.3.1 Phenylcyclohexylpiperidine

PCP is an acronym for phenylcyclohexylpiperidine, which is often shortened to phencyclidine, a name seemingly unrelated to the acronym. It is a Schedule II controlled substance that produces hallucinogenic and neurotoxic effects.

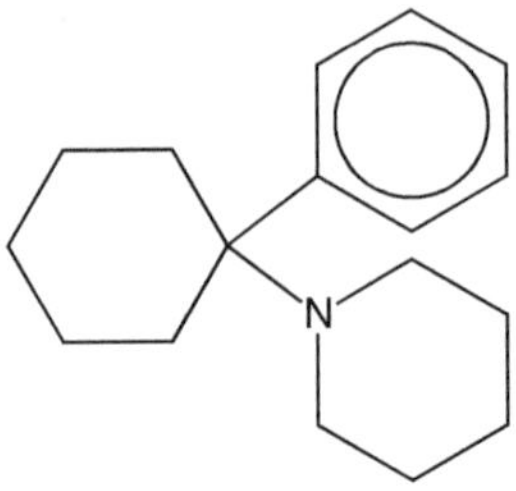

PCP

Structure 14.5

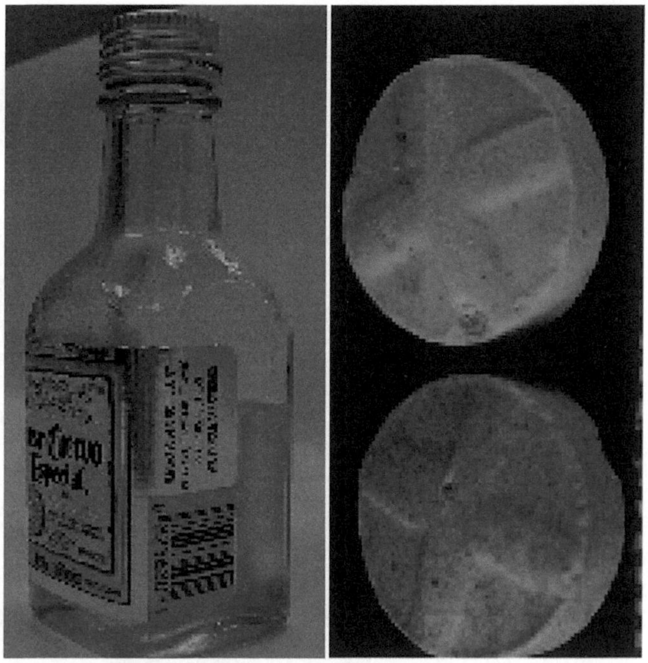

Fig. 14.5 PCP can exist in either pure base form (*left*) or pure hydrogen-chloride-salt form (*right*). The salt is typically precipitated from the base form using hydrogen chloride gas or isopropyl alcohol saturated with HCl.

PCP blocks neurological signals from the brain to the conscious mind, resulting in dream-like states of sensory deprivation and dissociation. It is a general depressant that produces euphoria, sedation, respiratory depression, analgesia, and amnesia. PCP is a yellow oil in its pure base form and appears as a white-tan crystals or powder in pure hydrogen-chloride-salt form (PCP hydrochloride) (Fig. 14.5). The primary effects typically last 8–10 h, but complete removal from the body can exceed 7 days.

PCP is an extremely powerful hallucinogen that alters the mind in an unpredictable fashion. It causes some users to become detached from reality, while others become animated. Intoxicated users exhibit behavioral disturbances driven by their delusions and hallucinations. This is often manifested in violent acts of self-injury (including suicide), attacks on others, or destruction of property. Toxic doses can produce a psychotic state with periodic schizophrenic episodes that can reoccur for months.

PCP encountered in the United States is exclusively produced in clandestine laboratories. It usually contains a number of contaminates, which taint the color of the product from tan to brown and produce a consistency ranging from powder to a gummy mass. PCP can be snorted, smoked, injected, or administered orally in either tablet or capsule form. It is typically obtained in either powder or liquid form, which is applied to a cigarette, joint, oregano, parsley, or mint and smoked.

Since its emergence as a drug of abuse in the late 1960s, PCP is universally recognized as the most dangerous of all synthetic hallucinogens. Street names for phencyclidine include PCP, angel dust, super grass, killer weed, embalming fluid, rocket fuel, wack, and ozone.

14.4 Analytical Methods

14.4.1 Visual Inspections

Visual inspections are an integral part of the identification process. Generally, they are used to determine subsequent testing methods, but it is very important to recognize their limitations. Tertiary amine controlled substances are submitted to forensic laboratories in many forms and shapes. Therefore, visual inspection should only be used to formulate a preliminary opinion on the basic nature of the substance. They are not a confirmatory method and should never be used to render a final judgment (Fig. 14.6).

14.4.2 Chemical Screening of Tertiary Amines

A comprehensive flowchart for the color screening of tertiary amines is illustrated in Fig. 14.7. The chart contains testing methods and results for commonly encountered controlled substances in this category.

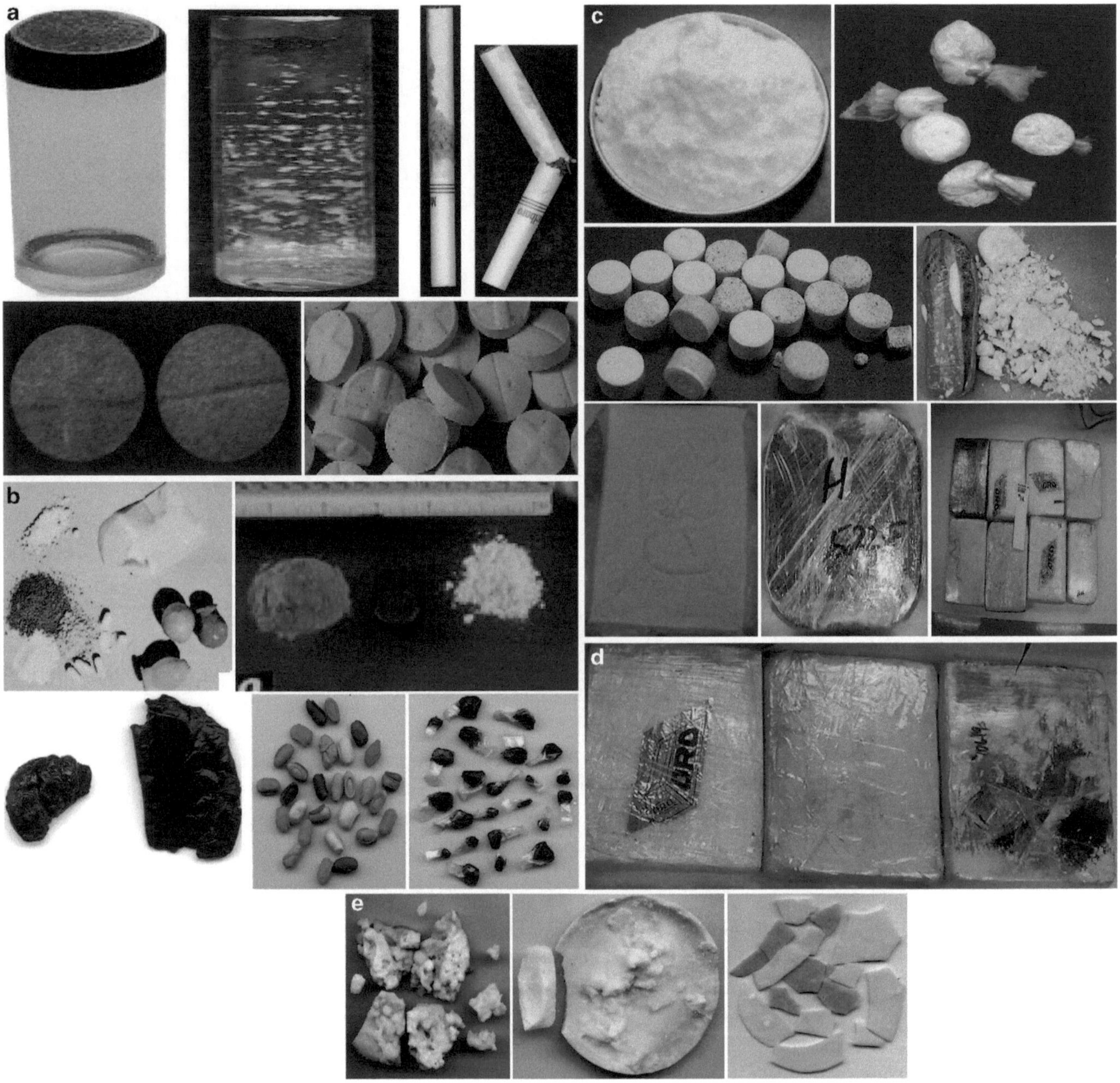

Fig. 14.6 Tertiary amines are often submitted to forensic laboratories in a variety of forms. Set (**a**) First Row: A jar (left) and plastic bag (center) containing the pure base form of PCP, a yellow oil. Cigarettes dipped in PCP (right), street name "a boat." Set (**a**) Second Row: Tablets of PCP. Set (**b**) First Row: Powdered heroin packaged in balloons and paper envelopes for distribution (left). Piles of powdered heroin (right), a penny is shown between the two piles for reference. Note the color difference between the two piles indicating the presence of contaminants. Set (**b**) Second Row: Black-tar heroin (left), black-tar heroin packaged for distribution (center and right). Set (**c**) First Row: Powdered cocaine (left), powdered cocaine packaged for distribution (right). Set (**c**) Second Row: Tablets (left) and bricks (right) of cocaine HCl. Set (**c**) Third Row: Bricks of cocaine HCl packaged for distribution. Set (**d**): Bricks of cocaine HCl packaged for distribution. Set (**e**): Forms of free-base cocaine, street name "crack".

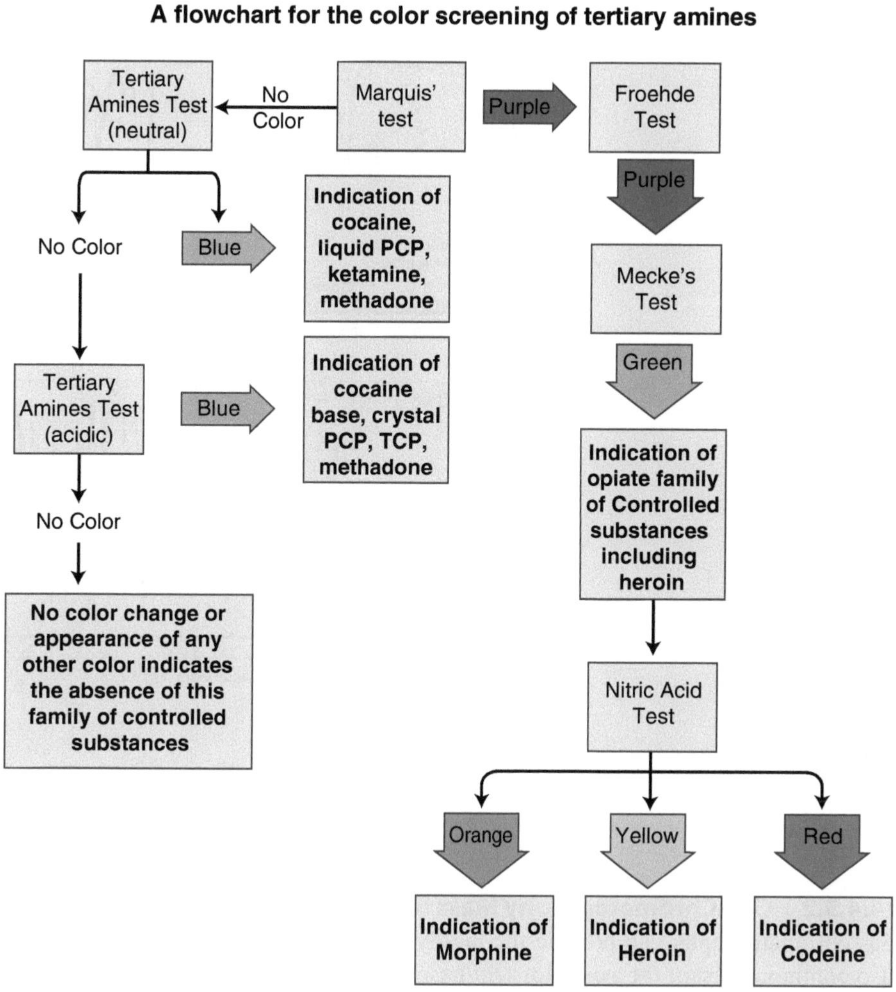

Fig. 14.7 A flowchart used for the color screening of frequently encountered controlled substances in the tertiary amine class. Less common drugs in this class could also be screened using these methods in different combinations.

14.4.3 Confirmatory Examination

14.4.3.1 Fourier Transform Infrared Spectroscopy

Fourier transform infrared spectroscopy (FTIR) is often used to identify controlled substances in the tertiary amine class. Surprisingly, the characteristic absorption bands result from the aromatic ring and not necessarily the amine functional group.

Typical spectra exhibit prominent bands in the low-frequency region at 600–900 cm^{-1} resulting from out-of-plane bending of aromatic C–H bonds. Bands at 1,585–1,600 cm^{-1} result from aromatic skeletal vibrations involving conjugated (alternating single and double bonds) carbon-to-carbon stretching within the ring. These are usually doublets but can show considerable variation, depending on the nature of ring substitutions. Supporting bands are observed as mixed multiplets in the 1,400–1,500 cm^{-1} region and result from similar skeletal vibrations. Distinctive aromatic C–H stretching bands occur at 3,000–3,100 cm^{-1}. Amine salts typically exhibit a broad band in the 2,400–2,700 cm^{-1} region, which is especially useful in differentiating cocaine HCl (salt) from its free-base form. Characteristic strong C–N stretching bands in the spectra of aromatic tertiary amines are observed at 1,250–1,300 cm^{-1}. These may be difficult to isolate from various adjacent bands but are usually distinguishable. Cocaine HCl and free-base cocaine are shown below as representative examples.

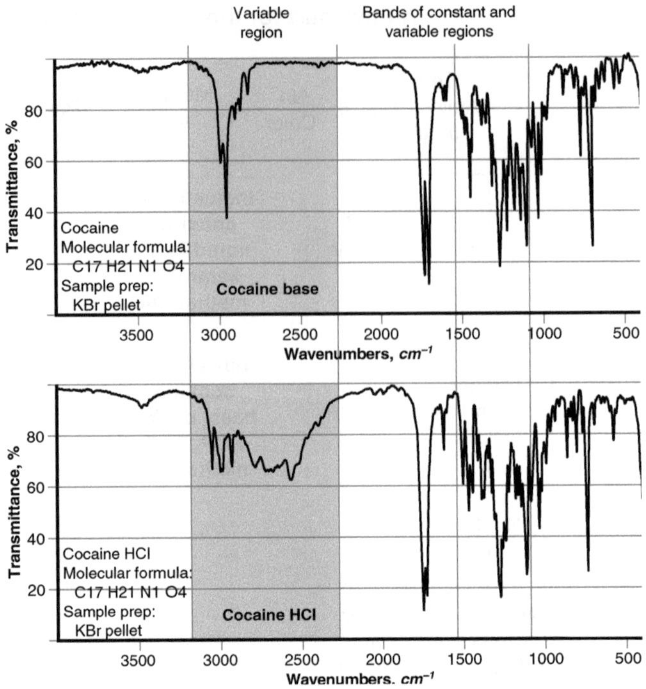

Structure 14.6

14.4.3.2 Gas-Chromatography Mass Spectrometry

Gas chromatography can provide the highly purified samples required for FTIR; however, less sensitive techniques are also commonly used, for example, acid/base and base extraction. GCMS provides highly reliable quantitative data in the form of GC retention rates and MS-fragmentation patterns, but it has limitations. The spectra of cocaine HCl and free-base cocaine are identical, and differentiation requires FTIR. Notwithstanding the previous, GCMS and FTIR are often complementary confirmatory methods used to provide definitive proof in the identification process (Fig. 14.8).

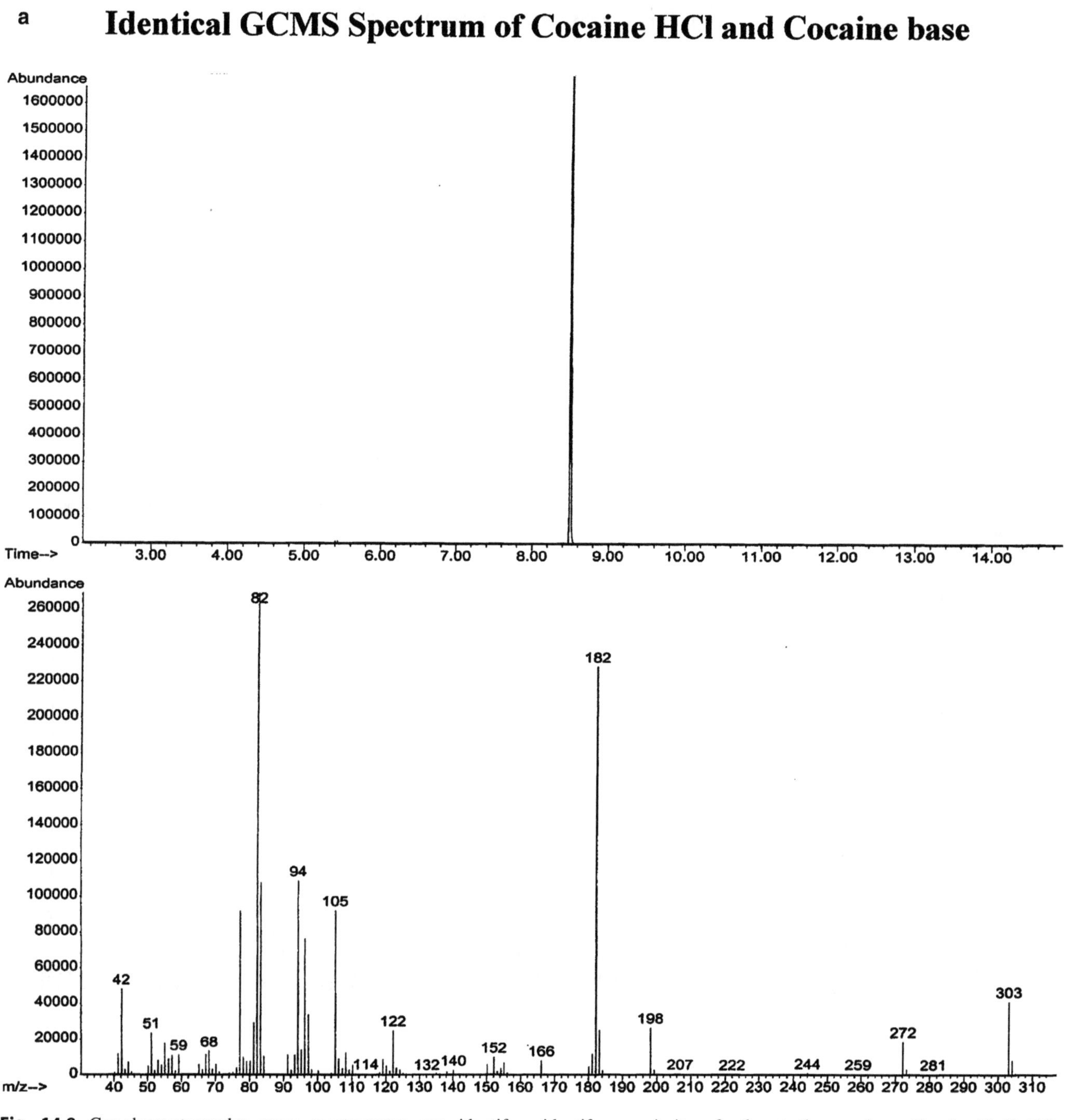

Fig. 14.8 Gas-chromatography mass spectrometry can identify cocaine ($C_{17}H_{21}NO_4$, molecular weight 303.3 g/mol) from other tertiary amines (note the M^+ peak at 303 in the above mass spectra), but it cannot differentiate the salt from the free-base form (note the single peak in the GC chromatogram). However, it is effectively used to identify a majority of other tertiary amines. Heroin ($C_{21}H_{23}NO_5$ molecular weight 369.4 g/mol) and PCP ($C_{17}H_{25}N$, molecular weight 243.4 g/mol) are shown above. Note the MS spectrum of heroin is complicated by the presence of a contaminant (two peaks in the GC chromatogram).

b

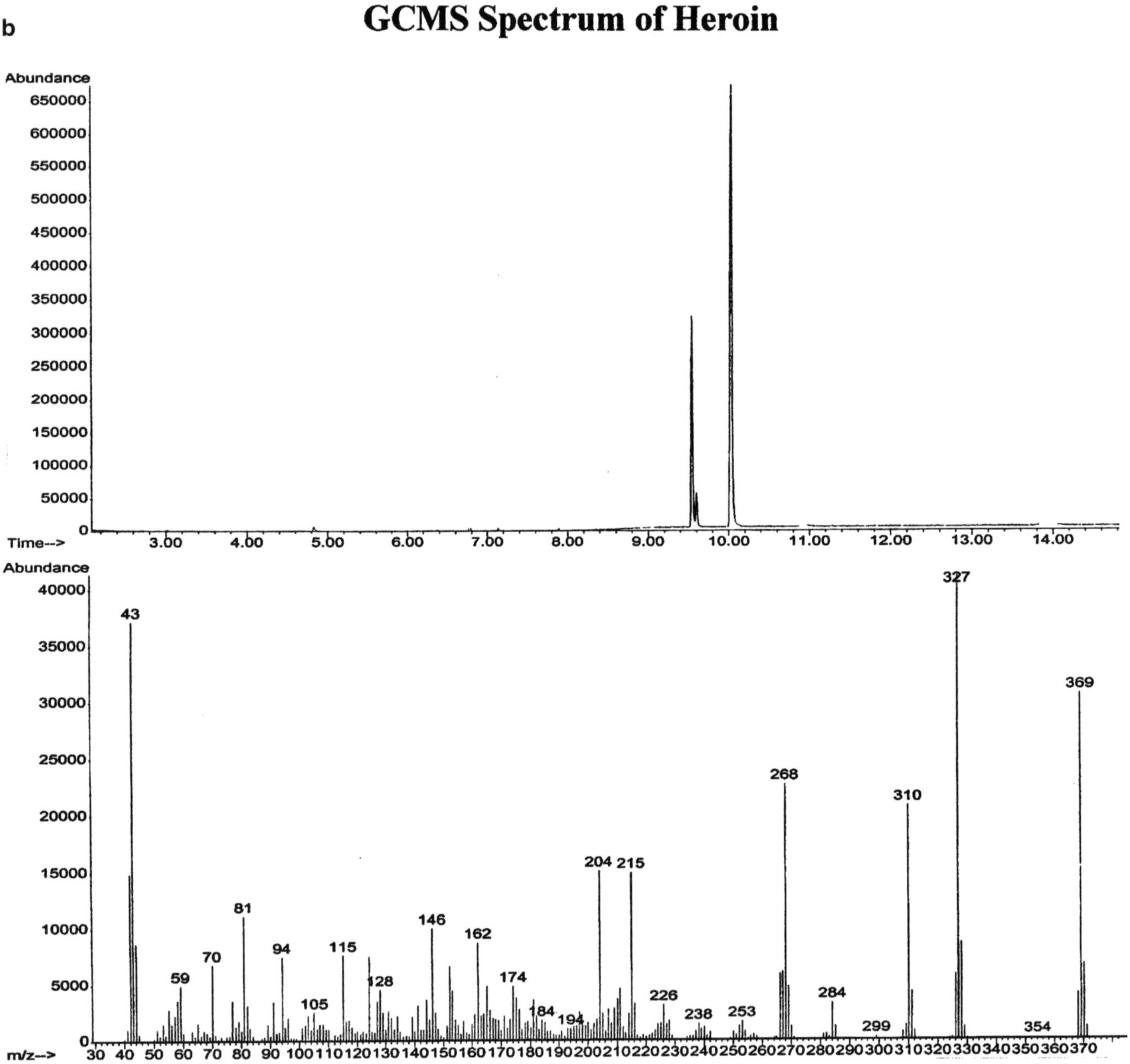

Fig. 14.8 (continued)

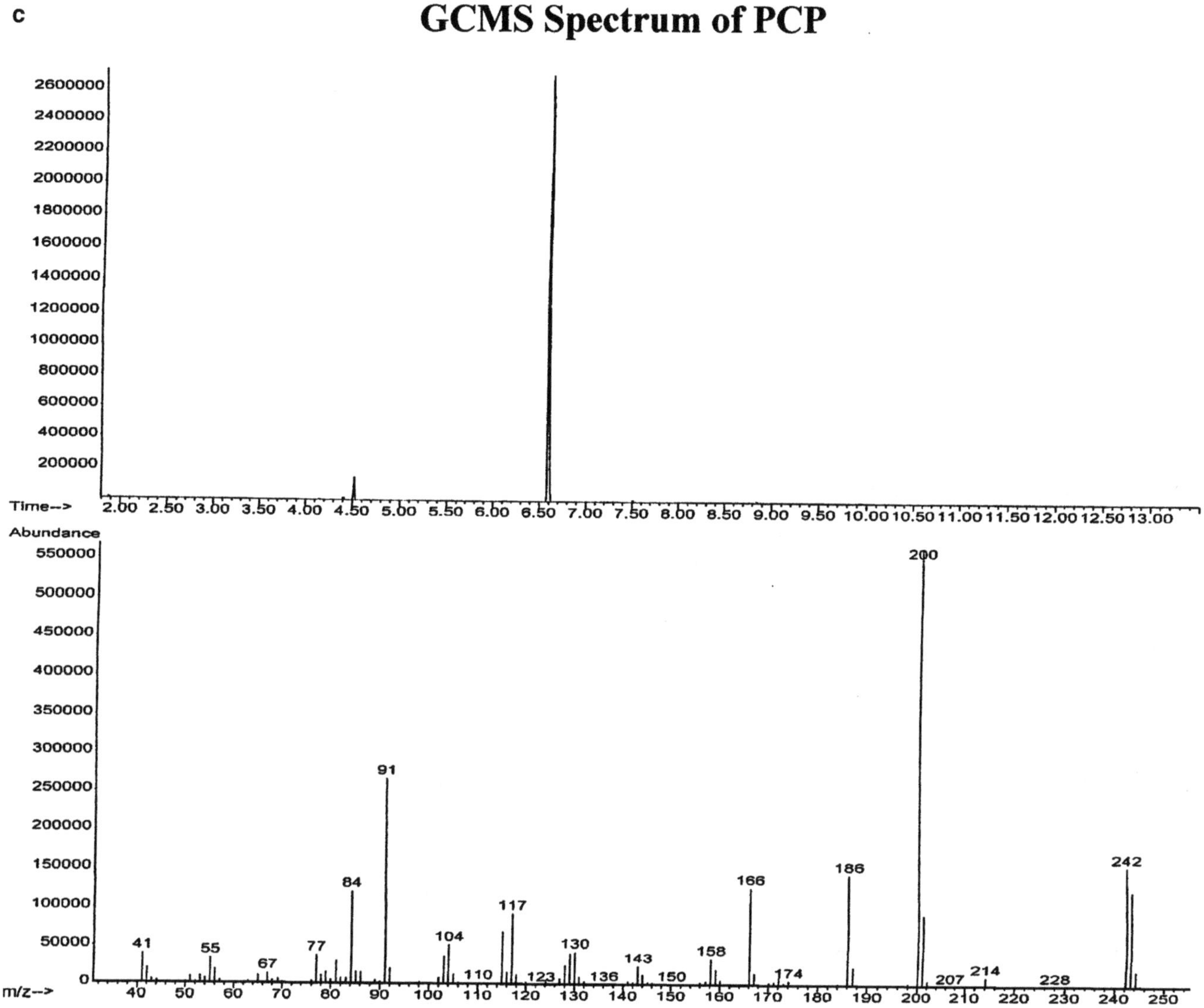

Fig. 14.8 (continued)

14.5 Questions

1. Draw the structure of a controlled substance in the tertiary amine class and circle the functional group.
2. How did crack cocaine get its name?
3. List three effects of cocaine use.
4. List two species of coca plants.
5. Describe latex from poppy plants.
6. List two naturally occurring alkaloids found in latex.
7. List three side effects of morphine use.
8. Please describe the structural difference between morphine and codeine to the jury.
9. Please explain to the jury why your analysis of suspected codeine revealed the presence of aspirin.
10. Why is the administration of codeine in a clinical setting closely monitored?
11. Describe the actions of morphine on the central nervous system in humans.
12. Briefly describe how poppy plants utilize morphine.
13. How is heroin often differentiated from other opiate alkaloids using visual examinations?
14. Describe how heroin abuse collapses veins.
15. What does PCP stand for?
16. Please describe to the jury the physical properties that are used to differentiate the base and salt forms of PCP.
17. Discuss the effects of PCP abuse.
18. Outline separate color-screening procedures that indicate the presence of heroin and PCP.
19. How is FTIR used to differentiate the salt and free-base forms of cocaine? Include a discussion of the prominent bands commonly produced from both forms.

Suggested Reading

Allen, A. C.; *et al.* The Cocaine Diastereomers. *J. Forensic Sci.* **1981**, 26, 12–26.

Audier, H. E. Mass Spectrum of Cocaine and Related Compounds. *Org. Mass Spectrum.* **1969**, 2, 83.

Audier, H. E.; Sozzi, G. Mass Spectrum of Heroin and Related Compounds. *Org. Mass Spectrum.* **1984**, 19, 150.

Bohm, B. A.; Ganders, F. R.; Plowman, T. Biosystematics and Evolution of Cultivated Coca (Erythroxylaceae). *Syst. Bat.* **1982**, 7, 121–133.

Bray, W.; Dallery, C. Coca Chewing and High Altitude Stress: A Spurious Correlation. *Curr. Anthropol.* **1983**, 24, 269–274.

Christian, D. R. Jr. Analysis of Controlled Substances. In *Forensic Science: An Introduction to Scientific and Investigative Techniques*, 3rd ed.; James, S. H.; Nordby, J. J., Eds.; CRC Press: Boca Raton, FL, 2009.

Deglin, J. H.; Vallerand, A. H. *Davis's Drug Guide for Nurses*, 11th ed.; F. A. Davis Company: Philadelphia, PA, 2009.

Duke, J. A.; Aulik, D.; Plowman, T. Nutritional Value of Coca. *J. Am. Chem. Soc.* **1975**, 87, 2920.

Gottlieb, A. *The Pleasures of Cocaine*. Andor Press: Berkeley, CA, 1976.

Manura, J. J.; Chao, J. M.; Saferstein, R. The Forensic Identification of Heroin. *J. Forensic Sci.* **1978**, 23, 44–56.

National Institute on Drug Abuse. Heroin. http://www.nida.nih.gov/infofacts/heroin.html (accesses September 2009).

National Institute on Drug Abuse. PCP/Phencyclidine. http://www.nida.nih.gov/DrugPages/PCP.html (accessed September 2009).

Office of National Drug Control Policy. Heroin Facts & Figures. http://www.whitehousedrugpolicy.gov/DrugFact/heroin/heroin_ff.html (accessed September 2009).

United Nations. *Recommended Methods for Testing Cocaine. Manual for Use by National Narcotics Laboratories*; ST/NAR/7; United Nations: New York, 1986.

United Nations. *Recommended Methods for Testing Opium, Morphine and Heroin. Manual for Use by National Drug Testing Laboratories*; ST/NAR/29/Rev.1; United Nations: New York, 1998.

U.S. Drug Enforcement Administration. Cocaine. http://www.justice.gov/dea/concern/cocaine.html (accessed September 2009).

U.S. Drug Enforcement Administration. Heroin. http://www.justice.gov/dea/concern/heroin.html (accessed September 2009).

Tryptamines

15

15.1 Introduction

Tryptamine is an aromatic heterocyclic organic compound derived from *indole*. Indole is found in several fragrances and is a common precursor to many pharmaceutical drugs. It consists of a characteristic two fused-ring system containing benzene and pyrrole, a five-membered conjugated diene ring containing nitrogen (heterocyclic ring).

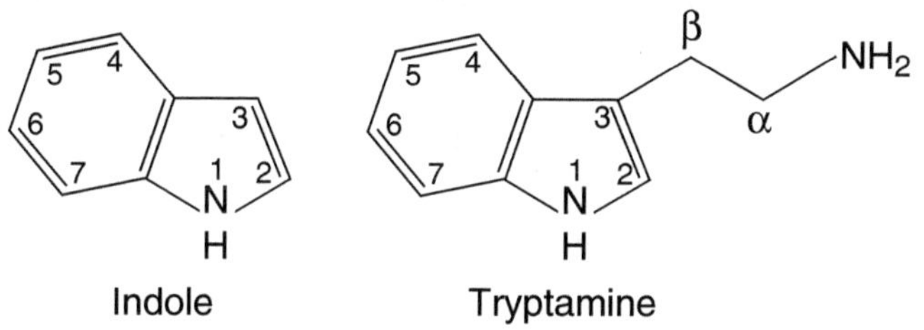

Structure 15.1

The most famous indole derivative is the amino acid tryptophan from which the name tryptamine is partially derived.

Tryptamine alkaloids are naturally occurring psychoactive hallucinogens found in plants, fungi (mushrooms), and animals. Most can also be synthetically produced through substitution at various positions on tryptamine.

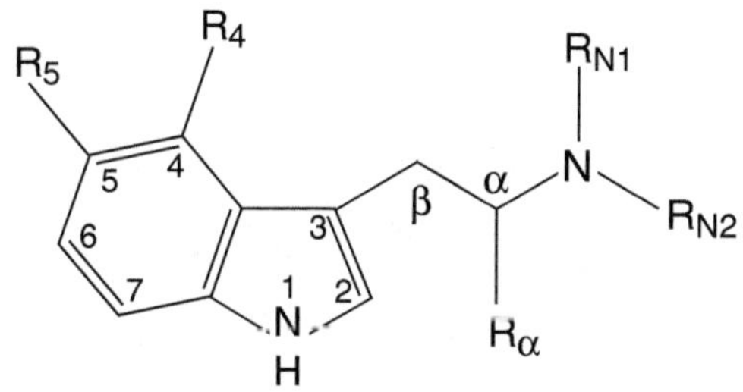

Structure 15.2

Tryptamine analogs, both natural and synthetic, produce psychedelic effects, including sensory deprivation, illusions, changes in perception, altered states of awareness, and periodic episodes of schizophrenia. Time becomes meaningless, and out-of-body experiences and extra-sensory-type perception occur with a blending of the senses. Some users lose touch with reality as the senses cease to function normally. There is a distortion of time and space as objects morph into other objects with great clarity and intense colors. There is a feeling of unexplained connectivity with space and objects resulting in complete unity with the universe. In this state, it is impossible for the user to distinguish conscious thought from the hallucination. For most, the emotional and mental impact of their experience is quite positive and enduring.

J.I. Khan et al., *Basic Principles of Forensic Chemistry*, DOI 10.1007/978-1-59745-437-7_15,
© Springer Science+Business Media, LLC 2012

15.2 Natural Tryptamines

15.2.1 Psilocin and Psilocybin (Psychoactive Mushrooms)

Psilocin and psilocybin are naturally occurring active agents in psychoactive "magic" mushrooms.

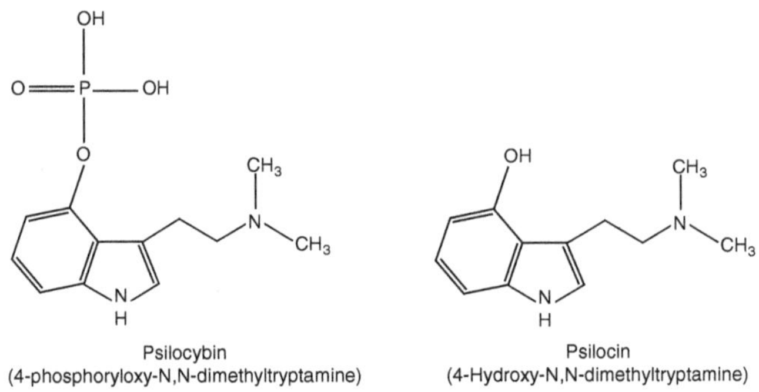

Structure 15.3

A number of tryptamines are classified as Schedule I hallucinogens and most are naturally occurring, but many, if not all, can be synthetically produced. Psilocybin (4-phosphoryloxy-*N*,*N*-dimethyltryptamine) and psilocin (4-hydroxy-*N*,*N*-dimethyltryptamine) are found in certain mushrooms indigenous to tropical and subtropical regions of South America, Mexico, and the United States (Fig. 15.1; Table 15.1).

Psilocin is structurally similar to serotonin, a neurotransmitter that regulates human mood, anger, and aggression.

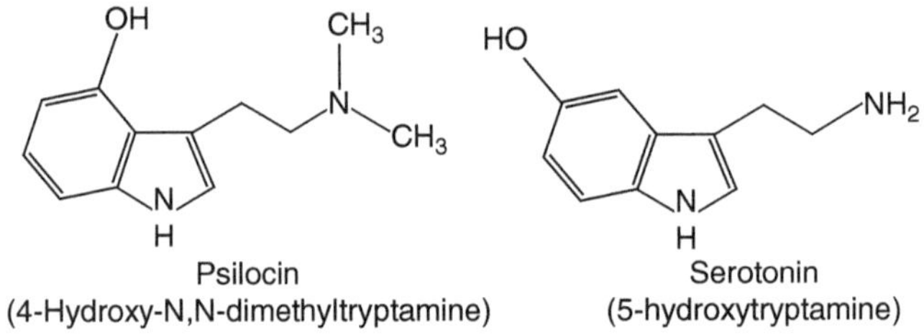

Structure 15.4

Fig. 15.1 Considerable variation exists in both the form and shape of "magic" mushrooms. Cultivated mushrooms (*left*) are harvested and dried (*right*) for distribution and ingestion. These hallucinogens have a documented history of abuse dating back to 500 BC and have been used for centuries in Native American cultures.

Table 15.1 Classification of psychoactive mushrooms

Kingdom	Fungi
Subkingdom	Mycota
Phylum	Basidiomycota
Class	Hymenomycetes
Order	Agaricales
Families	Strophariaceae, Bolbitiaceae, Coprinaceae, and Cortinariaceae
Genus	*Amanita*, *Conocybe*, *Panaeolus*, *Psilocybe*, and *Stropharia*
Species	*Psilocybe mexicana* (common) *Psilocybe baeocystis* (potent psilocybe) *Psilocybe caerulipes* (blue foot psilocybe) *Psilocybe coprophila* (dung-loving psilocybe) *Psilocybe cubensis* (common large psilocybe) *Psilocybe cyanescens* (bluing psilocybe) *Psilocybe pelliculosa* (conifer psilocybe) *Psilocybe semilanceata* (liberty cap) *Psilocybe stunzii* (Stunz's blue legs)
Psychoactive ingredients	Psilocin (4-hydroxy-*N*,*N*-dimethyltryptamine) Psilocybin (4-phosphoryloxy-*N*,*N*-dimethyltryptamine)
Contents	May contain psilocybin, psilocin, or both. The concentration varies from species to species

It mimics the effects of serotonin at receptor sites in the brain. Ingestion of psychoactive mushrooms is perceived as high levels of serotonin, resulting in psychedelic hallucinations. Surprisingly, psilocin does not normally affect dopamine receptors and only acts on serotonin receptors at high doses (10–50 mg). Psilocybin produces effects similar to psilocin because it is rapidly dephosphorylated (removal of phosphoryl group) to psilocin in the body.

Psilocin is generally considered more potent than psilocybin, but the effects produced by consuming preparations of dried or brewed mushrooms are unpredictable and are largely dependent on the type of mushroom, its age, and the preservation of the extract. There are many species of "magic" mushrooms that contain varying amounts of these tryptamines, as well as uncertain amounts of other biologically active agents. Consequently, the hallucinogenic activities, as well as the extent of toxicity produced, are often dangerously unpredictable.

15.2.2 Bufotenin

Bufotenin (5-hydroxy-*N*,*N*-dimethyltryptamine or 5-OH-DMT) is a psychedelic hallucinogen found in mushrooms (Amanita species), higher order plants (Leguminosae family), and in the venom and eggs of several species of toads in the *Bufo* genus.

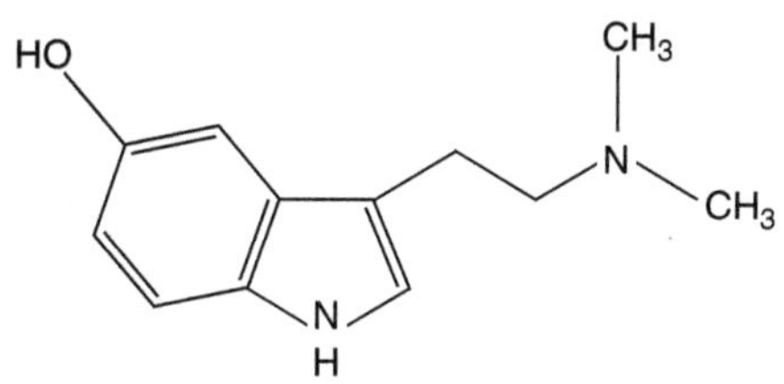

Bufotenin
(5-hydroxy-N,N-dimethyltryptamine)

Structure 15.5

Bufotenin was first isolated from *Bufo alvarius* in 1968 and the "psychedelic toad" gained immediate worldwide notoriety. It is most effective when collected, dried, and then smoked in a pipe. This hallucinogen has effects similar to psilocybin, DMT (*N,N*-dimethyltryptamine), and DET (diethyltryptamine).

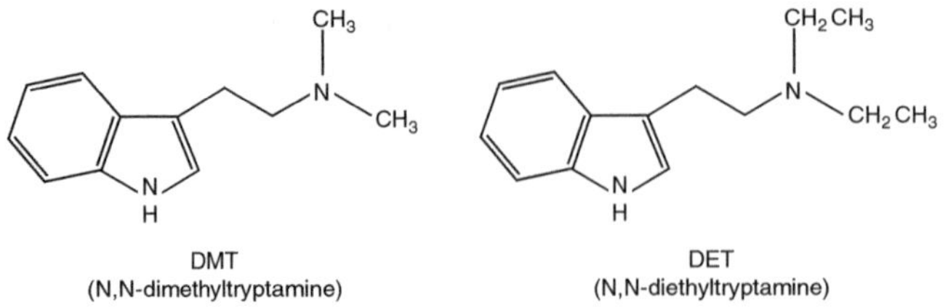

Structure 15.6

The name bufotenin is derived from *bufotoxin*, a general term referring to a number of chemical poisons secreted from the skin glands (parotoid glands) of Bufo toads. These multicellular glands produce a variety of biologically active toxins that vary from species to species. The term "*toad licking*" originated in the United States, but licking toads to "get high" is an urban myth. Bufotenin is rapidly metabolized by the body and loses a significant degree of activity when taken orally. However, its effects can be felt if taken in combination with other drugs that slow down or inhibit its metabolism. Generally, it is sniffed, smoked, or injected. The effective hallucinogenic dose in humans is 50–100 mg and the effects last for only 45–60 min. As a result, bufotenin-induced hallucinations are often called a "businessman's trip."

There are more than 200 different species of Bufo toads (Fig. 15.2). The most common is *Bufo alvarius*, found mainly in the Sonoran Desert in North America. The *Bufo marinus* is native to the Amazon region and the European species is called *Bufo vulgaris*. Historically, Bufo toads have been used throughout the world for centuries. The skin and venom have been used in medicines and hallucinogenic tribal rituals that date back to 2000 BC. Bufo and other toads were often included in many mystical and cultural charms for sexuality and rain fertility. They are also commonly used in burial ceremonies and are found in large numbers at burial sites in Mexico. Chinese healers have used Bufo venom for thousands of years in a potion called *Ch'an Su*. Bufo toads also have a role in the medicinal history of Nepal, Tibet, India, Germany, and Africa.

15.2.2.1 Yopo Seeds (*Anadenanthera peregrina*)

The use of *cohoba snuff*, a potent psychoactive compound made from the seeds of the yopo tree (*A. peregrina*), was first observed during Columbus' second journey to the New World (Fig. 15.3).

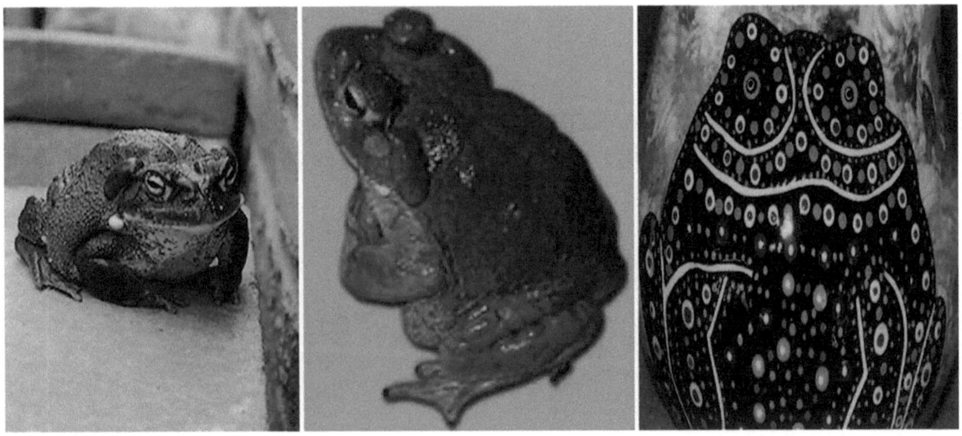

Fig. 15.2 Different species of Bufo toads. Bufotenin, a potent psychedelic hallucinogen, is secreted from glands located at the surface of the skin. "Toad-licking" was a popularized urban myth believed by many as a method of oral ingestion.

Fig. 15.3 Leaf (*left*) and seeds (*right*) from the yopo tree. Cohaba snuff is a powerful hallucinogen used in ancient tribal rituals. It is a mixture of crushed seeds, ash or lime, and calcinated shells.

The yopo is an elegant tree with fern-like leaves and wide branches. DMT and bufotenin are the main psychoactive components in the seeds and fruits of this plant. The stem, bark, and pods also contain numerous tryptamine analogs, that is, 5-methoxy-*N*,*N*-dimethyltryptamine, 5-methoxy-*N*-methyltryptamine, *N*-methyltryptamine, and bufotenin oxide.

15.2.3 Methoxy Derivatives

Methoxy-substituted ($-OCH_3$) tryptamines produce psychedelic effects similar to those commonly associated with most hallucinogens. The two most common derivatives are 5-methoxy-*N*,*N*-dimethyltryptamine (5MeODMT), a naturally occurring tryptamine found in plants, and 5-methoxy-*N*,*N*-diisopropyltryptamine (5MeODIPT or foxy methoxy), a synthetic analog.

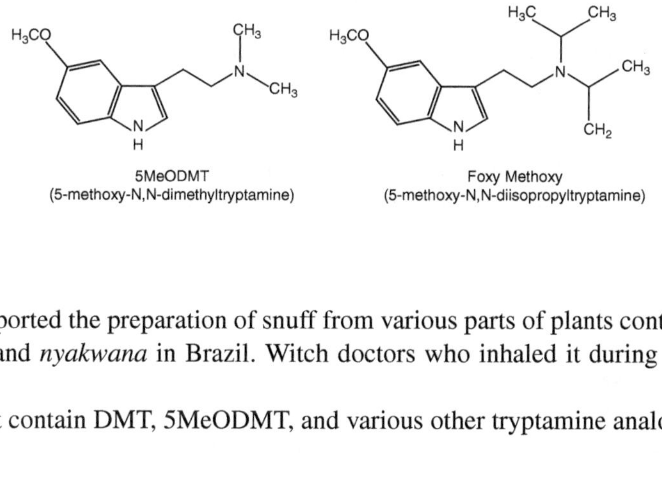

Structure 15.7

In 1909, Koch-Grunberg reported the preparation of snuff from various parts of plants containing 5MeODMT. This snuff was called *yato* in Columbia and *nyakwana* in Brazil. Witch doctors who inhaled it during ritualistic healing ceremonies called it *hakudufha*.

Below is a list of plants that contain DMT, 5MeODMT, and various other tryptamine analogs:

ACANTHACEAE
Justicia pectoralis: Leaves contain DMT.

AGARICACEAE
Amanita citrina: Plant contains DMT, 5MeODMT.
Amanita porphyria: Plant contains 5MeODMT.

AIZOACEAE
Delosperma sp. Plant contains DMT.

GRAMINEAE
Arundo donax: Plant contains 5-methoxy-*N*-methyltryptamine, leaves contain bufotenin, DMT, and flower contains *N*,*N*-dimethyltryptamine methohydroxide.
Phalaris arundinacea: Leaves and Plant contain DMT and 5MeODMT.
Phalaris tuberosa (*P. aquatica*): Leaves contain DMT and 5MeODMT.

Phragmites australis: Roots contain DMT.

LEGUMINOSAE

Acacia confusa: Stem contains DMT.

Acacia maidenii: Bark contains DMT.

Acacia nubica: Leaves contain DMT.

Acacia phlebophylla: Leaves contain DMT.

Acacia polyacantha subsp. *campylacantha*: Leaves contain DMT.

Acacia Senegal: Leaves contain DMT.

Acacia simplicifolia: Stem, bark, and leaves contain DMT.

Anadenanthera colubrina var. *cebil* (*Piptadenia macrocarpa*): Stem and pods contain DMT, seeds contains bufotenin and DMT

Anadenanthera excelsa: Stem and pods contain DMT.

Anadenanthera peregrina: Stem, bark of stem, pods, and leaves contain DMT, 5MeODMT. Bark of tree contain 5MeODMT, 5-methoxy-*N*-methyltryptamine, and *N*-methyltryptamine. Plant contains bufotenin. Fruit of the plant contains bufotenin oxide, DMT, and *N*,*N*-dimethyltryptamine oxide.

Desmanthus illineonsis: Roots and root bark contain DMT.

Desmodium caudatum: Stem and roots contain DMT.

Desmodium gangeticum: Plant (without flowers) contains DMT and 5MeODMT.

Desmodium gyrans: Leaves and roots contain DMT and 5MeODMT.

Desmodium pulchellum: Plant contains DMT and 5MeODMT.

Desmodium racemosum: Plant contains 5MeODMT.

Despedeza bicolor va. *Japonica*: Leaves and bark of roots contain DMT and 5MeODMT. Bark of tree also contains 5-methoxy-*N*-methyltryptamine and *N*-methyltryptamine. Plant contains bufotenin. Fruit contains bufotenin oxide, DMT, and *N*,*N*-dimethyltryptamine oxide.

Mimosa hostilis: Roots of the plant contain DMT.

Mimosa scabrella: Bark of the stem contains DMT.

Mimosa tenuiflora: Bark of the stem contains DMT and 5MeODMT.

Mucuna pruriens: Leaves, seeds, stem, and roots contain DMT and 5MeODMT.

Petalostylis labicheoides var. *casseoides*: Leaves and stem contain DMT.

MALPIGHIACEAE

Banisteriopsis muricata (*B. argentea*): Stem and leaves contain DMT.

Diplopterys cabrerana (*Banisteriopsis rusbyana*): Leaves contain DMT and 5MeODMT.

MYRISTICACEAE

Iryanthera ulei: Bark of stem contains 5MeODMT.

Osteophloem platyspermum: Bark of stem contains DMT and 5MeODMT.

Virola calophylla: Plant contains DMT and 5MeODMT, bark contains 5MeODMT, DMT, and *N*-methyltryptamine.

Virola carinata: Leaves contain DMT.

Virola divergens: Leaves contain DMT.

Virola elongata: Leaves and bark of the stem contain DMT and 5MeODMT.

Virola melinonii: Bark contains DMT.

Virola multinervia: Roots and bark contain DMT and 5MeODMT.

Virola pavonis: Leaves contain DMT.

Virola peruviana: Bark of stem contains DMT and 5MeODMT.

Virola rufula: Bark, stem, roots, and leaves contain DMT and 5MeODMT.

Virola sebifera: Bark contains DMT.

Virola theiodora: Plant contains DMT and 5MeODMT.

Virola venosa: Leaves and roots contain DMT and 5MeODMT.

OCHNACEAE

Testulea gabonensis: Bark and roots contain DMT

POLYGONACEAE

Erigonum sp.: Plant contains DMT

RUBIACEAE
Psychotria carthaginensis: Leaves contain DMT and *N*-methyltryptamine.
Psychotria viridis (*P. psychotriaefolia*): Leaves contain DMT and *N*-methyltryptamine.

RUTACEAE
Dictyoloma incanescens: Leaves contain 5MeODMT.
Dutaillyea drupacea: Leaves contain 5MeODMT.
Dutaillyea oreophila: Leaves contain 5MeODMT.
Evodia rutaecarpa: Leaves contain 5MeODMT.
Limonia acidissima: Stem contains 5MeODMT.
Pilocarpus organensis: Leaves contain 5MeODMT.
Vepris ampody: Leaves contain DMT.
Zanthoxylum arborescens: Leaves contain DMT.
Zanthoxylum procerum: Leaves contain DMT.

15.3 Synthetic Tryptamines

Most synthetic analogs of tryptamine are preferred by users over naturally occurring forms because they can be taken orally to produce hallucinogenic effects. Natural analogs are rapidly degraded into inactive forms prior to absorption by the enzyme monoamine oxidase. They are only active when taken in combination with monoamine oxidase inhibitors (MAOIs), a dangerous alternative. MAOIs are potent antidepressants commonly used to treat depression. The most significant risk associated with MAOI treatment is their ability to interact with over-the-counter medications, prescription drugs, illicit drugs, and even some dietary supplements. These interactions are often unpredictable and can be life threatening. Synthetic derivatives of tryptamine usually contain alkyl substitutions on, or near, nitrogen. These substituted molecules are generally unaffected by monoamine oxidase and the drug is rapidly absorbed by the body. Commonly abused synthetic analogs include DET (*N,N*-diethyltryptamine), foxy methoxy (5-methoxy-*N,N*-diisopropyltryptamine), and AET (α-ethyltryptamine). DET is derived from DMT and is generally considered less potent, while AET is an analog of foxy methoxy.

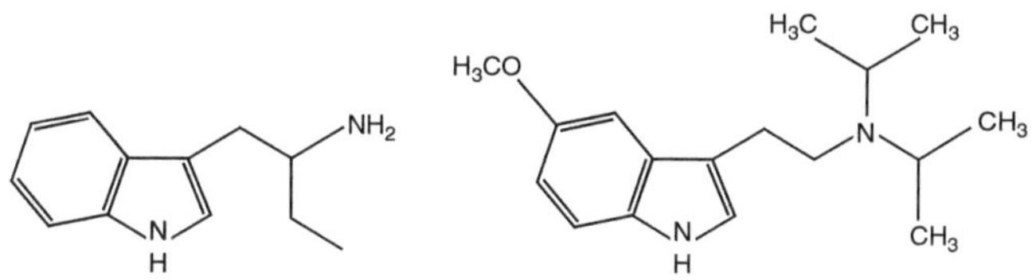

AET
(α-ethyltryptamine)

Foxy Methoxy
(5-methoxy-N,N-diisopropyltryptamine)

Structure 15.8

In recent years, a number of synthetic tryptamine analogs have been surfacing in the illicit drug market. They are often produced outside the United States and imported using legitimate scientific names under the label "for scientific research purpose only." Many law-enforcement agencies at border crossings were unaware of the connection between these unfamiliar names and illegal drug activity. Therefore, they often did not question a small amount of drugs crossing the border "for research purposes." Recently, the Drug Enforcement Administration, border patrol, and other courier services have been alerted to this innovative form of smuggling.

15.4 Analytical Methods

15.4.1 Visual Identification

The visual identification of psychoactive mushrooms is complicated by the fact that many come in various sizes, shapes, and physical forms (Fig. 15.4). Color and odor are usually the only characteristics that differentiate psychoactive mushrooms

Fig. 15.4 Identification of psychoactive mushrooms using only visual inspection is nearly impossible due to extreme variations in physical appearance across all species. Psilocin and psilocybin mushrooms (*top left*), bufotenin toadstools (*top right*), and dried psilocin/psilocybin for distribution (*bottom*).

from others. Therefore, visual identification of hallucinogenic mushrooms is very simple.... do not do it. An experienced forensic chemist can usually differentiate various mushrooms quite effectively. However, identification by odor or color is not an accepted confirmatory method and is not recognized in a court of law.

15.4.2 Chemical Screening Tests

15.4.2.1 Weber Test

The Weber test was first introduced by Weber State College in 1984. It has since become the preferred color-screening method used to indicate the presence of psilocin and psilocybin in mushrooms.

Weber Test: Reagent 1

- Freshly prepared 0.1% (m/v) Fast Blue B or Diazo Blue B (*O*-dianisidine, tetrazotized) solution in water.

Note: Do not make more than 1 ml or the amount of solution you would likely use in approximately 1 h.

Weber Test: Reagent 2

- Concentrated hydrochloric acid.

Dry the mushrooms at a low temperature (~45°C) overnight. High temperatures will degrade the sample and psychoactive components. Place 1–2 mg of dried, ground sample in the well of a clean spot plate, add 1 drop of reagent 1 and note color, then add 1 drop of reagent 2 and note color.

Results

- Red color after the addition of reagent 1, followed by a blue-green color after the addition of reagent 2 indicates the presence of psilocin or psilocybin.

Methanol extraction for GCMS or thin-layer chromatography (TLC) is a common subsequent method. It is worth noting that not all laboratories require or utilize the Weber test as a preliminary screening method. The choice varies according to the particular operating procedures of the forensic laboratory.

15.4.3 Extraction of Psilocin and Psilocybin from Mushrooms

1. Dry mushrooms at low temperature (45°C) in an oven overnight.
2. Grind the dried mushrooms into a fine powder using an agate mortar.
3. Place ~1 g of the powder into a test tube.
4. Add 1 ml of methanol and vortex the mixture for 1 min.
5. Allow mixture to stand undisturbed until larger particles settle.
6. Remove solution and transfer into a micro-test tube. The solution will appear slightly cloudy because of tiny, undissolved particulate.
7. Centrifuge solution for 1 min. The particulate will pack at the bottom of the micro-test tube, producing a clear, light-yellow solution on top.
8. Carefully transfer the clear, light-yellow solution into a separate micro-test tube without disturbing the pellet.
9. Perform confirmatory examination on solution using TLC or GCMS, or both. The recovered volume of the extract is only about 0.2 ml, so use it sparingly.

15.4.4 Thin-Layer Chromatography

Thin-layer chromatography is a confirmatory method commonly used to identify psilocin and psilocybin extracted from psychoactive mushrooms. Suspected samples are run against known standards and identification is achieved using component R_f-values and spot color comparisons.

1. Draw a reference line in pencil across the width of a standard TLC plate ~1.5 cm up from the bottom. Generally, a total of four samples (unknowns + standards) can be run effectively on a standard plate. A larger plate would be required to run a greater number of samples. The simplest form of TLC contains one sample and one reference.
2. Mark two dots on the reference line ~2 cm apart. Label one dot reference standard and the other with the appropriate case number or identifier.
3. Spot the reference standard (usually psilocin and psilocybin in methanol) and sample (solution from methanol extraction) on their respective dots using separate capillary pipettes.
4. Place the running solvent in the developing chamber. Typically, acetic acid/water/1-butanol in ratios 1:1:2, respectively.
5. Carefully place the TLC plate in the developing chamber with the samples down toward the running solvent. Make sure that the level of the solvent is below the reference line. Cover the beaker.
6. Capillary action will force the running solvent up the plate. When the solvent stops eluting, or approaches the top of the plate, remove the plate. Immediately draw a line in pencil across the solvent front. Allow a few minutes for the plate to dry.
7. Spray color-enhancing reagent on plate. Typically, Van Urk's reagent (1% *p*-dimethylaminobenzaldehyde + 10 ml of HCl + 90 ml ethanol).

15.4.4.1 Interpretation of Thin-Layer Chromatography Results

Colored spots with identical retention factors (R_f) will appear in columns containing unknown(s) and standard(s). Psilocin is smaller than psilocybin and therefore travels further up the plate. Also, psilocin appears as a blue-colored spot with Van Urk's reagent, while psilocybin is pink. The R_f factors and color comparisons provide definite proof in the identification of both components (Fig. 15.5).

15.4.5 Gas-Chromatography Mass Spectrometry

15.4.5.1 Examination of Mushrooms

GCMS is used to analyze the light-yellow solution from methanol extraction without dilution. Unfortunately, differentiation of psilocin from psilocybin is not routinely achieved using this technique because psilocybin is easily converted to psilocin in the instrument. However, both are potent psychoactive drugs and the identification of psilocin using GCMS is often extended to include psilocybin as well. Therefore, the results are often reported as "the mushrooms contain psilocin/psilocybin" and the

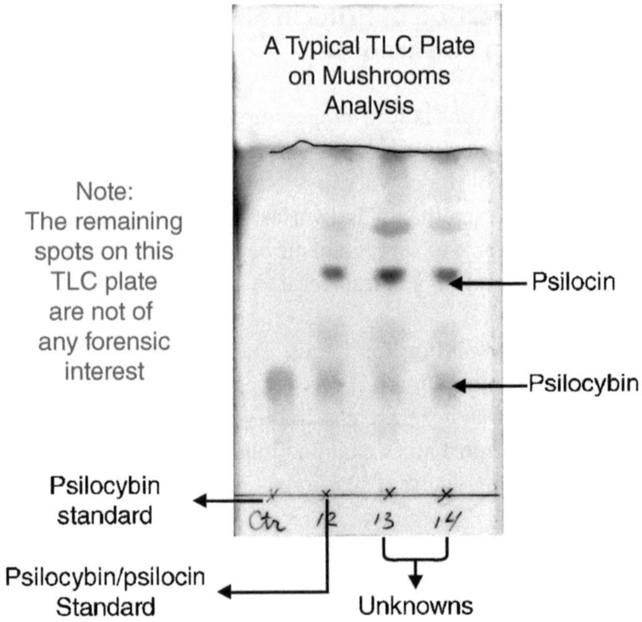

Fig. 15.5 A typical TLC plate on mushroom analysis.

presence of psilocybin is confirmed by TLC analysis. Psychoactive mushrooms also contain a large number of noncontrolled organic molecules and biomolecules that are soluble in methanol. These substances can complicate the GC chromatogram and MS spectrum (Fig. 15.6).

15.4.5.2 Examination of Bufotenin and Synthetic Tryptamines

Psilocin and psilocybin toadstools (mushrooms) are usually brown in color and have a characteristic odor. The entire mushroom is covered in what appears to be a greenish powder residue. By contrast, bufotenin toadstools are normally quite colorful and vary in both size and shape (Fig. 15.7). Analytical techniques used to isolate and identify bufotenin are the same as those used for psilocin and psilocybin.

Several species of plants contain various tryptamine analogs (Fig. 15.8). Plant material submitted for analysis of bufotenin, natural methoxy derivatives, or other tryptamine analogs are generally first extracted with methanol and then analyzed using GCMS (Figs. 15.9 and 15.10).

It is worth noting that forensic chemists are not experts in mycology (study fungi) or botany (study plants). Their primary responsibility is to determine whether a particular fungus or plant contains controlled substances and to isolate and identify such components. Therefore, the formal classification of plants or mushrooms should be avoided during any examination of evidence.

Synthetic tryptamine analogs are commonly submitted to forensic laboratories as highly concentrated solutions. Most are identified using ether extraction followed by GCMS analysis. Solid forms of the analogs (if submitted) are usually extracted with methanol prior to GCMS (Fig. 15.11).

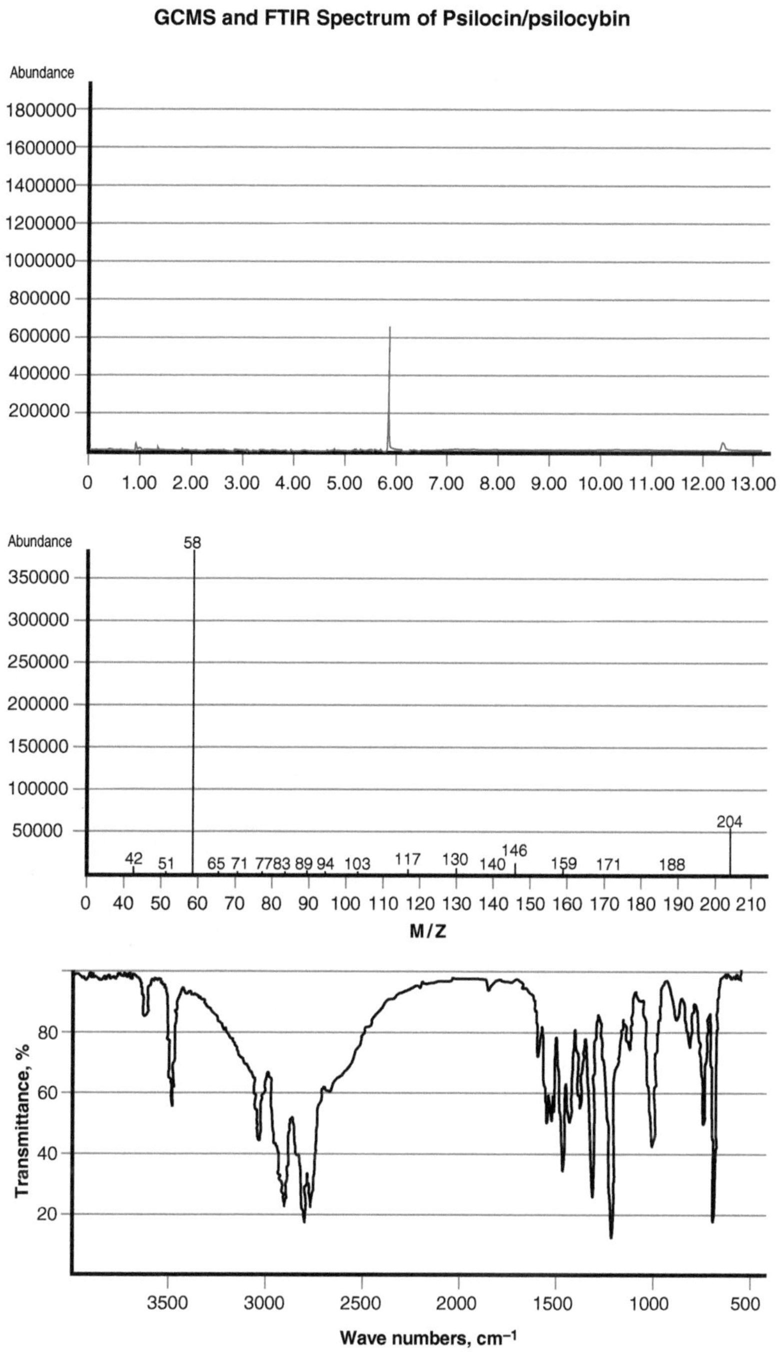

Fig. 15.6 Spectral data for psilocin and psilocybin. The GC chromatogram will always show a single peak because psilocybin is converted into psilocin in the instrument (*top*). Typical mass spectra for psilocin ($C_{12}H_{16}N_2O$, 204.3 g/mol), note M^+ peak at 204 (*middle*). The IR spectrum shows a typical free hydroxyl group sharp band at 3,500 cm^{-1} and a very broad band in the 2,800–3,200 cm^{-1} region resulting from a coupling of hydroxyl vibrations to ring C–H stretching. Note the aromatic supporting bands at 1,500, 1,250, and 1,000 cm^{-1}.

Fig. 15.7 Examples of bufotenin toadstools in their natural environment (*second photo*), recently harvested (*first and third*), and dried forms for distribution (*fourth*).

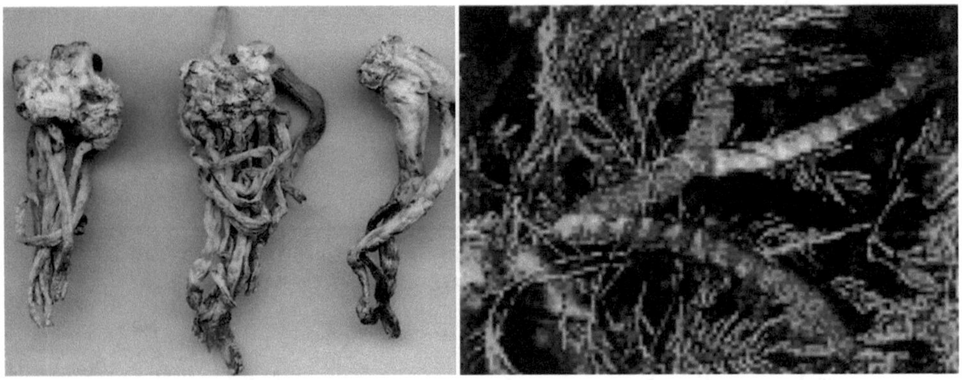

Fig. 15.8 Tryptamine analogs are naturally occurring in a vast number of plant species. Cultivated plants (*right*) are harvested and dried for distribution as illicit drugs (*left*).

Mass Spectrum and FTIR Spectrum of Bufotenine

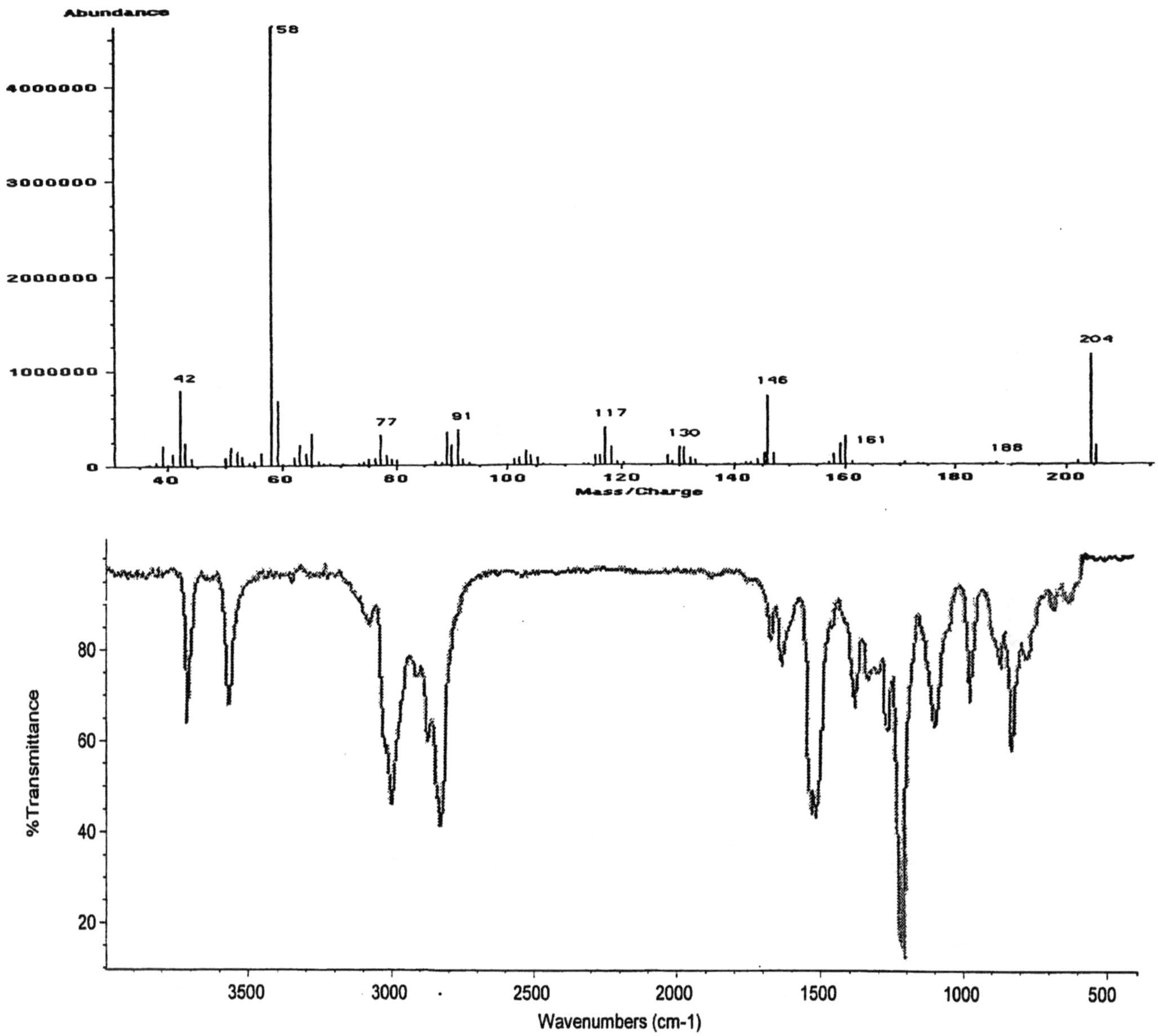

Fig. 15.9 Spectral data for bufotenin, also spelled bufotenine ($C_{12}H_{16}N_2O$, 204.3 g/mol). The mass spectrum (*top*) of bufotenin is very similar to psilocin; this is expected for positional isomers (differ only in OH location). The FTIR spectrum (*bottom*) is less complicated because the coupling interactions between the hydroxyl group and the heterocyclic ring (nitrogen containing ring) are drastically reduced by transferring the OH group from position 4 (psilocin) to position 5 (bufotenin) on the aromatic ring.

GCMS Spectrum and FTIR Spectrum of Methoxy

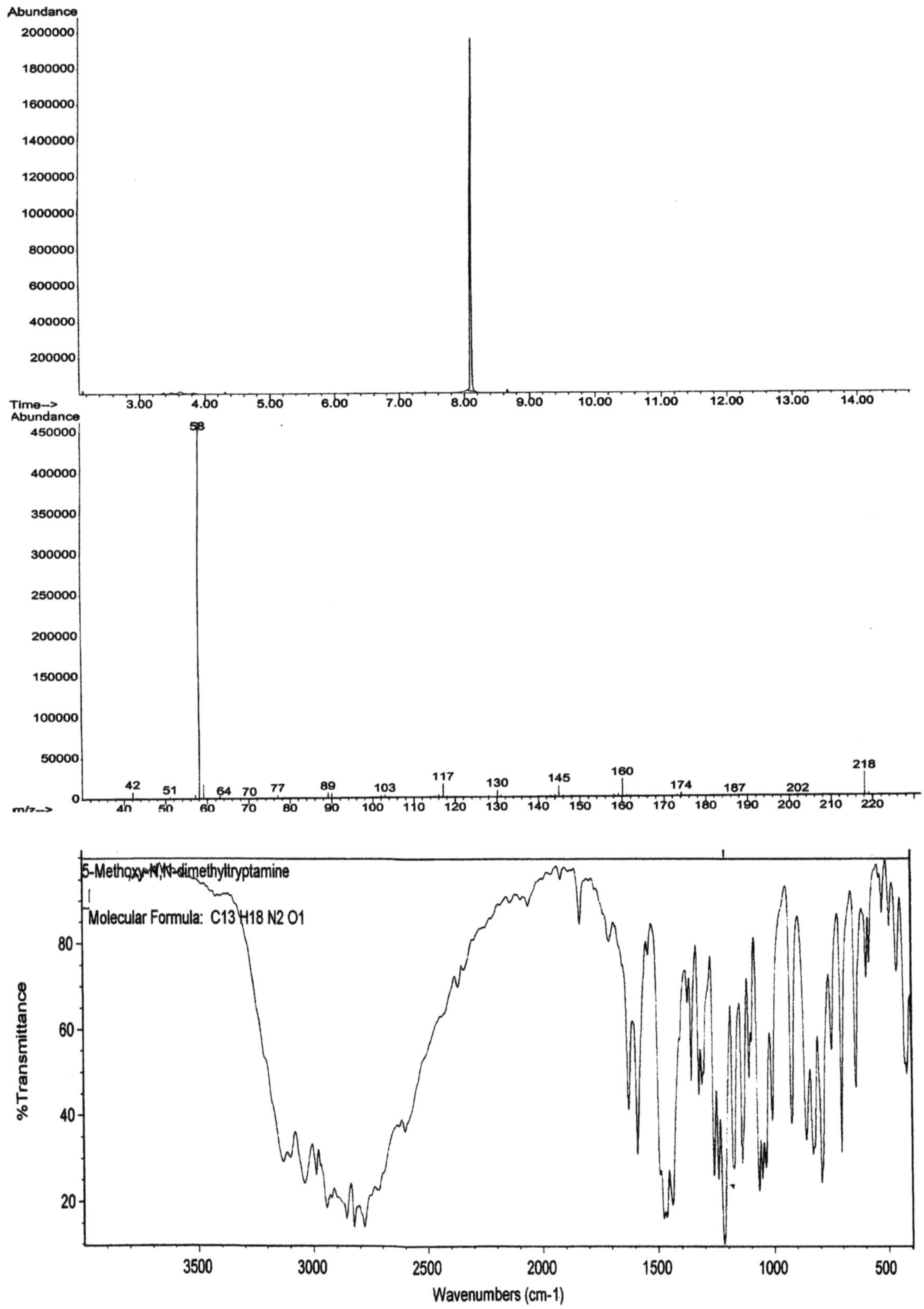

Fig. 15.10 Spectral data for 5-methoxy-*N,N*-dimethyltryptamine also known as methoxy or 5OHDMT ($C_{13}H_{18}N_2O$, 218.3 g/mol). Note the absence of absorption bands above 3,300 cm^{-1} in the FTIR spectrum (*bottom*). There is no hydroxyl group present, but notice the band broadening effect of methoxy (–OCH_3) in 2,600–3,200 cm^{-1} region.

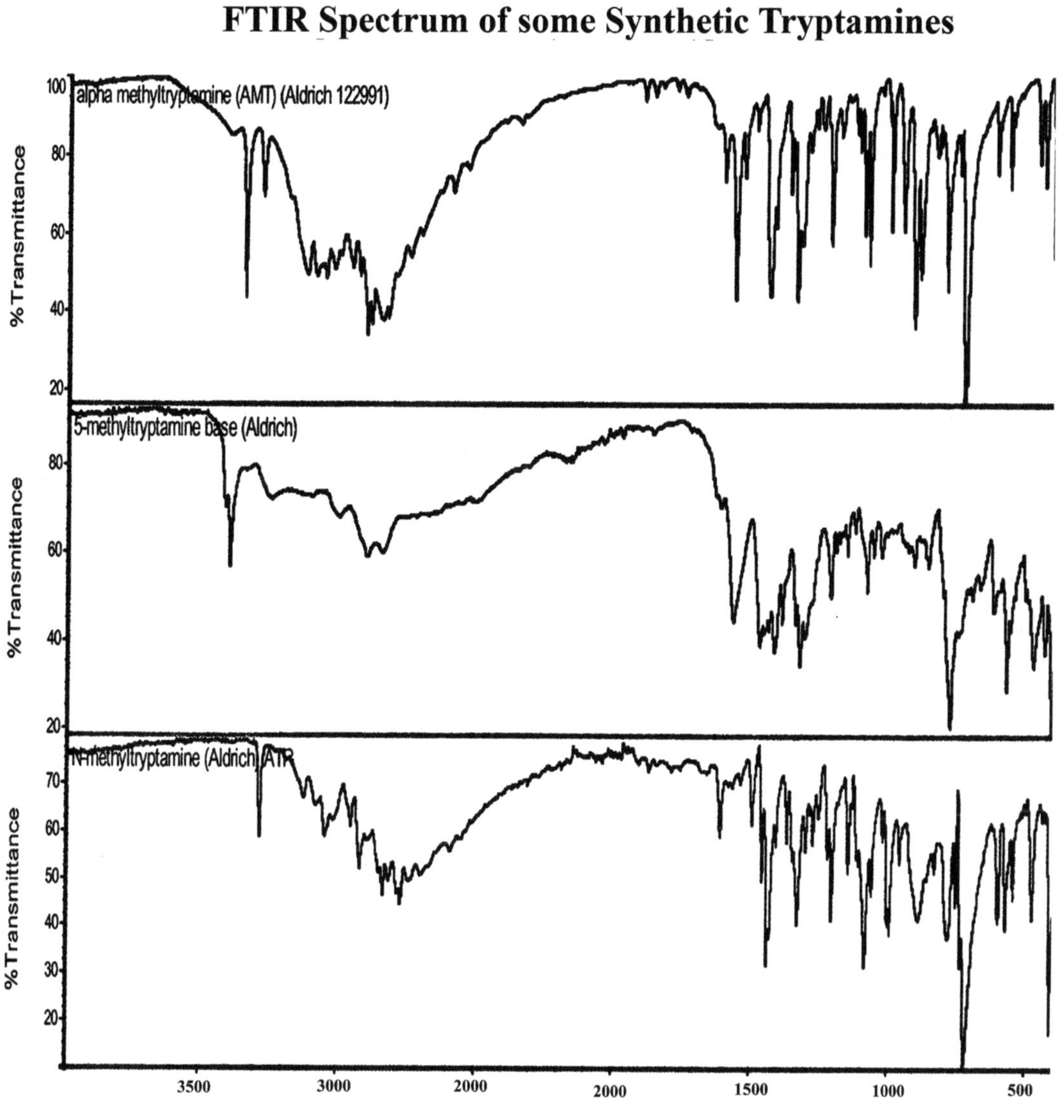

Fig. 15.11 FTIR spectra of the synthetic tryptamines α-methyltryptamine (AMT), 5-methyltryptamine base, and *N*-methyltryptamine.

15.5 Questions

1. Draw indole.
2. Draw tryptamine and label the positions commonly substituted to produce hallucinogenic drugs.
3. Naturally occurring tryptamines are commonly found in which kingdoms?
4. Briefly explain how psilocin produces hallucinogenic effects.
5. Briefly describe the structural differences between psilocin and bufotenin to members of the jury.
6. List the tryptamine analogs contained in *Despedeza bicolor va. Japonica* and *Anadenanthera peregrine*.
7. Explain to members of the jury why synthetic analogs of tryptamine are often abused more than natural forms.
8. Give two examples of synthetic tryptamines.

9. Briefly describe an instance when visual examination of psychedelic mushrooms would be an acceptable confirmatory method.
10. Describe a positive Weber test.
11. Describe the procedure of methanol extraction to members of the jury.
12. How is psilocin differentiated from psilocybin using TLC?
13. Is it possible to differentiate psilocin from psilocybin using GCMS? Explain.
14. Is fungi (mushroom) classification part of forensic analysis? Explain.
15. What is the difference between a toadstool and a mushroom?

Suggested Reading

Arora, D. *Mushrooms Demystified*, 2nd ed.; Ten Speed Press: Berkely, CA, 1986.

Burke, A. A.; *et al. Derivatization of Bufotenine*. American Academy of Forensics: New York, 1997.

Glennon, R. A.; Dromey, R. G. Drug Disrimination: Applications to Drug Abuse Research. *Anal. Chem.* **1987**, 48, 1464.

National Drug Intelligence Center. Foxy Fast Facts. http://www.usdoj.gov/ndic/pubs6/6440/index.htm (accessed September 2009).

National Drug Intelligence Center. Psilocybin Fast Facts. http://www.usdoj.gov/ndic/pubs6/6038/index.htm#illegal (accessed September 2009).

National Institute on Drug Abuse. Research Report: Hallucinogens and Dissociative Drugs. http://www.drugabuse.gov/PDF/RRHalluc.pdf (accessed September 2009).

Ott, J. *Ayahuasca Analogs*. Jonathan Ott Books: Occidental, CA, 1994.

U. S. Department of Justice, Drug Enforcement Administration. *Symposium on Tryptamine in DEA Southwest lab in California.* Drug Enforcement Administration: Washington, DC, 2004.

U. S. Drug Enforcement Administration. Psilocybin & Psilocyn and Other Tryptamines. http://www.usdoj.gov/dea/concern/psilocybin.html (accessed September 2009).

16 Anabolic Steroids

16.1 Introduction and History

Steroid hormones are a group of powerful chemical messengers that regulate a significant number of physiological functions, including muscle and bone growth, sexual development and reproduction, carbohydrate metabolism, and immune system response functions. All steroids contain a *sterane* core, a characteristic four-fused ring system containing three cyclohexane rings and one cyclopentane ring. The rings are labeled A through D and each carbon is numbered to identify its location in sterane.

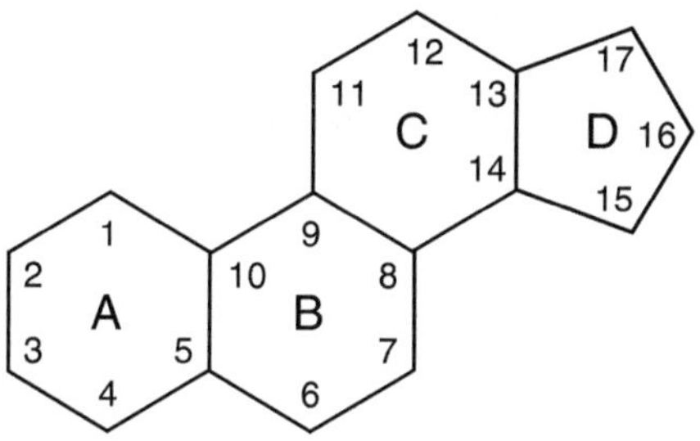

Sterane
(Cyclopentanoperhydrophenathrene)

Structure 16.1

Steroid hormones are derived from cholesterol, the infamous steroid implicated in a variety of cardiovascular diseases. Cholesterol is actually classified as a *sterol* due to the presence of a hydroxyl group (OH) at carbon three of sterane.

It is a vital component of cellular membranes where it is used to establish proper membrane permeability and fluidity.

J.I. Khan et al., *Basic Principles of Forensic Chemistry*, DOI 10.1007/978-1-59745-437-7_16,
© Springer Science+Business Media, LLC 2012

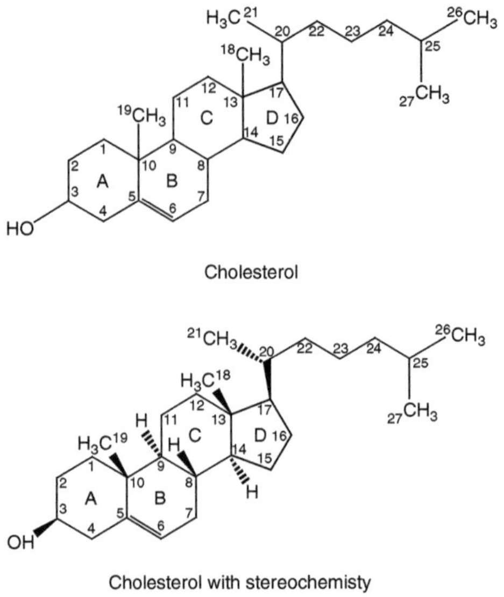

Structure 16.2

16.2 Naturally Occurring Steroid Hormones

There are five basic classes of naturally occurring steroid hormones: progestagens, mineralocorticoids, glucocorticoids, estrogens, and androgens. The primary function of *progestagens* (progestins) is to maintain the uterus during pregnancy. They have a basic skeleton containing 21 carbons and are produced from cholesterol by cleaving the hydrocarbon chain attached at carbon 17.

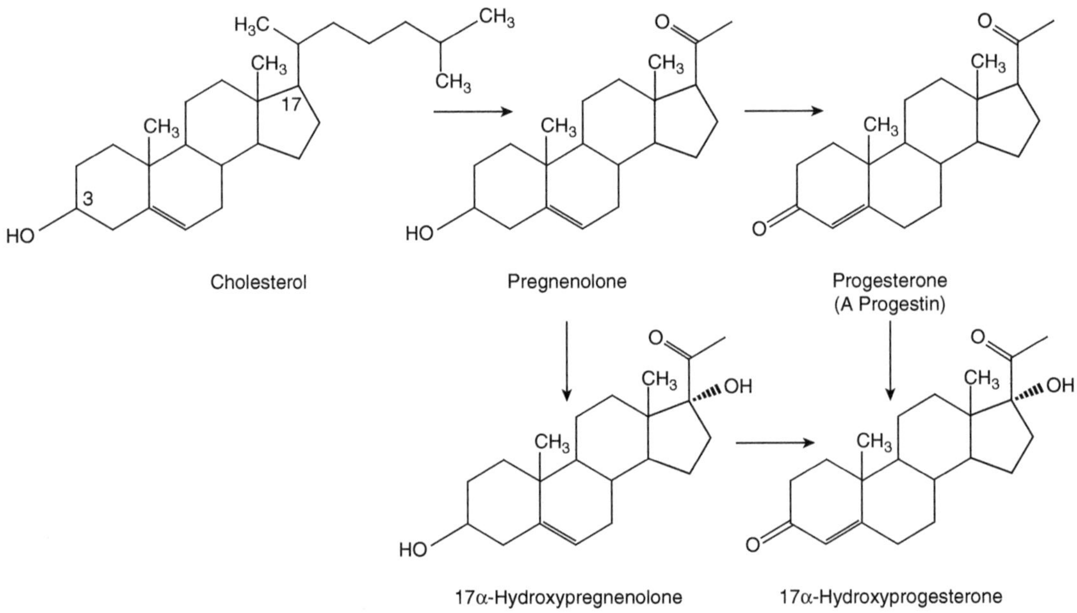

Structure 16.3

Oral contraceptives, i.e.; birth control pills, contain progesterone (a progestin) in combination with various estrogens. These hormones mimic the effects of pregnancy, which prevent the release of eggs from the ovaries.

Mineralocorticoids and *glucocorticoids* are synthesized in the adrenal cortex from progesterone. Mineralocorticoids primarily regulate reabsorption and excretion of sodium (a mineral), potassium (a mineral), and water by the kidneys. Glucocorticoids regulate carbohydrate, protein, and lipid metabolism, as well as inflammatory reactions and "stress-coping" mechanisms.

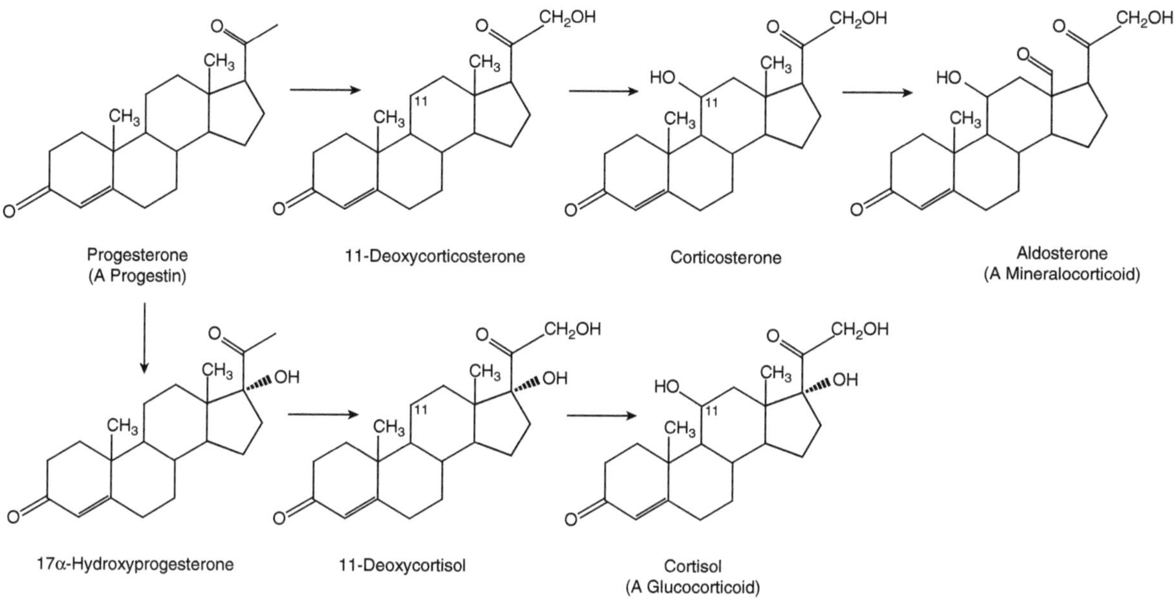

Structure 16.4

Androgen is a general term used to describe steroid hormones that control or regulate the development and maintenance of male characteristics. It is most often associated with testosterone and its natural or synthetic derivatives. They usually have a basic skeleton containing 19 carbons and are common precursors in the production of *estrogen*. Most are surprised by the fact that estrogen does not have a specific structure. It is routinely (and rather loosely) used to define a group of hormones that produce female characteristics including estradiol, estrone, and estriol, or any combination of the three.

Androgens are the only class of controlled or regulated steroids and are thus the only steroids of forensic interest.

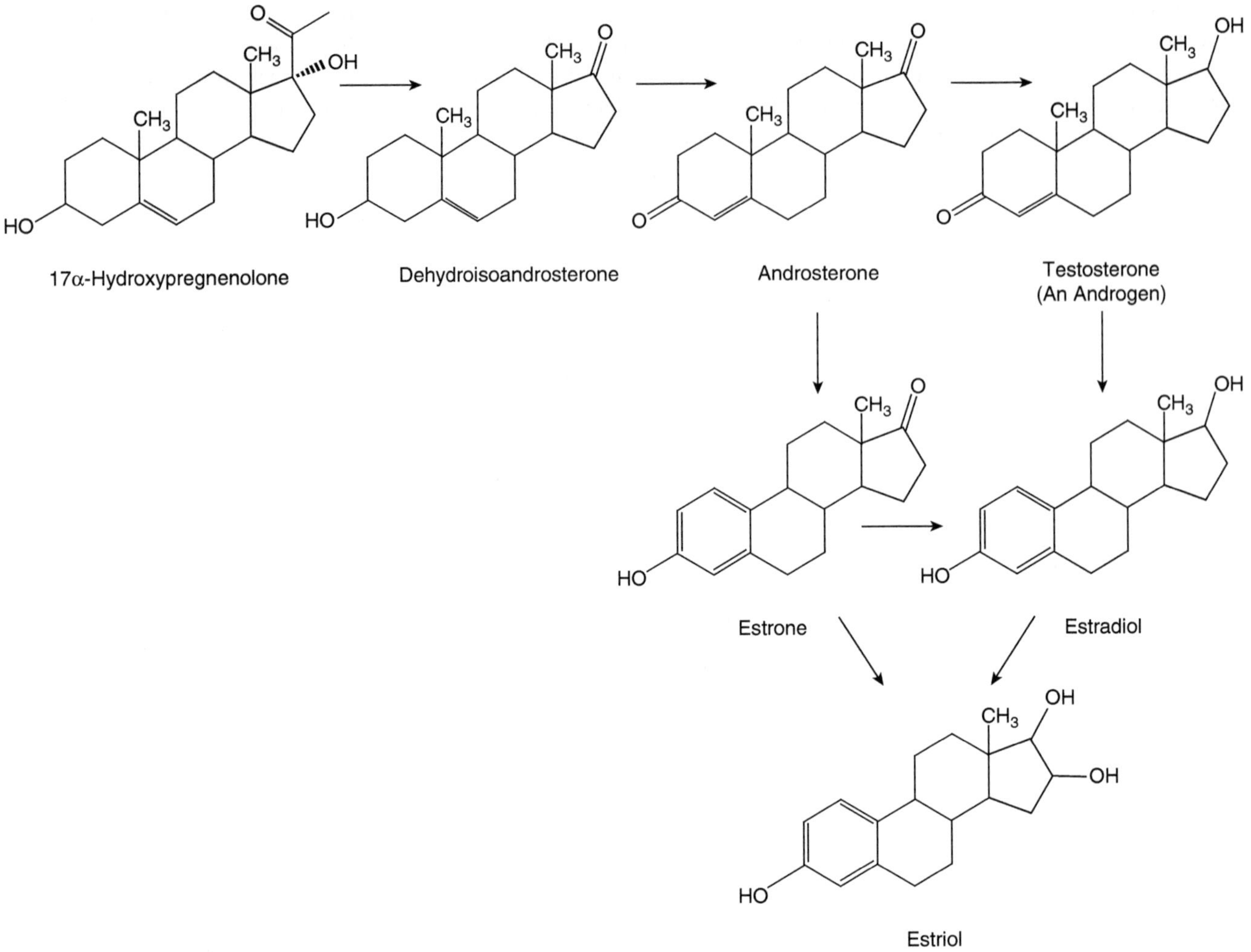

Structure 16.5

16.3 Anabolic Steroids

16.3.1 General Structure

Although the term "steroid" defines a large number of biological molecules in the *lipid* class, its use in popular culture references "anabolic steroids" almost exclusively. Anabolic steroids are synthetic derivatives of testosterone, a naturally occurring hormone in the body that regulates various functions including skeletal muscle growth (anabolic effects) and the development of male sexual characteristics (androgenic effects).

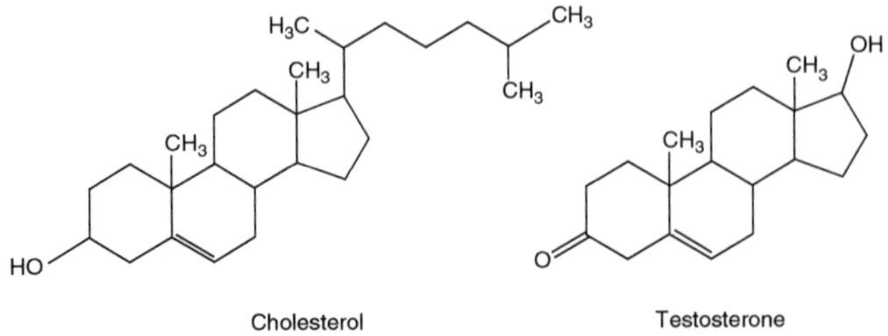

Structure 16.6

Anabolic steroids were originally developed in the late 1930s to treat hypogonadism, a condition in which the testes do not produce enough testosterone for normal growth, development, and sexual functioning. They were later commonly used to treat delayed puberty and specific types of impotence.

16.3.2 Physical and Psychological Effects

A particularly important effect of anabolic steroid use is the stimulation of protein synthesis, which results in (among other things) the growth of skeletal muscle and bone tissue. This increase in muscle mass and bone density enhances physical performance and endurance, a clear advantage in competitive sports. Steroid abuse by professional athletes has become so widespread that it often affects the outcome of sporting competitions. Currently, most steroids are banned in every major amateur and professional sport. In an effort to maintain the integrity of sporting competitions, sanctioning bodies require periodic testing for "doping" and impose serious penalties on those not in compliance.

Steroid use has damaging side effects, including elevated blood pressure, severe acne, liver damage (where it is metabolized), harmful changes in the composition of cholesterol (LDL to HDL ratios, low- and high-density lipoproteins), and alterations in heart structure (abnormal thickening of left ventricle) resulting in arrhythmia, congestive heart failure, and heart attack. There are also male- and female-specific side effects. In females, abnormal hair growth, deepening of the voice, enlarged clitoris, and decreased or absent menstrual cycles can occur. In males, testicular atrophy, development of breast tissue, enlarged Adam's apple, and temporary infertility may result.

Psychological effects are primarily related to mood disturbances or disorders that result in depression and overly aggressive or violent behavior. Chronic users may also develop symptoms of dependency and withdrawal.

16.3.3 Methods of Administration

Anabolic steroids are available in capsules, tablets, gel, patches, and injections. The level of effectiveness is directly related to the rate of absorption. Although oral administration is the most convenient method, injection is the most effective. Steroids taken by mouth are rapidly converted into inactive metabolites resulting in limited absorption. Dermal patches exhibit slow or limited absorption because active steroids must first permeate the skin prior to absorption. Injection is the preferred method of administration because steroids are directly introduced into deep muscle tissue.

16.3.4 Nomenclature of Anabolic Steroids

The nomenclature of androgen-based steroids follows the same general procedures as those developed for the naming of organic compounds (Chap. 4). Saturated hydrocarbons are members of the *alkane* class and contain only carbon–carbon single bonds. Unsaturated hydrocarbons are members of the *alkene* class and contain at least one carbon–carbon double bond.

The names for anabolic steroids can be divided into three basic parts. The first part, "*androst*," indicates the presence of the basic skeletal structure of an androgen, usually testosterone. The second part describes saturation or degree of unsaturation on the androgen skeleton. In these cases, *androst* precedes the suffix "*ane*" for saturated androgens (*androstane*) and the suffix "*ene*" for unsaturated forms (*androstene*). Some steroids contain more than one carbon–carbon double bond. In these cases, the suffix "*ene*" is amended to "*adiene*" or "*atriene*" depending on the number of double bonds present. For example, androst*adiene* defines an androgen molecule with two points of unsaturation (two carbon–carbon double bonds). Points of unsaturation usually occur in the rings of sterane and are rarely found in substituted groups attached to the characteristic four-fused ring system. They are located using the first carbon in the double bond.

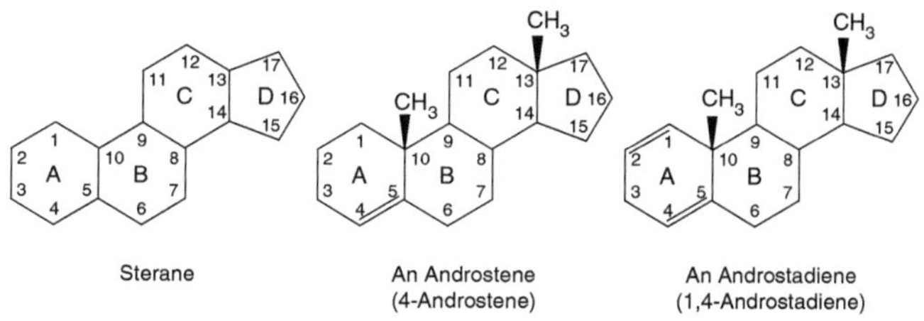

Structure 16.7

The last part of the name describes characteristic functional groups attached to sterane. Androgens containing *ketone* functional groups end in "*one*," while those containing *hydroxyl groups* (–OH, alcohols) end in "*ol*." Ketone groups are typically found at carbon 3 and hydroxyl groups are commonly substituted at carbon 17. The rigid structure of the four-fused ring system is planar, allowing only substitutions above or below the plane. The positional substitutions above the plane are defined as beta (β) substitutions and those below the plane are alpha (α). Androstane containing a ketone at carbon 3 would be correctly named 3-androst*anone*, which is commonly shortened to simply androstanone. The name of androstene containing the same substitution would correctly locate both the double bond and the ketone group, for example, 4-androst*en*-3-*one*, usually shortened to just androstenone. A hydroxyl group substituted at carbon 17 of androstenone, above the plane of the ring (β position), would change the name to 4-*androsten*-17β-*ol*-3-*one*.

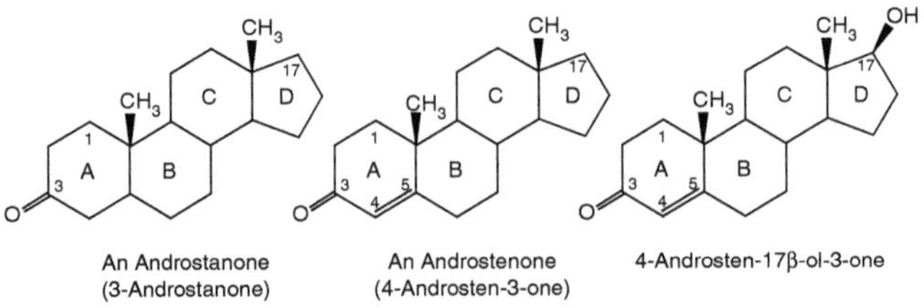

Structure 16.8

Careful examination of the name would be interpreted as a carbon–carbon double bond from carbon 4 to carbon 5 (4-androsten), an OH up at carbon 17 (17β-*ol*-), and a ketone functional group at carbon 3 (3-*one*). This systematic method is used to explicitly define the structure of steroids. Unfortunately, most natural and synthetic anabolic steroids are commonly referenced using generic names.

16.3.5 Frequently Encountered Steroids

The possession or sale of anabolic steroids without a valid prescription is illegal. Androgenic (promoting masculine characteristics) anabolic (tissue-building) steroids are Schedule III controlled substances defined as "any drug or hormonal substance chemically and pharmacologically related to testosterone (except estrogen, progestins, and corticosteroids) that promotes muscle growth." Street names include arnolds, gym candy, pumpers, roids, and juice. Some of the most commonly abused anabolic steroids are listed below.

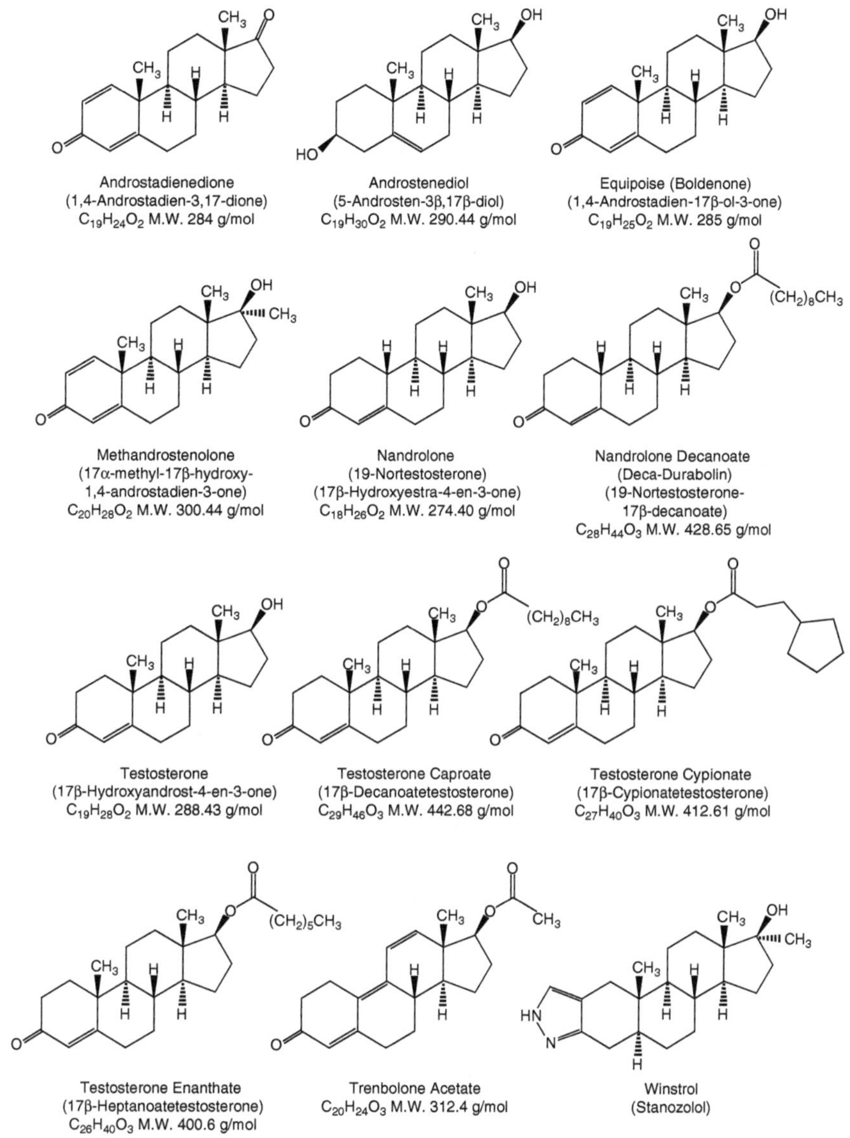

Structure 16.9

16.4 Analytical Methods

16.4.1 Visual Inspections

The illicit steroid market is inundated with an unknown number of genuine, chemically modified, mixed, and even counterfeit steroids, available in forms ranging from injections, capsules, tablets, dermal patches, ointments, and lotions. It is virtually impossible to gain even the slightest insight on the identity of a steroid using visual inspection. Mexico, Denmark, Australia, and the Netherlands are a few of the worldwide suppliers of anabolic steroids (Fig. 16.1).

16.4.2 Gas Chromatography Mass Spectrometry

Presumptive color-screening tests for the identification of anabolic steroids do not exist. FTIR analysis is often of little value due to the size and complexity of steroid molecules. Therefore, GCMS is the preferred technique when examining this group of controlled substances. Methanol is the solvent of choice; a small amount of suspected steroid liquid or solid is dissolved (or mixed) in methanol and the resulting solution is analyzed.

16.4.3 Mass Spectra of Commonly Encountered Steroids

The mass spectra of several steroids commonly submitted to forensic laboratories for analysis are shown below. Most structures are provided in previous pages for comparison and parent ion peak (M^+) identification (Fig. 16.2a–g).

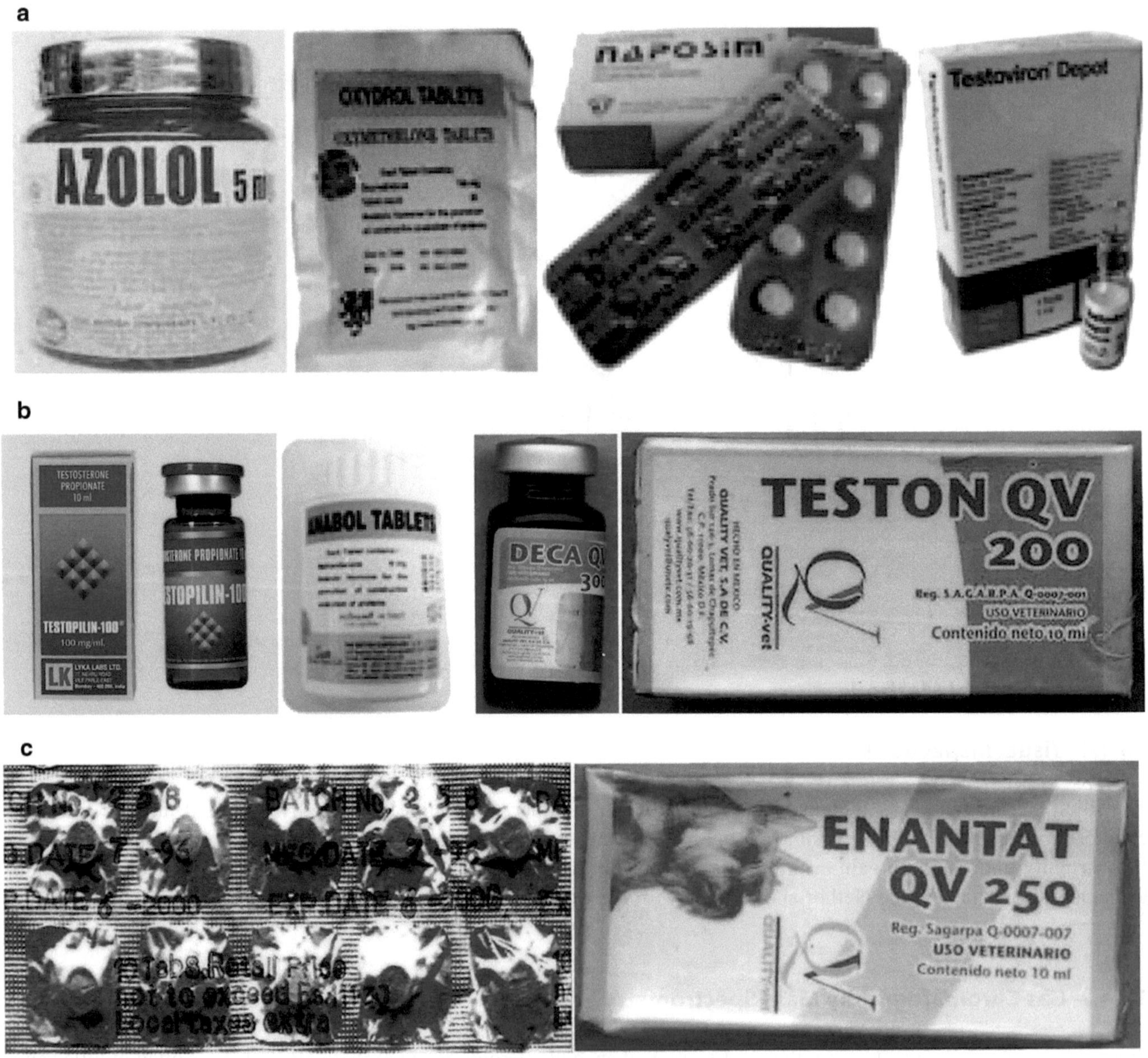

Fig. 16.1 Representative forms of steroids packaged for distribution.

d

e

Fig. 16.1 (continued)

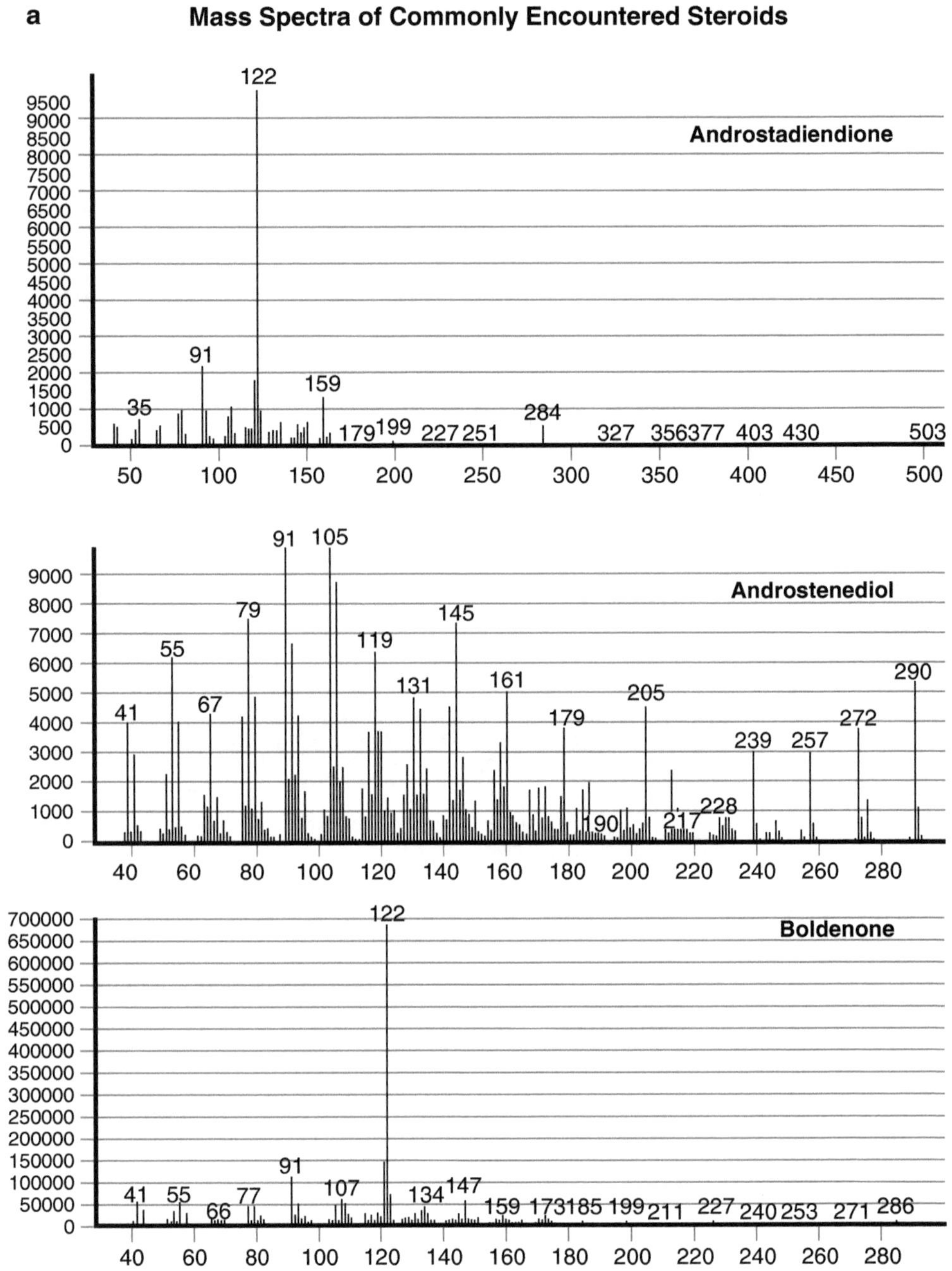

Fig. 16.2 The MS spectra for steroids commonly submitted to forensic laboratories are shown. Correlate steroid structural formulas shown in struct. 16.9 to their corresponding spectra and identify M^+ peaks. Note: Struct. 16.9 does not provide a structure for all spectra.

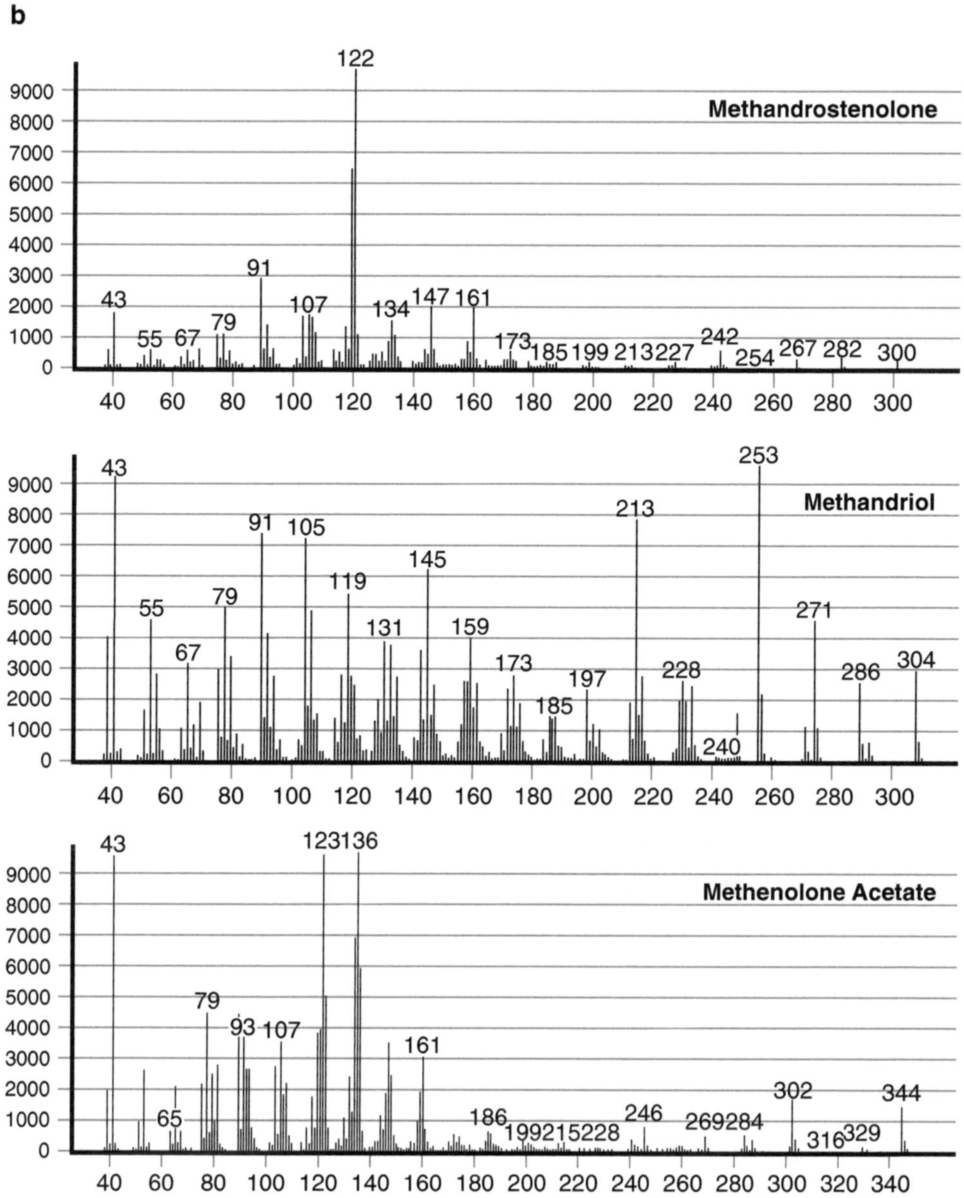

Fig. 16.2 (continued)

c

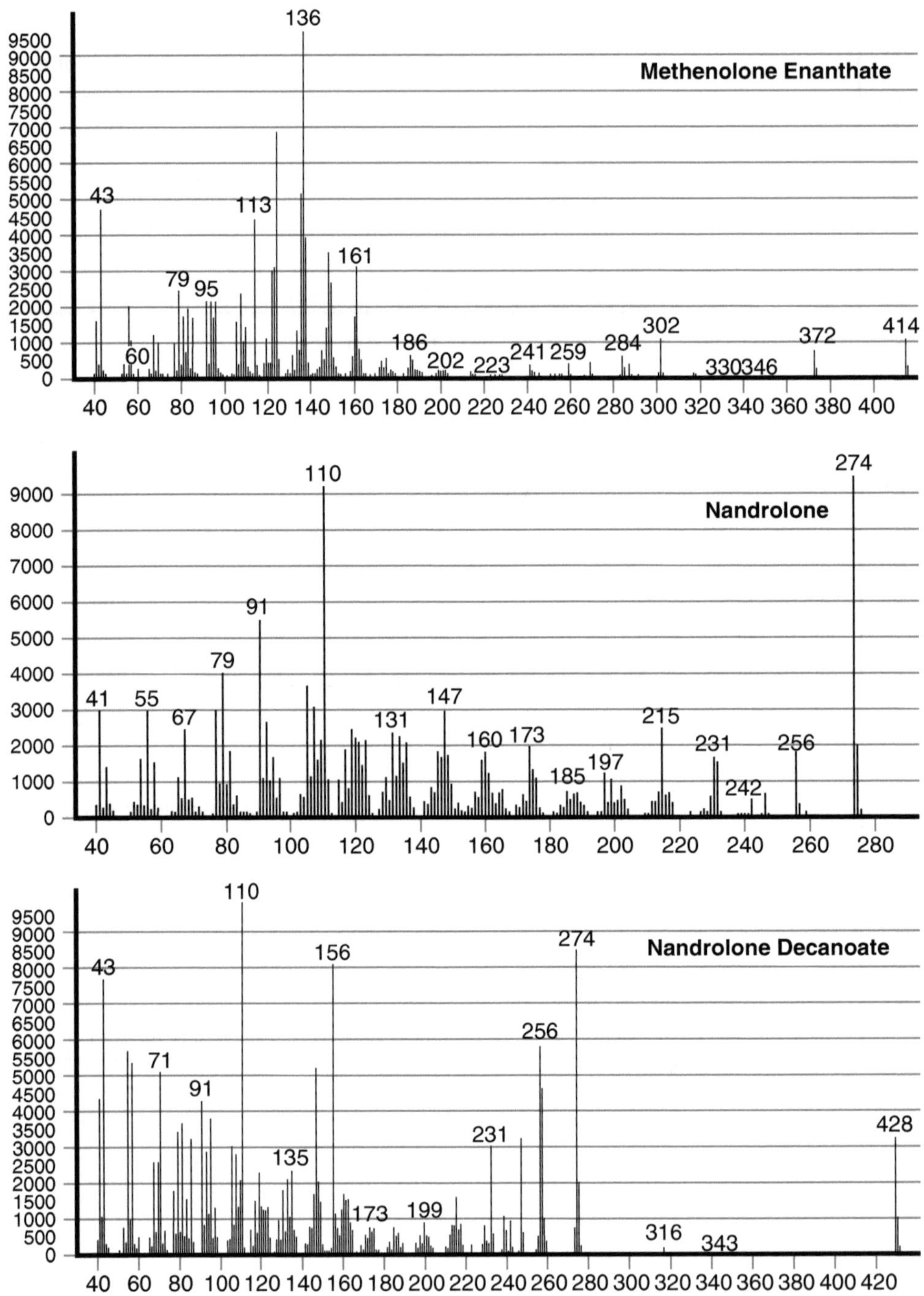

Fig. 16.2 (continued)

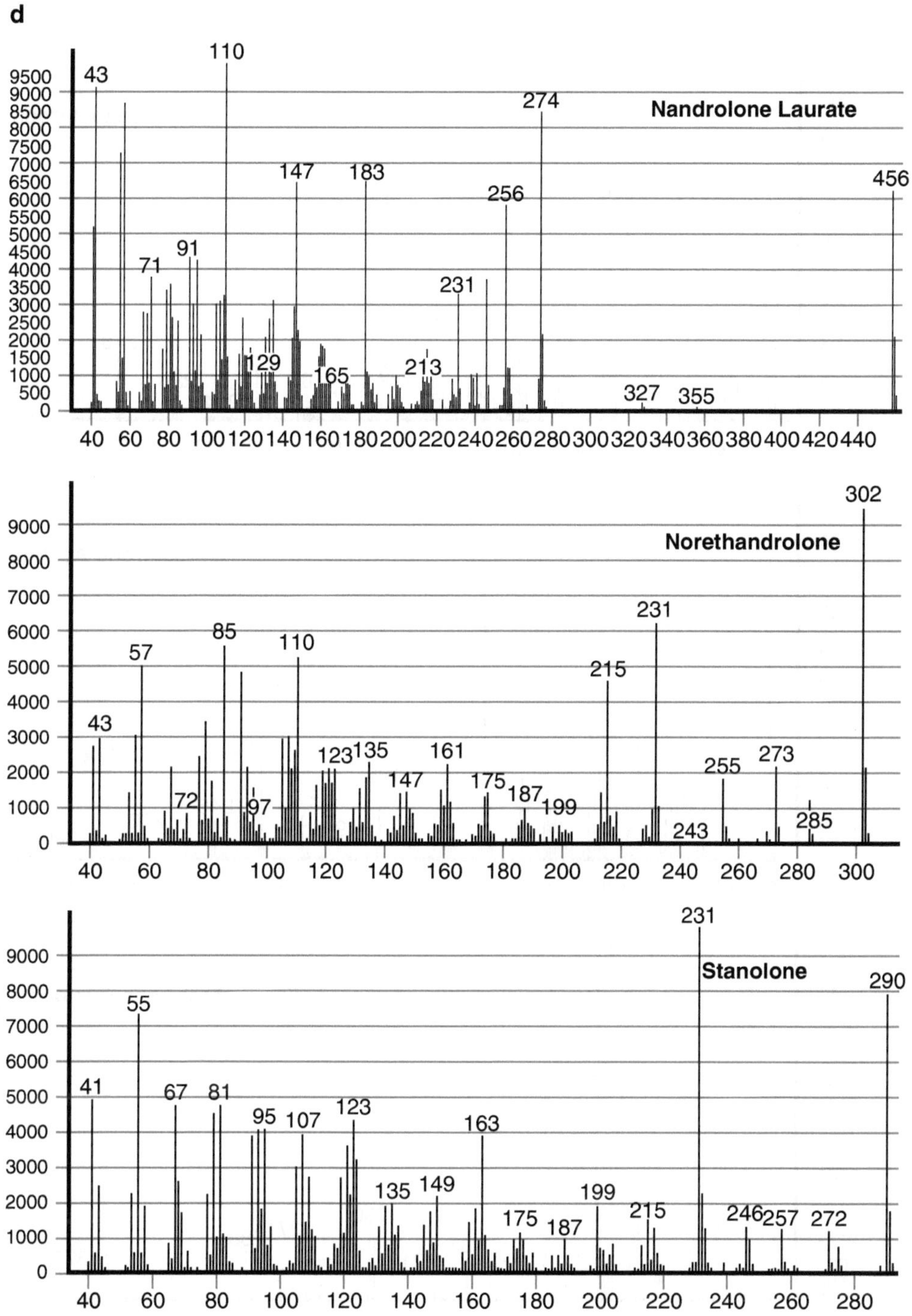

Fig. 16.2 (continued)

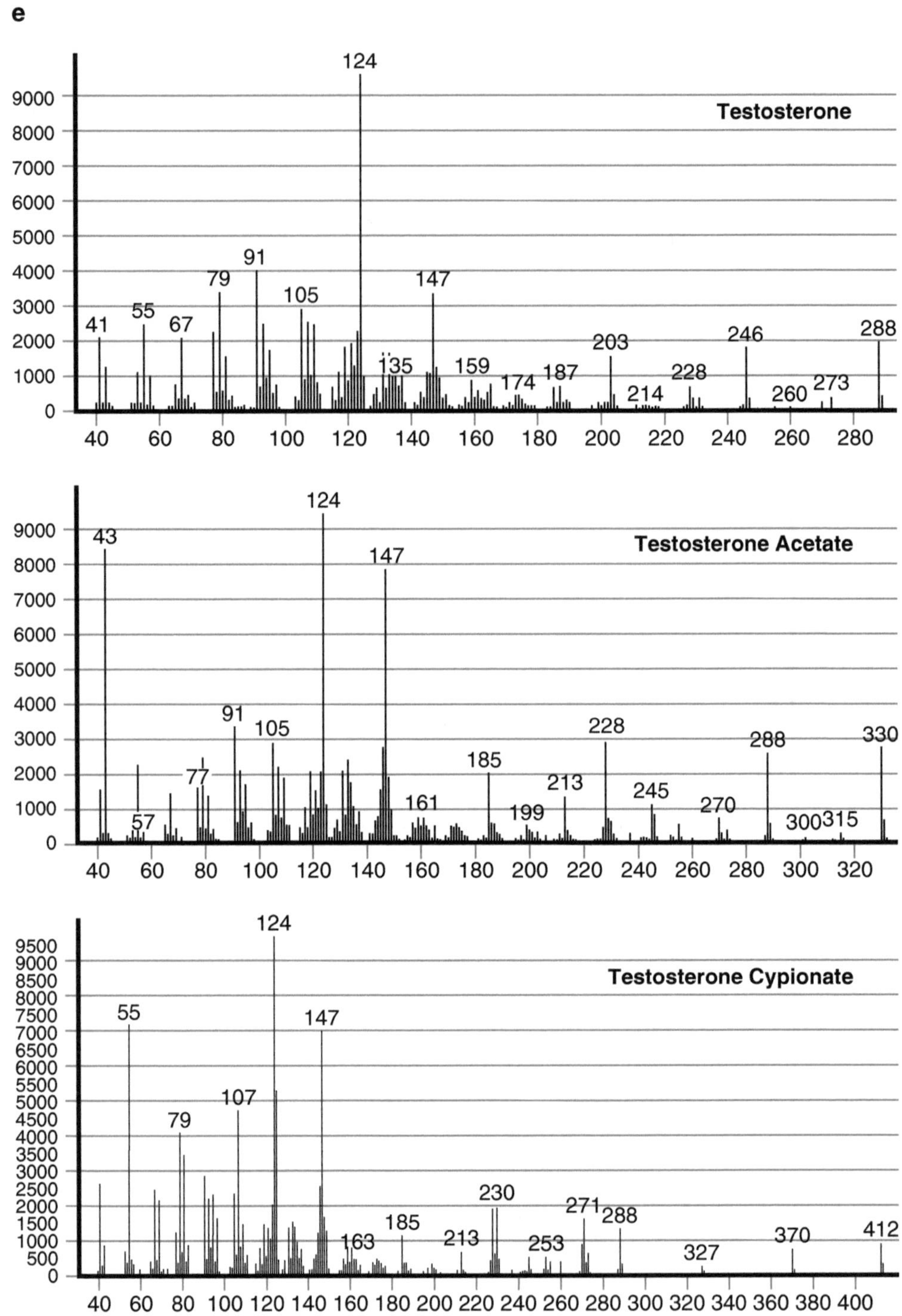

Fig. 16.2 (continued)

f

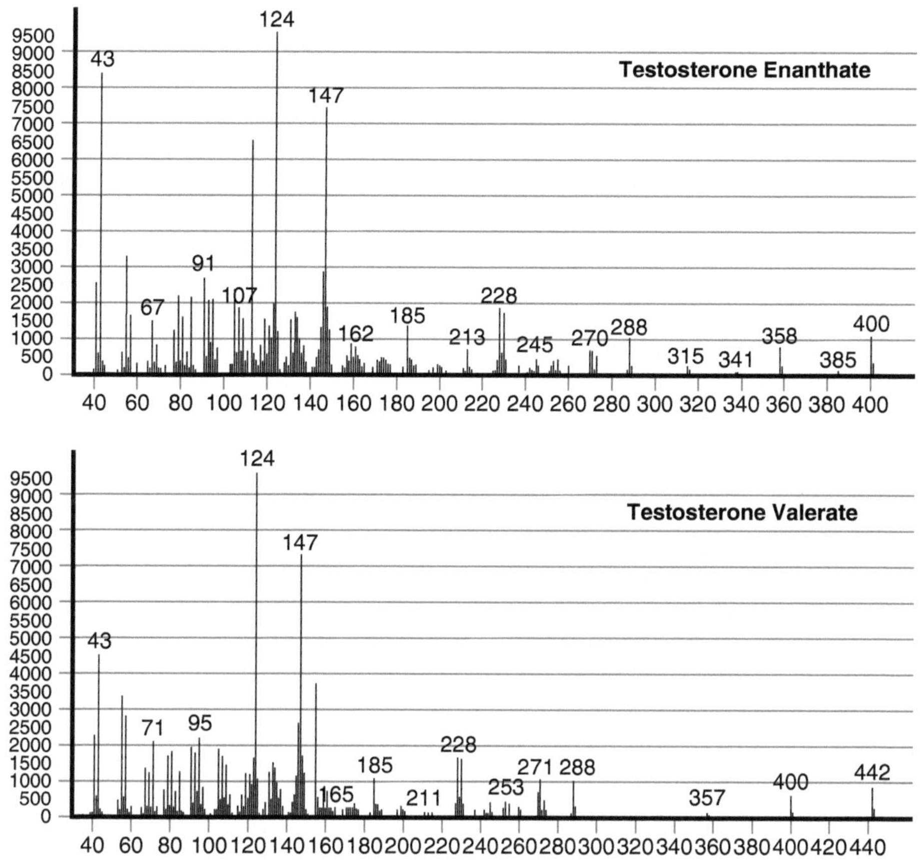

g

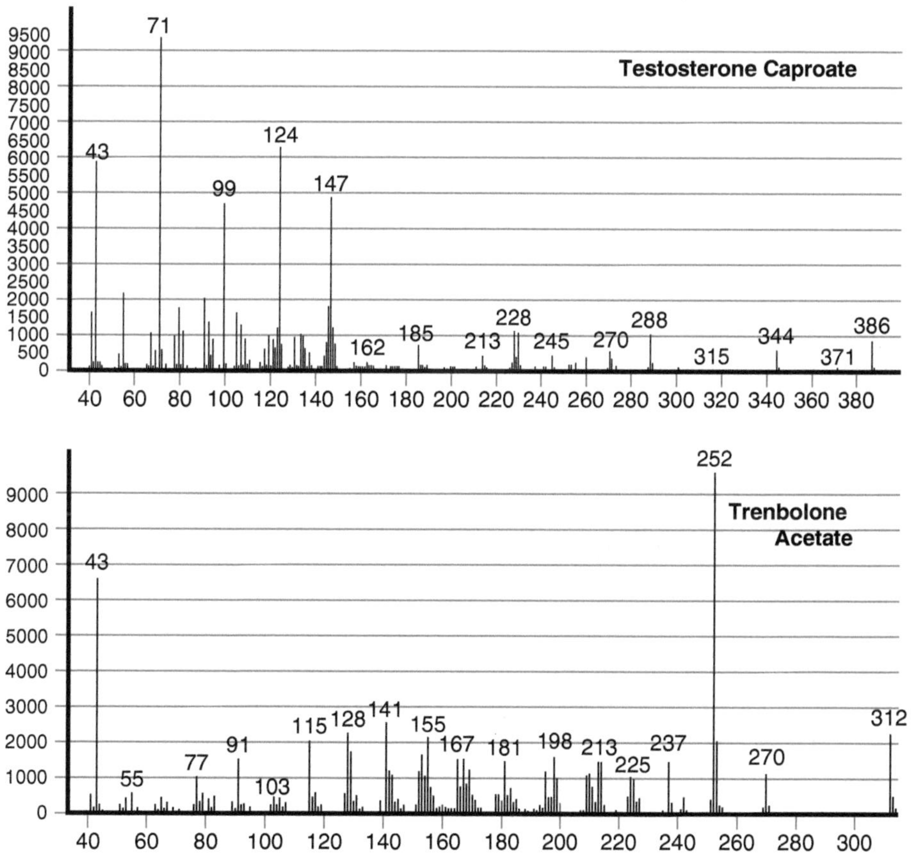

Fig. 16.2 (continued)

16.5 Questions

1. Describe the function of steroid hormones to members of the jury.
2. Briefly discuss the five classes of steroid hormones.
3. Define anabolic steroid to members of the jury and briefly explain how they differ from other steroids.
4. List four damaging side effects of steroid abuse.
5. Explain to the jury why injection of steroids is generally more effective than tablets.
6. Interpret the name 17α-methyl-17β-hydroxy-1,4-androstadien-3-one to members of the jury. How is this different from 17α-methyl-1,4-androstadien-17β-ol-3-one?
7. How are androgenic anabolic steroids defined and classified according to the Controlled Substance Act?
8. Draw the structure of the following steroids.
   (a) 4-Androstenediol (4-androsten-3β,17β-diol)
   (b) 5-Androstenediol (5-androsten-3β,17β-diol)
   (c) 5-Androstenedione (5-androsten-3,17-dione)
   (d) 5-Androsten-3β,17α-diol.
9. List three street names for anabolic steroids.
10. Outline a chemical screening test commonly used in the identification of steroids.
11. Why is FTIR generally ineffective in steroid examination?
12. Determine the molecular ion peak (M^+) for each of the following using the MS spectra in the chapter. Provide the chemical name in each example.
   (a) Boldenone
   (b) Nandrolone
   (c) Nandrolone decanoate
   (d) Testosterone
   (e) Testosterone cypionate
13. Choose one of the MS spectra shown in Fig. 16.2 that does not have a structural formula in struct. 16.9. Research your chosen steroid and report your results to the class.

Suggested Reading

Allen, M.; Rahkila, P. Anabolic-Androgenic Steroid Effects on Endocrinology and Lipid Metabolism in Athletes. *Sports Med.* **1988**, 6, 327–332.

American College of Sports Medicine. Position Stand on the Use of Anabolic-Androgenic Steroids in Sports. *Med. Sci. Sports Exerc.* **1987**, 19, 534–539.

American Medical Association, Council on Scientific Affairs. Medical and Non-Medical uses of Anabolic-Androgenic Steroids. *JAMA*. **1990**, 264, 2923–2927.

Bahrke, M. S.; Yesalosk, C. E.; Wright, J. E. Psychological and Behavioral Effects of Endogenous Testosterone Levels and Anabolic-Androgenic Steroids Among Males: A Review. *Sports Med.* **1990**, 10, 303–337.

ChemNet. Global Chemical Network. http://us.chemnet.com/ (accessed September 2009).

Haupt, H. A. Anabolic Steroids and Growth Hormone. *Am. J. Sports Med.* **1993**, 21, 468–474.

United States Olympic Committee. *USOC Drug Education Handbook*. USOC: Colorado Springs, 1989.

U. S. Department of Justice Drug Enforcement Administration. A Dangerous and Illegal Way to Seek Athletic Dominance and Better Appearance. http://www.deadiversion.usdoj.gov/pubs/brochures/steroids/public/index.html (accessed September 2009).

U. S. Department of Justice Drug Enforcement Administration. *Symposium on Steroids in DEA Southwest Labs in California, USA*. U. S. Drug Enforcement Administration: Washington, DC, 2005.

Voet,D.; Voet, J.G. *Biochemistry*, 2nd ed.; John Wiley & Sons, Inc.: New York, 1995; p 703.

Wilson, J. D. Androgen Abuse in Athletes. *Endocr. Rev.* **1988**, 9, 181–199.

Zaretski, I.; Zollinger, M.; Seibl, J. Mass Spectrometry of Steroids. *Mass Spectrum*. **1976**, 20, 649–656.

17 Miscellaneous Controlled Substances

17.1 Introduction

This chapter focuses on a variety of controlled substances that cannot be accurately classified using the methods developed in previous chapters. Most contain more than one functional group and, although they produce effects similar to drugs previously discussed, their structures prohibit classification as phenethylamines, tertiary amines, tryptamines, or steroids. This group includes barbiturates, fentanyl, γ-hydroxybutyric acid (GHB), ketamine, and lysergic acid diethylamide (LSD).

17.2 Barbiturates

All barbiturates are derived from barbituric acid, a heterocyclic six-membered ring containing two nitrogen atoms and three carbonyl groups (carbon–oxygen double bonds). The nature of the group substituted at position 5 of barbituric acid is typically responsible for the effects produced by different types of barbiturates.

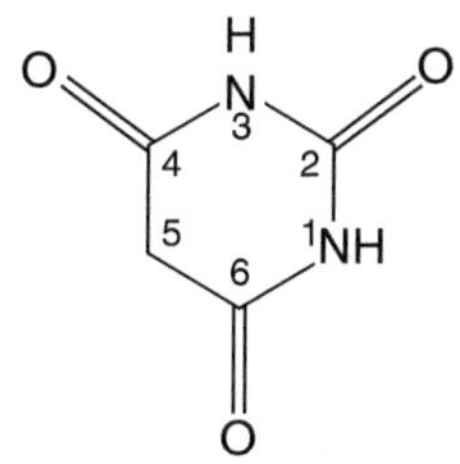

Barbituric Acid

Structure 17.1

Barbiturates are commonly used as sedatives, hypnotics, anesthetics, and anticonvulsants. They act as central nervous system depressants, producing effects ranging from mild sedation to coma. Barbiturates are usually classified into groups termed ultra-short acting, short acting, intermediate acting, or long acting, based on their rate of onset and duration of clinical effects.

Ultra-short-acting barbiturates produce anesthetic effects almost immediately after intravenous injection. The schedule IV drug methohexital (Brevital) and the schedule III drugs thiamylal (Surital) and sodium thiopental (Sodium Pentothal) fall into this category and have common applications in medicine. Abusers of barbiturates often prefer members of the schedule II short- and intermediate-acting forms such as amobarbital (Amyta), pentobarbital (Nembutal), and secobarbital (Seconal). Schedule III short- and intermediate-acting barbiturates include butalbital (Fiorina) and butabarbital (Butisol).

J.I. Khan et al., *Basic Principles of Forensic Chemistry*, DOI 10.1007/978-1-59745-437-7_17,
© Springer Science+Business Media, LLC 2012

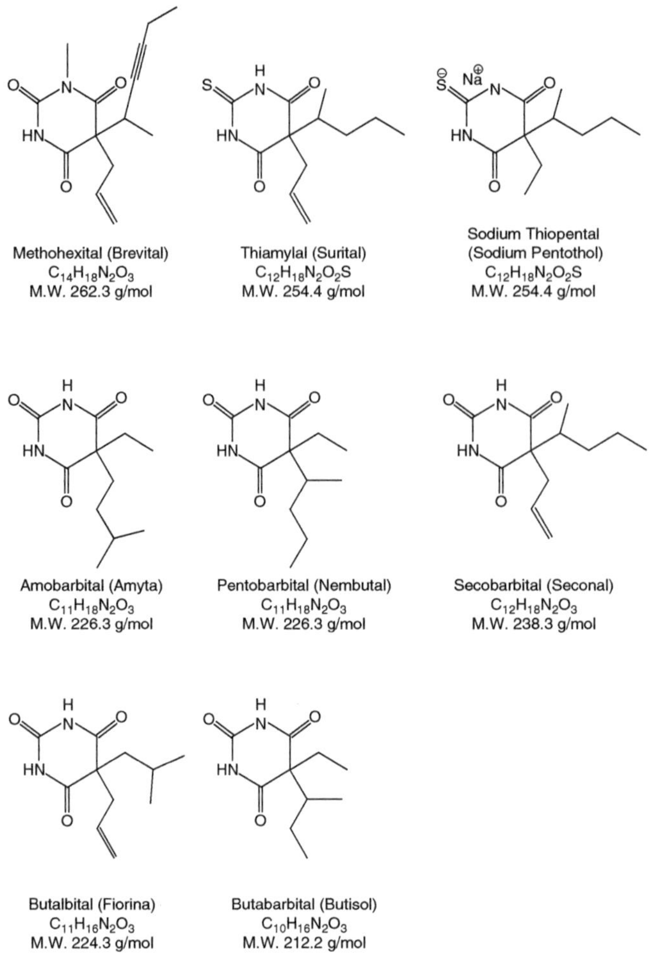

Structure 17.2

The onset of activity associated with intermediate types is usually 15–40 min after oral ingestion and the effects may last up to 6 h. These durations are useful in the treatment of insomnia and preoperative sedation. Veterinarians often use low doses of pentobarbital (short/intermediate acting) and cyclopal (long acting) for anesthesia and high doses for euthanasia.

The schedule IV long-acting barbiturates include phenobarbital (Luminal) and mephobarbital (Mebaral). With an onset time of approximately 1 h and a 12-h duration of effectiveness, these drugs are ideally suited for daytime sedation and the treatment of seizure disorders (Fig. 17.1).

Fig. 17.1 Barbiturate capsules and tablets. Barbiturates are commonly used in combination with muscle relaxants for capital punishment by lethal injection. Contrary to popular belief, sodium pentothal-"truth serum" (a barbiturate) does not force people to tell the truth. However, the effects reduce inhibitions, which may predispose individuals to be caught off-guard during questioning.

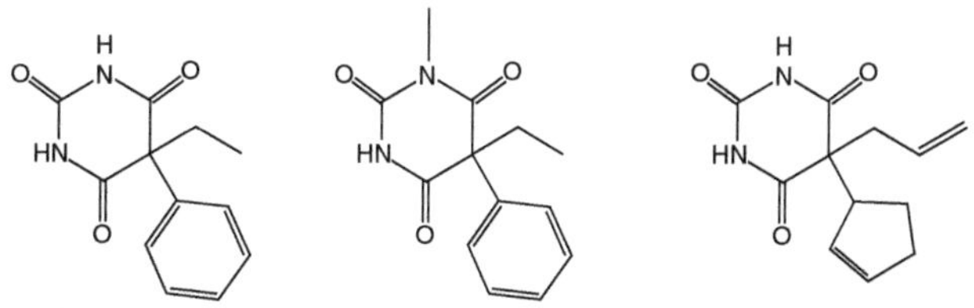

Phenobarbital (Luminal) $C_{12}H_{12}N_2O_3$ M.W. 232.2 g/mol

Mephobarbital (Mebaral) $C_{13}H_{14}N_2O_3$ M.W. 246.2 g/mol

Cyclopal (Allylcyclopentenyl Barbituric Acid) $C_{12}H_{14}N_2O_3$ M.W. 234.3 g/mol

Structure 17.3

In the very near future, it is highly likely that barbiturates will be found only in the illicit drug market. Their use in common medical practice is declining because many treatments choose to use benzodiazepines instead. This decision is based primarily on the fact that the same clinical affects can be achieved using benzodiazepines without the high risk of lethal overdose associated with barbiturates.

Street names for barbiturates include barbs, bluebirds, downers, goofballs, tooties, and yellow jackets.

17.3 Fentanyl

Fentanyl (*N*-phenyl-*N*-(1-phenethyl-4-piperidinyl) propanamide) is extensively used as an anesthetic and analgesic in operating rooms and intensive care units under the trade name Sublimaze.

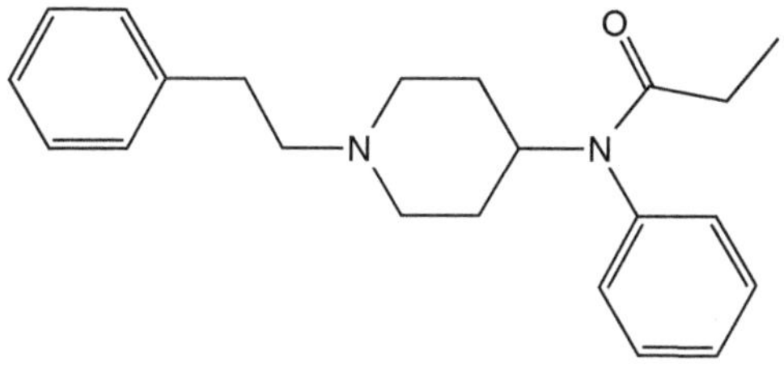

Fentanyl $C_{22}H_{28}N_2O$ M.W. 336.5 g/mol

Structure 17.4

It is a schedule II opioid that is nearly 100 times more potent than morphine and is particularly effective in treating chronic pain associated with various forms of cancer. Actiq, a "lollypop" form of fentanyl citrate, is often used in cancer treatments because no known opioid is more effective in reducing cancer pain. Duragesic is a fentanyl transdermal patch commonly used in long-term pain management and Carfentanil, the most potent analog of fentanyl, is roughly 10,000 times stronger than morphine. It has applications in veterinary medicine where it is often used to sedate or immobilize large animals. Fentanyls are potent drugs that have an extremely high potential for lethal overdose. They are generally sold as tablets, transdermal patches, lollypops, and injections (Fig. 17.2).

Fentanyl is potentially hazardous to handlers, transporters, and examiners due to its high potency and susceptibility to dermal absorption. Therefore, additional safety measures are highly recommended when in direct contact with this drug, i.e.; field collection and laboratory analysis.

Illicit use of fentanyls first appeared in the mid-1970s and 12 different analogs have been currently identified by the DEA. The clinical effects of these derivatives are very similar (indistinguishable) to those of heroin; in fact, many fentanyl overdoses have been incorrectly attributed to heroin. The "China White" forms of fentanyl refer to any derivative produced in clandestine laboratories. They are classified as schedule I controlled substances and a particularly famous form is α-methyl-fentanyl (AMF).

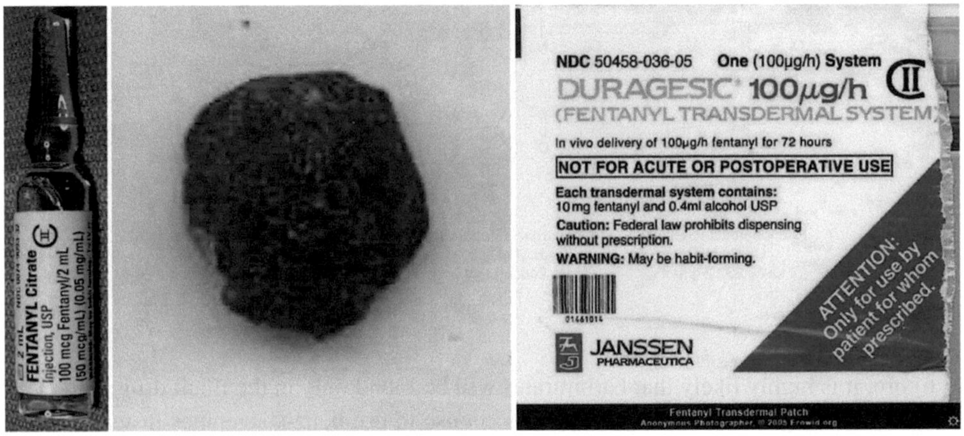

Fig. 17.2 Fentanyl is commonly available in tablets, patches, lollypops, and injections. This extremely potent narcotic has a high potential for lethal overdose. Carfentanil, or the "China White" variant 3-methylfentanyl, is believed to be the active ingredient in Kolokol-1, a highly classified Russian chemical weapon.

AMF was the first "designer drug" developed entirely in clandestine laboratories for sale as an illicit drug rather than as a product of legitimate scientific research. It produces longer-lasting effects because it is relatively more resistant to metabolic degradation.

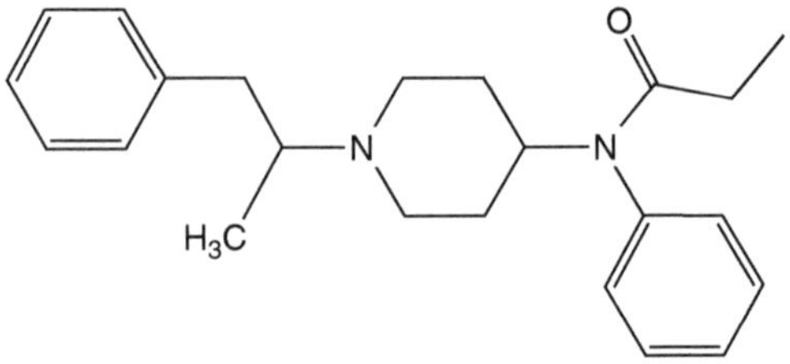

α-Methylfentanyl
$C_{23}H_{30}N_2O$
M.W. 350.5 g/mol

Structure 17.5

Almost immediately, ten different analogs of AMF appeared in the illicit drug market. The emergence of such a wide class of novel narcotic drugs had a significant impact on the government's approach to classifying illicit drugs. It prompted the implementation of the Federal Analog Act which, for the first time, attempted to control entire families of drugs based on structural similarities as opposed to listing each drug individually.

Heroin and cocaine are commonly mixed ("cut") with fentanyl to increase the quantity for sale or to compensate for low-grade or impure forms. This combination has claimed the lives of many unsuspecting addicts who unknowingly overdose on the potent, highly toxic mixture.

17.4 *Gamma*-Hydroxybutyric Acid: γ-Hydroxybutyric Acid

GHB is a naturally occurring substance found in the central nervous system in small concentrations.

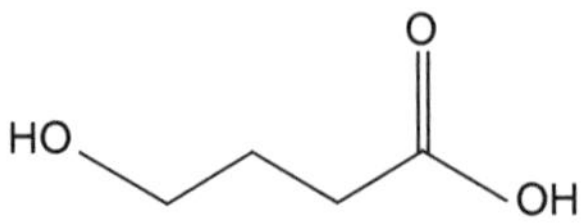

GHB
(γ-Hydroxybutyric Acid)
$C_4H_8O_3$
M.W. 104.1 g/mol

Structure 17.6

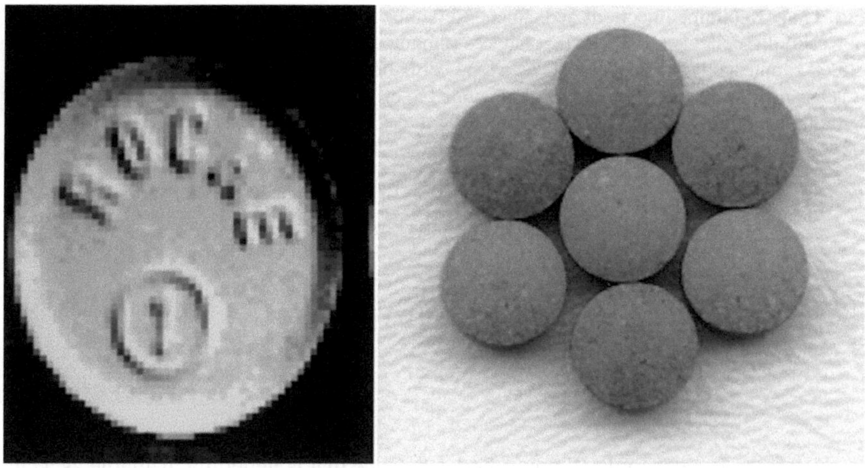

Fig. 17.3 Common forms of gamma-hydroxybutyric acid (GHB). The popular children's toy "Aqua-Dots" was recalled and cancelled when it was discovered that 1,4-butanediol (which metabolizes to GHB) was used in the bead manufacturing process.

It is a neuroprotective nutrient whose principal activity is to guard against neuron degradation. At recreational doses of 500–3,000 mg, it is a central nervous system depressant that induces euphoria.

In recent years, GHB has emerged as a significant drug of abuse throughout the United States. It is typically an odorless, colorless liquid, and is also available as a tablet (Fig. 17.3). The production of GHB in clandestine laboratories is a relatively simple process and local operators usually handle its distribution. It is usually mixed with alcohol for abuse and has become popular among teens and young adults at dance clubs, bars, gyms, and "raves." Also, it is routinely used by bodybuilders for its anabolic effects. In low doses, GHB causes drowsiness, dizziness, nausea, and visual disturbances. In high doses, unconsciousness, seizures, severe respiratory depression, and coma can occur. GHB overdose usually requires emergency room treatment, including intensive care for respiratory depression and coma.

GHB gained notoriety as a "date-rape" drug and has been implicated in a number of drug-related sexual assaults. It is very difficult to establish the involvement of GHB in such cases because its detection in bodily fluids is virtually precluded by the fact that it is quickly eliminated from the body (within 24 h). Also, the rapid onset of CNS depressant effects commonly leaves the victim with little memory of the details of the attack. Thus, a majority of cases are unreported, and, those that are reported, are often unsubstantiated.

Street names of GHB include lquid ecstasy, scoop, easy lay, Georgia home boy, grievous bodily harm, liquid X, and goop.

17.5 Ketamine

Ketamine (ketamine hydrochloride) is primarily used as a tranquilizer in veterinary medicine, but it also has legitimate uses in humans as well. It is often part of the treatment for pediatric burns and has various applications in dentistry and experimental psychotherapy. The hydrochloride salt is marketed as Ketalar for medicinal use and Ketaset to veterinarians.

Ketamine is a *dissociative* anesthetic that produces hallucinogenic effects similar to PCP.

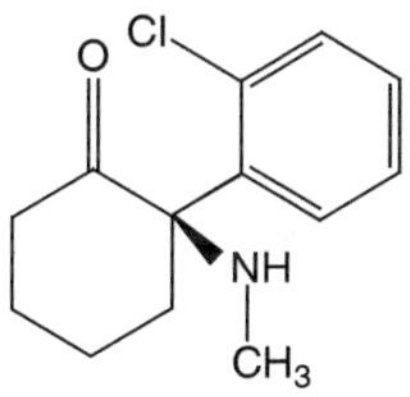

Ketamine HCl
$C_{13}H_{16}ClNO$
M.W. 237.7 g/mol

Structure 17.7

It blocks signals to the conscious mind from various parts of the brain and severely impairs all senses, especially sight, balance, and sense of time. Ketamine is characterized as a "date-rape drug" and is typically abused as a "club drug" distributed at raves and parties.

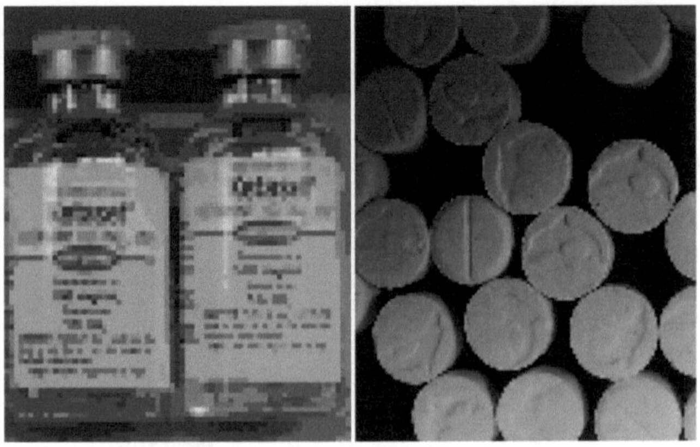

Fig. 17.4 Ketamine injections and tablets. Ketamine is a powerful dissociative anesthetic that induces (among other effects) amnesia. It is considered a "date rape" drug because it is has no odor or taste and can be easily added to drinks without detection.

Ketamine is a schedule III controlled substance that is generally available as a clear liquid or white (or off-white) powder. The powder can be snorted or compressed into tablets for oral administration (Fig. 17.4). The powder or liquid can also be added to tobacco or marijuana and smoked. Ketamine is most effective when injected either IM (intramuscular) or IV (intravenous). The powder form of ketamine is strikingly similar to cocaine and crystal methamphetamine and careful examination is required to avoid incorrect identification. Ketamine is sometimes sold as counterfeit MDMA (ecstasy) or as a mixture with other drugs such as ephedrine and caffeine. Most illicit ketamine comes from diverted legitimate suppliers or theft, primarily from veterinary clinics.

Ketamine produces effects similar to PCP (phencyclidine or "Angel Dust"), including numbness, loss of coordination, sense of invulnerability, muscle rigidity, aggressive or violent behavior, slurred or blocked speech, exaggerated sense of strength, and a blank stare. Ketamine causes depression of respiratory functions, but, unlike most hallucinogens, the central nervous system is generally unaffected and cardiovascular function is maintained. The anesthetic properties of ketamine can relieve tension and anxiety, but inadvertent self-injury to the user is cause for great concern.

Ketamine produces a general state of dissociation that can render the user comatose. It is typically characterized by a sense of detachment from one's body and the external world. The effects usually last an hour but can easily extend to 4–6 h. The effects of chronic use may take as long as 2 years to subside and flashbacks are common up to 1 year after use. Low doses (25–100 mg) produce psychedelic effects quickly, while large doses (above 200 mg) can cause convulsions resulting in oxygen deprivation in the brain and muscles. Long-term effects include tolerance and a high potential for both physical and psychological dependence.

Some street names for ketamine include K, Ket, special K, vitamin K, vit K, kit kat, Keller, Kelly's day, green, blind squid, cat valium, purple, special la coke, super acid, and super C. Slang for the "ketamine high" include "K-hole," "K-land," "baby food," and "god."

17.6 Lysergic Acid Diethylamide

Lysergic acid diethylamide (LSD) is a powerful psychedelic hallucinogen classified as a schedule I controlled substance.

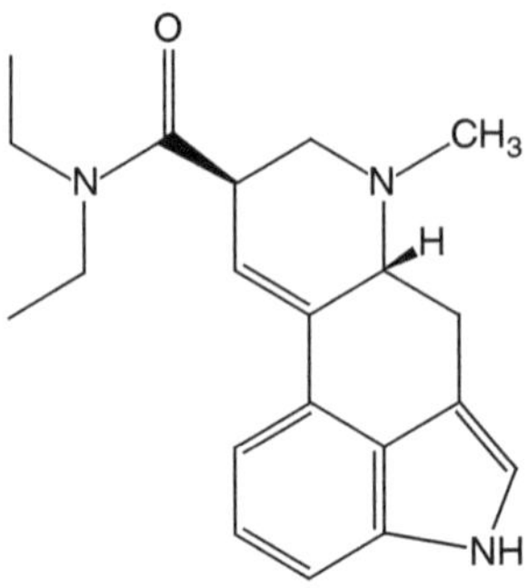

LSD
(Lysergic Acid Diethylamide)
$C_{20}H_{25}N_3O$
M.W. 323.4 g/mol

Structure 17.8

Fig. 17.5 Common forms of LSD for distribution. Contrary to popular belief, LSD is not considered addictive because it does not produce a compulsive need for the drug. Users can develop tolerance however, requiring progressively higher doses of the drug in order to achieve the desired affects.

Fig. 17.6 Ergot-infected rye. Note the *dark* sclerotia extending from the base. Thousands of pounds of ergot sclerotia are harvested each year from Midwestern rye farms and used in the production of various prescription drugs.

It is a naturally occurring alkaloid found in *ergot*, a fungus (*Claviceps purpurea*) that infects the grains of rye and related grasses. It is also synthetically produced from the schedule III precursors lysergic acid and lysergic acid amide, and the List I chemicals ergotamine and ergonovine. A particularly dangerous synthetic form of LSD, known as LSD-25, is derived from natural psychoactive alkaloids found in ergot. It is roughly 4,000 times stronger than mescaline and is one of the most powerful psychoactive drugs in existence.

LSD is available in tablet, capsule, and liquid form. It is commonly blotted on decorative paper and cut into individual squares that are sold and ingested orally (Fig. 17.5). Naturally occurring LSD (ergine) is also found in the seeds of two species of Mexican morning glory vines that are ingested by Native Americans as part of cultural medicinal and religious rituals.

Ergot forms a dark, compact, fungal mass called a sclerotium where the grain would normally develop. These pellet-like structures can be seen in an infected grain spike typically extending out from the bracts (glumes) (Fig. 17.6). The sclerotia are the source of potent alkaloids found in *Claviceps purpurea*. In late spring, when rye plants are in bloom, the sclerotia from the previous year's crop produce stalked ascocarps resembling microscopic fungal fruiting bodies. The head of each ascocarp contains several embedded perithecia, which contain numerous saclike asci, each with eight ascospores. The ascospores infect the young, developing grains (ovaries) of rye plants, eventually replacing them with purplish-black sclerotia. Because it produces ascospores within saclike asci, *Claviceps* is placed in the fungal class Ascomycetes.

The study of ergot fungus has led to several important medical discoveries. In 1935, the alkaloid ergonovine, which causes strong muscular contractions, was isolated from ergot and used to induce labor and control hemorrhaging. Also, the alkaloid ergotamine has been used extensively to relieve migraine headaches by causing constriction of blood vessels.

17.7 Analytical Methods

17.7.1 Visual Identification

The visual identification of this group of controlled substances is difficult because most samples are commonly submitted to forensic laboratories in many shapes, sizes, and physical forms (Fig. 17.7).

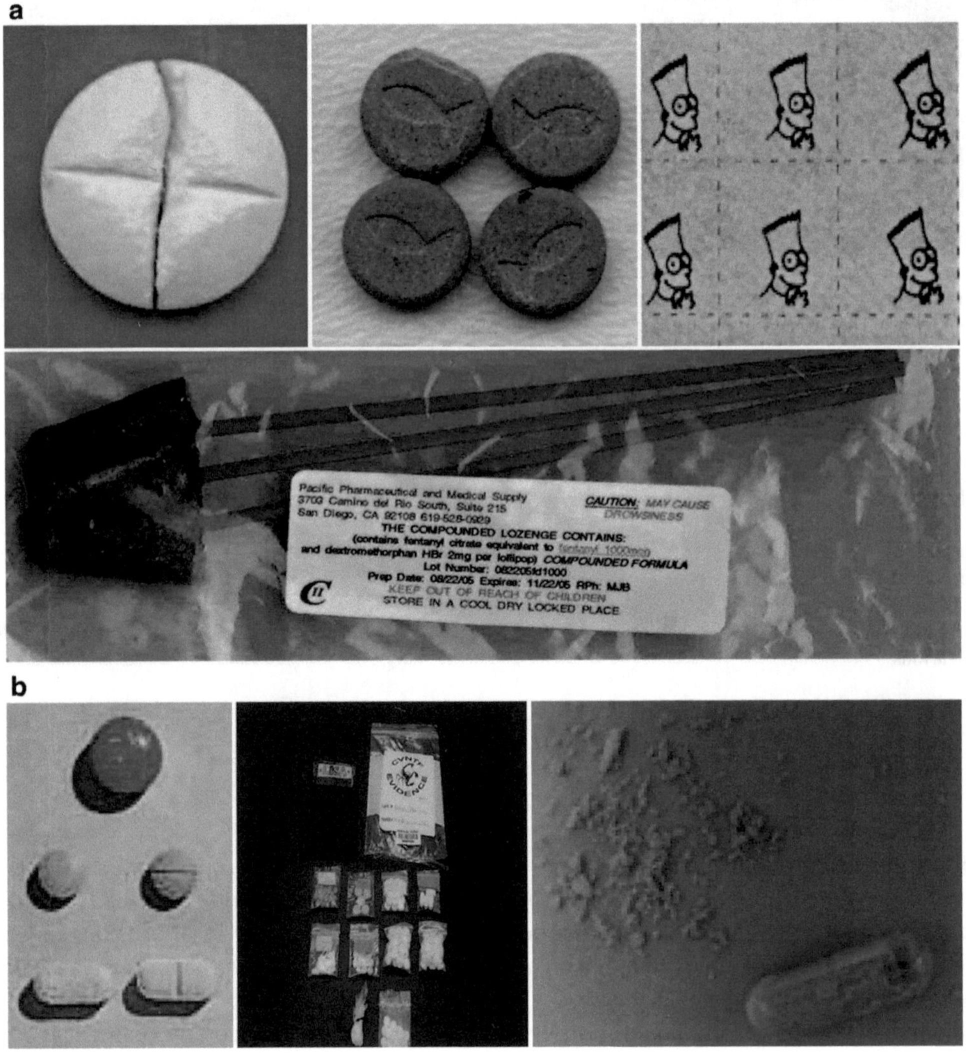

Fig. 17.7 Common forms of controlled substances submitted to forensic laboratories for analysis. (**a**) LSD tablets (*top left and center*) and designer paper blotted with liquid LSD (*top right*). Fentanyl lollypops (*middle*), (**b**) barbiturate tablets (*bottom left*) and capsules (*bottom right*), and *gamma*-hydroxybutyric acid (*bottom center*).

17.7.2 Chemical Screening Tests

The comprehensive flowchart illustrates the results of color-screening tests commonly performed on suspected samples of LSD and barbiturates. The chart also contains tests used on a variety of other controlled substances. Ketamine is indicated by the formation of blue color using the neutral tertiary amine test. GHB is indicated using one of the following tests: cromic acid test (5 M H_2SO_4 saturated with cromic acid) produces a blood red-brown color, ferric chloride test (distilled water saturated with Iron(III) chloride-$FeCl_3$) produces a rust brown-red color, and cobalt nitrate test (1% cobalt(II) nitrate-$Co(NO_3)_2$) produces a violet color (Fig. 17.8).

17.7.3 Gas-Chromatography Mass Spectrometry

GCMS is typically used to identify the controlled substances in this chapter. Sample preparation is performed using acid–base, base, and methanol extraction. For direct extraction of suspected LSD, the paper tabs, gel tabs, or tablets are cut into small pieces and added to the extracting solvent. The mixture is vortexed and the resulting solution is transferred (without debris) into an autosampler vial for GCMS. Typical GCMS results for a variety of controlled substances are shown below (Fig. 17.9a–h).

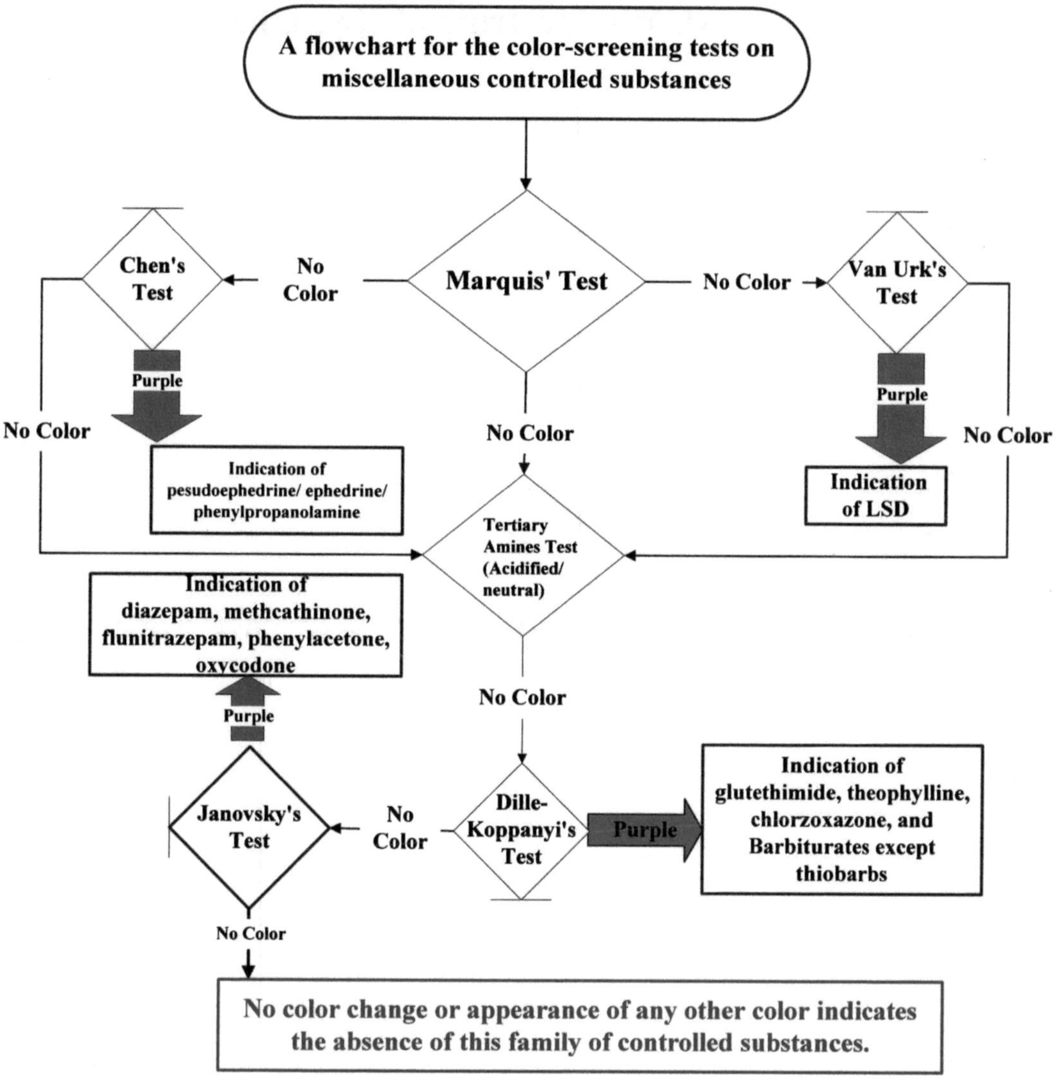

Fig. 17.8 Results of presumptive color tests used to screen miscellaneous controlled substances.

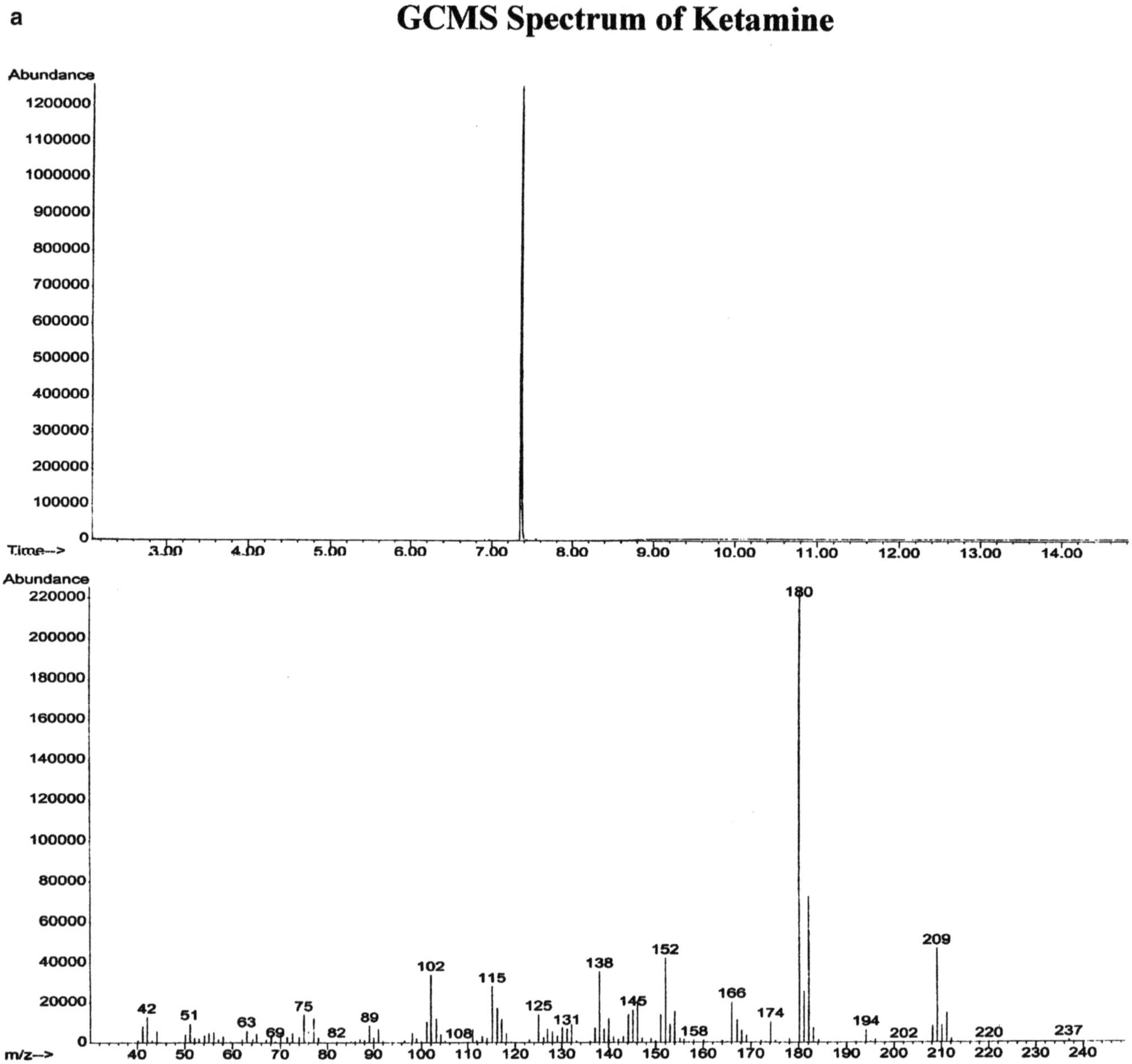

Fig. 17.9 Representative GCMS spectra of miscellaneous controlled substances.

b

GCMS Spectrum of Demerol

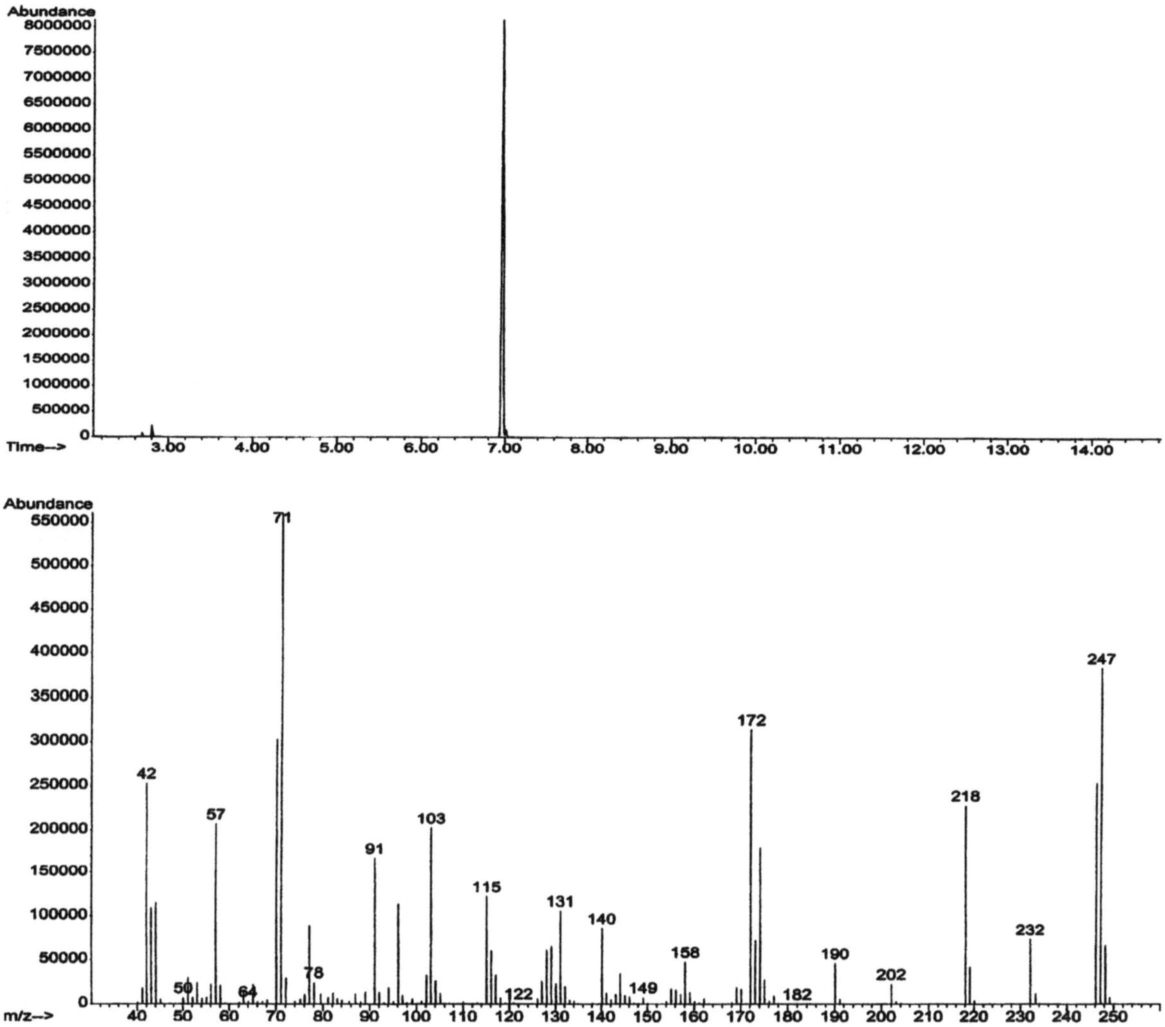

c

MS Spectra of GBL/GHB

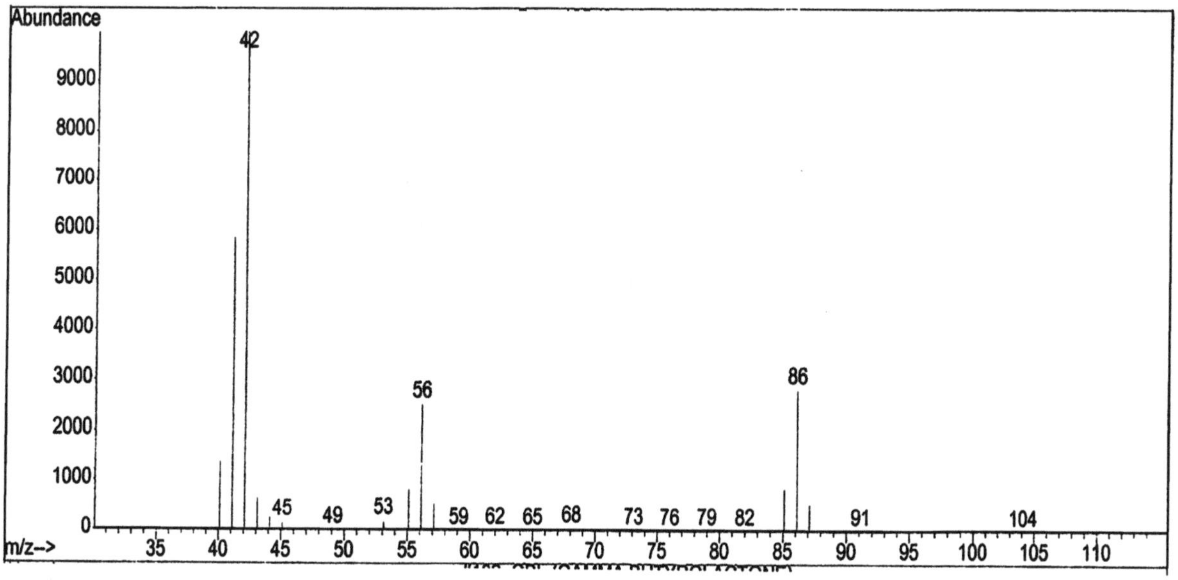

Fig. 17.9 (continued)

d

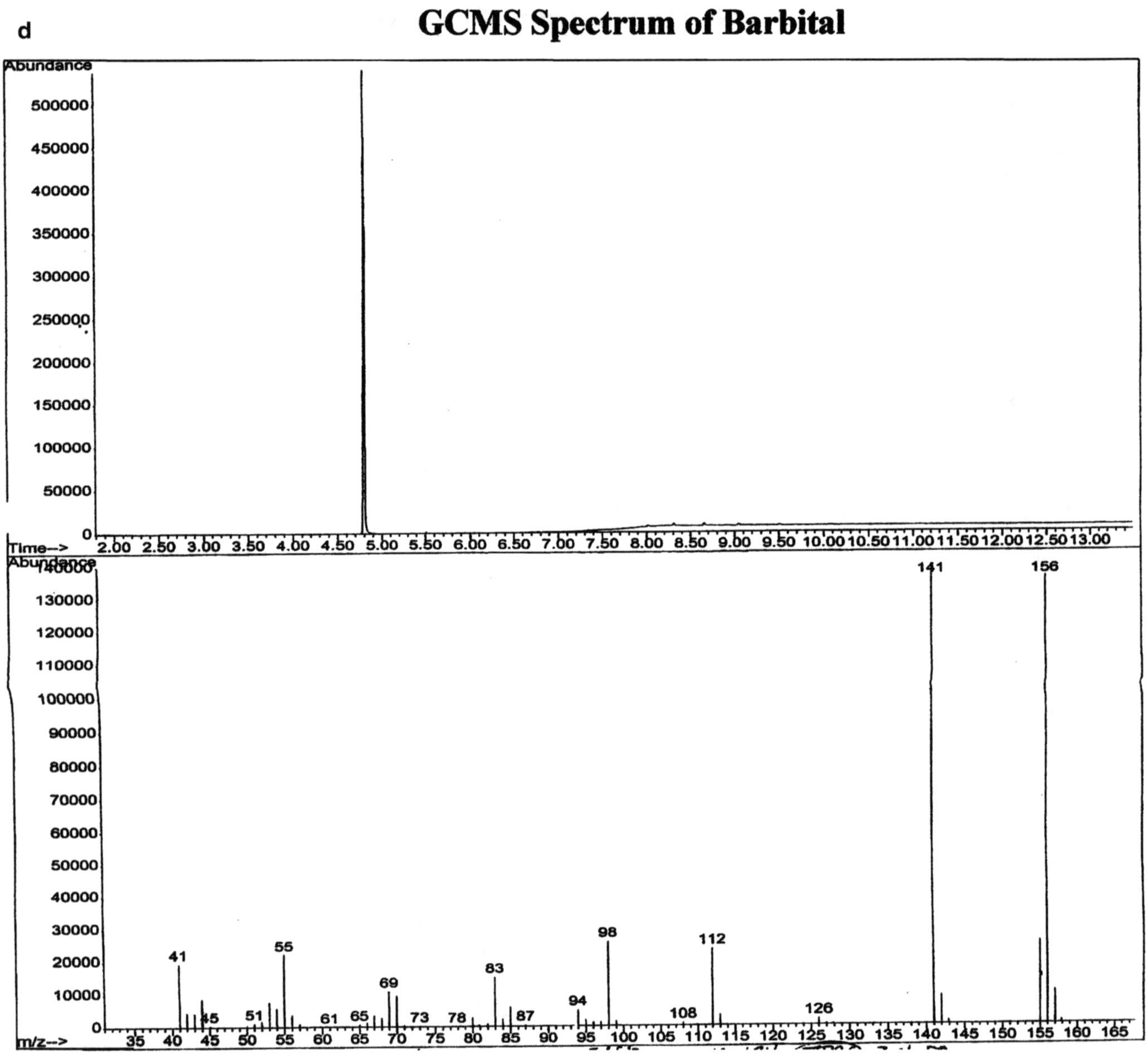

Fig. 17.9 (continued)

e **GCMS Spectrum of Allyl-cyclopentenyl-Barbiteric Acid**

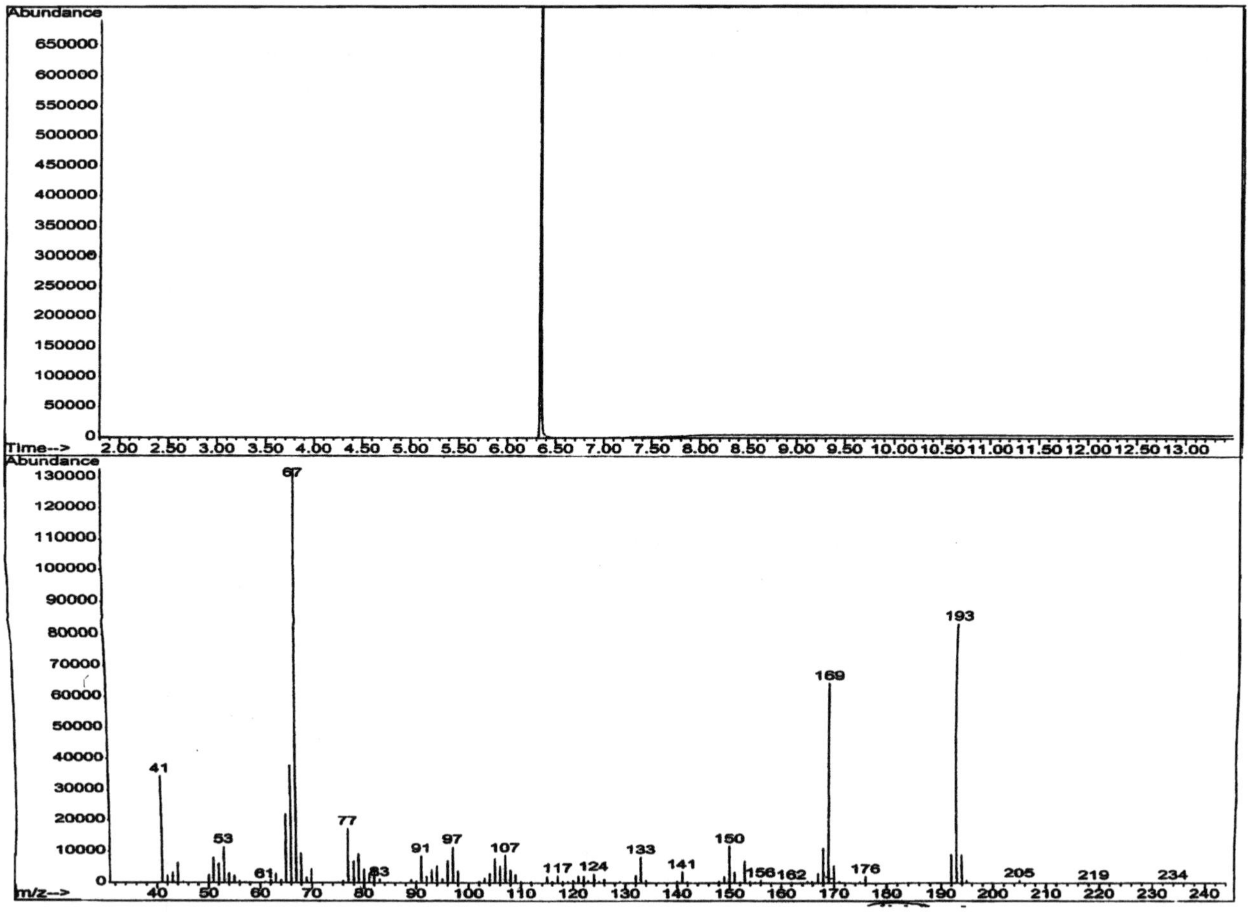

f **GCMS Spectrum of Secobarbital**

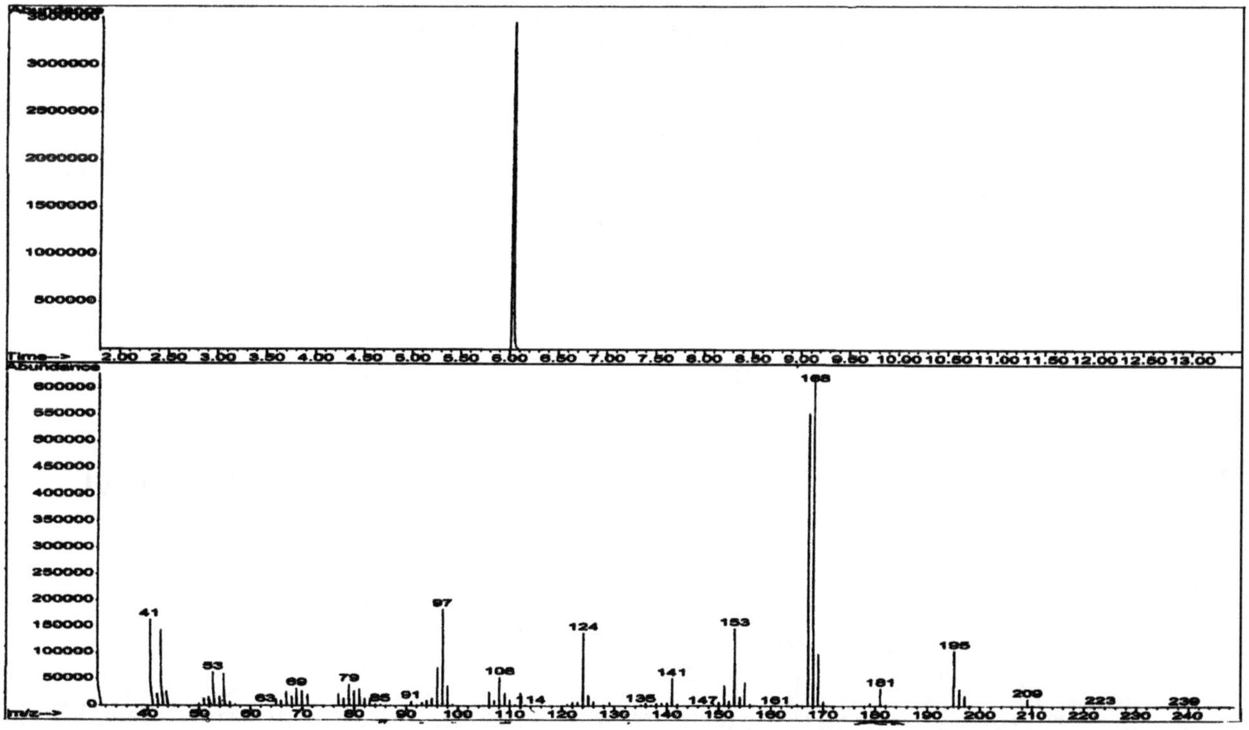

Fig. 17.9 (continued)

g

GCMS Spectrum of LSD

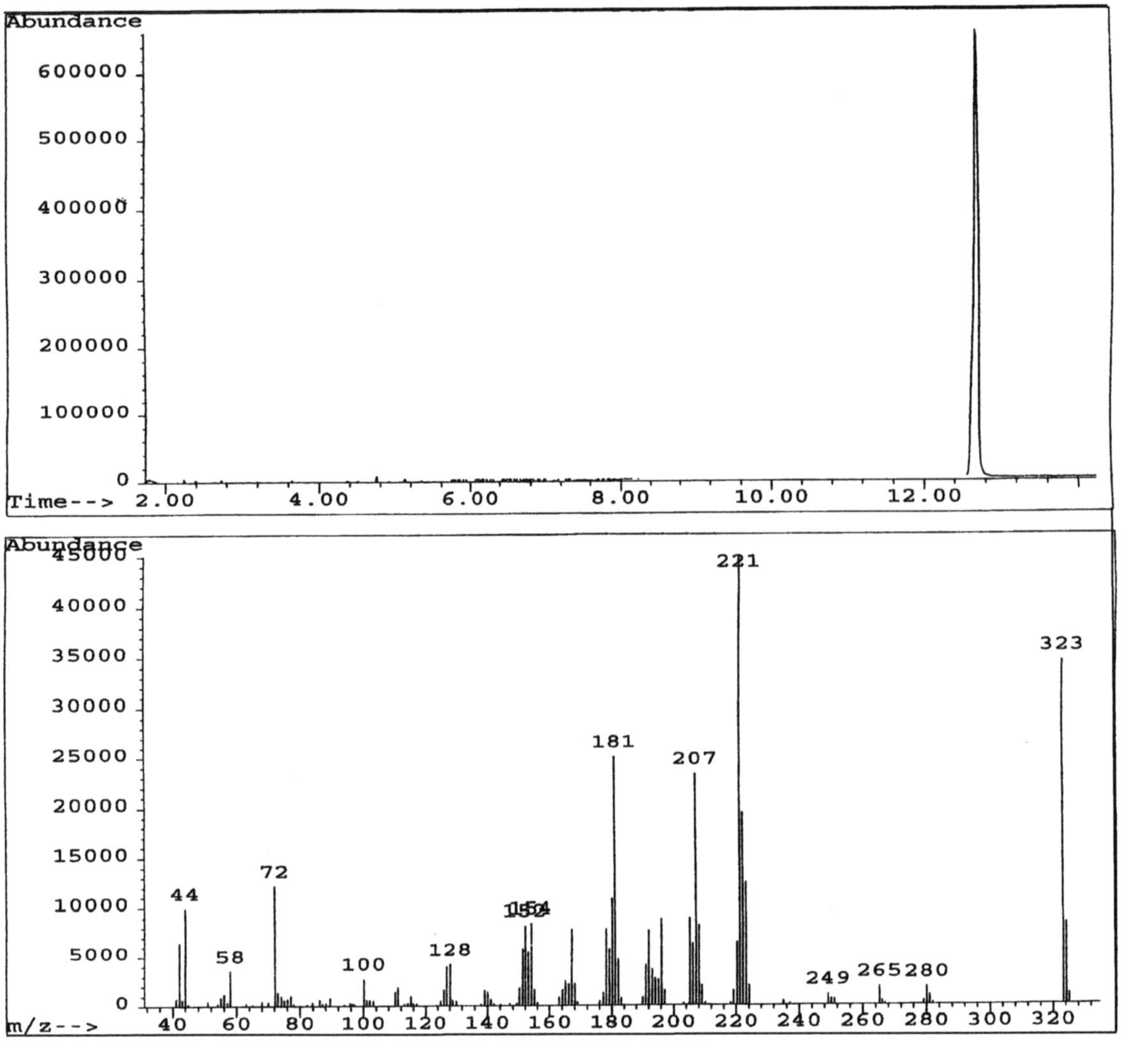

h

MS Spectra of Fentanyl

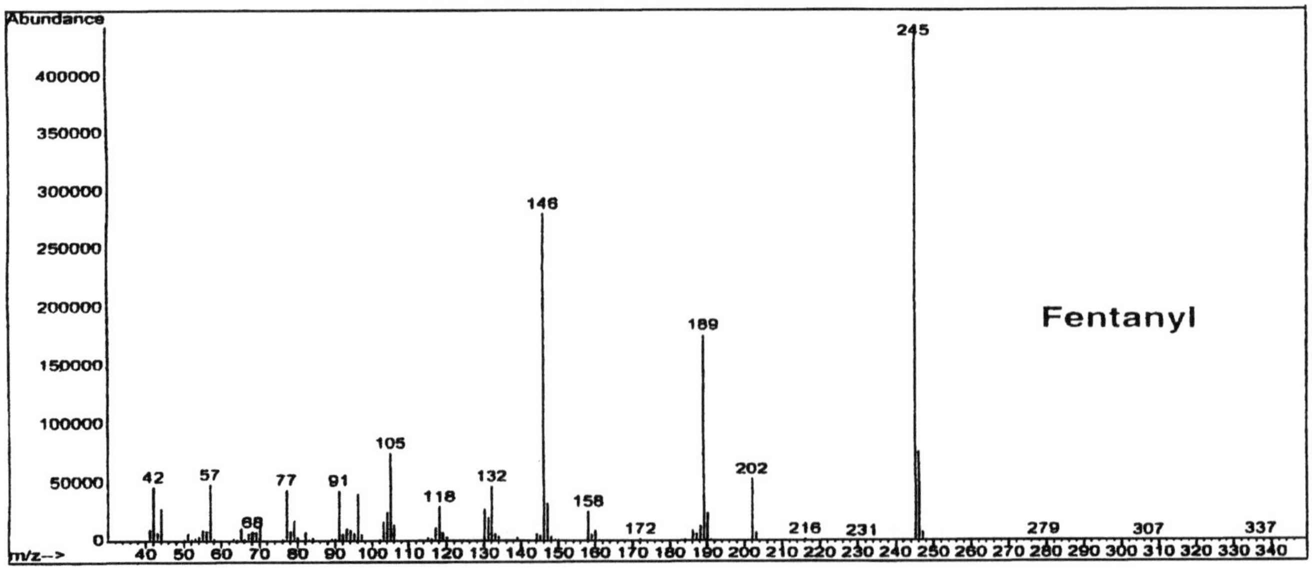

Fig. 17.9 (continued)

17.8 Questions

1. Draw the structure of barbituric acid and label position five.
2. List three effects of barbiturate use.
3. Please explain to the jury how barbiturates are usually classified.
4. Give two examples of barbiturate use in veterinary medicine.
5. Cite a reason why barbiturates will eventually be replaced by benzodiazepines.
6. Draw the structure of fentanyl and list two legitimate uses.
7. What is the most potent derivative of fentanyl?
8. Why is fentanyl potentially hazardous in forensic investigators?
9. Discuss the significance of α-methylfentanyl (AMF).
10. Draw the structure of *gamma*-hydroxybutyric acid (GHB).
11. Briefly explain to the jury the physiological role of GHB in the body.
12. List three effects of GHB use.
13. GHB is classified as a "date-rape" drug. Explain to the jury why most sexual assault cases are unreported.
14. Draw the structure of ketamine and explain the term "dissociative anesthetic."
15. What is the most effective method to administer ketamine?
16. What is a natural source for LSD?
17. What is the most potent derivative of LSD and how is it commonly sold in the illicit drug market?
18. Outline a series of color-screening tests that would indicate the presence of LSD and barbiturates.
19. What color-screening test could be used on GHB?
20. Identify the molecular ion peak (M^+) and the base peak in the MS spectrum of the following:
    (a) Ketamine
    (b) Demerol
    (c) GHB
    (d) Barbital
    (e) LSD
    (f) Fentanyl

Suggested Reading

Andera, K. M.; Evans, H. K.; Wojcik, C. M. Microchemical Identification of Gamma Hydroxy-Butyrate (GHB). *J. Forensic Sci.* **2000**, 45, 665–668.

ChemNet. Global Chemical Network. http://us.chemnet.com/ (accessed September 2009).

Hambling, D. Pentagon's New Drug Weapons. http://www.wired.com/dangerroom/2007/08/drugs-r-us/?commenter (accessed September 2009).

Henderson, G. et. al. Designer Drugs: The California Experience. In *Clandestinely Produced Drugs, Analogues, and Precursors.* U.S. Department of Justice, Drug Enforcement Administration: Washington, DC, 1989.

Lewis, W. H.; Elvin-Lewis, M. P. F. *Medical Botany*. John Wiley & Sons: New York, 1977.

Martin, W. R. The Nature of Opiate and LSD Receptors: Structural Activity Relationship Implications. In *Quantitative Structure Activity Relationships of Analgesics, Narcotic Antagonists, and Hallucinogens*; Barnett, G.; Trsic, M.; Willette, R. E., Eds.; National Institute of Drug Abuse: Rockville, MD, 1994, pp. 60–69.

National Drug Intelligence Center. Ketamine. http://www.usdoj.gov/ndic/pubs4/4769/index.htm (accessed September 2009).

Palenik, S. *Particle Atlas of Illicit Drugs*; Walter McCrone Associates: Chicago, 1974.

Shafer, J. Designer Drugs. *Science*. **1990**, 85, 60–67.

United Nations. *Recommended Methods for Testing Barbiturate Derivatives Under International Control. Manual for Use by National Narcotics Laboratories*; ST/NAR/18; United Nations: New York, 1989.

United Nations. *Recommended Methods for Testing LSD. Manual for Use by National Narcotics Laboratories*; ST/NAR/17; United Nations: New York, 1989.

U. S. Department of Justice Drug Enforcement Administration. *Symposium on Fentanyl in DEA Southwest Labs in California, USA*. U. S. Drug Enforcement Administration: Washington, DC, 2007.

U. S. Drug Enforcement Administration. Barbiturates. http://www.usdoj.gov/dea/concern/barbiturates.html (accessed September 2009).

U. S. Drug Enforcement Administration. Drug Scheduling. http://justice.gov/dea/pubs/scheduling.html (accessed September 2009).

U.S Drug Enforcement Administration. Fentanyl. http://www.usdoj.gov/dea/concern/fentanyl.html (accessed September 2009).

U. S. Drug Enforcement Administration. GHB. http://www.usdoj.gov/dea/concern/ghb_factsheet.html (accessed September 2009).

U. S. Drug Enforcement Administration. LSD. http://www.justice.gov/dea/concern/lsd.html (accessed September 2009).

Part IV

Clandestine Laboratory Operations

18 Clandestine Operations: Synthetic Methods, Hazards, and Safety

18.1 Introduction

It is not our intention to provide a detailed reference on the synthetic methods used to manufacture illegal drugs. Consequently, critical information on specific steps and reagent amounts is intentionally omitted from selected procedures. This has little, if any, affect on either the continuity or depth of the topic under discussion. The reactions represent a basic survey of those procedures commonly used in clandestine operations. The primary focus of the remaining chapters is on the identification and forensic examination of illicit methamphetamine production.

The enforcement of laws and government policies regulating illegal production of controlled substances in the United States is the responsibility of the Drug Enforcement Agency (DEA). The illicit production of drugs in clandestine laboratories represents a significant challenge to law-enforcement agencies around the world. This is complicated by the fact that drug policies and enforcement resources vary from country to country. For example, analogs of tryptamine are legally manufactured in India and China and imported into the United States, using loopholes in import regulations.

The illicit production of phencyclidine (PCP) in clandestine laboratories is a relatively simple process. However, the required solvents, precursors, and chemicals are restricted and regulated in the United States and many other countries. As a result, PCP is usually smuggled into the United States from outside sources or the required chemicals are smuggled in under different labels.

The clandestine production of γ-hydroxybutyric acid (GHB) and cocaine base is common in the United States, whereas 3,4-methylenedioxymethamphetamine (MDMA), 3,4-methylenedioxyamphetamine (MDA), lysergic acid diethylamide (LSD), ketamine, and tryptamines are typically imported under technical names from Latin America, India, or China. The synthesis of these drugs requires advanced technical knowledge of chemical procedures and elaborate chemical glassware which, in most cases, preclude clandestine manufacturing. Surprisingly, illicit production of amphetamine is rarely encountered in the United States despite the fact that it can be produced using a relatively simple process.

18.2 Clandestine Operations

18.2.1 Synthesis of Cocaine

Ground coca (*Erythroxylum coca*) leaves are mixed with water and lime (calcium oxide) to produce a pulp. Kerosene (or other hydrocarbon solvent) is added with constant and vigorous stirring. The mixture is filtered and weak sulfuric acid is added to the filtrate (solution); the pulp is discarded. The resulting two-layer mixture (biphasic) is mixed thoroughly. The cocaine is extracted into the aqueous layer (acidic layer), which is removed and treated with ammonium chloride to precipitate (solid) the cocaine as coca paste. The solvent is evaporated, and the coca paste is dissolved in dilute sulfuric acid.

J.I. Khan et al., *Basic Principles of Forensic Chemistry*, DOI 10.1007/978-1-59745-437-7_18,
© Springer Science+Business Media, LLC 2012

Fig. 18.1 Native coca leaves (*left*) are harvested for cocaine production (*right*).

Potassium permanganate (oxidizing agent) is added to remove (or minimize) cinnamoyl derivatives in the final product. The mixture is stored at room temperature for several hours. The addition of ammonia converts the coca paste into cocaine base, which precipitates out and is separated by filtration. The free-base form is either dried for direct use or converted into cocaine HCl by adding a solution of ether, HCl, and acetone.

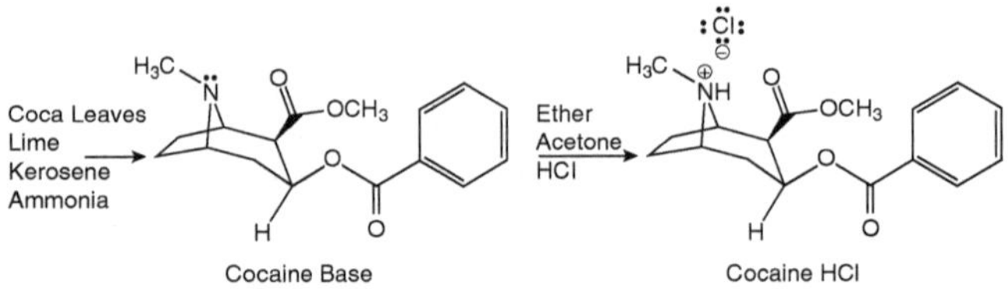

Structure 18.1

The cocaine salt is then filtered out and dried for use (or abuse) (Fig. 18.1).

18.2.2 Synthesis of Fentanyl

There are several methods commonly used to synthesize fentanyl, and each is largely dependent on the nature of the starting material. We have chosen a four-step method for illustration which takes slightly longer than 48 h to complete but is performed at room temperature. In the first step, *N*-phenethyl-4-piperidone is produced via reductive amination using phenylacetaldehyde, 4-piperidone, and triacetoxyborohydride ($NaBH(AcO)_3$) in triethyamine (Et_3N). The solution is mixed for 24 h, and then, aniline, glacial acetic acid, and triacetoxyborohydride are added to produce 4-Anilino-*N*-phenylethyl-4-piperidine. The solution is mixed for an additional 24 h. Propionyl chloride is added dropwise over a period of 2 h to convert 4-Anilino-*N*-phenylethyl-4-piperidine into soluble fentanyl. Addition of a strong base, such as sodium hydroxide (NaOH), crystallizes the fentanyl, which is separated by filtration and dried (Fig. 18.2).

18.2.3 Synthesis of γ-Hydroxybutyric Acid

The hydrolysis of γ-butyrolactone (GBL) using NaOH results in ring cleavage producing the sodium salt of GHB. The salt is distilled under slightly acidic conditions to produce GHB. A saturated solution of sodium hydroxide in 40% ethanol is placed

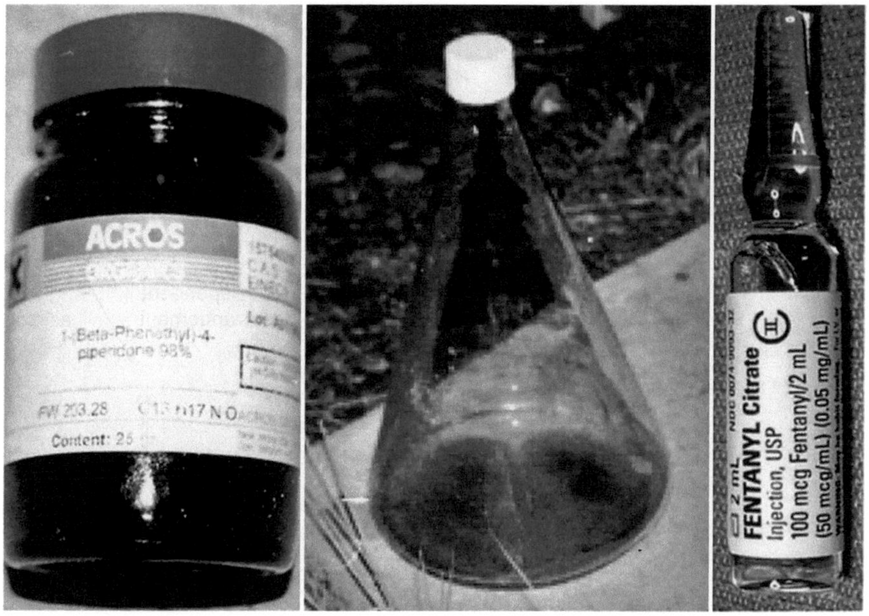

Fig. 18.2 Precursor reagent (*left*) and typical reaction vessel (*center*) used in the clandestine manufacturing of fentanyl. Fentanyl is also available by legal prescription; a typical 2 ml vial of the citrate derivative is shown above (*right*).

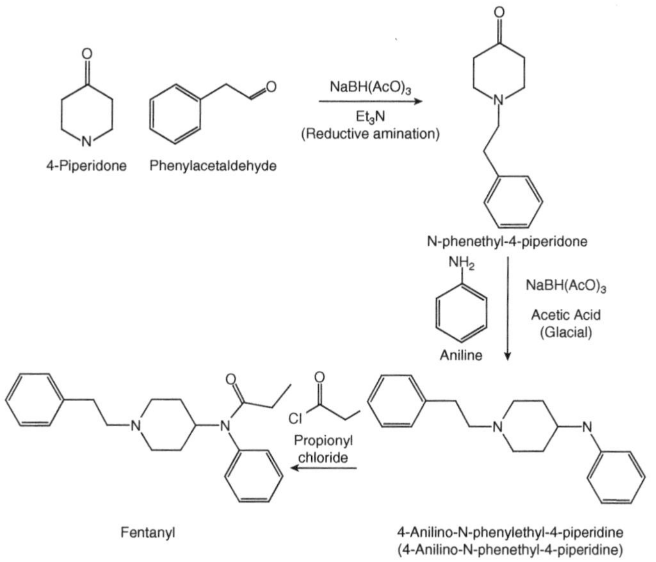

Structure 18.2

in a two- or three-neck round-bottomed flask fitted with a condenser. The flask is placed on a heating mantle and the condenser water is turned on. The mixture is heated until boiling and γ-butyrolactone (GBL) is slowly added. When the pH of the mixture has decreased to 7, distillation is performed and a clear solution of GHB is collected. The pH must be closely monitored because GHB will spontaneously rearrange into GBL under basic conditions (above pH 7) (Fig. 18.3)

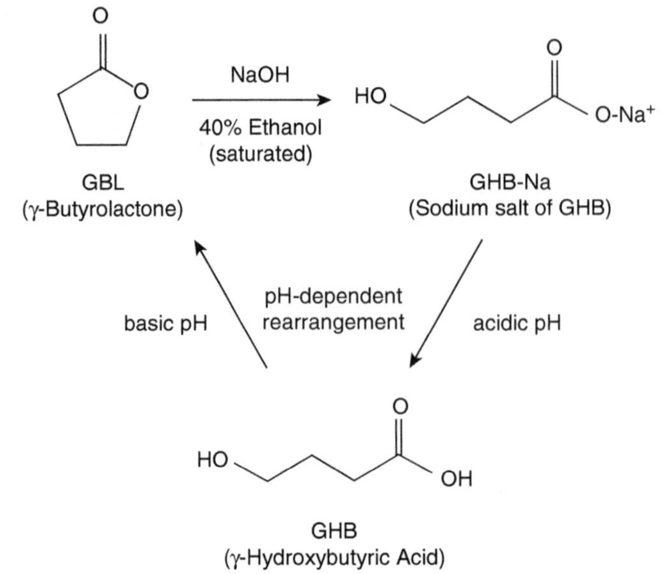

Structure 18.3

Fig. 18.3 Common household items are often used to manufacture illicit drugs in clandestine laboratories. The above apparatus was used to produce illicit *gamma*-hydroxybutyric acid (GHB).

18.2.4 Synthesis of Heroin

Raw opium is extracted from opium poppy (*Papaver somniferum*) by first dispersing the poppy in water, followed by the addition of lime (calcium oxide). The heterogeneous solution is mixed for several hours. The solid plant material is removed by filtration and ammonium chloride is added to the filtrate; the solid material is discarded. The addition of ammonium chloride increases the pH of the solution to about 9, which causes the precipitation of the free-base form of morphine. The mixture is separated using filtration and concentrated HCl is added to the solid to convert morphine base into morphine HCl. The mixture is heated to evaporate the solvent. At this stage, the resulting solid residue may be purified

Fig. 18.4 Opium poppy is a natural source of morphine, a precursor in the synthesis of heroin and codeine (*left*). The latex contains morphine and is obtained by slicing the poppy (*right*).

for morphine-related medicinal purposes. Subsequent addition of acetic anhydride and heating converts the isolated morphine salt into heroin HCl. In some instances, clandestine operators convert heroin into codeine (Fig. 18.4).

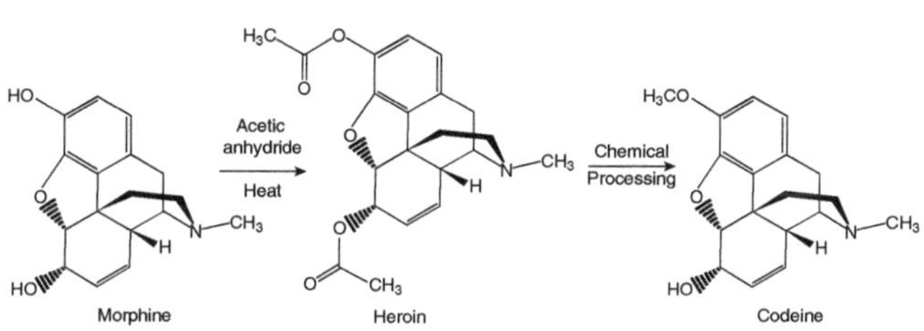

Structure 18.4

18.2.5 Synthesis of Lysergic Acid Diethylamide

LSD is produced from an ergot alkaloid isolated from certain species of morning glory or Hawaiian Baby Woodrose plants. Fermentation of *Claviceps purpurea* or *Aspergillus clavatus* is another technique used to produce the required alkaloid precursor.

The conversion of ergot alkaloid into lysergic acid is performed by refluxing the alkaloid in a mixture of water, alcohol, and potassium hydroxide. The resultant mixture is acidified to precipitate lysergic acid, which is isolated using filtration. The solid lysergic acid is suspended in a mixture of acetonitrile and trifluoroacetic anhydride at low temperature (−20°C) for 2 h. Diethylamine is added, and the solvent is evaporated under vacuum, producing a solid residue of LSD (Fig. 18.5).

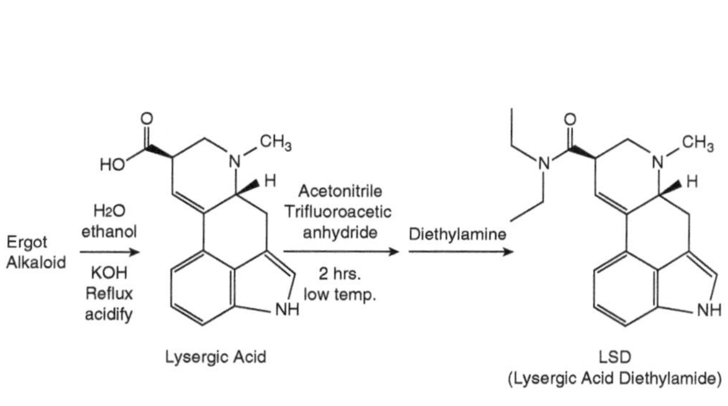

Structure 18.5

Fig. 18.5 Ergot of rye is a natural source of precursors used in the clandestine manufacturing of LSD. The black ergot component is clearly visible in the photo (*left*). LSD is commonly sold as a liquid adsorbed to the surface of designer paper (*center*). A typical apparatus used to produce LSD (*right*).

18.2.6 Synthesis of 3,4-Methylenedioxymethamphetamine

A solution of isosafrole in acetone is added dropwise to a mixture of 30% hydrogen peroxide in formic acid producing an exothermic reaction. The products are sensitive to thermal degradation; therefore, the rate of isosafrole/acetone addition is carefully monitored to ensure reaction temperatures do not exceed 40°C. This generally takes an hour and external cooling may be necessary. The solution is stirred continuously for 16 h. Volatile components are removed under vacuum to yield a deep red residue, which is dissolved in methanol and subsequently treated with 15% H_2SO_4. The solution is heated for 3 h using a steam bath and then cooled to room temperature. The mixture is extracted with ethanol and the combined extracts washed twice, first with water, and then with dilute NaOH. The ethanol is removed under vacuum and the resulting liquid is distilled (2.0 mmHg/108–112°C) to recover 3,4-methylenedioxyphenylacetone as a pale-yellow oil. This intermediate is a critical component in the synthesis of most methylenedioxy derivatives.

Thin pieces of cut aluminum are added to distilled water containing mercury(II) chloride. The reaction proceeds for 15–30 min. and is characterized by the evolution of small bubbles, formation of a light-gray precipitate, and the appearance of occasional silvery spots on the surface of the aluminum. The water is removed by decantation and the solid is washed three times with fresh distilled water. A solution of methylamine hydrochloride in warm water is added to the resulting solid, followed by the successive addition of isopropyl alcohol, 25% NaOH, and 3,4-methylenedioxyphenylacetone (from above). The exothermic reaction is kept below 60°C with occasional immersion in cold water. When thermally stable, the mixture is cooled to room temperature and the insoluble components settle to the bottom as a gray sludge, which is separated by filtration. The solvent is removed from the filtrate under vacuum, and the resulting solid residue is dissolved in water followed by treatment with HCl. The mixture is extracted with methylene chloride, and the solvent is removed from the combined extracts under vacuum. The resulting amber oil is distilled (0.4 mmHg/100–110°C), and the distillate is dissolved in isopropyl alcohol. Concentrated hydrochloric acid is added, followed by anhydrous diethyl ether, which precipitates MDMA as fine white crystals (Fig. 18.6).

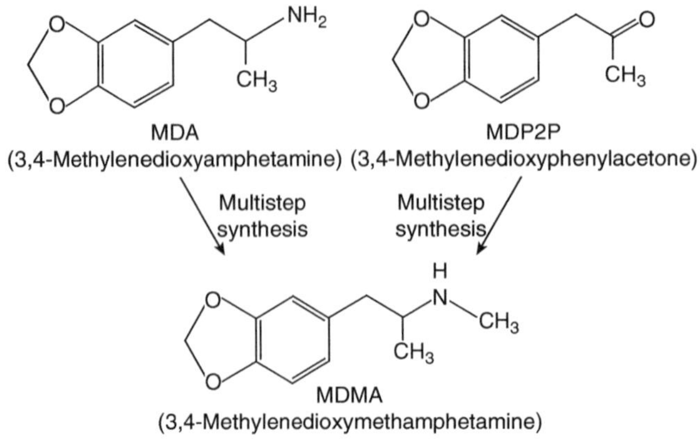

Structure 18.6

Fig. 18.6 The synthesis MDMA (ecstasy) requires specialized glassware and advanced knowledge of chemical techniques (*left*). The degree of hydration in the final product is affected by heating at various stages of the synthetic procedure. Hydrated and anhydrous MDMA (*right*).

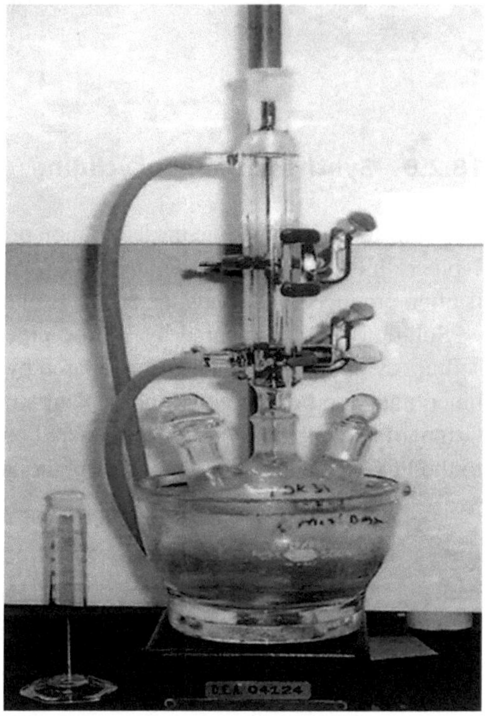

Fig. 18.7 A typical flask and condenser set-up used in the clandestine synthesis of methcathinone. The sequence of acid and base additions is critical to the overall success of the manufacturing process.

18.2.7 Synthesis of Methcathinone

Methcathinone is also known as Cat, Jeff, or ephedrone. Ephedrine, acetic acid, potassium permanganate, and methylene chloride are mixed in either a two- or three-neck round-bottomed flask fitted with a condenser (Fig. 18.7). The solution is mixed for half an hour and sodium hydrogen sulfite is added to precipitate out manganese(IV) oxide, which is separated by filtration. Sodium hydroxide (6 M) is added to the filtrate until the mixture is basic. The organic layer (methylene chloride) is removed and 0.5 M sulfuric acid is added until acidic. The aqueous layer (acidic layer) is removed and sodium bicarbonate is added until basic. The mixture is extracted with methylene chloride and HCl gas is bubbled through the combined extracts to precipitate methcathinone HCl.

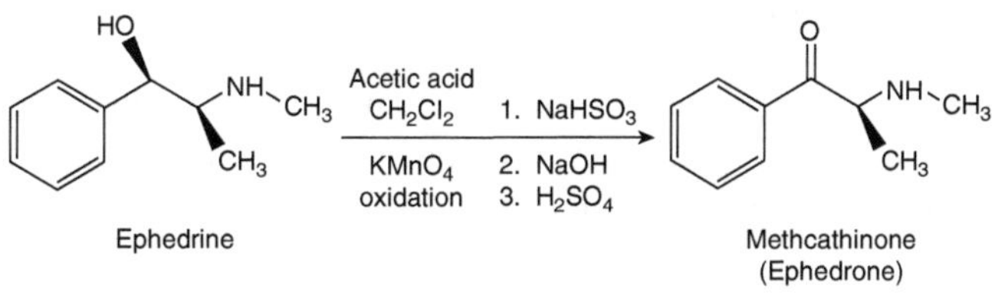

Structure 18.7

Fig. 18.8 The term "bucket method" is derived from the practice of using 5-gallon buckets as reaction vessels in the production of illicit PCP. It is the preferred method used by clandestine operators.

18.2.8 Synthesis of Phencyclidine

The "bucket method" is the most common process used in clandestine laboratories to synthesize PCP (Fig. 18.8). It requires 1-piperidinocyclohexanecarbonitrile (PCC) as a precursor, but this chemical is also a controlled substance. Therefore, PCC is often produced by mixing the products of separate reactions involving piperidine and cyclohexanone.

Solution "A" is prepared by adding piperidine to a solution of HCl and ice-cold water, followed by the addition of sodium cyanide. Solution "B" is prepared by mixing cyclohexanone and sodium metabisulfite. The two solutions are combined and stood overnight with constant stirring. The solid PCC is recovered, washed with ice-cold water, and dried. A Grignard reagent is produced from phenyl bromide, solid magnesium (turnings), and anhydrous ether, which is then added to the PCC. This mixture stands overnight, and the final product (PCP) is extracted using petroleum ether.

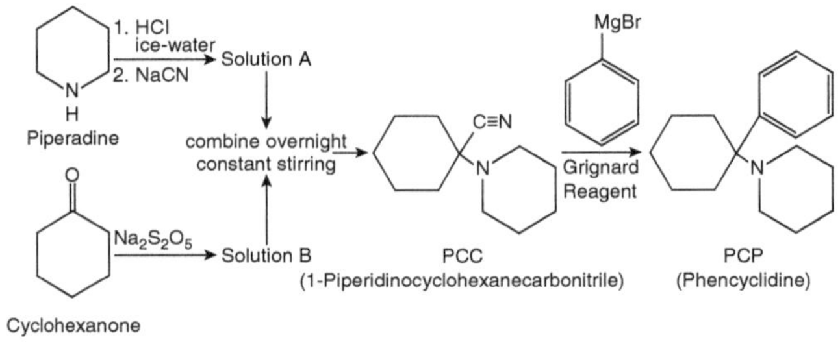

Structure 18.8

18.2.9 Synthesis of *N,N*-Dimethyltryptamine

There are several methods commonly used to prepare tryptamine and tryptamine analogs. One of the simplest procedures for DMT production utilizes 5-methoxyindole as a precursor. A mixture of anhydrous ether, 5-methoxyindole, and oxalyl chloride is chilled and subsequently reduced using lithium aluminum hydride ($LiAlH_4$). Heating the mixture to reflux produces DMT (Fig. 18.9).

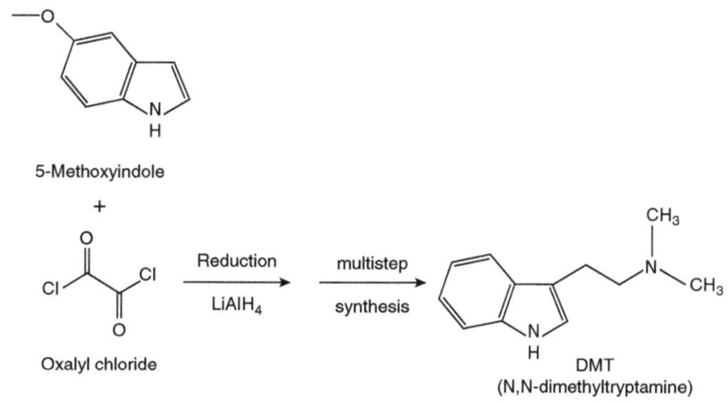

Structure 18.9

Fig. 18.9 Apparatus used to produce tryptamine and tryptamine analogs (*left*). Note the round-bottomed flask fitted with an automatic stirring device and separatory funnel. Purified DMT is a fine, white powder (*right*).

18.3 Synthesis of Methamphetamine: The Clandestine Operation of Choice

A vast majority of clandestine laboratories produce illicit methamphetamine. It has become a popular alternative to cocaine, heroin, and PCP synthesis because most precursors, solvents, and chemicals are readily available. Pseudoephedrine, although highly regulated, is a well-known over-the-counter decongestant used as a precursor in methamphetamine production. As a result, it has been removed from shelves and is currently sold only after consultation with a pharmacist. In an effort to circumvent these restrictions, clandestine operators often use the ephedra plant as a source of precursors. Additional chemicals required to convert pseudoephedrine into methamphetamine are contained in common household items that can be obtained at most hardware stores.

The synthetic procedures used to produce illicit methamphetamine are shockingly simple. Generally, a basic recipe and ordinary household items are all that is required. Simple cookware, glass jars, and other household utensils can be used during the "cooking" process, and filtration is accomplished using coffee filters (Fig. 18.10). Clandestine operations, however, often utilize slightly more elaborate set-ups.

Although there are several ways to manufacture methamphetamine, the cold and hot methods – described below – are the two most frequently used.

18.3.1 Cold Method

The cold method is a Birch reduction sometimes called the Nazi method. As the name implies, the cold method does not require a heat source to initiate the reaction. Lithium or sodium metal, usually obtained from common batteries, is used as a catalyst to

Fig. 18.10 The regulation of common household items is impossible by anyone's standards. Ordinary cookware, glass containers, and household products are effectively used to produce illicit methamphetamine.

reduce the hydroxyl group on pseudoephedrine. In the process, small pieces of lithium or sodium metal are mixed with pseudoephedrine, and ammonia is slowly added with consistent and vigorous stirring. This part of the process is exothermic and potentially hazardous. Methamphetamine is subsequently extracted from the mixture using organic solvents. HCl gas is bubbled through the organic extract to precipitate a usable form of methamphetamine.

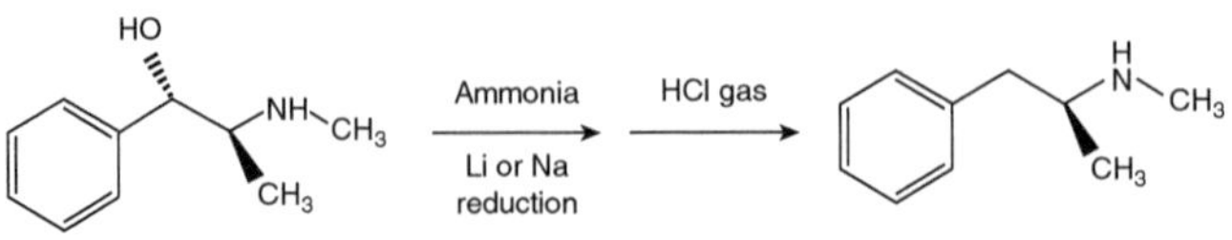

Pseudoephedrine *d*-Methamphetamine

Structure 18.10

Fig. 18.11 Hydrogen gas is stored in metal containers resembling propane tanks (*left*). The gas is used in the last step of illicit methamphetamine production. Large Erlenmeyer flasks are usually found at clandestine crime scenes (*right*).

The cold method is preferred by clandestine operators in rural areas of the United States because ammonia is readily available in most fertilizers (Fig. 18.11).

18.3.2 Hot Method

In the hot method, also known as the catalytic reduction of ephedrine/pseudoephedrine, a metal catalyst provides a surface for the reduction of the hydroxyl (OH) group on ephedrine/pseudoephedrine. There are two common methods for accomplishing reduction. The first uses palladium (Pd) or platinum (Pt) on activated charcoal (carbon) as the metal catalyst and hydrogen gas as the reducing agent. The second uses red phosphorus (Red P) as the metal catalyst and hydroiodic acid (HI, strong acid) as the reducing agent. This is the preferred method used in the United States and is generally known as the red phosphorus–hydroiodic method (Red-P/HI) (Fig. 18.12).

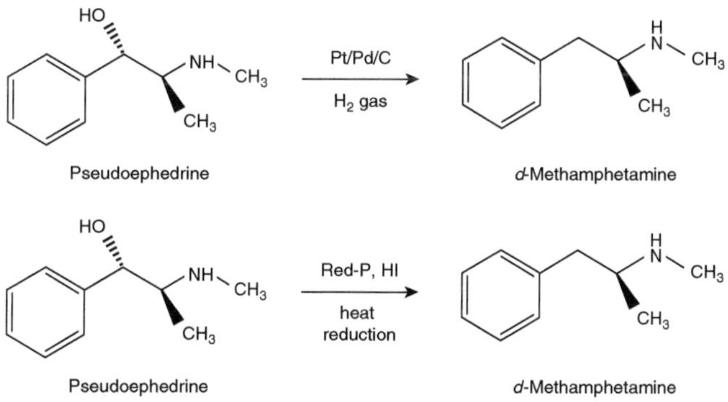

Structure 18.11

Fig. 18.12 Hot method of methamphetamine synthesis.

The Federal Government has passed strict laws regulating the sale and possession of pseudoephedrine-containing tablets. As a result, clandestine operators are using precursors in ephedra to evade these regulations. When ephedra is used, regardless of the method, a mixture of products is formed.

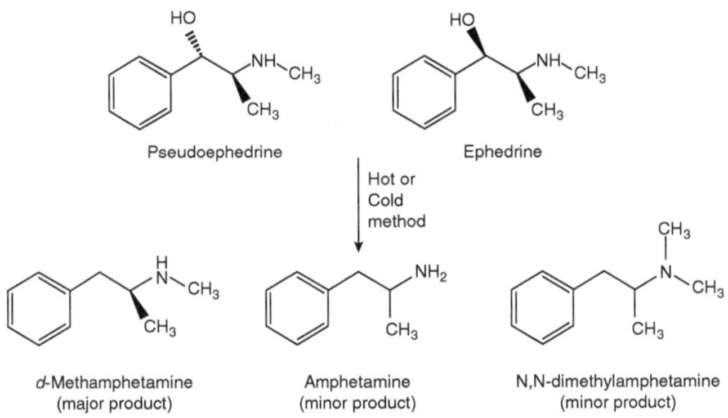

Structure 18.12

18.4 Potential Hazards Associated with Clandestine Operations

Clandestine operations vary from site to site; some are small, while others may be large. They may be conducted in a garage, a backyard, or at remote locations. Regardless of site size or location, all are hazardous to operators, the environment, and law-enforcement-response teams.

Many clandestine operators are unaware of the potential hazards associated with illicit-drug manufacturing. Their lack of scientific knowledge is inherently dangerous to themselves, their households, their neighbors, and the environment. All clandestine lab sites are potentially hazardous due to the necessary presence of various organic solvents, corrosive acids and bases, and sources of heat or flames. A list of common hazards is below.

1. *Chemical reactivity hazard*: Clandestine operations require the use of acids and bases, which are both corrosive and caustic. Acid–base reactions are very rapid and, in some cases, extremely violent. They are usually handled improperly and stored carelessly at clandestine sites, which increases the risk associated with these chemicals.
2. *Fire hazard*: Many chemicals used in clandestine operations, such as denatured alcohol, charcoal lighter fluid, and Coleman fuel, are easily ignited and have caused damage to both property and operators.
3. *Explosive hazard*: Chemical reactions conducted near open flames are extremely dangerous. Many clandestine operations have been discovered because the operator produced an explosion that destroyed the lab.
4. *Exposure hazard*: Volatile organic chemicals and HCl gas are frequently used in clandestine operations. Fumes from these known carcinogens are often present in the facility and surrounding areas. Clandestine operators, their family members, their associates, and anyone in the immediate area are routinely exposed to these hazardous vapors. Phosphene and phosgene are two deadly gases liberated at various stages of methamphetamine production. They have claimed the lives of several unsuspecting operators.

18.5 Safety Considerations

Clandestine investigation is the responsibility of specialized, highly trained response teams. Members wear maximum safety gear upon initial entry and, after assessing the hazard, wear appropriate safety gear while processing the scene (Fig. 18.13).

18.6 Role of the Forensic Chemist at Clandestine Lab Sites

The responsibilities of forensic chemists at clandestine-laboratory crime scenes are outlined below. In addition, they may also have specific, agency-related assignments.

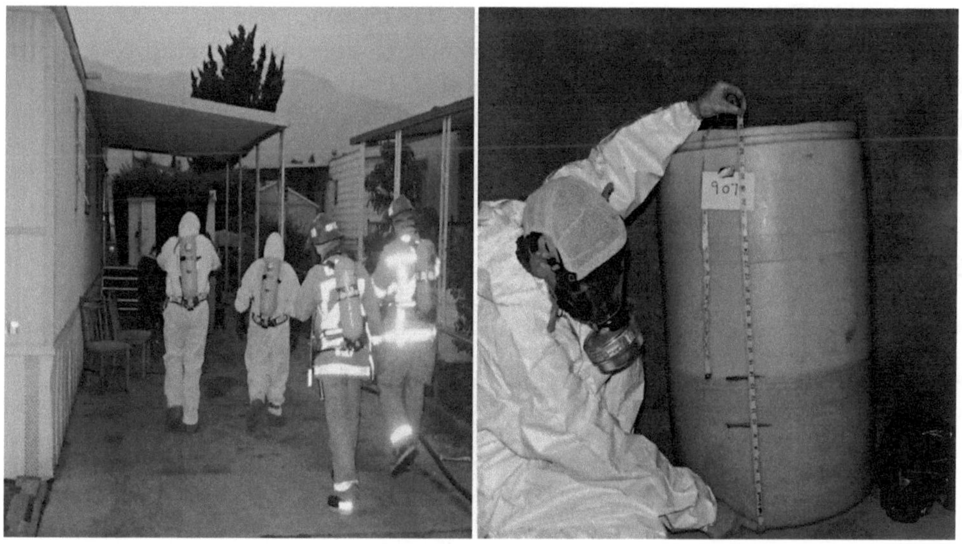

Fig. 18.13 Clandestine laboratories are extremely dangerous and often unpredictable. The safety equipment worn by response team members varies at different stages of the investigation. Initial entry (*left*) and crime-scene processing (*right*).

18.6.1 Advisory

The primary responsibility of a forensic chemist at an alleged clandestine site is to determine whether clandestine operations actually exist. They assess reagents, chemicals, equipment, and glassware to formulate a preliminary opinion. This is critical in establishing probable cause for law-enforcement action and also important in the arraignment process. When rendering an opinion, it is vital to establish whether the site is a current manufacturing or processing facility, or one used in the past. Differentiation between the two is usually based on experience and common sense. For example, the presence of biphasic liquids or moist filter paper stained with red phosphorous indicates current manufacturing and processing, whereas dry red phosphorous, empty jars of iodine, and used, but empty glassware, indicates past activity.

18.6.2 Evidence Collection

If the site is determined to be an active facility, environmental-safety measures are implemented for the processing of evidence. The forensic chemist identifies relevant evidence and appropriate samples are collected for examination. The evidence is documented and sealed at the scene, and a chain of custody is established for each sample. Transportation of the evidence to the laboratory is policy dependent; some agencies prefer to transport, while others may defer this responsibility to the forensic chemist.

18.7 Questions

1. Can you please explain to the jury the method used to manufacture cocaine?
2. Outline the procedure for the clandestine production of cocaine.
3. What is the role of propionyl chloride and sodium hydroxide in fentanyl synthesis?
4. Why must the pH be closely monitored in the production of γ-hydroxybutyric acid?
5. Draw the reaction for the synthesis of γ-hydroxybutyric acid.
6. Describe the procedure for synthesizing heroin to members of the jury.
7. Briefly describe the role of acid addition in LSD production.
8. What is the key intermediate in the synthesis of most methylenedioxy compounds?
9. Why is hydrogen-chloride gas used in methcathinone synthesis?
10. Explain the term "bucket method" to members of the jury.
11. Outline the method used to produce PCP in clandestine laboratories.
12. How is the Grignard reagent produced?
13. What is the common precursor to DMT?
14. Describe the events resulting in the use of ephedra by clandestine operators.
15. Why is methamphetamine the most common drug produced in clandestine labs?
16. Outline the hot and cold methods used to produce methamphetamine.
17. List and describe two hazards associated with illegal-drug manufacturing.
18. Briefly describe the role of a forensic chemist at clandestine-lab sites.
19. Discuss two ways to differentiate past clandestine sites from active sites.
20. Why is establishing "probable cause" important to clandestine investigation?

Suggested Reading

Allen, A. C.; Kiser, W. O. Methamphetamine from Ephedrine: Chloroephedrine and Aziridine. *J. Forensic Sci.* **1987**, 32, 953–962.

Augustine, R. L. Toxic Effects of Impurities in Methamphetamine from Cladestine Labs. Present at DEA Conference, New York, 1979.

Baum, R. M. New Variety of Street Drugs Pose Growing Problems. *Chem. Eng. News.* **1985**, 9, 7–16.

Beagle, J. Q. Synthesis and Effects of PCP Analogs. http://www.erowid.org/archive/rhodium/chemistry/pcp/ (accessed October 2009).

Casale, J. F.; Klien, R. F. X. Illicit Production of Cocaine. *Forensic Sci. Rev.* **1993**, 5, 95–107.

Dal Cason, T. A. An Evaluation of the Potential for Clandestine Manufacturing of 3,4-Methylenedioxyamphetamine (MDA) Analogs and Homologs. *J. Forensic Sci.* **1970**, 35, 675–697.

Davenport, T. W.; Allen, A. C.; Cantrell, T. S. Synthetic Reductions in Clandestine Amphetamine and Methamphetamine Laboratories: A review. *Forensic Sci. Int.* **1989**, 42, 183–199.

Dos Res, L. J.; Maes, R. A. Piperazine-Like Compounds: A New Group of Designer Drugs-of-Abuse on the European Market. *Forensic Sci. Int.* **2001**, 121, 47–56.

Henderson, G. et al. Designer Drugs: The California Experience. In *Clandestinely Produced Drugs, Analogues, and Precursors.* U.S. Department of Justice, Drug Enforcement Administration: Washington, DC, 1989.

Metsger, L.; Bittner, S. Autocatalytic Oxidation of Ethers with Sodium Bromate. *Tetrahedron.* **2000**, 56, 1905–1910.

Sanguinetti, V. R.; Angelo, A.; Frank, M. R. GHB: A Home Brew. *Am. J. Drug Alcohol Ab.* **1997**, 23, 637–642.

Shulgin, A. Phenethylamines I have Known and Loved: Synthesis of MDMA; MDM; ADAM; Ecstacy; 3,4-Methylenedioxy-N-Methylamphetamine. http://www.mdma.net/mdma.html (accessed October 2009).

Shulgin, A.T.; Shulgin, A. DMT Synthesis from Indole. http://deoxy.org/dmt-tihkal.htm (accessed October 2009).

Siegfried. Synthesis of Fentanyl. http://www.opioids.com/fentanyl/synthesis.html (accessed October 2009).

Soine, W. H. Clandestine Drug Synthesis: Heroin. *Med. Res. Rev.* **1986**, 6, 41–74.

United Nations. *Guidelines for the Import and Export of Drug and Precursor Reference Standards for Use by National Drug Testing Laboratories*; United Nations Publication: New York, 2007.

U. S. Department of Justice, Drug Enforcement Administration. *Synthetically Manufactured Drugs of Abuse*, Drug Enforcement Administration: Washington, DC, 1986.

Evidence Identification and Collection 19

19.1 Clandestine Operations: A Forensic Analogy

Clandestine manufacturing of illicit drugs is a multistep synthetic process that can be accurately compared to preparing a dinner from a cookbook. First, the raw materials listed in the "recipe" are purchased and separately prepared. Next, the appropriate cookware is gathered and set up. The ingredients are combined and cooked according to the instructions and the raw materials are processed into the final product. Impurities are often added to illicit drugs as "cutting agents" to increase the overall amount of product, which has the effect of increasing profit margins for the operator. Ultimately, the final product is served and consumed.

19.2 Signs of Clandestine Operations

The evidence associated with clandestine operations is highly variable and depends on a number of factors such as site, method of manufacturing, type of controlled substance, and quantity produced. These must be considered when formulating a preliminary opinion and collecting evidence for examination. For example, wall stains and carpets stains are photographed and/or collected for examination to prove that clandestine manufacturing was performed (Fig. 19.1).

It is not a crime to possess acids, bases, organic solvents, plastic tubing, coffee grinders, coffee filters, coffee pots, and pseudoephedrine tablets. However, a clandestine operation is suspected if:

1. Coffee grinders contain a white solid residue.
2. Coffee pots contain acidic, basic, or organic solvents instead of coffee.
3. There is evidence of red phosphorus or white solid binders present (Fig. 19.2).
4. An excessive number of empty pseudoephedrine blister packs are found on site.
5. Plastic tubing is found connected to plastic or glass bottles (Fig. 19.3).
6. Large quantities of Red Devil Lye (NaOH), acetone, charcoal lighter fluid, denatured alcohol, rubbing alcohol, tincture-iodine, red phosphorous, muriatic acid (HCl), sulfuric acid, iodine crystals, or hypophosphorous acid are found at the scene.
7. Two-phase (biphasic) liquids are present.

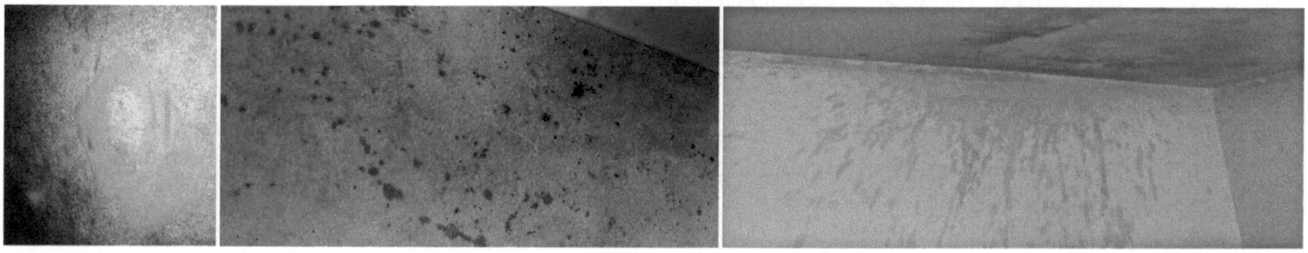

Fig. 19.1 Stains are important evidence in clandestine investigation. Specific types and patterns are often used to indicate the presence of active clandestine manufacturing.

J.I. Khan et al., *Basic Principles of Forensic Chemistry*, DOI 10.1007/978-1-59745-437-7_19,

© Springer Science+Business Media, LLC 2012

Fig. 19.2 Red phosphorus and white solid binders are strong evidence of clandestine activity. Red phosphorus is commonly used in methamphetamine production and binders are common "cutting agents".

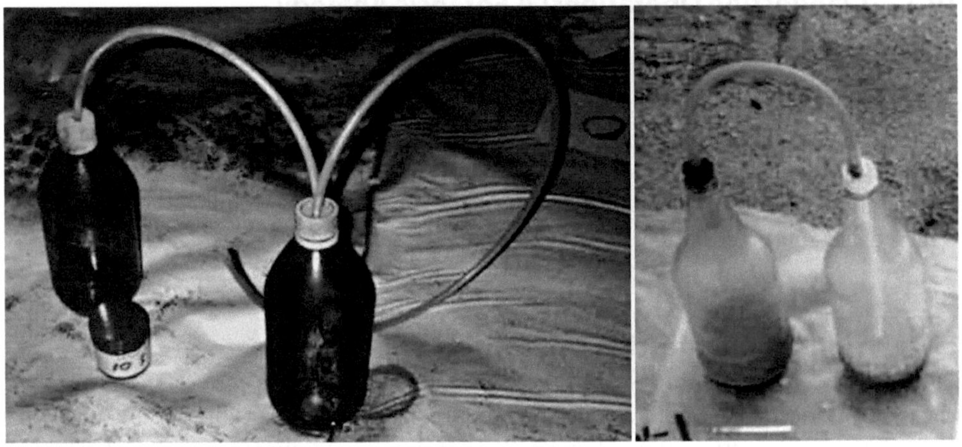

Fig. 19.3 A variety of clandestine procedures require gas-generating systems. Operators typically produce primitive systems using common tubing and readily available glass or plastic containers.

19.3 Identification of Related Evidence

Differentiating evidence from common household items at clandestine sites is a delicate process requiring great care. This task is often complicated by the fact that many chemicals used in clandestine manufacturing have legitimate domestic uses. Although the presence of the individual items listed below does not necessarily indicate illegal activity, their collective presence at an isolated site is usually a compelling argument for illicit drug production.

1. *Solvents*: Rubbing alcohol, methanol, denatured alcohol, Coleman fuel, charcoal lighter fluid, and acetone (Fig. 19.4).
2. *Elements*: Iodine crystals, red phosphorus, lithium, or sodium (Fig. 19.5).
3. *Precursors*: Ephedrine/pseudoephedrine/ephedra plant.
4. *Chemicals*: Muriatic acid, sulfuric acid, hypophosphorous acid, hydroiodic acid, sodium hydroxide (Red Devil Lye), sodium chloride (table salt), or ammonia (Fig. 19.6).
5. *Miscellaneous items*: Coffee filters, plastic tubing, glass jars, gloves, stained cloths, and biphasic liquids.
6. *Equipment*: Coffee grinders, electric skillets, stoves, electric scales, coffee pots, heating mantles, and household cookware (Fig. 19.7).

Propane gas tanks are fairly common in most households. However, clandestine operators modify the cylinders for storage of corrosive agents such as hydrogen chloride gas or ammonia gas. Therefore, the presence of corroded cylinders could be an indication of a clandestine operation (Fig. 19.8).

The presence of a small quantity of the items listed above with obvious signs of manufacturing and processing is required for prosecution. However, large quantities are an indication of illegal use even if found without supporting evidence. For example, a single bottle of denatured alcohol, or a few packs of pseudoephedrine tablets, would be common in most households, but a 10-gallon drum of alcohol, or a large number of empty bottles of pseudoephedrine, is a clear indication of abuse (Fig. 19.9).

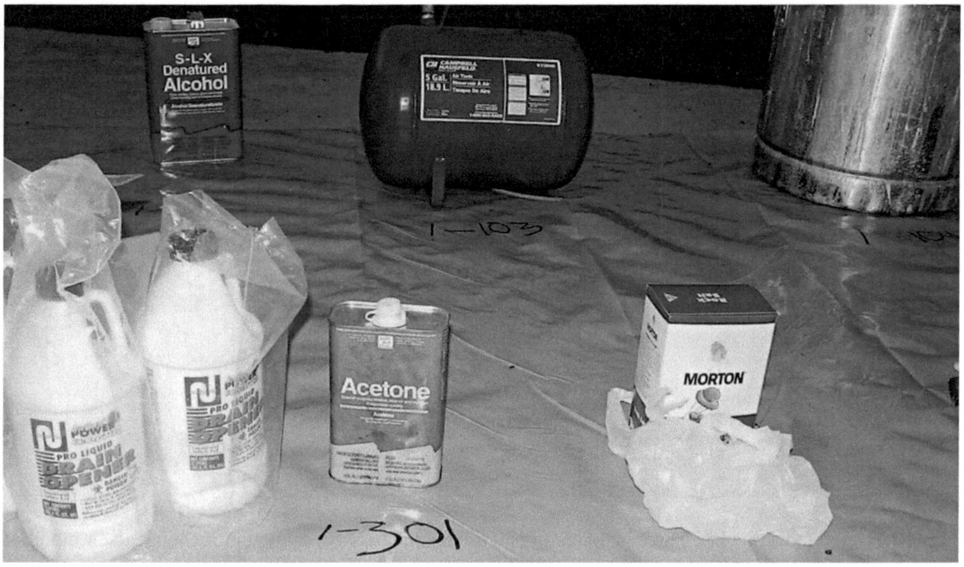

Fig. 19.4 The possession of moderate supplies of common household products is not illegal. The amount of chemicals in the photo would not be sufficient to conclusively indicate illegal activity unless supported by more definitive evidence.

Fig. 19.5 The discovery of catalysts such as red-phosphorus, lithium, or sodium at clandestine sites is generally considered conclusive evidence of illicit drug manufacturing. These chemicals have no legitimate household uses.

Fig. 19.6 Vast amounts of acids, bases, and organic solvents used in clandestine drug manufacturing. This evidence is compelling, despite the conspicuous absence of hardware used in production.

Fig. 19.7 A clandestine lab. There is no legitimate use for the excessive amount of chemicals or modified heat sources.

Fig. 19.8 A common propane tank (*left*) and one modified for HCl gas (*right*). HCl and ammonia gas are corrosive reagents used in illicit drug production that change the outward appearance of the tank.

19.4 Solutions Frequently Encountered at Clandestine Sites

Solutions are an integral part of illicit drug manufacturing and are often found at clandestine crime scenes. The amount discovered is often irrelevant because generally no legitimate domestic use of these mixtures exists. They serve no purpose other than manufacturing and processing. A representative list of solutions commonly associated with methamphetamine production is given in the following:

1. White cloudy solutions of alcohol or water with, or without, sludge.
2. Red phosphorus, hydroiodic acid, water (with or without pseudoephedrine).
3. Red phosphorus, hydrochloric acid, water, and iodine (with or without pseudoephedrine)
4. Hypophosphorous acid, iodine, and water (with or without pseudoephedrine)
5. Pseudoephedrine, ammonia.
6. Two-layer (biphasic) liquids.

Fig. 19.9 Chemicals and manufacturing setups are not necessarily required for successful prosecution. The individual presence of large quantities of solvents (*top left*), empty packaging (*top right*), or manufacturing hardware (*bottom*) is strong evidence of clandestine manufacturing.

Biphasic liquids are particularly incriminating. The only one commonly found in most households is oil and vinegar salad dressing. A biphasic liquid discovered at a scene is either salad dressing or a step in the processing of methamphetamine. A coffee pot filled with a biphasic liquid is not a familiar sight in the average household (Fig. 19.10).

19.5 Clandestine Production of Methamphetamine

As previously stated, illicit production of methamphetamine is the most common clandestine operation and catalytic reduction of pseudoephedrine is often the preferred method. Although there are several ways to reduce the hydroxyl group on pseudoephedrine, the focus here is on the "Red Phosphorus-HI Method," the most frequently encountered process in the United States.

There are typically four steps in the clandestine production of methamphetamine using the Red Phosphorus-HI Method.

19.5.1 Extraction of Pseudoephedrine from Cold Tablets (Step I)

The first step is the extraction of pseudoephedrine from cold tablets with intent to manufacture methamphetamine. Clandestine operators obtain bulk quantities of pseudoephedrine-containing cold tablets from pharmacies or convenience stores. The tablets are ground to fine powder using coffee grinders and added to an organic solvent, typically denatured alcohol (Fig. 19.11). The heterogeneous mixture is stirred vigorously in a suitable container.

Fig. 19.10 The requirement of solutions in the production of many types of illicit drugs makes them a valuable source of definitive proof of clandestine manufacturing.

Fig. 19.11 The purchase of bulk solvents (*left*) and pseudoephedrine-containing cold tablets (*center and right*) by clandestine operators often initiates clandestine crime investigation.

Pseudoephedrine readily dissolves in alcohol, whereas the insoluble binder (normally starch) usually settles at the bottom of the container. The alcohol solution is transferred to another container and the solvent is evaporated at room temperature or on a stove. The recovered pseudoephedrine is collected and further processed in the second step. It is worth noting that extracting pseudoephedrine from cold tablets using water is also very common.

19.5.2 Manufacturing of Methamphetamine (Step II)

The extracted pseudoephedrine is mixed with hydroiodic acid, red phosphorus, and water in a round-bottomed flask (Fig. 19.12). This mixture is boiled ("cooked") for a few hours using a stove or heating mantle. The addition of heat to the mixture causes the reduction of pseudoephedrine to methamphetamine. It is very important to add water continuously to prevent boiling to dryness. The absence of water generates phosphine, an extremely toxic gas that is lethal at very low concentrations.

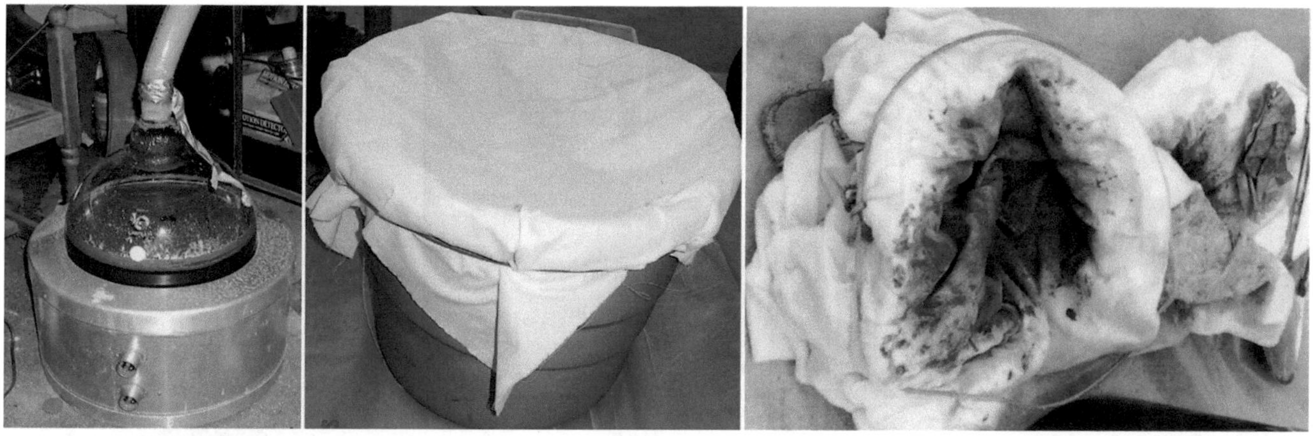

Fig. 19.12 Large round-bottom flasks and heating mantles are often used in the second step of methamphetamine synthesis (*right*). Large container prepared for red phosphorous addition (*center*); the same container after addition (*right*). The residual stains are often discovered at clandestine sites and collected as evidence.

19.5.3 Processing of Methamphetamine (Step III)

Red Devil Lye (NaOH) and an organic solvent, typically Coleman fuel, are added to the solution from step II and mixed thoroughly. The free base form of methamphetamine is extracted into the organic layer (Coleman fuel) and removed. Table salt and sulfuric acid are mixed to create a primitive HCl gas generator. The HCl gas is bubbled through the organic layer converting methamphetamine base into methamphetamine-HCl. The salt form (methamphetamine-HCl) is insoluble in the organic layer and precipitates out. The recovered methamphetamine is usable at this stage, but its dark physical appearance is somewhat less appealing to abusers. Therefore, clandestine operators often proceed to the icing stage (Fig. 19.13)

19.5.4 Icing of Methamphetamine (Step IV)

Methamphetamine produced in step III is dissolved in acetone and the mixture is chilled in a refrigerator, producing ice-like crystals of methamphetamine (Fig. 19.14).

19.6 Collection of Evidence

Definitive proof of at least one step in the Red Phosphorus-HI Method is required to convict clandestine operators. As expected, the probability of conviction increases dramatically if more steps are proven. Representative evidence collected to prove various steps is illustrated below:

Samples collected to prove extraction of pseudoephedrine (step I) (Fig. 19.15).

Samples collected to prove methamphetamine production (step II) (Fig. 19.16).

Samples collected to prove processing of methamphetamine (step III) (Fig. 19.17).

Samples collected to prove icing (step IV) (Fig. 19.18).

It is always prudent to collect more than one piece of supporting evidence for a particular step because a single item may produce negative results upon examination. Evidence collection is the sole responsibility of members of the clandestine investigation response team. Always remember, clandestine operators have never been exonerated because too much evidence was collected.

Fig. 19.13 Chemicals used in the third step of methamphetamine synthesis. Red-Devil Lye is a drain cleaner used in clandestine operations as a source of NaOH (*top left*). Coleman fuel (*top center*) is the preferred organic solvent. Methamphetamine-HCl prior to icing (*top right*) is usable, but lacks the physical appearance that appeals to most abusers. Common table salt (*bottom left*) is used to make HCl gas-generators (*bottom right*).

Fig. 19.14 Methamphetamine crystals produced after the icing stage.

Fig. 19.15 Large containers of white powder are commonly found at clandestine lab sites. This type of evidence is collected to prove stage I of methamphetamine production (the extraction of pseudoephedrine from cold tablets).

Fig. 19.16 Heating mantles are frequently used as a heat source in the production of methamphetamine. The raw materials are often stored in a variety of large containers.

Fig. 19.17 Bilayer liquids are strong evidence of methamphetamine processing. The right examples are commonly found at clandestine laboratories and are collected to prove stage III of methamphetamine production.

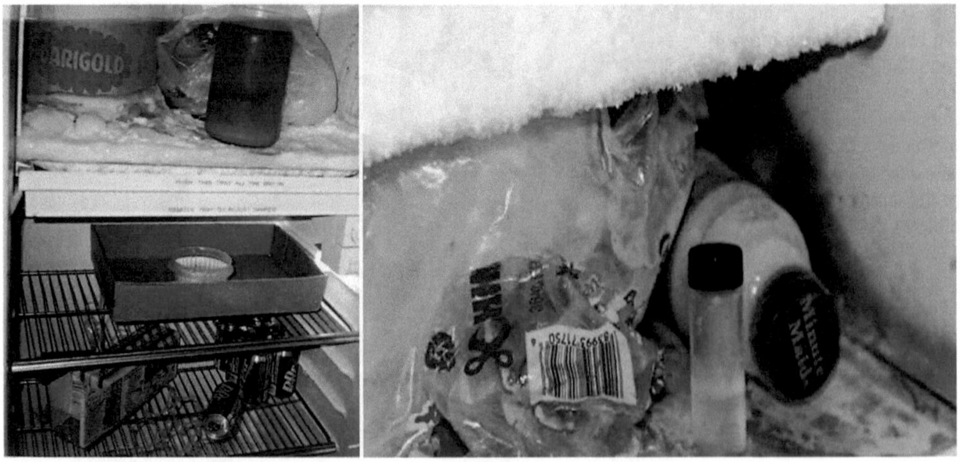

Fig. 19.18 Examples of evidence collected to prove the icing stage of methamphetamine production.

Fig. 19.19 Remnants of past clandestine operations. The inert nature of water makes it an ideal solvent for recovering residue left in glassware.

19.7 Collection of Washes

In some cases, clandestine laboratories are discovered after the production process has been performed. The inactive site is generally clean, and very little remains except for a few pieces of empty glassware left behind. Under these circumstances, fingerprints and washes from the empty glassware are critical in building a case.

The collection of wash is a delicate matter and should be taken seriously. Methanol is generally the solvent of choice because it dissolves a variety of different substances. Unfortunately, it can also produce chemical changes that alter the natural pH of the dissolved residue. Subjecting evidence to the possibility of change prior to examination is unacceptable under any circumstances. If a wash using methanol contains methamphetamine, it can only indicate that the glassware contained methamphetamine. It cannot be stated, beyond a reasonable doubt, that the glassware was used in the manufacturing or processing of methamphetamine. Therefore, collecting evidence using methanol washes is generally not a sound forensic practice because it may be disputed during trial.

It is extremely important to maintain the integrity of the clandestine samples prior to testing. To achieve this, water washes are highly desirable for several reasons (Fig. 19.19). First, the dissolved sample is well preserved with no chemical changes. Second, water has no affect on the pH of the dissolved residue. Acids and bases are used extensively in methamphetamine production and their presence can be easily detected in unaltered residues. If, at the time of examination, a water wash is acidic or basic and contains methamphetamine, the analyst can say, beyond a reasonable doubt, that the glassware was used in the manufacturing or processing of methamphetamine. Also, fingerprints, if found, can incriminate individual(s) in the clandestine manufacturing of methamphetamine and add supporting evidence for prosecution.

19.8 Questions

1. List four indications of clandestine manufacturing.
2. List three types of related evidence that supports the existence of clandestine operations.
3. Explain to members of the jury how the simple observation of a common propane tank indicated illegal activity.
4. List three solutions commonly found at clandestine sites.
5. Please explain to members of the jury how the possession of common household items indicates their use in clandestine manufacturing (hint: quantities).
6. Outline the steps used in the Red Phosphorus-HI Method.
7. Why is water addition in Step II critical?
8. Why is sulfuric acid used in Step III?
9. What is a common source of NaOH used in methamphetamine production?
10. What is the purpose of "icing" methamphetamine?
11. Why is it prudent to collect multiple samples of evidence?
12. Describe the evidence which is typically collected to prove step IV.
13. Describe how samples are collected using washing.
14. Explain to members of the jury why water is preferred over methanol when washing empty glassware.
15. Why is it generally unacceptable to use methanol as a wash solvent?

Suggested Reading

Christian, D. R. Jr. *Forensic Investigation of Clandestine Laboratories*; CRC Press: Boca Raton, FL, 2005.

Davenport, T. W.; Allen, A. C.; Cantrell, T. S. Synthetic Reductions in Clandestine Amphetamine and methamphetamine Laboratories: A Review. *Forensic Sci. Int.* **1989**, 42, 183–199.

Goddard, K. *Clandestine Operations*. Presented at 37th Semiannual Seminar of the California Association of Criminalists, Newport Beach, CA, 1971.

Examination of Clandestine Evidence 20

20.1 Introduction

The sole purpose of evidence collected from clandestine sites is to prove that operators are engaged in the illegal production of controlled substances. The discovery of illicit drugs packaged for distribution is clearly compelling evidence, but successful prosecution does not necessarily require the presence of final product at the crime scene. It is quite common to convict operators using evidence that proves *intent* to produce and distribute illicit drugs. This broadens the scope of evidence examination to include the following objectives:

Objective 1: Identification of chemicals, solvents, precursors, or elements related to illicit production. The presence of final product is not required.

Objective 2: Identification of solutions frequently associated with illicit production. The presence of final product is not required.

Objective 3: Identification of synthetic steps used in illicit production. The presence of final product is critical depending on which stage is under investigation.

Although the main focus of this chapter is on the Red Phosphorus-HI Method, the science and examination techniques discussed have applications in the analysis of most drugs and trace evidence. Specific reagents may change or sample preparation techniques may be modified, but the concepts are interchangeable.

20.2 Examination of Evidence to Prove Extraction (Step I) (Fig. 20.1)

20.2.1 Evidence Type

(a) Raw chemicals and bulk solvents.
(b) Glass or plastic containers containing clear or cloudy heterogeneous solutions with solid material settled at bottom. The solution or solid may be colored depending on the binder used in the tablets. This evidence is an indication of *current* clandestine activity.
(c) Moist coffee filters containing a white or colorful residue indicate separation of pseudoephedrine from binder. This evidence is an indication of *current* clandestine activity. Dry filters are an indication of past activity.
(d) A coffee grinder with traces of white or colorful solid residue.

J.I. Khan et al., *Basic Principles of Forensic Chemistry*, DOI 10.1007/978-1-59745-437-7_20,

© Springer Science+Business Media, LLC 2012

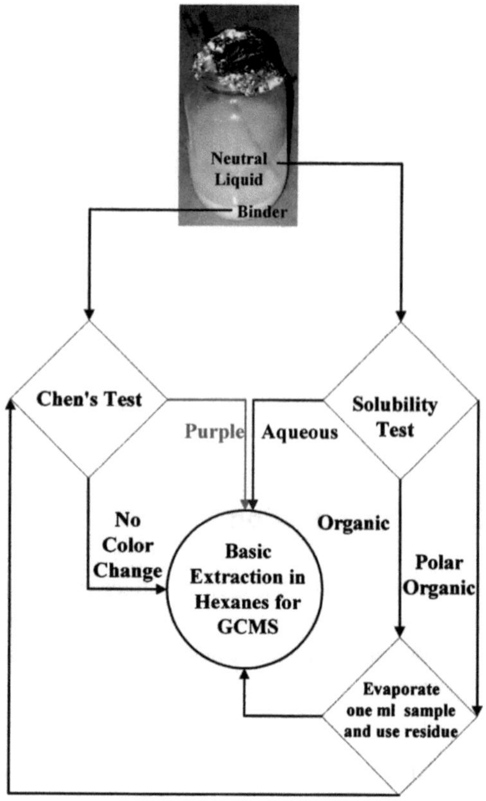

Fig. 20.1 Summary of clandestine evidence examination to prove step I.

20.2.2 Examination

1. Test the pH of the solution; it should be neutral.
2. Place a small amount of the solid or liquid sample in two separate test tubes for solubility testing.
3. Add 0.5 ml of chloroform to one test tube and 0.5 ml of water to the other. If the sample mixes in chloroform and water, the operator used alcohol in the extraction. If the sample is insoluble in chloroform and mixes in water, the operator used water in the extraction.
4. Evaporate the solvent and perform a base extraction on the solid residue for gas-chromatography mass spectrometry (GCMS).
5. Perform Chen's test on the binder, if present. Purple is positive for ephedrine/pseudoephedrine.
6. Perform base extraction in hexane on the binder for identification using GCMS.

20.3 Examination of Evidence to Prove Manufacturing of Methamphetamine (Step II) (Fig. 20.2)

20.3.1 Evidence Type

Elements: Red phosphorus (P), iodine (I_2), sodium (Na), or lithium (Li).
Acids: Hydroiodic acid (HI), hydrochloric acid (HCl), or hypophosphorous acid (H_3PO_2)
Precursors: Pseudoephedrine
Products: Methamphetamine, amphetamine, *N,N*-dimethylamphetamine
By-products: Azidridines, substituted naphthalenes, phenyl-2-propanone (P2P), or other transient products.

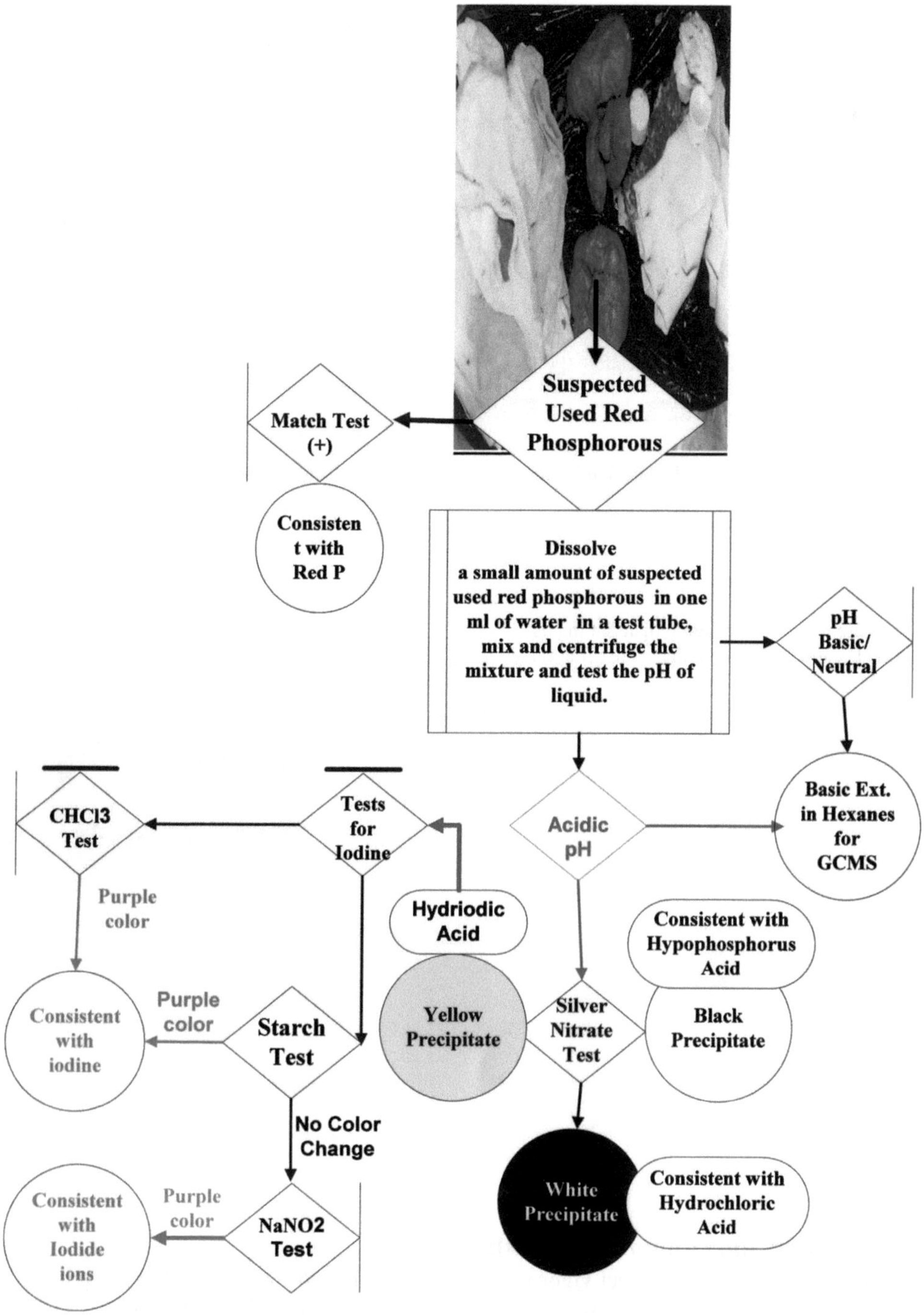

Fig. 20.2 Summary of clandestine evidence examination to prove step II.

20.3.2 Examination

20.3.2.1 Identification of Elements

Forensic laboratories often identify elements using their chemical and physical properties. Confirmatory methods using a scanning electron microscope are also quite common.

20.3.2.2 Flame-Test for Red Phosphorus

Red phosphorus is the substance that ignites when the tip of a match is rubbed against a course striker. The *flame-test* is a confirmatory method used to identify red phosphorus. A small portion of the suspected red solid is placed on a clean glass slide and exposed to a flame. If the substance is red phosphorus, it will ignite generating a small cloud of white smoke (Fig. 20.3).

Fig. 20.3 Samples of red phosphorus recovered from clandestine crime scenes. The presence of this catalyst is a clear indication of methamphetamine production.

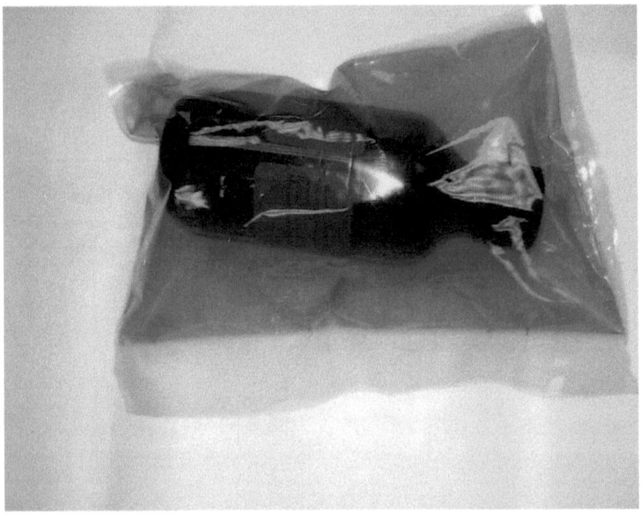

Fig. 20.4 Iodine crystals are found in packaging material that is discolored by sublimation because iodine containers are often sealed improperly.

20.3.2.3 Tests for Sodium and Lithium (Nazi Method)

Sodium and lithium are group I metals that react vigorously with water, including water in the air (humidity). Consequently, they must be stored in an environment devoid of water, often kerosene or any number of thick viscous oils. It is highly likely that metals recovered from clandestine sites in kerosene (or other oil) are sodium or lithium. For confirmation, a small portion of suspected metal is placed in a test-tube and water is added. Sodium ignites with a yellow flame and lithium ignites with a red flame. Unlike most metals, these are relatively soft and can be easily cut with a knife.

20.3.2.4 Identification of Iodine

1. Sublimation: Iodine crystals (I_2) characteristically sublime, that is, they convert directly from solid to vapor without appearing in the liquid phase. The process of sublimation normally taints iodine packaging material yellow (Fig. 20.4).
2. Starch Test: Acidified iodine turns purple when starch is added. Dissolve a few crystals of suspected iodine in 0.25 ml of acetic acid and add a small amount of starch. The solution turns purple if iodine is present.
3. Chloroform Test: Acidified iodine turns a chloroform layer purple. Dissolve a few crystals of suspected iodine in 0.25 ml of acetic acid and add 0.25 ml of chloroform. The organic layer (chloroform) will turn purple in the presence of iodine.

Table 20.1 Examination results and interpretation

Evidence	Testing results	Interpretations
Cloudy white/colorful solutions	Pseudoephedrine	Extraction of pseudoephedrine (step I)
Biphasic/organic/acidic/ basic solutions	Pseudoephedrine	Extraction of pseudoephedrine (step I)
Red solid	Red phosphorus, iodine, HCl/HI/ hypophosphorous acid	Mixture used in the manufacturing of methamphetamine (step II)
Red solid	Red phosphorus, iodine, HCl/HI/ hypophosphorous acid, pseudoephedrine	Mixture used in the manufacturing of methamphetamine (step II)
Red solid	Red phosphorus, iodine, HCl/HI/ hypophosphorous acid, pseudoephedrine, methamphetamine	Manufacturing of methamphetamine (step II)
Biphasic/organic/acidic/ basic solutions	Methamphetamine	Processing of methamphetamine (step III)
Polar-organic solvents	Methamphetamine	Processing/icing of methamphetamine (steps III or IV)
Stain wash[a]	Acidic or basic pH, HCl/HI/iodine, pseudoephedrine, methamphetamine	Manufacturing or processing of methamphetamine (steps II or III)
Water wash	Acidic or basic pH, pseudoephedrine, methamphetamine	Manufacturing or processing of methamphetamine (steps II or III)
Methanol wash	Pseudoephedrine, methamphetamine	None

[a]Stains from clothing, plastic tubing, filter papers, carpets, floors, or walls

Fig. 20.5 Silver cations form insoluble ionic compounds when combined with a variety of anions. The color of the precipitate is used to identify acids recovered from clandestine sites.

4. To confirm the presence of iodine in red phosphorus, place a small amount of sample in a test tube and add 1 ml of water. Mix, centrifuge, and test the pH of the solution. If it is acidic, add chloroform and interpret results as described in test 3. If it is not acidic, divide the sample into two equal portions and add acetic acid to both. Add chloroform to one and starch to the other. Interpret results as described in tests 2 and 3 above (Table 20.1).
5. An additional test confirms the presence of iodide (I^-), the ionic form of iodine. Perform the starch test as described in test 2. If negative, add sodium nitrite ($NaNO_2$). The solution turns purple in the presence of iodide ions.

20.3.2.5 Identification of Acids: Silver Nitrate Test

1. Test the pH of the suspected solution.
2. If acidic, add one or two drops of 5% silver nitrate and observe the color of the precipitate (Fig. 20.5):

White precipitate: Hydrochloric acid (HCl)
Black precipitate: Hypophosphorous (H_3PO_2)
Yellow precipitate: Hydroiodic acid (HI)

20.3.3 Confirmatory Examination

Precursors, products, and transient by-products collected in pure form are identified using GCMS. Impure samples require extraction prior to GCMS analysis.

1. Place a small amount of sample in a test tube and add 1 ml of water.
2. Mix and centrifuge.
3. Add 1 ml of 6 M NaOH.
4. Extract in hexane for GCMS analysis.

20.4 Examination of Evidence to Prove Processing of Methamphetamine (Step III) (Fig. 20.6)

20.4.1 Evidence Type

Chemicals: Organic solvents, two-layer (biphasic) solutions, acidic and/or basic solutions.
Products: Methamphetamine, amphetamine, *N*,*N*-dimethylamphetamine
By-products: Azidridines, substituted naphthalenes, phenyl-2-propanone (P2P), or other transient products

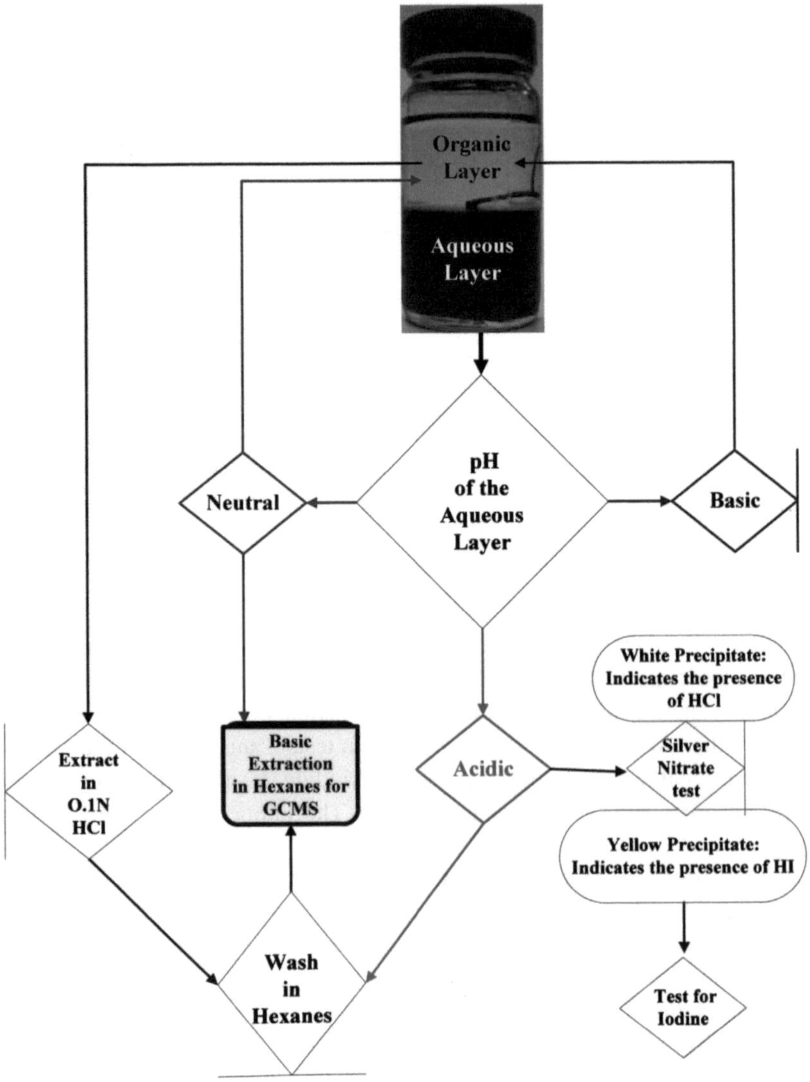

Fig. 20.6 Summary of clandestine evidence examination to prove step III/IV.

Fig. 20.7 The discovery of biphasic solutions at clandestine sites is strong evidence of methamphetamine processing. Acidic aqueous layers often contain methamphetamine.

20.4.2 Examination of Biphasic Solutions

First, test the pH of both layers of the solution. The organic layer is usually neutral, but the aqueous layer may be acidic, basic, or neutral. In general, the examination procedure is determined by the pH of the aqueous layer.

20.4.2.1 Examination of Acidic Layers

The pH of the aqueous layer is confirmed using the silver nitrate test as described previously (Fig. 20.5). Methamphetamine is highly soluble in acidic solutions. If the aqueous layer is acidic, it is tested directly for the presence of methamphetamine (Fig. 20.7).

1. Wash approximately 1 ml of the acidic layer three times with 1 ml portions of hexane to extract organic contaminants from the aqueous layer. This step is necessary for a clear GC.
2. Add 6 M NaOH to the acidic layer until the solution is basic.
3. Extract methamphetamine with hexane for GCMS.

20.4.2.2 Examination of Basic Layers

If the aqueous layer is basic, the organic layer is tested for the presence of methamphetamine.

1. Add 1 ml of 0.1 M HCl to 1 ml of the organic layer. If methamphetamine is present, it will transfer to the acidic aqueous layer.
2. Follow the examination procedure for acidic layers above. Wash the acidic layer three times with 1 ml portions of hexane, add 6 M NaOH until basic, extract in hexane for GCMS (Fig. 20.8).

20.4.2.3 Examination of the Neutral Layers

When the aqueous layer is neutral, both the organic and aqueous layer must be tested for the presence of methamphetamine.

1. Remove 1-ml samples of each layer for separate analysis.
2. Add 6 M NaOH to each sample until basic.
3. Extract both samples with hexane for GCMS (Fig. 20.9).

20.5 Examination of Evidence to Prove Icing (Step IV)

Collect and examine any solutions containing acetone, especially those discovered in a refrigerator or freezer. Acetone is a polar-organic solvent with a neutral pH and distinctive odor. Evaporate the solvent, perform color-screening tests on the residue, and perform base extraction in hexane for the GCMS (Fig. 20.10).

Fig. 20.8 A jar containing a biphasic solution recovered from a clandestine crime scene. Organic layers are tested for methamphetamine when the pH of the aqueous layer is above 7 (basic).

Fig. 20.9 A biphasic solution containing a neutral aqueous layer. In this case, both layers are tested for methamphetamine.

Fig. 20.10 Iced methamphetamine crystals. Acetone is usually a critical source of evidence because it is required in the icing stage of methamphetamine production.

20.6 Examination of Stains (Fig. 20.11)

Characteristic stains related to the clandestine production of methamphetamine are usually red, brown, or yellow in color. They are often found on fabric, clothing, plastic tubing, filter paper, carpet, floors, and walls (Fig. 20.12).

1. Collect a small sample of the stain and dissolve in 1 ml of water. Test pH of the solution.
2. If basic, perform extraction with hexane for GCMS.
3. If acidic, test for acid type using the silver nitrate test.
4. Test the acidic solution for the presence of iodine using starch test or chloroform test.
5. Perform base extraction in hexane for GCMS.

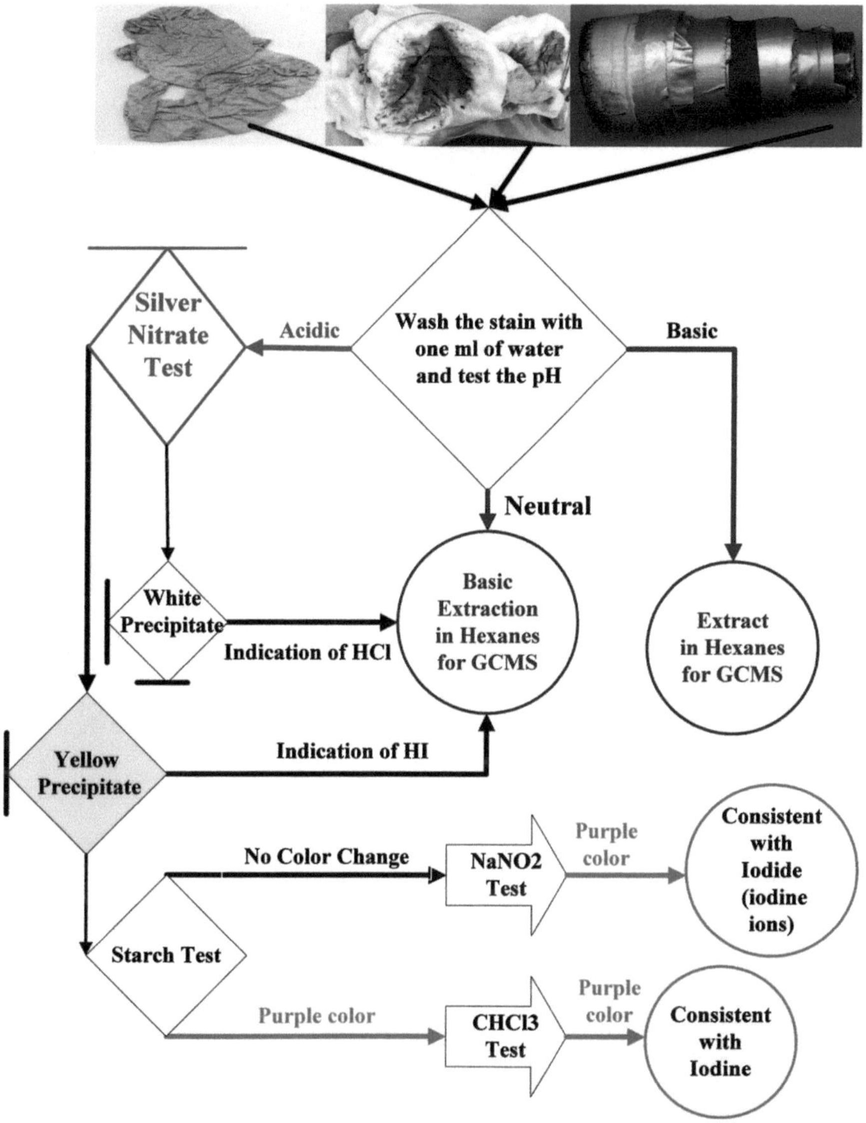

Fig. 20.11 Summary of forensic examination of stain evidence.

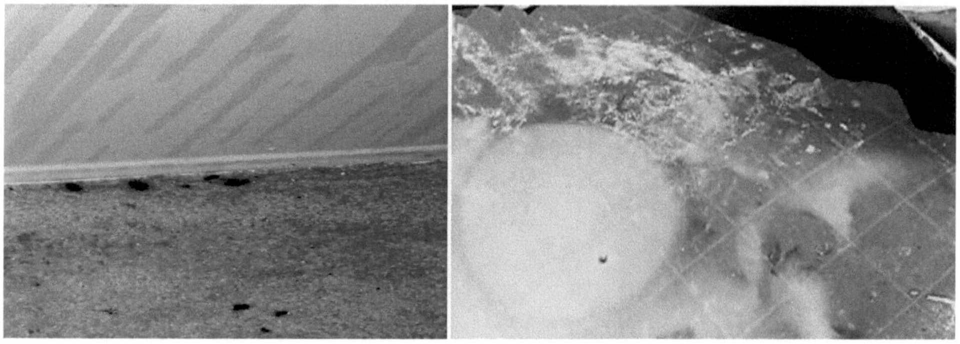

Fig. 20.12 Stains characteristic of methamphetamine production. The patterns are highly variable but the colors are usually *red*, *brown*, or *yellow*. Stains can be found on anything used in the production process or on contaminated material or fabric in the local vicinity.

20.7 Examination of Washes

Washing is typically performed using methanol or water; however, water is usually preferred as discussed in Chap. 19. The procedure for examining washes is illustrated in the flowchart in Fig. 20.13, regardless of solvent (Fig. 20.13).

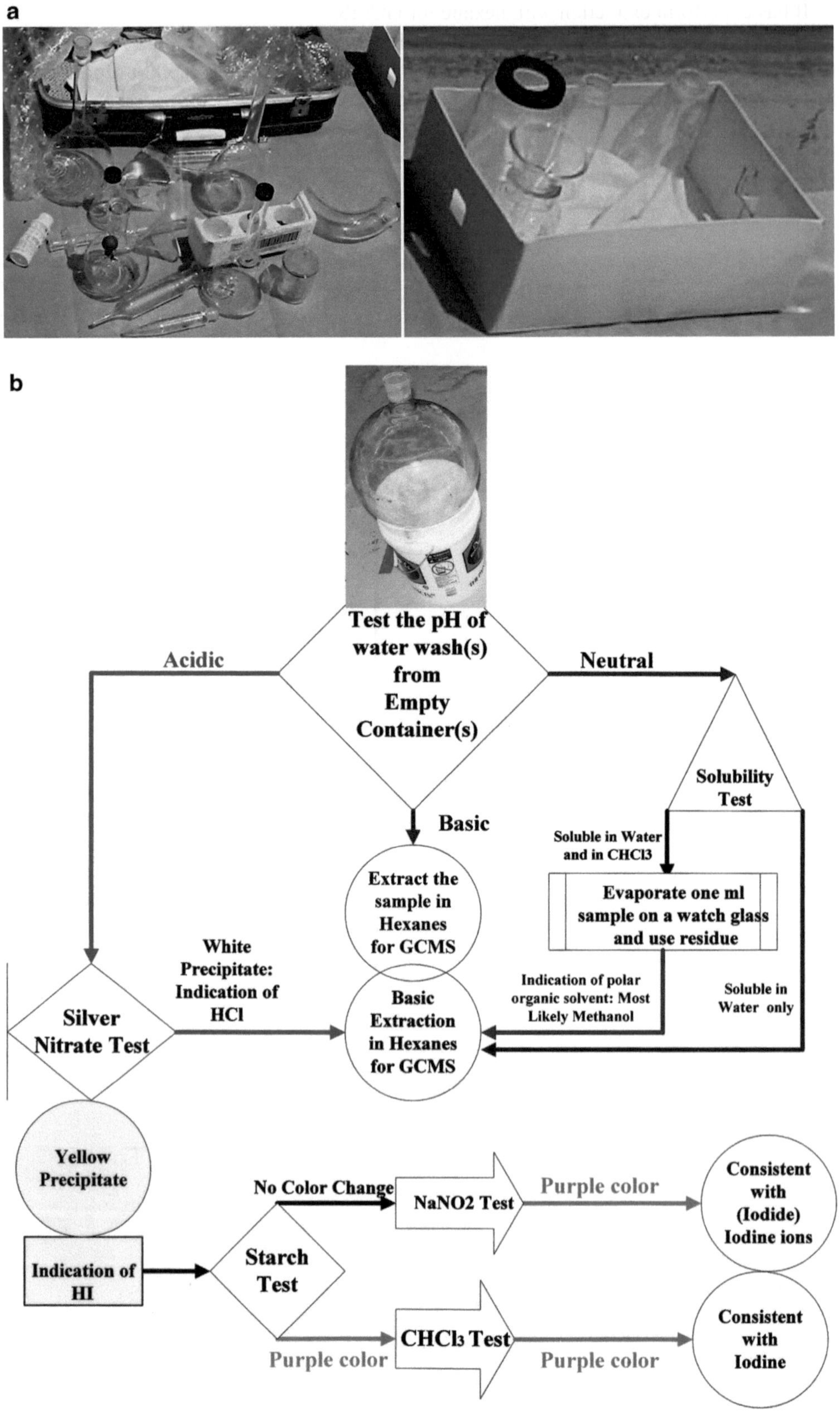

Fig. 20.13 (**a**) Glassware collected from clandestine sites is tested for the presence of methamphetamine using washes. (**b**) The residue is dissolved in water and the resulting solution is tested using the flowchart.

20.8 Determining Methods of Methamphetamine Production

Criminal prosecution requires proof that individuals engaged in the production of illicit drugs. Although the actual method used is often of little concern, identifying the process is necessary to prove unlawful activity.

The simple observation of chemicals and equipment discovered at the scene is one way to identify the method used to produce methamphetamine. For example, the Nazi method would be suspected if sodium (Na) or lithium (Li) is discovered with a conspicuous absence of heat sources (cold method), whereas the Red-P/HI method would be implicated by the presence of red phosphorus, hydroiodic acid, and heat sources (hot method).

A significantly more reliable way to determine the method of production is through the use of analytical results. Individual methods yield a set of distinctive by-products that are used to identify the process. For example, Red-P/HI reduction produces a set of by-products that are different from those found in Nazi reduction. In addition, reduction methods using ephedrine or pseudoephedrine as precursors generate different by-products than those using the ephedra plant (Figs. 20.14–20.16).

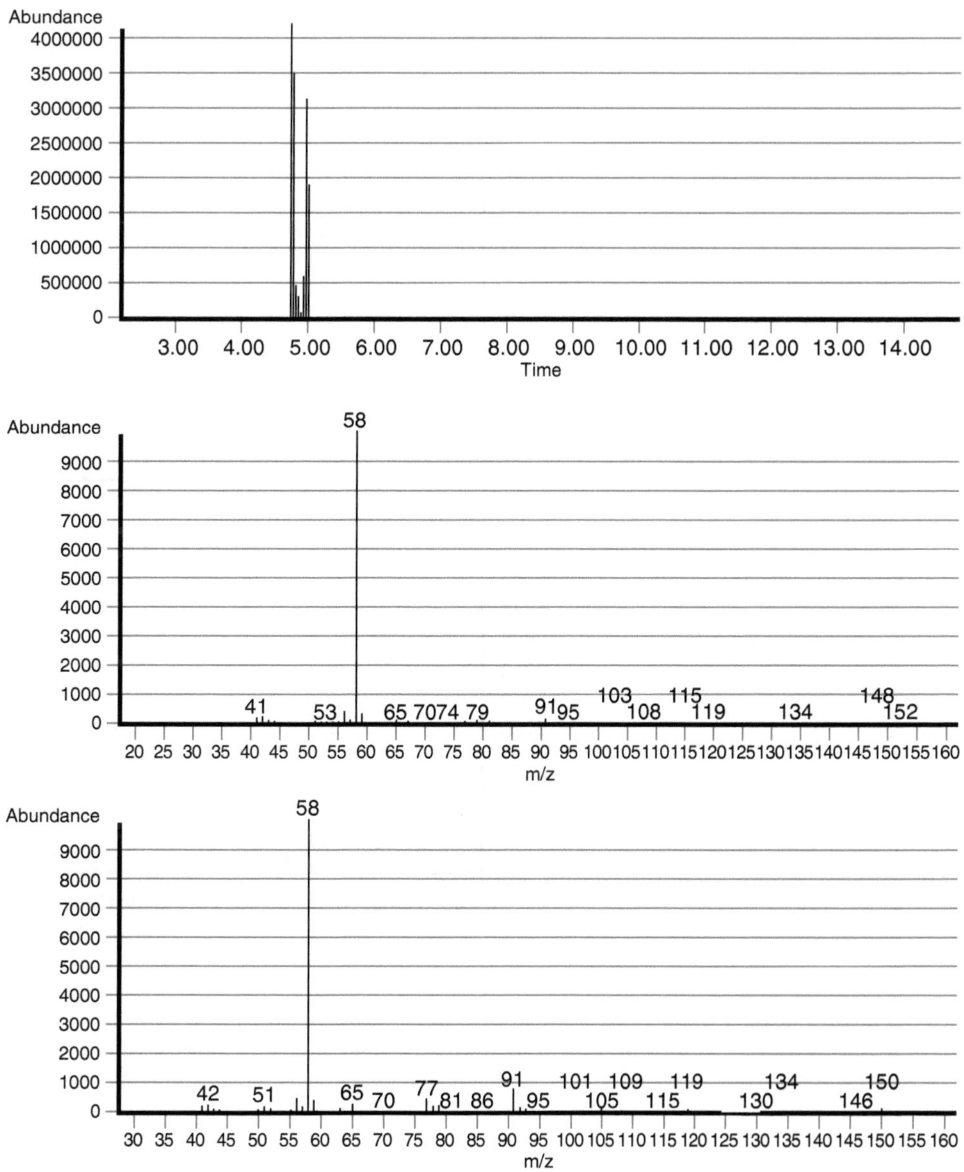

Fig. 20.14 Signature GCMS results of the Nazi method.

Fig. 20.15 Signature GCMS results of the HI-red phosphorus method.

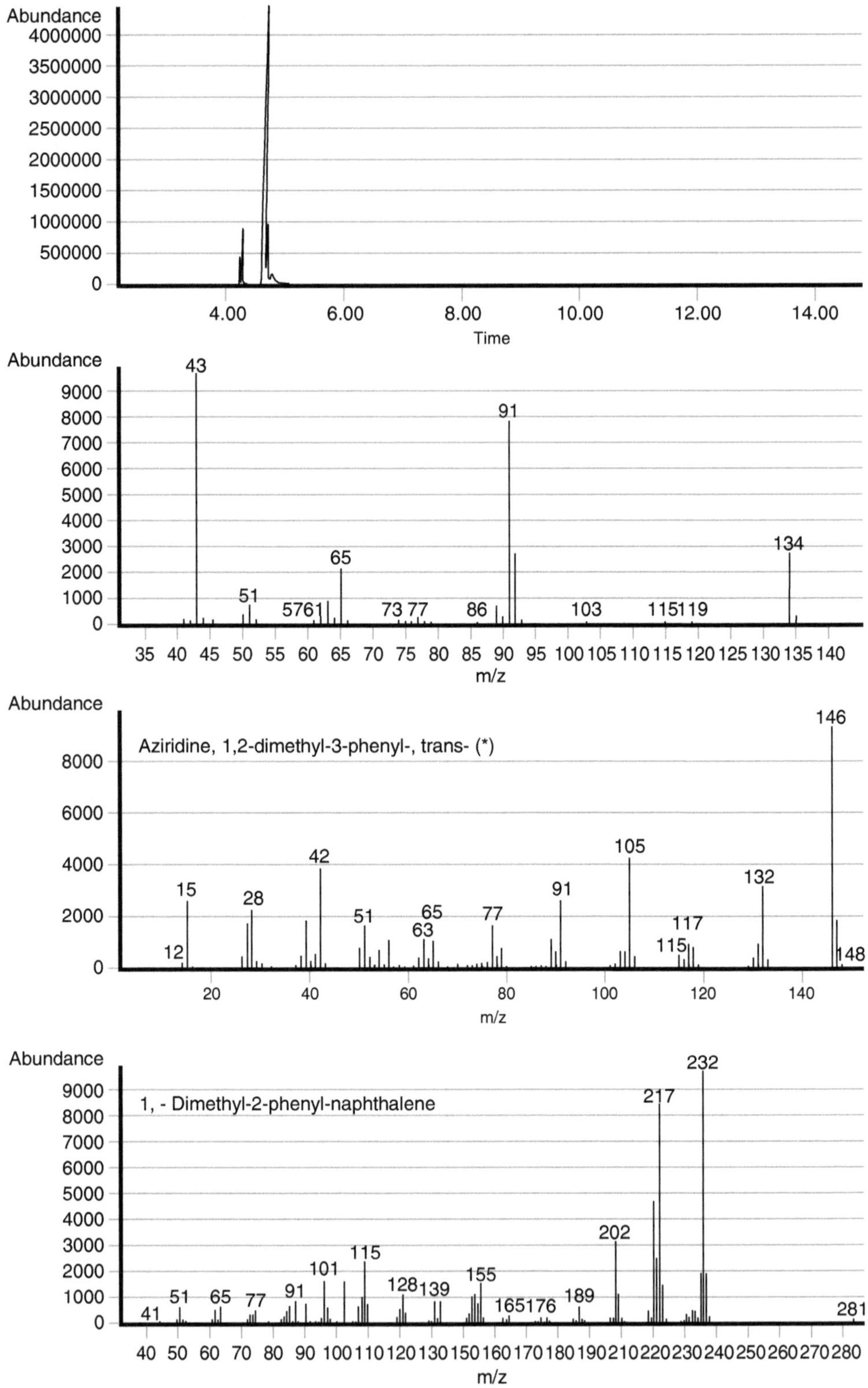

Fig. 20.16 Signature GCMS results when the Ephedra plant is used as a precursor.

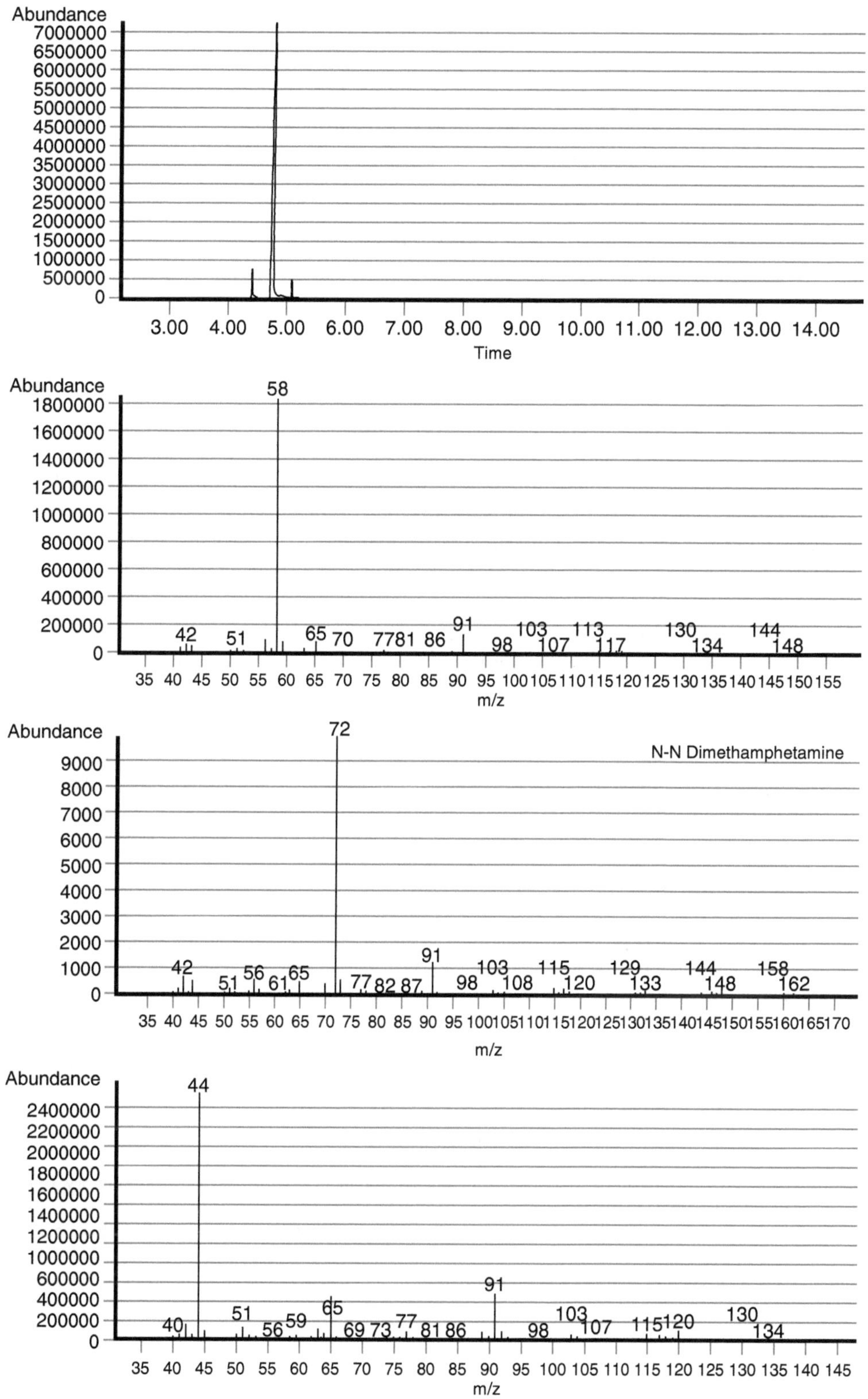

20.9 Questions

1. List the three objectives of evidence examination.
2. Is it possible to convict clandestine operators who have not actually produced illicit drugs? Explain.
3. List two types of evidence indicating current clandestine activity.
4. Please explain to members of the jury how chloroform is used in the examination of evidence to prove extraction of pseudoephedrine (step I).
5. Why are the solutions used in step I often colored?
6. Please explain to members of the jury why Chen's test is used on binders.
7. List the acids and elements commonly used in methamphetamine production?
8. Please explain to members of the jury how you identified red phosphorus, sodium, and lithium.
9. Why is sodium and lithium stored in kerosene?
10. Describe the starch and chloroform test. How is it used in forensic investigation?
11. How is the anionic form of iodine (iodide) tested?
12. Outline the procedure used to identify acids.
13. Which step of methamphetamine production uses biphasic solutions? What is the procedure to test for methamphetamine when a basic aqueous layer is found?
14. What solvent is used in the icing stage of methamphetamine?
15. What are the characteristic colors of stains commonly associated with methamphetamine production and where are they found?
16. A stain was tested with the following results: washed with water, pH tested acidic, silver nitrate produced yellow solid, starch test was purple, and chloroform test was purple. Identify the two components (refer to Fig. 20.12).
17. What are two observations that indicate the use of the Red-P/HI method?
18. How is GCMS used to identify the process used in methamphetamine production?
19. Using the GCMS data provided, differentiate the Red-P/HI method from the Nazi method (hint: GC chromatogram).
20. Using the GCMS data, describe how methamphetamine produced from the ephedra plant is different from that produced using other reductive methods?

Suggested Reading

Allen, A. C.; Kiser, W. O. Methamphetamine from Ephedrine: Chloroephedrine and Aziridine. *J. Forensic Sci.* **1987**, 32, 953–962.

Audier, E.; Millet, A.; Sozzi, G. Mass Spectra of Amphetamine and Related Compounds. *Org. Mass Spectrum.* **1984**, 19, 522–523.

California Department of Justice. *Technical Procedures for Controlled Substance Analysis.* California Department of Justice: Sacramento, 2006.

Christian, D. R. Jr. Analysis of Controlled Substances. In *Forensic Science: An Introduction to Scientific and Investigative Techniques*, 2nd ed.; James, S. H., Nordby, J. J., Eds.; CRC Press: Boca Raton, FL, 2005.

Christian, D. R. Jr. *Forensic Investigation of Clandestine Laboratories*; CRC Press: Boca Raton, FL, 2005.

Davenport, T. W.; Allen, A. C.; Cantrell, T. S. Synthetic Reductions in Clandestine Amphetamine and methamphetamine Laboratories: A Review. *Forensic Sci. Int.* **1989**, 42, 183–199.

Goddard, K. *Clandestine Operations.* Presented at 37th Semiannual Seminar of the California Association of Criminalists, Newport Beach, CA, 1971.

McKibben, T. *et al.* Analysis of Inorganic Components Found in Clandestine Drug Laboratory Evidence. *J. Clandestine Lab. Investigating Chemists Assoc.* **1995**, 5, 19–33.

Rasmussen, K. E.; Knutsen, P. Techniques for the Detection and Identification of Amphetamines and Amphetamine-Like Substances. *UN Bull Narcotics.* 1985, 37, 95–112.

United Nations Division of Narcotic Drugs. *Rapid Testing Methods of Drugs of Abuse.* United Nations: New York, 1988.

Laboratory Manual

Preface

The goal of this laboratory manual is to provide the student with an enjoyable experience that is both informative and challenging. It is our hope that the thrill of "seeing" theoretical principles "come to life" in the laboratory will enhance your overall understanding, as well as stimulate and develop critical thinking skills. Although there has always been considerable debate over the most effective methods of instruction; many agree that lecture topics supported by practical exercises are a proven model to create a successful learning environment. We embrace the spirit of this model in the *Forensic Chemistry Laboratory Manual*. Our approach is to correlate laboratory exercises to the theoretical and investigative principles of forensic chemistry. This will provide the student with valuable hands-on experience while adding clarity and continuity to lecture topics. This laboratory manual was written within the framework of each of the following areas.

Level and Audience

The *Forensic Chemistry Laboratory Manual* covers the laboratory component of a one semester class in forensic chemistry. It is not designed to be a stand-alone laboratory manual. It was specifically written to complement *Basic Principles in Forensic Chemistry*, the required text for a one semester class offered as part of our forensic certificate program. The course requires no prerequisite and is designed for students with little, if any, background in chemistry or forensics. The laboratory exercises are designed to provide practical experience in forensic investigative techniques. Emphasis is on the development of proper technique, handling of evidence, and interpretation of data and results. Although there is brief exposure to more sophisticated chemical principles, it is not the main focus of the manual. It is possible to perform complex procedures and reliably interpret results without an in-depth understanding of the complex reaction mechanisms involved.

Forensic Investigation

Investigative techniques are developed using evidence and test results from actual case studies. Students learn to exercise due diligence in the formulation of hypotheses, preparation of courtroom testimony, and presentation of results. "Moot" courts are used to develop proper courtroom demeanor, i.e., giving testimony, presenting evidence, jury interaction, etc. In addition, students are exposed to proper format and writing techniques typically used in the submission of case reports.

Stockroom Preparation

It was important to develop experiments that require chemicals and laboratory equipment that is both inexpensive and readily available.

Safety

Forensic chemical analysis is often performed by highly trained scientists in a controlled environment. Consequently, a few procedures used in forensic investigation have been intentionally omitted. These omissions may be based on reagent cost or availability, lack of analytical instrumentation or specialized glassware, or safety concerns when working with potentially dangerous chemicals. In these few cases, data and/or test results are provided for interpretation and presentation purposes only. Strict adherence to all safety procedures is highly stressed.

J.I. Khan et al., *Basic Principles of Forensic Chemistry*, DOI 10.1007/978-1-59745-437-7,
© Springer Science+Business Media, LLC 2012

To The Student

We wish you success as you begin your journey into forensic chemistry. This manual was specifically designed to illustrate principles and techniques commonly used in forensic investigation. All too often, students fail to realize (or appreciate) the importance of practical laboratory experience and its relationship to theoretical principles. Lost in the topics presented in lecture are the long hours scientists spend in the laboratory developing and proving these theories. As you perform the experiments in this manual, you will learn proper experimental technique and develop an appreciation of the correlation that exists between theory and practice. We hope that your laboratory experience is enjoyable and informative.

Strict adherence to safety procedures will create a relatively safe and hazard free laboratory environment. It is the responsibility of each student to contribute to this safe environment by following all safety rules and regulations. The following list of safety procedures should be followed at all times. Your laboratory may have specific safety rules and practices, in addition to those below, that will be thoroughly explained by your laboratory instructor.

Wear approved safety glasses or goggles at all times
If you have contact lenses, nonvented goggles are required.

Prepare for lab
Read the experiment carefully and be aware of potential hazards before coming to lab.

Dress for lab
No loose fitting cloths, no shorts, no open-toed shoes, no tank tops. Tie long hair back to prevent contact with an open-flame. Lab coats or aprons are highly recommended and may be available in lab.

No food or drinks are allowed in the laboratory
Chemicals may adhere to food or liquids and may cause illness. If you take a break to eat, wash your hands thoroughly.

Know the location and proper use of all safety equipment
Survey the lab and locate all exits, safety showers, fire extinguishers, fire blankets, eye wash facilities, emergency gas shut-off valves, emergency phones, etc.

No unauthorized experiments
Closely follow the instructions given in this manual. Do not deviate from the procedures or techniques explained.

Practice proper laboratory behavior at all time
Do not take unnecessary risks. Playing or "horsing around" in lab will not be tolerated and will result in your expulsion.

Handle all chemicals properly
Never taste chemicals or inhale chemical vapors.
Avoid direct contact of chemicals with skin.
Never pour excess chemicals back into the original container.
Your instructor will advise you in the proper disposal of waste material.

Keep your work area neat and organized
Do not clutter your work area with excessive chemicals, glassware, and books.

Smoking is not permitted in the laboratory

Report all accidents to your instructor
Report all accidents, no matter how small, to your laboratory instructor. This information may be used to further develop and/or refine existing safety procedures.

If you have questions ASK YOUR INSTRUCTOR

Your instructor is a trained professional who is very familiar with the procedures performed in each experiment. If you have questions or require clarification, do not hesitate to ask.

The above represents a list of minimum safety precautions that should be followed to create a safe laboratory environment. Following these procedures will not guarantee a safe, accident-free environment, nor are they intended to represent a complete list of all safety rules and regulations. The possibility of accident and/or injury is always present in the lab; however, strict adherence to proper safety procedures at all times will minimize the risk for such occurrences.

Laboratory Manual Table of contents

Experiment	Topic
1	Introduction and Safety
2	Forensic/Scientific Investigation and Atomic Structure
3	Properties of Elements
4	Mixtures and Compounds
5	Chemical Formulas and Nomenclature
6	Solubility
7	Molecular Geometry
8	Organic Chemistry and Functional Groups
9	Microcrystallography
10	Chemical Extraction
11	Chromatography
12	Interpretation of GCMS Spectra
13	IR Spectroscopy
14	Examination of Marijuana (moot)
15	Examination of Controlled Substances: Primary and Secondary Amines (moot)
16	Examination of Controlled Substances: Tertiary Amines and Opiates (moot)
17	Examination of Controlled Substances: Tryptamines (moot)
18	Examination of Anabolic Steroids (moot)
19	Examination of Miscellaneous Controlled Substances (moot)
20	Clandestine Manufacturing of Methamphetamine (moot)

Experiment # 2

Name ______________________________

Forensic/Scientific Investigation and Atomic Structure

Reference: Chapters 1 and 2

Objectives: Students will gain practical experience using the scientific method to develop conclusions. Students will become familiar with atomic structure and writing electron configurations for ground state neutral atoms and ions.

Introduction:

Scientific discoveries are usually the result of a systematic approach to a good idea or an unexplained observation. Although many variations of this "systematic" approach exist, it often involves a stepwise process called the *scientific method.* The *scientific method* is a procedure used to develop technical theories and generally includes four phases; *observation, hypothesis, experimentation, and theory*. It begins with the observation of some type of unexplained phenomenon. A possible cause of the observation is proposed during the hypothesis phase. Experiments are then specifically designed in order to prove the hypothesis. If experimental results do not support the hypothesis another possibility is considered and tested. If experimental results confirm the hypothesis, and are consistently reproducible, a formal explanation is developed and subsequently offered as a theory. The theory is then presented to the scientific community where it may be accepted or rejected. If accepted, it may become a *principle* or a *law*.

Part: A: Forensic/ Scientific Investigation

Observation: Different types of cell phone ring tones.

Hypothesis: No cell phones are currently set to identical ring tones

Experiment: Each student that has a cell phone will play their ring tone. Record the results.

Theory:

Part B: Atomic Structure

Complete the following table

Symbol	# Protons	# Neutrons	# Electrons	Atomic Mass
H				1
	19		19	
K^+				39
B		6		11
	5		2	
Cl			17	
Cl^-	17			

Part C: Electron Configuration

Write the electron configuration for each of the following.

Mg (atomic mass = 24)

Mg^{2+} (atomic mass = 24)

C (atomic mass = 12)

N^{3-} (atomic mass = 14)

Ca (atomic mass = 40)

Cl^{-} (atomic mass = 35.45)

H (atomic mass = 1)

Experiment # 3

Name ____________________________

Properties of Elements

Reference: Chapter 3

Objectives: Students will distinguish the difference between a chemical property and a physical property. Students will gain practical experience in the use of properties to identify elements and compounds.

Materials: 1M hydroiodic acid (HI), 1M hydrochloric acid (HCl), 0.3M hypophosphorous acid (H_3PO_2), 0.5M sulfuric acid (H_2SO_4), 1M acetic acid ($HC_2H_3O_2$), 3% silver nitrate solution ($AgNO_3$), 1% iodine solution in water, chloroform ($CHCl_3$), starch (sugar), sodium nitrite ($NaNO_2$), test tubes, matches, and wax pencils.

Introduction:

There are fundamental properties associated with all forms of matter. These *distinguishing characteristics* may be physical or chemical in nature, and are commonly used to identify and classify a particular substance. A *physical property* is anything that can be measured or observed without changing chemical composition. The melting point and boiling point of water are examples of physical properties because these temperatures can be measured without changing the chemical composition of water. A *physical change* is a change in the state of matter, but not its chemical composition. There are three accepted states of matter; solid, liquid, and gas (although some would argue *plasma* is also a state). Other physical properties commonly used in the forensic identification of elements and compounds are: color, odor, density, solubility, conductivity, and sublimation.

Chemical properties are a measure of the ability of a substance to produce new substances, or simply, a measure of the reactivity of a substance. *Chemical changes* are transformations that produce products chemically and physically different from the starting material. A solution containing silver nitrate will produce a white precipitate (solid) in the presence of chloride ions and a yellow precipitate in the presence of iodide ions. These observations illustrate the chemical changes that result from the reactivity (chemical properties) of silver nitrate. Physical and chemical properties are commonly used to identify elements and compounds in the field of forensic science. Consequently, these properties may be used to support or reject specific parts of an investigation.

Part A:

Clean five test tubes and label each with the name of an acid shown below. Place 10 drops (approx. ½ ml) of the corresponding acid into each of the labeled test tubes. Add 1–2 drops of 3% silver nitrate solution ($AgNO_3$) to each test tube and observe the results. If a precipitate (solid) forms, record the color below next to the corresponding acid. If no precipitate is observed, write "none."

Acid	Color of Precipitate
Hydroiodic acid	
Hydrochloric acid	
Hypophosphorous acid	
Sulfuric acid	
Acetic acid	

Conclusion:

Part B:

Strike a matchstick on the following surfaces and record your observations.

Nature of Surface	Ignite (yes/no)
Wood	
Cement	
Metal	
Plastic	
Paper	
Course sandpaper (matchbox)	

Conclusion:

Part C:

Clean three test tubes and label each with a reagent shown below. Place ten drops (approx. ½ ml) of 1% iodine solution in each test tube and add the corresponding reagent. Record your observations.

Reagent	Add	Observation
Starch	Approx. "½ pea size" of starch	
Chloroform	1 ml chloroform	
Acetic acid	1 ml acetic acid	

Conclusion:

Experiment # 4

Name________________________________

Mixtures and Compounds

Reference: Chapters 1, 2, and 3

Objective: Students will observe common properties of mixtures and compounds.

Materials: methanol, DI water, sugar, salt, analytical balance, watch glass, and oven.

Introduction:

Elements and compounds may exist as pure substances or as mixtures. *Pure substances* contain only one component and have the same composition throughout, i.e., pure gold, pure sugar, pure water, etc. *Mixtures* contain two or more pure substances and may be homogeneous or heterogeneous. *Homogeneous mixtures* have the same composition and properties throughout. However, they are not pure substances because they contain more than one component. *Heterogeneous mixtures* have distinctly different properties within the mixture; water and sand would be an example. In any *binary solution* (a solution that contains only two components), the *solvent* is the component present in greatest amount and the *solute* is the component present in least amount.

The following mixtures will be provided. Classify each mixture by circling homogeneous or heterogeneous.

Solution #1	Sugar in water (sat.)	homogeneous or heterogeneous
Solution #2	Salt in water (sat.)	homogeneous or heterogeneous
Solution #3	Sugar in methanol (sat.)	homogeneous or heterogeneous
Solution #4	Salt in methanol (sat.)	homogeneous or heterogeneous

Part A:

Clean and dry four watch glasses and label each 1, 2, 3, or 4. Weigh each empty watch glass on an analytical balance and record the mass in the table below under "watch glass." Be sure to weigh the watch glasses after they are labeled! Place 1.0 ml of the corresponding solutions above on each of the labeled watch glasses, i.e., place 1.0 ml of solution #1 on watch glass labeled 1, etc. Using the same analytical balance that was used to weigh the empty watch glasses, carefully weigh the watch glasses containing each solution. Determine the mixture mass for each solution by subtracting the mass of the empty watch glass from the mass of the watch glass containing solution. Record the mass of each mixture in the table below under "mixture mass." Save the watch glasses containing each solution for Part B. Clean and dry two small test tubes and place one test tube into a small beaker. Place the test tube/beaker on the analytical balance and *tare* the balance (zero the balance with the test tube/beaker on the pan). Place 1.0 ml of water into the test tube and record the mass in the table below under Solvent Mass for water. Repeat the procedure using the other test tube and 1.0 ml of methanol. Record the mass below under Solvent Mass for methanol (water is the solvent in solutions 1 and 2, methanol is the solvent in solutions 3 and 4). Subtract the solvent mass from the mixture mass and record the difference in the table below under "solute mass."

Watch Glass	Mixture Mass	Solvent Mass	Solute mass
#1-		Water-	
#2-		Water-	
#3-		Methanol-	
#4-		Methanol-	

What is the mass of sugar in mixture #1?
What is the mass of salt in mixture #2?
What is the mass of sugar in mixture #3?
What is the mass of salt in mixture #4?

Did the mixture mass exceed the solvent mass in any solution? If so, explain.

Did the solvent mass exceed the mixture mass in any solution? If so, explain.

Did the solvent mass equal the mixture mass in any solution? If so, explain.

Explain your observed results using your knowledge of homogeneous and heterogeneous mixtures.

Part B:

Carefully place the watch glasses containing each solution in the oven and evaporate the solvent to dryness. When evaporation is complete, weigh each watch glass and record the mass in the table below under "watch glass/residue." Determine the mass of the residue by subtracting the mass of the empty watch glass (measured in Part A) from the mass of the watch glass/residue. The residue is the actual mass of solute contained in each solution. Record the residue mass in the table below under "solute mass (actual)."

Watch Glass/Residue	Solute Mass (actual)

What is the actual mass of sugar in mixture #1?
What is the actual mass of salt in mixture #2?
What is the actual mass of sugar in mixture #3?
What is the actual mass of salt in mixture #4?

Conclusion: (Hint: Definition of solution; did any of the 1.0 ml solutions actually contain 1.0 ml of solvent?)

Experiment # 5

Name ______________________________

Chemical Formulas and Nomenclature

Reference: Chapters 2 and 3

Objective: Students will gain experience writing chemical formulas for ionic compounds. Students will learn formal procedures used to name ionic and covalent compounds.

Introduction:

Substances are either elements or compounds. A *compound* is a substance that consists of two or more elements bonded together in a specific way. The forces that hold atoms together in a compound are called *chemical bonds.* An *ionic bond* involves the *transfer* of electrons from a *metal* to a *nonmetal.* A *covalent bond* consists of a pair of electrons *shared* between two *nonmetals*.

Ions and the Octet Rule

Atoms are electrically neutral because they have an equal number of electrons and protons. An atom can be converted into a charged particle called an *ion* by losing or gaining one or more electrons. The loss of electron(s) from a neutral atom produces a *positively* charged ion called a *cation* (pronounced cat-ion). The gain of electron(s) by a neutral atom produces a *negatively* charged ion called an *anion* (pronounced an-ion).

Generally, the charge on an ion can be predicted from the position of the element on the periodic table. The metals (on the left-hand side of the table) lose electrons to form cations. The Group IA elements lose ONE electron to achieve an octet and take a charge of 1 positive. This is correctly written using a superscript "+" attached to the upper right side of the elemental symbol, i.e., Na^+. Notice that the number "1" is *not* written when the cation carries a positive *one* charge. The Group IIA elements lose TWO electrons and take a charge of 2 positives. This is correctly written using a superscript "2+" attached to the upper right side of the elemental symbol, i.e., Mg^{2+}. When cations carry a charge greater than one, the number is written first, followed by the sign. The Group IIIA elements lose THREE electrons and take a charge of 3 positives which is written as a superscript "3+", i.e., Al^{3+}.

The nonmetals (on the right-hand side of the table) gain electrons to form anions. The Group VIIA elements gain ONE electron to achieve an octet and take a charge of 1 negative. This is written using a superscript "-" attached to the elemental symbol, i.e., Cl^-. Once again, the "1" is *not* written. The Group VIA elements gain TWO electrons and take a charge of 2 negatives which is written as a superscript "2-", i.e., O^{2-}. The Group VA elements gain THREE electrons and take a charge of 3 negatives which is written as superscript "3-", i.e., N^{3-}. Some transition metals and metals in Group IVA have variable charges (more than one positive ion is possible). See table below.

Some common ions and their location on the periodic table.

IA	IIA											IIIA	IVA	VA	VIA	VIIA	
H^+																	
Li^+	Be^{2+}													N^{3-}	O^{2-}	F^-	
Na^+	Mg^{2+}											Al^{3+}		P^{3-}	S^{2-}	Cl^-	
K^+	Ca^{2+}						Fe^{2+} Fe^{3+}	Co^{2+} Co^{3+}	Ni^{2+}	Cu^+ Cu^{2+}	Zn^{2+}					Br^-	
Rb_+	Sr^{2+}									Ag^+			Sn^{2+} Sn^{4+}			I^-	
Cs^+	Ba^{2+}										Hg_2^{2+} Hg^{2+}		Pb^{2+} Pb^{4+}				

Writing Formulas for Ionic Compounds

Ionic compounds are electrically neutral. Therefore, when writing formulas, the cations (positive) and anions (negative) must combine to produce a net charge of zero. In the formula, the cation (metal) is always written first, followed by the anion (nonmetal). The number and types of each element must be clearly shown in the formula; the type of element is indicated using the elemental symbol, and the number of each element is indicated using a subscript attached at the lower right side of the symbol. The number "1" is not written in cases requiring only a single element. Formulas for ionic compounds are called *formula units*.

The correct ratio required to produce a net charge of zero when Na^+ ions combine with Cl^- ions is one to one because one Na^+ cancels one Cl^-. Therefore, the formula is NaCl. Notice this is not written Na_1Cl_1.

The correct ratio when Na^+ ions combine with O^{2-} ions is two to one because two Na^+ are required to cancel one O^{2-}. The 2 atoms of Na are indicated in the formula using a subscript "2" directly attached to Na. The formula is Na_2O.

The correct combining ratio when Na^+ ions and P^{3-} ions combine is: Na_3P (three to one).

Practice Examples:

Write the formula for the ionic compound that is formed when each of the following pairs of ions interact:

a) K^+ and S^{2-}
b) Mg^{2+} and O^{2-}
c) Ca^{2+} and I^-
d) Li^+ and N^{3-}
e) Al^{3+} and S^{2-}

Solution

(a) The cation has a charge of 1+ because K is a member of Group IA. The anion has a charge of 2- because S is member of Group VIA. Thus, two positive ions (2+) are required for each negative ion (2-) to produce an electrically *neutral* formula unit.

The formula is $\mathbf{K_2S}$.

(b) The cation has a charge of 2+ and anion has a charge of 2-. The ratio is 1:1. The formula is **MgO**.

(c) The cation has a charge of 2+ and anion has a charge of 1-. Two negative ions are required for each positive ion. The formula is $\mathbf{CaI_2}$.

(d) The cation has a charge of 1+ and anion has a charge of 3-. Three positive ions are required for each negative ion. The formula is $\mathbf{Li_3N}$.

(e) The cation has a charge of 3+ and anion has a charge of 2-. Two positive ions are required for three negative ions. Here, the lowest common factor of 3+ and 2- is 6 (without sign). The formula is $\mathbf{Al_2S_3}$.

Naming Ions

The names of cations and anions are determined by a system developed by the International Union of Pure and Applied Chemistry (IUPAC).

Metals That Form Only One Type of Positive Ion

Elements in Groups IA, IIA, IIIA, and some transition elements form only one type of cation. For these ions, the name of the cation is the elemental name of the metal followed by the word "ion." Cations can now be differentiated from their corresponding neutral forms using specific names. For example, K is potassium (neutral form) and K^+ is potassium ion (cationic form).

Na^+	sodium ion	K^+	potassium ion	Mg^{2+}	magnesium ion
Al^{3+}	aluminum ion	Ag^+	silver ion	Zn^{2+}	zinc ion

Metals That Form Two Different Positive ions

Metals in Group IVA, and most transition metals, form more than one type of cation and the charge must be included in the name. For these ions, the name of the cation is the elemental name of the metal followed by a Roman numeral in parentheses, *with no space* after the name. The Roman numeral indicates the positive charge on the ion. Technically, the names do not end with the word "ion," although some still prefer to include it.

Sn^{2+}	tin(II)	Sn^{4+}	tin(IV)
Pb^{2+}	lead(II)	Pb^{4+}	lead(IV)
Cu^{+}	copper(I)	Cu^{2+}	copper(II)
Fe^{2+}	iron(II)	Fe^{3+}	iron(III)
Co^{2+}	cobalt(II)	Co^{3+}	cobalt(III)
Hg_2^{2+}	mercury(I)	Hg^{2+}	mercury(II)

Naming Anions

Anions are named by replacing the last part of the elemental name with the suffix *–ide*, and adding the word "ion". Anions can now be differentiated from their corresponding neutral forms using specific names. For example, S is sulfur (neutral form) and S^{2-} is sulf*ide* (anionic form).

F^{-}	fluoride ion	Cl^{-}	chloride ion	Br^{-}	bromide ion	I^{-}	iodide ion
O^{2-}	oxide ion	S^{2-}	sulfide ion	N^{3-}	nitride ion	P^{3-}	phosphide ion

Polyatomic Ions

A *polyatomic ion* is an ion that contains two or more elements. It is recommended that you memorize the names and formulas of the following polyatomic ions:

NH_4^{+}	ammonium	SO_3^{2-}	sulfite
CN^{-}	cyanide	SO_4^{2-}	sulfate
OH^{-}	hydroxide	HSO_3^{-}	hydrogen sulfite
$C_2H_3O_2^{-}$	acetate	HSO_4^{-}	hydrogen sulfate
CrO_4^{2-}	chromate	PO_3^{3-}	phosphite
$Cr_2O_7^{2-}$	dichromate	PO_4^{3-}	phosphate
MnO_4^{-}	permanganate	HPO_4^{2-}	hydrogen phosphate
NO_2^{-}	nitrite	ClO^{-}	hypochlorite
NO_3^{-}	nitrate	ClO_2^{-}	chlorite
CO_3^{2-}	carbonate	ClO_3^{-}	chlorate
HCO_3^{-}	hydrogen carbonate	ClO_4^{-}	perchlorate

The common name for HCO_3^{-}, HSO_3^{-}, and HSO_4^{-} are bicarbonate, bisulfite, and bisulfate respectively.

Naming Ionic Compounds

Binary ionic compounds containing metals that form only one type of positive ion

These compounds contain only *two types* of elements; a metal ion and a nonmetal ion. Note: binary refers to element types, not total number of atoms. For example; $MgBr_2$ contains 3 total atoms, 1 Mg and 2 Br's, but it contains only two types, Mg and Br. Also, recall all elemental symbols begin with capital letters, if you have a binary compound, your formula contains only two capital letters. The metal is always named first using the elemental name of the metal. The nonmetal is named second using the anionic name of the nonmetal (elemental name modified with the suffix *–ide*).

Practice Examples:

Name the following binary ionic compounds:

NaCl	$MgBr_2$	AlP	K_2S	SrF_2	ZnI_2

Solution

NaCl	sodium chloride	K_2S	potassium sulfide
$MgBr_2$	magnesium bromide	SrF_2	strontium fluoride
AlP	aluminum phosphide	ZnI_2	zinc iodide

Binary ionic compounds containing metals that form two different positive ions

These compounds are essentially named using the same procedure developed for metals forming only one type of cation. The distinction is that the charge on the cation must be written as a Roman numeral in parentheses immediately after (with no space) the metal name.

Practice Examples:

Name the following binary ionic compounds:

$FeBr_3$	CoF_2	SnO	PbI_4	HgS	Cu_3P

Solution

$FeBr_3$	iron(III) bromide	PbI_4	lead(IV) iodide
CoF_2	cobalt(II) flouride	HgS	mercury(II) sulfide
SnO	tin(II) oxide	Cu_3P	copper(I) phosphide

Ionic compounds containing polyatomic ions

Identifying compounds containing polyatomic ions is somewhat simplified by the fact that all elemental symbols begin with capital letters. If you identify more than two capital letters in the formula, your compound contains a polyatomic ion that you must immediately recognize (the value of memorizing!). Naming these compounds is simply based on your familiarity with the polyatomic ions. The cation is named first using its elemental name, followed by the name of the polyatomic ion.

Practice Examples:

Name the following polyatomic ionic compounds:

$Ca(NO_3)_2$	$ZnSO_4$	NH_4CN
Li_3PO_4	Na_2CO_3	$Mg(HCO_3)_2$

Solution

$Ca(NO_3)_2$	calcium nitrate
$ZnSO_4$	zinc sulfate
NH_4CN	ammonium cyanide
Li_3PO_4	lithium phosphate
Na_2CO_3	sodium carbonate
$Mg(HCO_3)_2$	magnesium hydrogen carbonate

Compounds containing only nonmetals (molecular compounds)

This type of compound contains covalent bonds, so the concept of cations and anions is somewhat obscure and does not necessarily apply. Regardless, they are essentially named using the previously developed procedure for naming binary compounds containing a metal. The first element is named using its elemental name and the second is named using its anionic name. The difference is the use of Greek prefixes attached to each name, which indicate the number of each element present in the formula. There is one important exception; the prefix "mono" is never attached to the name of the first element in the formula. Let us look at SF_6 as an example. You should immediately notice that S and F are nonmetals because both are located in the nonmetal section of the periodic table (right side). This observation should immediately trigger "prefixes" in your mind. SF_6 is sulfur hexafluoride because the formula indicates one sulfur (note the absence of the subscript "1" attached to S) and 6 fluorides (subscript 6 attached to F). This is not monosulfur hexafluoride because "mono" is never used with the first element. It is possible, however, to attach all other prefixes to the name of the first element, i.e., N_2O_5 is dinitrogen pentoxide. The Greek prefixes are listed below, note that all end in a vowel. When the prefix ends with an "a" or "o", and the elemental name begins with an "a" or "o", the vowel of the prefix is usually dropped to simplify pronunciation. Notice in our N_2O_5 example above, the name of the anion is pentoxide, not pentaoxide.

Greek prefixes

1 (mono-)	2 (di-)	3 (tri-)	4 (tetra-)	5 (penta-)
6 (hexa-)	7 (hepta-)	8 (octa-)	9 (nona-)	10 (deca-)

Practice Examples:

Name the following binary molecular compounds:

N_2O_5	CO_2	P_4S_3	XeF_6	ICl	NH_3	I_4O_9	CO	H_2O	H_2O_2

Solution

N_2O_5	dinitrogen pentoxide	NH_3	nitrogen trihydride (ammonia)
CO_2	carbon dioxide	I_4O_9	tetraiodine nonoxide
P_4S_3	tetraphosphorous trisulfide	CO	carbon monoxide
XeF_6	xenon hexafluoride	H_2O	dihydrogen monoxide (water)
ICl	iodine monochloride	H_2O_2	dihydrogen dioxide

Exercises

Inspect the periodic table and list ALL elements with a ONE letter abbreviation.

Symbol	Name	Period number	Group number	metal, nonmetal, or metalloid

Formulas of Ionic Compounds

Complete the following table with the formula of the compound.

	Br^-	O^{2-}	NO_3^-	PO_4^{3-}
Na^+				
Mg^{2+}				
Al^{3+}				
Pb^{4+}				
NH_4^+				
Fe^{2+}				
Fe^{3+}				

Ionic Compounds

Write the correct name for each of the following ionic compounds.

Formula	Name
$Pb(HCO_3)_4$	
Al_2S_3	
$LiHSO_4$	
$Zn_3(PO_4)_2$	
CoF_3	
$Ca(CN)_2$	
SnO_2	
Na_2CrO_4	
K_2CO_3	
Cu_3P	
$Sr(OH)_2$	
$(NH_4)_2HPO_4$	
Hg_2Cl_2	

Formula	Name
$BaSO_4$	
$Sn(NO_3)_2$	
$AgClO_3$	
$Cu(HSO_3)_2$	

Molecular Compounds

Name each of the following compounds.

Formula	Name
BBr_3	
Br_3O_8	
CI_4	
C_3O_2	
Cl_2O_7	
IF_5	
I_2O_5	
NCl_3	
N_2O_5	
OF_2	
P_4S_3	
P_4S_9	
P_4O_{10}	
SF_6	
S_2Cl_2	
SiS_2	
$SiBr_4$	
XeF_2	
XeO_4	
XeF_6	

Experiment # 6

Name ______________________________

Solubility

Reference: Chapter 3

Objective: Students will test the solvent properties of various liquids to observe and understand the chemical nature of solubility and miscibility.

Materials: acetone, chloroform, ammonia, methanol, water, test tubes, and pipettes.

Introduction:

Water is a common solvent in many solutions and substances like salt and sugar readily dissolve in water. Any substance that dissolves appreciatively in a specified solvent is said to be *soluble* in that solvent. Technically, the term *solubility* refers to a quantitative maximum amount of substance that can dissolve in a given volume of solvent at a specific temperature. The ability of a substance to dissolve in a particular solvent depends on the identity of both the solvent and the substance; the general rule is "likes dissolve likes." Water is a polar covalent molecule and, as a solvent, can dissolve similar molecules (polar covalent). The polarity of water is responsible for its remarkable solvent properties and explains why ionic compounds (i.e., NaCl) and polar covalent compounds (i.e., sucrose, ammonia) are soluble, whereas nonpolar molecules (i.e., organic, gasoline, oils) are not. Terms such as soluble, slightly soluble, insoluble, and solubility are used to describe the ability of a substance to dissolve in a solvent. Intuitively, we associate the term "dissolving" with solids and liquids; however, a liquid may also be soluble in another liquid. For example, when 25.0 ml of ether is added to 25.0 ml of water, the resulting total volume is <u>not</u> 50.0 ml, in fact, it is slightly less (approx. 48.5 ml). This is the result of solubility; ether and water are slightly soluble in one another and consequently, the volumes are not additive. The solubility of one liquid in another is difficult to determine and is usually not readily observed upon mixing. For this reason, liquids are often characterized using *their ability to mix with other liquids* rather than their *solubility* in other liquids. The degree of mixing between two liquids is described using the terms *miscible* and *immiscible*. Two liquids are miscible (soluble) if a uniform solution results after mixing (i.e., water and ammonia). Two liquids are immiscible (insoluble) if two distinct layers form after mixing (i.e., oil and water). Water is miscible in polar liquids and immiscible in nonpolar (organic) liquids.

Part A:

Add 1.0 ml of each of the following reagents to four separate, clean, and dried test tubes: chloroform, ammonia, methanol, water. Add 1.0 ml of acetone to each test tube and mix. Observe the results and characterize the liquids as miscible (**M**) or immiscible (**I**). Record your results in the table below under the corresponding reagents. Clean and dry three test tubes and add 1.0 ml of each of the following reagents to separate test tubes: ammonia, methanol, water. Add 1.0 ml of chloroform to each test tube and mix. Record your results in the table below under the corresponding reagents. Clean and dry two test tubes and add 1.0 ml of each of the following reagents to separate test tubes: methanol, water. Add 1.0 ml of ammonia to each test tube and mix. Record your results in the table below under the corresponding reagents. Clean and dry one test tube and add 1.0 ml of methanol and 1.0 ml of water to the test tube and mix. Record your results in the table below under the corresponding reagents.

Reagents	Chloroform	Ammonia	Methanol	Water
Acetone				
Chloroform	--------------			
Ammonia	--------------	------------		
Methanol	--------------	------------	------------	

Identify the organic liquids:

Identify the inorganic liquids:

What liquids are polar?

What liquids appear to be non-polar?

List the liquid-pairs that are miscible:

List the liquid-pairs that are immiscible:

Are immiscible liquids soluble in one another? Briefly explain.

Are miscible liquids soluble in one another? Briefly explain.

Experiment # 7

Name ______________________________

Molecular Geometry

Reference: Chapter 3

Objective: To illustrate the geometric structures of simple molecules and to demonstrate the relationship between bonding and molecular geometry.

Materials: Molecular model kits.

Introduction:

In recent years, advancements in research and technology have provided precise information on molecular geometry, i.e., bond distances, angles, and energies. Structural theory has advanced far beyond the simple electron dot representation and now rests securely on the foundations of quantum and wave mechanics. Although problems involving only simple molecules can be solved through rigorous mathematical calculations, approximations such as Valence Bond Theory (VBT) and Molecular Orbital Theory (MOT) are very successful in providing results that agree favorably with experimental measurements. This exercise will utilize Valence Bond Theory to illustrate the geometry of a variety of simple molecules. This will be accomplished through the use of model kits that show the correct angles formed between atoms in the molecule. The first covalent bond formed between any two atoms is always a sigma-bond (σ-bond). This type of bond has electrons distributed symmetrically about the bond axis and is used to define the bond axis. Additional bonds (double or triple) formed between the same two atoms will be pi-bonds (π-bonds). These bonds are perpendicular to the defined σ-bond and do not influence geometry. It is the σ-bonds, and any lone pairs of electrons occupying sigma hybrid orbitals, that determine molecular geometry.

Model kits are a necessary and integral part of the study of molecular geometry. They are tools that allow students to transcend the inherent difficulties that arise from visualizing a three dimensional structure on a two dimensional piece of paper. A complete understanding of the capabilities and limitations of model kits is essential in their successful use as a visual aid. Open your model kits and carefully read the instructions. Inventory the various pieces contained in your kit and pay special attention to the color codes used to designate specific atoms. It is important that you take the time now to familiarize yourself with all the components contained in your kit. This will allow you to concentrate on the structure or concept currently under study and prevent wasted time (and confusion) that may result from constantly referring back to the instructions. A table is included at the end of the lab for reference.

Methane (CH_4)

Construct a model of methane using your model kit. Locate a tetrahedral center (a carbon atom) and attach four rods (bonds) in each hole of the atom. Attach 4 balls (hydrogen atoms) of the same color to each rod extending from your structure.

Sketch the model using the solid/dashed wedge convention and name the geometry.

Write the structural formula.

Write the condensed structural formula.

What are the H–C–H bond angles?__________________

How many σ-bonds are on the central carbon?__________________

Identify the two planes present in this molecule. Do the planes divide the molecule into equal halves?_______________

How many atoms, including carbon, are in the same plane?_________________

Ammonia (NH_3)

Construct a model of ammonia using your model kit.

Sketch the model, including the lone pair of electrons, and name the geometry.

Write the structural formula.

Write the condensed structural formula.

What is the H–N–H bond angle?__________________
How many σ-bonds are on the central nitrogen?__________________
How many lone pairs are on the central nitrogen?__________________
Are any atoms in ammonia in the same plane?__________________
Suggest a reason why the H–N–H bond angles in ammonia differ from the H–C–H bond angles observed in methane.

Water (H_2O)

Construct a model of water using your model kit.

Sketch the model, including lone pairs of electrons, and name the geometry.

Write the structural formula.

Write the condensed structural formula.

What is the H–O–H bond angle? ____________________
How many σ-bonds are on the central oxygen? _________________
How many lone pairs are on the central oxygen? ______________________
How many atoms, including oxygen, are in the same plane? __________________
Why is the H–O–H bond angle slightly smaller than the bond angles observed in ammonia and methane?

What conclusions can be made on how lone pairs affect molecular geometry?

Sulfur Hexafluoride (SF_6)

Construct a model of sulfur hexafluoride using your model kit.

Sketch the model and name the geometry.

Write the structural formula.

Write the condensed structural formula.

What are the F–S–F bond angles? ____________________
How many σ-bonds are on the central sulfur? _________________
How many lone pairs are on the central sulfur? ____________________
How many atoms, including sulfur, are in the same plane? ___________________

Carbon Dioxide (CO_2)

Construct a model of carbon dioxide using your model kit.

Sketch the model and name the geometry.

Write the structural formula.

Write the condensed structural formula.

What is the O–C–O bond angle? ___________________
How many σ-bonds are on the central carbon? _____________________
How many π-bonds are on the central carbon? ___________________
How many lone pairs are on the central carbon? ____________________
How many atoms are in the same plane? ________________________

Lone pairs	Bonded atoms[a]	Geometry	Molecular Shape	Examples
0	2	Linear		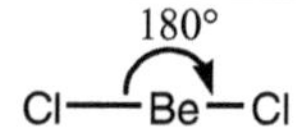 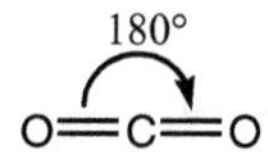
0	3	Trigonal Planar		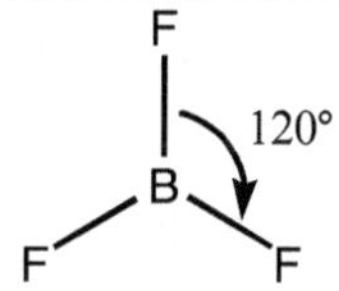
0	4	Tetrahedral	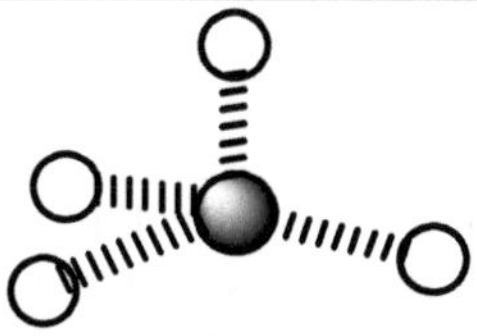	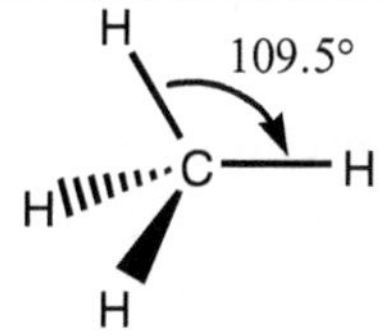
1	3	Trigonal Pyramidal	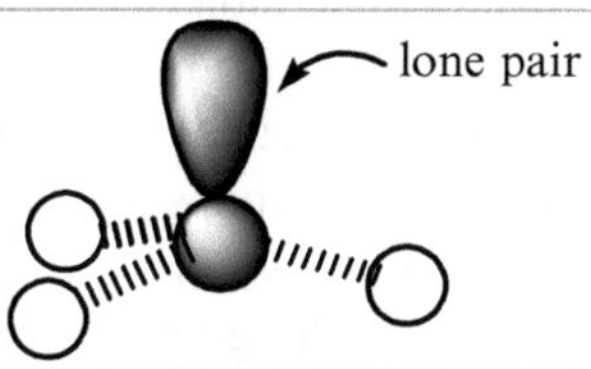	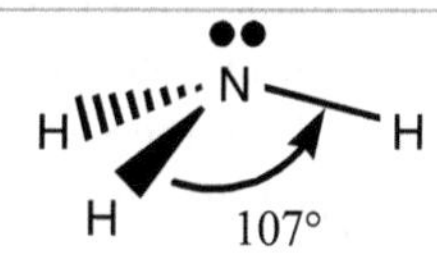
2	2	Angular (bent)	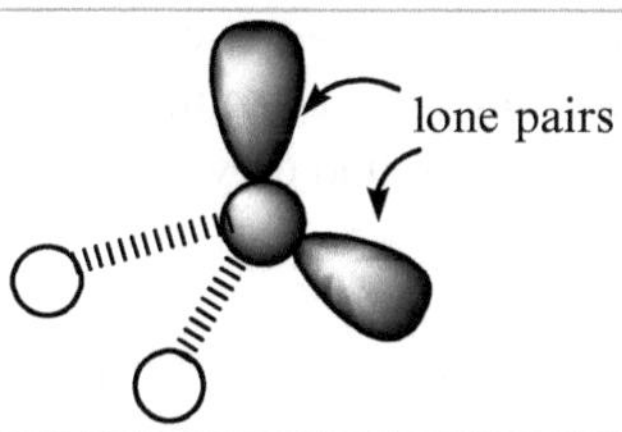	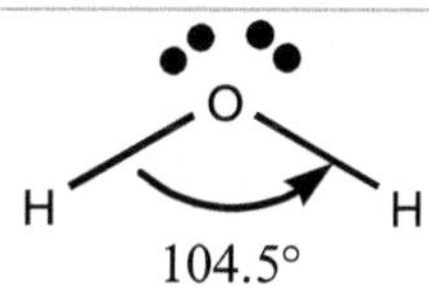
0	5	Trigonal Bipyramidal	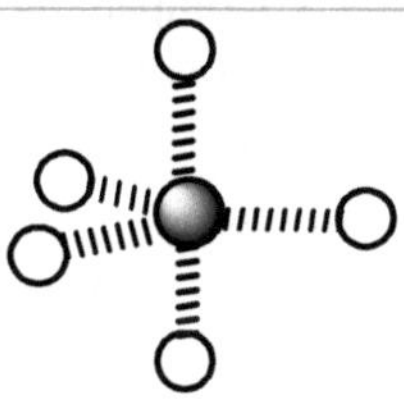	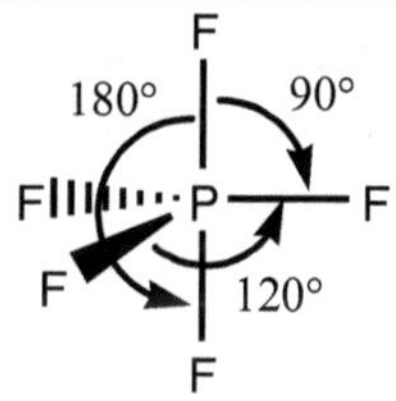
0	6	Octahedral	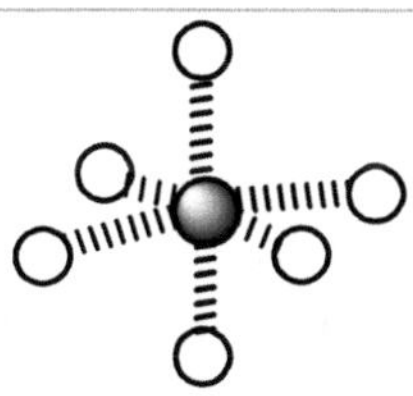	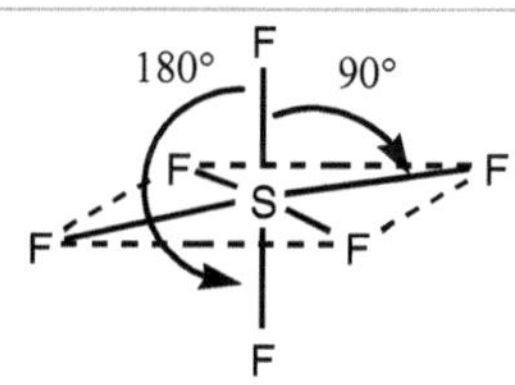

aTechnically this should read # of σ-bonds. A single bond contains 1 σ-bond and 0 π-bonds, double bonds contain 1 σ- and 1 π-bond, triple bonds contain 1 σ- and 2 π -bonds. Therefore, when counting bonded atoms to determine molecular geometry, count single bonds as 1, double bonds as 1, and triple bonds as 1. See CO_2 example above.

Experiment # 8

Name ______________________________

Organic Chemistry and Functional Groups

Reference: Chapter 4

Objective: Students will gain experience in the basic recognition of functional groups present in a variety of organic molecules.

Materials: Molecular model kits.

Introduction:

Organic chemistry is the study of the properties, structure, and function of compounds containing carbon. Although the defining element in organic molecules is carbon, it is common practice to define them by the obligate presence of both carbon and hydrogen. This does not mean that organic molecules contain only carbon and hydrogen; elements such as nitrogen, oxygen, sulfur, phosphorous, and chlorine may also be present. The study of organic chemistry does not involve the individual study of the vast number of organic compounds. Instead, organic compounds are divided into broad classes based on the presence of a *functional group*. Functional groups are atoms, groups of atoms, or common structural features used to classify organic molecules. In general, functional groups will react in a unique, predictive manner and this chemical behavior is similar in all compounds containing a specific group. It is possible, and quite common indeed, to have more than one functional group on a single molecule. In these cases, the molecule will exhibit chemical and physical properties of all groups present. The study of functional groups is the most effective and efficient approach to the study of organic chemistry. A wide range of functional groups can be found on various types of controlled substances.

Part A:

Use model kits to construct the following molecules using their chemical formulas. Rotate each structure in space and observe the geometry from different perspectives. Compare the structures of different molecules and determine some factors that influence geometry, i.e., number and types of bonds present, bond orientation on central atom, number of atoms in the molecule, etc. Draw the structures of each molecule.

Alkanes:

Name	Formula	Structure
Methane	CH_4	
Ethane	C_2H_6	
Propane	C_3H_8	
Butane	C_4H_{10}	

Alkenes:

Name	Formula	Structure
Ethene	C_2H_4	
Propene	C_3H_6	
1-Butene	C_4H_8	
2-Butene	C_4H_8	

1-Butene and 2-Butene are positional isomers because they have the same molecular formula and differ only in the location of the double bond.

Alkynes:		
Name	Formula	Structure
Acetylene	C_2H_2	
(Ethyne)		
Propyne	C_3H_4	
1-Butyne	C_4H_6	
2-Butyne	C_4H_6	

Note: acetylene does not contain a double bond despite its "ene" ending.

Alcohols:		
Name	Formula	Structure
Ethanol	C_2H_5OH	
Propanol	C_3H_7OH	
Isopropanol	C_3H_7OH	
(Isopropyl alcohol)		
1-Butanol	C_4H_9OH	
2-Butanol	C_4H_9OH	

Aldehydes:		
Name	Formula	Structure
Formaldehyde	HCHO	
(Methanal)		
Acetaldehyde	CH_3CHO	
(Ethanal)		
Propanal	CH_3CH_2CHO	
Butanal	$CH_3CH_2CH_2CHO$	

Note: The location of the aldehyde functional group will always be carbon # 1.

Carboxylic Acids:		
Name	Formula	Structure
Formic Acid	HCOOH	
Acetic Acid	CH_3COOH	
(Ethanoic Acid)		
Propanoic Acid	CH_3CH_2COOH	
Butanoic Acid	$CH_3CH_2CH_2COOH$	

Others:		
Name	Formula	Structure
Chloroform	$CHCl_3$	
Benzene	C_6H_6	
Cyclohexane	C_6H_{12}	

Draw structures for the following:

1-bromo-2,2-dichloropentane

2-methyl-3-heptene

1-chloro-1-pentanol

3-isopropyloctane

3-methylbutanal

4-chloropentanoic acid

Name the following:

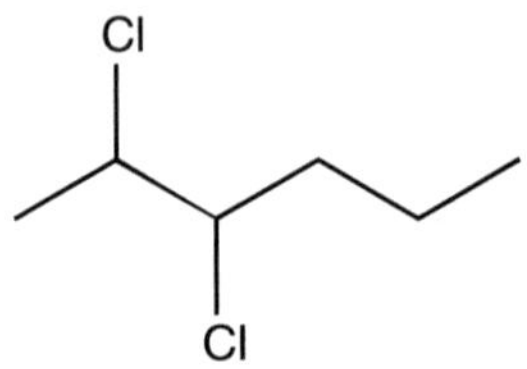

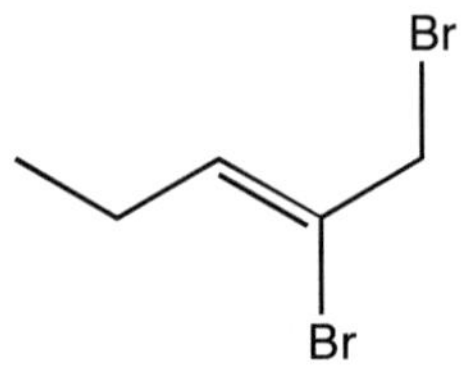

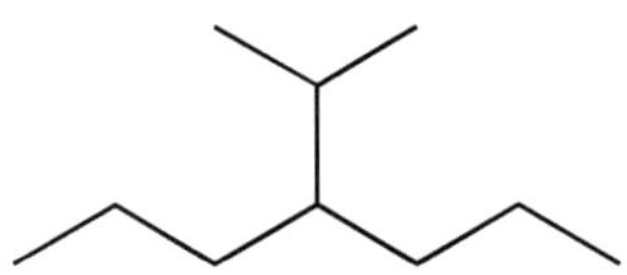

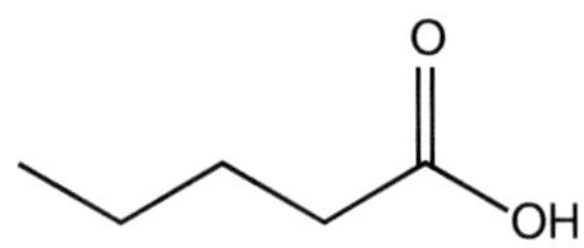

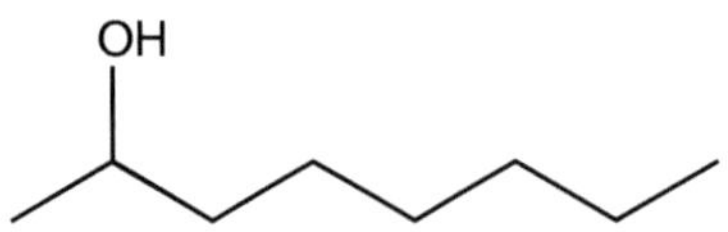

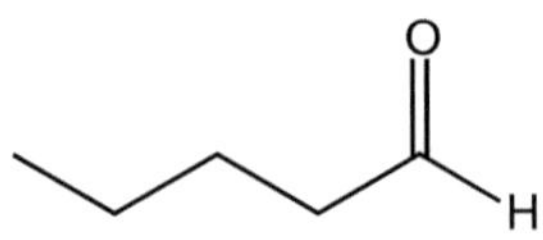

Part B:

Circle and name all functional groups present in each of the following compounds. A table of functional groups is provided on the last page of the lab.

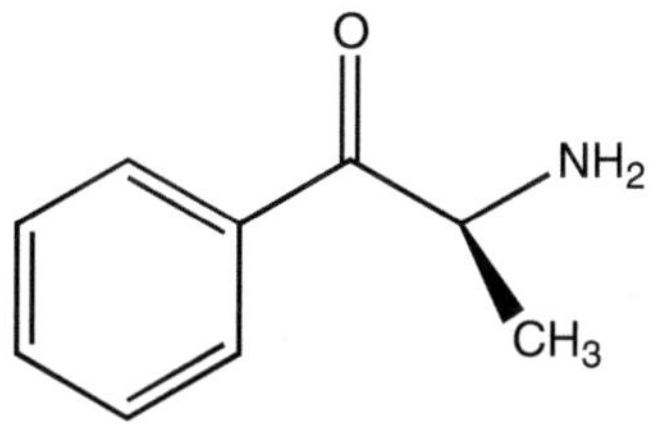

Cathinone

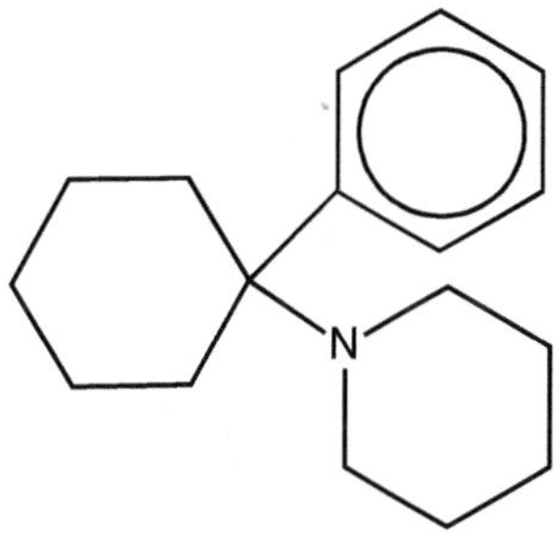

PCP

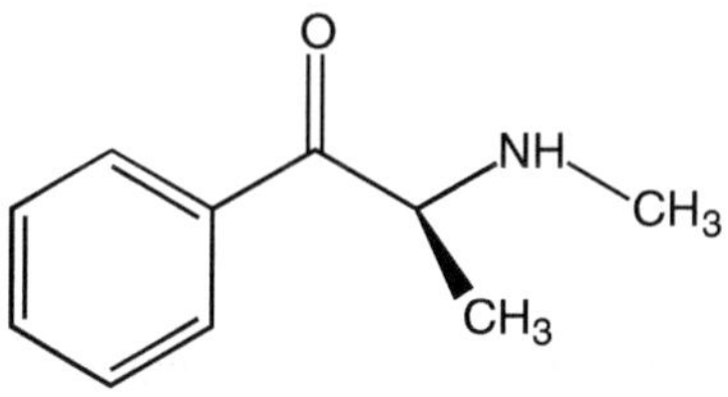

Methcathinone

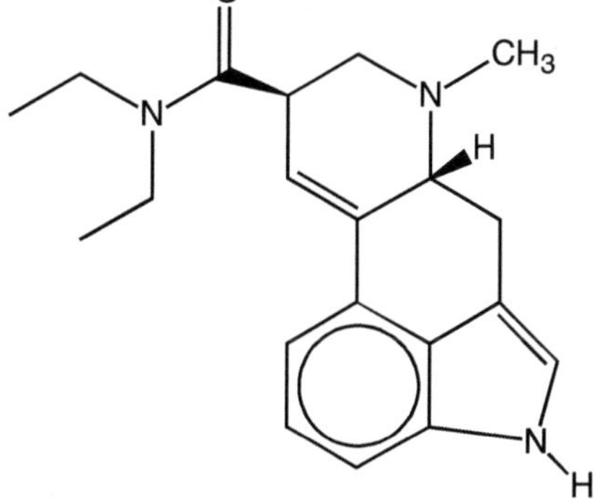

LSD

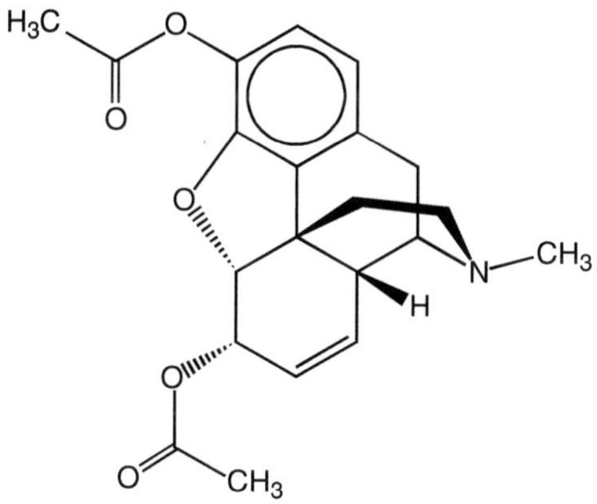

Heroine

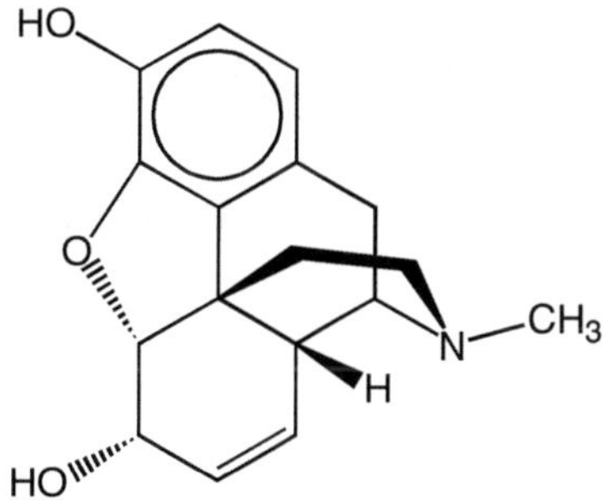

Morphine

Part C:

The structure below is tryptamine when all R-groups are hydrogen (H). Substitute the groups indicated in the table below for each corresponding "R" and name the resulting compounds (reference Chap. 15).

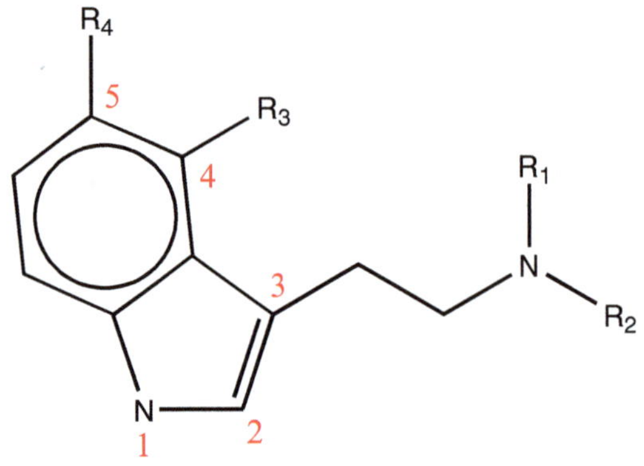

R_1	R_2	R_3	R_4	Name
H	H	H	H	Tryptamine
CH_3	CH_3	H	H	
CH_3	CH_3	H	OH	
CH_3	CH_3	OH	H	
CH_3	CH_3	PO_4^{3-}	H	
C_2H_5	C_2H_5	H	H	
CH_3	CH_3	H	OCH_3	

Part D:

Use your model kit to construct pseudoephedrine as shown below. Replace the "OH" functional group with "H" and draw the resulting structure. Name the new compound (reference Chap. 13).

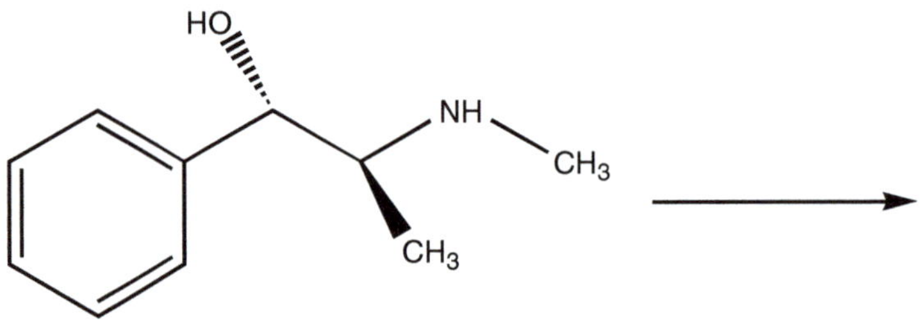

Pseudoephedrine

Class	Functional Group	IUPAC ending
Alkane	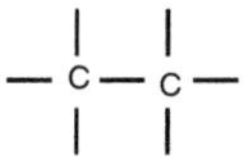	"-ane"
Alkene	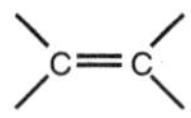	"-ene"
Alkyne	—C≡C—	"-yne"
Alcohol	R—OH	"-ol"
Ketone	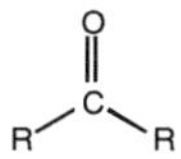	"-one"
Aldehyde	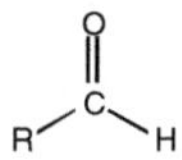	"-al"
Carboxylic Acid	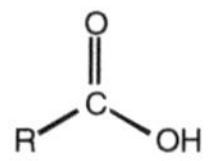	"-oic acid"
Ester	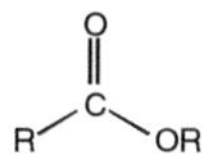	
Nitro Compounds	$R{-}NO_2$	
Amines	$R{-}NH_2$ 1° 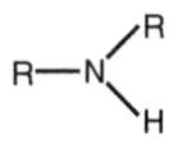 2° 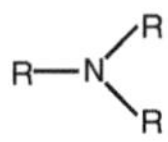 3°	"-amine"

Experiment # 9

Name ______________________________

Microcrystallography

Reference: Chapter 8

Objective: Students will gain experience in the forensic identification of elements using microcrystalline technique.

Materials: Solid form and 5% solutions of the following: NH_4Cl, NaCl, LiCl, ammonium molybdate (AMM-hydrate or anhydrous), and NaOH.
5% solutions of the following: $Mg(C_2H_3O_2)_2$, $Mg(NO_3)_2$, Na_2HPO_4 and $KMnO_4$ (or alternatively, $NaMnO_4$).
Acids: 5M HCl and 5M $HClO_4$ (perchloric acid).
Others: KCl (solid), $CaCO_3$ (sat. solution), and MgO (sat. solution), Microscope, microscope slides, droppers.

Solutions: Prepare 5% (m/v) solutions by dissolving 5 g of solid in 100 ml total volume. Alternatively, methanol may be used as solvent, however water forms slightly more stable crystals. $KMnO_4$ should be prepared fresh. If $NaMnO_4$ is used instead of $KMnO_4$, you will need to add a few drops of H_2SO_4 (5% v/v) to solution to increase solubility.

Part A:

Technique 1: Mount a slide on the microscope, place a few crystals of NaCl on the slide using a small spatula, and focus the solid in the microscope. Add a drop of Na_2HPO_4 solution directly on the solid and observe crystal formation using the microscope without mixing. Record your results below. Repeat with other solids and solutions listed below.

Solid	Solution	Crystals (Yes/No)	Sketch Crystals (if yes)
NaCl	Na_2HPO_4		
NaOH	$Mg(NO_3)_2$		
KCl	$HClO_4$		
AMM	HCl		
LiCl	$KMnO_4$		

Conclusion:

Part B:

Technique 2: Mount a slide on the microscope, place a drop of Na_2HPO_4 solution on the slide, and focus the drop in the microscope. Add a few crystals of NH_4Cl to the drop and, without mixing, observe crystal formation using the microscope. Record your results below. Repeat with other solutions and solids listed below.

Solids	Solutions	Crystals (Yes/No)	Sketch Crystals (if yes)
NH_4Cl	Na_2HPO_4		
NaOH	$Mg(NO_3)_2$		
KCl	$HClO_4$		
AMM	HCl		
LiCl	$KMnO_4$		

Conclusion:

Part C: Micro Test Technique

Technique 3: Mount a slide on the microscope, place a drop of solution-I on the slide and focus the drop in the microscope. Add a drop of solution-II to first drop and, without mixing, observe crystal formation using the microscope. Record your results below. Repeat with other solutions listed below.

Solutions-I	Solutions-II	Crystals (Yes/No)	Sketch Crystals (if yes)
NH_4Cl	Na_2HPO_4		
$CaCO_3$	$Mg(C_2H_3O_2)_2$		
MgO	$HClO_4$		
$Mg(NO_3)_2$	HCl		
MgO	$KMnO_4$		

Conclusion:

Part D: Macro Test Technique

Technique 4: Mount a slide on the microscope, place a drop of solution-I on the slide, and focus the drop in the microscope. Add a drop of solution-II to first drop and, without mixing, observe crystal formation using the microscope. Record your results below. Repeat with other solutions listed below.

Solution I	Solution II	Crystals (Yes/No)	Sketch Crystals (if yes)
NaOH	$HClO_4$		
NaCl	$HClO_4$		
AMM	Na_2HPO_4		
KMnO4	Na_2HPO_4		
LiCl	Na_2HPO_4		

Conclusion:

Experiment # 10

Name ______________________________

Chemical Extraction

Reference: Chapter 9
Objective: Students will gain practical experience using forensic extraction techniques to isolate solids from mixtures.
Materials: $NaHCO_3$ (sodium bicarbonate, saturated solution), 0.1 M HCl, starch, aspirin, acetaminophen, H_2SO_4 (concentrated), formaldehyde, 1% iodine solution acidified with glacial acetic acid, chloroform, hexane, vortex, spot plates, 5 ml test-tubes, droppers, and centrifuge.

Prepare a sample mixture containing 100 parts starch:1 part aspirin: 1 part acetaminophen, provide each group with a 1 g sample of the mixture.

Part I: Extraction:

1) Place about 0.25 g of sample in a 5 ml test tube labeled "A."
2) Add 1.0 ml (approx. 20 drops) of 0.1 M HCl.
3) Vortex the mixture and centrifuge.
4) Remove the top liquid using a dropper and place in a clean 5 ml test tube labeled "B." Be careful not to disturb the solid pellet at the bottom of test tube "A".
5) Save test tube "A" for Part-II.
6) Add a few drops of $NaHCO_3$ (sat.) to test tube "B" and vortex the mixture.
7) Add 1.0 ml of hexane and vortex the mixture.
8) Carefully transfer the organic layer using a dropper to a clean test tube labeled "C." Save test tube "C" for Part-II. The organic layer is hexane and should be the top layer, but check using solubility in DI water.
9) Add 1.0 ml chloroform to test tube "B" and vortex the mixture.
10) Carefully transfer the organic layer using a dropper to a clean test tube labeled "D." Save test tube "D" for Part-II. The organic layer is chloroform ($CHCl_3$) and should be the bottom layer, but check using solubility in DI water.

Part II: Screening Tests:

Perform the screening tests below on the samples contained in test tubes A (solid pellet), B (aqueous layer), C (organic layer, hexane), and D (organic layer, chloroform). The best results are obtained when the tests are run side-by-side as opposed to in sequence (reference Chap. 7). Starch test is positive when blue color appears with iodine. Concentrated sulfuric acid and formaldehyde produce red color with aspirin.

Test 1: Reference

Place a small sample (1/2 pea size) of the original mixture of starch, aspirin, and acetaminophen mixture (100:1:1) into each of two clean, separate wells of a spot plate. These wells will be your reference, so do not empty or clean after the tests are performed.

Starch Test:

Add one drop of 1% iodine solution to the first well containing the sample and record the color change: _______________. Do not empty or clean the well.

Aspirin Test:

Add one drop of formaldehyde and three drops of concentrated sulfuric acid to the second well containing the sample and record the color change: ____________________________. Do not empty or clean the well.

Test 2:

Place a small amount of solid from test tube "A" into each of two clean, separate wells of the spot plate. Repeat the tests above and record your results. Compare the intensity of the colors produced to the colors in the reference.

Result of starch test:____________________________

Result of aspirin test:____________________________

Test 3:

Place 3 three drops of solution in test tube "B" into each of two clean, separate wells of the spot plate. Evaporate the solvent and repeat the tests above. Record your results below and compare the relative intensity of the colors produced to the colors in the reference.

Result of starch test:____________________________

Result of aspirin test:____________________________

Test 4:

Place 3 three drops of solution in test tube "C" into each of two clean, separate wells of the spot plate. Evaporate the solvent and repeat the tests above. Record your results below and compare the relative intensity of the colors produced to the colors in the reference.

Result of starch test:____________________________

Result of aspirin test:____________________________

Test 5:

Place 3 three drops of solution in test tube "D" into each of two clean, separate wells of the spot plate. Evaporate the solvent and repeat the tests above. Record your results below and compare the relative intensity of the colors produced to the colors in the reference.

Result of starch test::____________________________

Result of aspirin test:____________________________

Summarize your conclusions based on the test results:

Experiment # 11

Name ______________________________

Chromatography

Reference: Chapter 10

Objective: Students gain practical experience using paper chromatography to separate the components of a variety of mixtures.

Materials: Red, blue, green, brown, and black felt tip pens, chromatographic paper, 1-butanol, 1-propanol, saran wrap.

Introduction:

Chromatography is a technique that utilizes a stationary phase and a mobile phase to separate the components of a mixture. The separation process is based on the fact that a given component (molecule) will exhibit a higher affinity for one of the phases and will either move in the mobile phase or remain in place on the stationary phase. Affinity is influenced by several factors, but size and polarity often play key roles. In this experiment, paper chromatography will be used to separate the components of ink commonly found in felt-tipped pens. The mixture (ink) will be spotted on the chromatographic paper (stationary phase) and placed in a developing chamber containing the mobile phase (solvent mixture). The mobile phase will migrate up the paper and contact each of the ink mixtures. The components in the ink that are soluble in the mobile phase will move up the stationary phase (paper) at different rates depending on size and polarity. The components that are not soluble will remain in place on the stationary phase. There are many types of chromatographic techniques and several factors must be considered when choosing a particular method. Thin-layer chromatography (TLC), for example, is simple, quick, and relatively inexpensive; however, it provides only qualitative results. High-performance liquid chromatography (HPLC) is more complex and requires sophisticated instrumentation, but it provides extremely accurate quantitative results. Other forms of chromatography include column chromatography, liquid chromatography (LC), and gas chromatography (GC).

Procedure:

1. Place 15.0 mL of 1-propanol, 15.0 mL of 1-butanol, and 15.0 mL of deionized water in a clean, dry 600 ml beaker. Cover the top of the beaker tightly with saran wrap. A rubber band may be used to secure the plastic wrapping. Lightly swirl the beaker to mix the contents. This will be used as the developing chamber.
2. Obtain a 10 cm × 15 cm piece of chromatographic paper. *Handle the chromatographic paper using the edges at all times.* Touching the paper with bare hands may deposit oils and dirt that will affect your results.
3. Using a ruler and a lead pencil, draw a straight line across the entire 10 cm width, 1.5 cm up from the bottom edge of the paper. *Do not use a pen to draw this line.*
4. Mark five small dots on the line with a pencil at 1.5 cm intervals.
5. Spot each dot with a different colored pen and allow the spots to dry.

6. Remove the saran wrap cover on the developing chamber and carefully place the chromatographic paper in the beaker with the edge nearest the samples placed downward in the solvent (handle the paper using the edges only). Important: Be sure that the samples are NOT immersed in the solvent. Replace the saran wrap cover.

7. The developing solvent will begin to migrate up the paper. Allow the system to stand undisturbed until the solvent has run close to the top of the paper, or for a period of 60 min. Do not allow the solvent to reach the top of the paper.

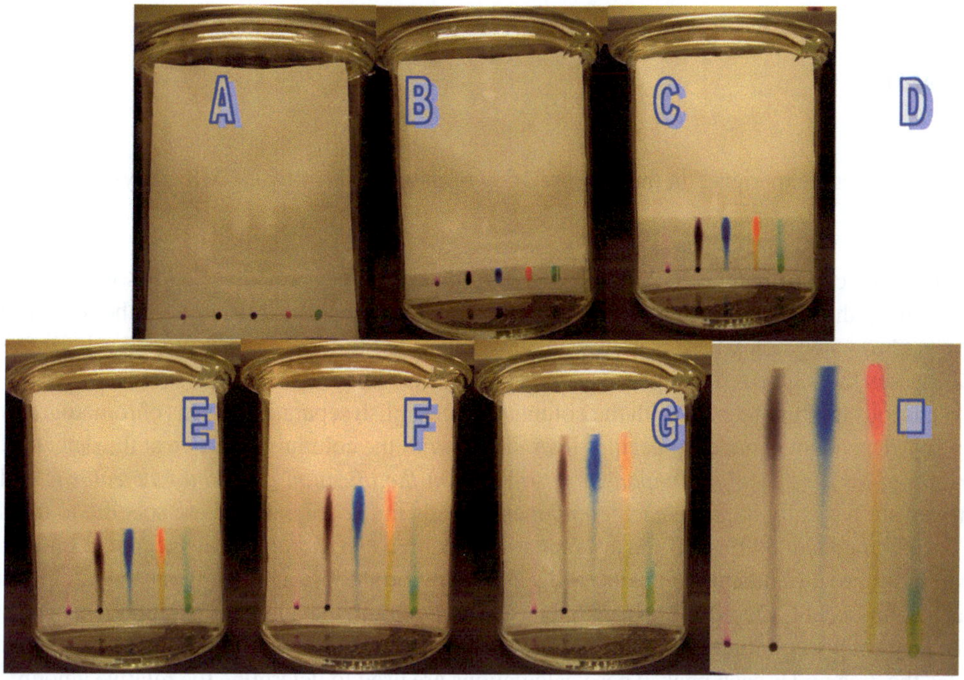

8. Remove the paper from the developing chamber and draw a line across the solvent front using a pencil. Allow the paper to dry.

9. Answer the following questions:

(a) What colors have properties similar to the solvent?

(b) What colors have properties that are different from the solvent?

(c) What pens contain components that are similar?

(d) How many components are contained in the red ink?

(e) How many components are contained in the blue ink?

(f) How many components are contained in the black ink?

(g) What components appear to be common in the pens tested?

(h) Do you believe your results would be different if ink pens from another manufacturer were used? Explain.

Experiment # 12

Name ______________________________

GCMS Interpretation

Reference: Chapter 10
Objective: Students will gain experience in the interpretation of data provided by GCMS analysis.

Introduction:

Gas chromatography (GC) can be combined with mass spectrometry (MS) to provide scientists with an extremely valuable technique used to identify the components present in a gaseous mixture (GCMS). A sample of the mixture under analysis is injected into the gas chromatograph where it is immediately vaporized in the injection port. A carrier gas (mobile phase, usually N_2 or He) passes the vaporized sample through a column contained in the GC (stationary phase). The individual components in the gaseous mixture interact with the column to accomplish separation. In gas chromatography, the separation process is usually size dependent; smaller molecules interact less with the column and arrive at the detector faster than larger ones. The *time required for a particular component to travel from the injection port to the detector* is called the *retention time*. Retention times are basically the time each component spends in the column and are specific to that component. The detector records each component eluting off the column as a peak on a *chromatogram* at specific retention times. Although a GC does not routinely provide quantitative data, it does give important information about the components in a gaseous mixture. Since each component is recorded as a peak on the chromatogram, the number of components present in the mixture is easily determined by simply counting the number of observable peaks. Most modern GCs are automated and contain software that calculates the area under each peak. These calculations are used to determine the relative amounts of each component. For example, if the area under one peak is twice the area under another, that component is present in twice the quantity. The exact amounts are not known, but the relative concentrations are. Retention times can be used to determine the relative size of each component present. Peaks observed at low retention times represent smaller molecules, while those with greater times represent larger ones. Once again, the actual size of each component is not known, but the relative sizes are easily determined. The identity of each component is determined through mass spectrometry. The separated components from GC are passed into a mass spectrometer coupled directly to the GC. In the mass spectrometer, each component is bombarded with a beam of high-energy electrons that fragment the molecule into positively charged ions. The fragmented ions are recorded on a *mass spectrum* by mass (specifically charge/mass). The *base peak* is the most intense (highest) and represents the *base ion*; usually the most stable ion formed (the one formed the easiest). The base peak is assigned 100% and the intensities of the other peaks are recorded as percentages of the base peak. Tables identifying ions of specific mass are used to reassemble the molecule from the ion fragments. The base peak and the molecular ion peak often play key roles in the identification process. The *molecular ion* is formed when a single electron is ejected from the molecule by the imposed high-energy beam. This peak commonly represents the mass of the molecule under investigation. In some cases, the molecular ion may be the base peak; however, this is not a requirement and should not be considered a standard.

Part A:

Interpret the gas chromatogram below and answer the following questions.

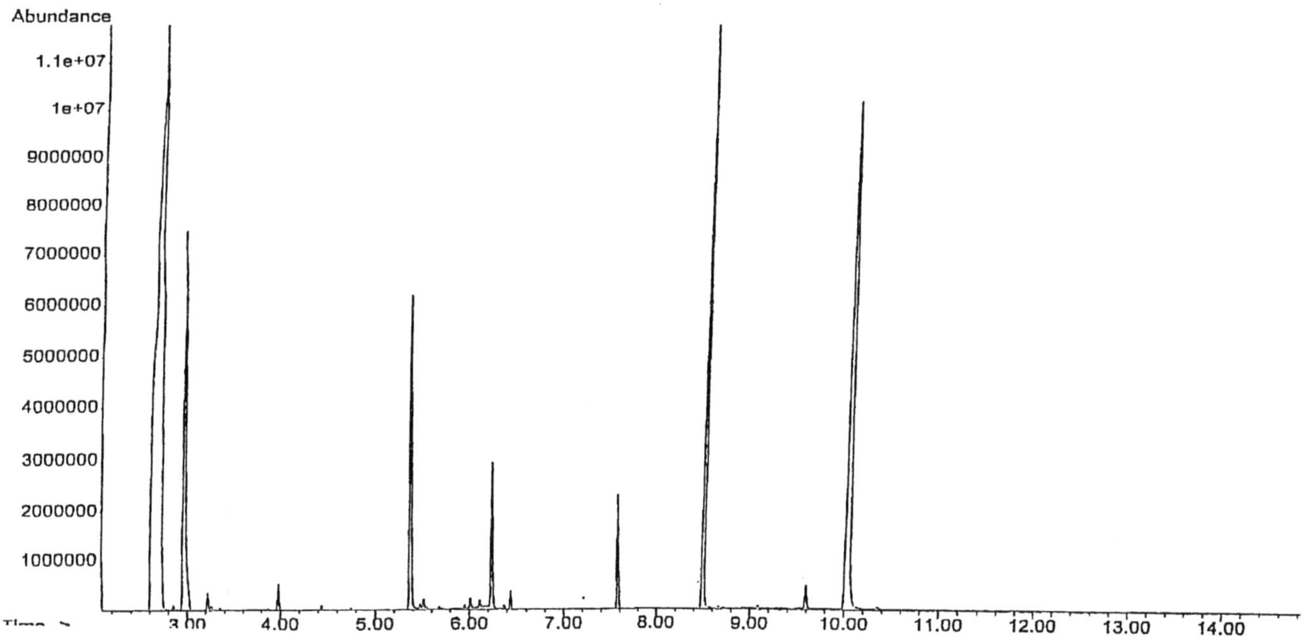

1. Identify each component in the mixture using retention times. How many different components are present? (Hint: peaks below Abundance 1000000 are not considered components).

2. Identify the peak representing the component of smallest mass.

3. Identify the peak representing the component of greatest mass.

4. Predict the peak representing the molecule present in greatest abundance.

5. Predict the peak representing the molecule present in least abundance.

Part B:

Identify each of the following using the MS data provided. You will need to reference additional information from your text.

Name	Molecular Ion (m/z)	Base Ion (m/z)	Other Prominent Ions
1)	135	044	91, 92, 65, 120, 134, 77
2)	303	082	182, 83, 77, 94, 105
3)	369	327	268, 204, 310, 315
4)	237	180	182, 209, 152, 138
5)	243	200	91, 242, 243, 186

Part C:

List the mass (m/z) of the major ion fragments formed from each the following. Draw the structure of each fragment.

1) Methamphetamine:

2) MDMA:

3) Psilocin/psilocybin:

4) THC:

5) Pseudoephedrine:

Part D:

Draw the ion fragments represented in the MS data provided for each of the following molecules.

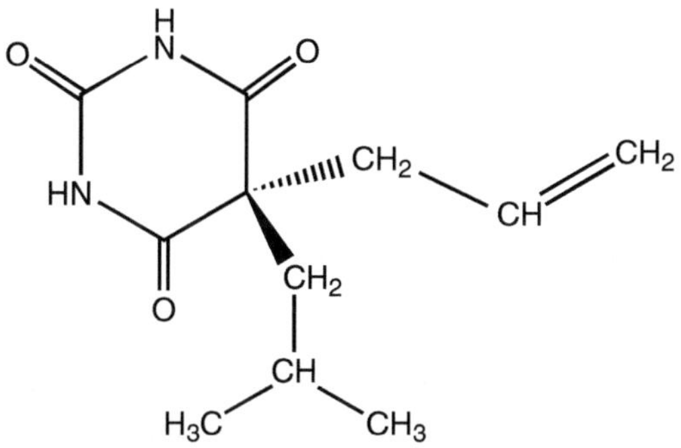

Molecular weight = 244
Base peak = 168
Prominant peaks:
97, 124, 141, 153, 167, 181

Butalbital

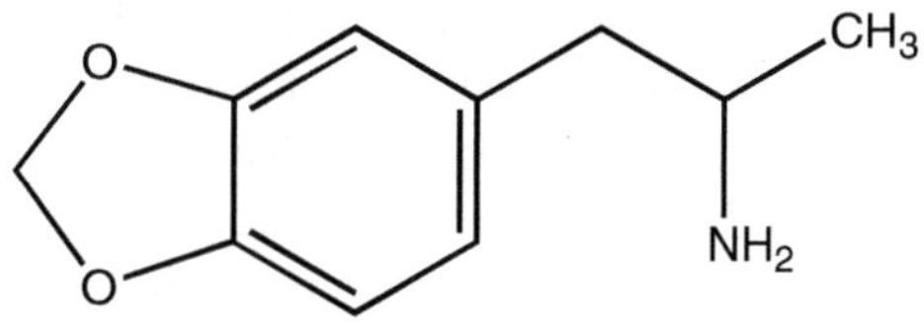

Molecular weight = 179
Base peak = 44
Prominant peaks:
51, 77, 81, 135, 136

MDA

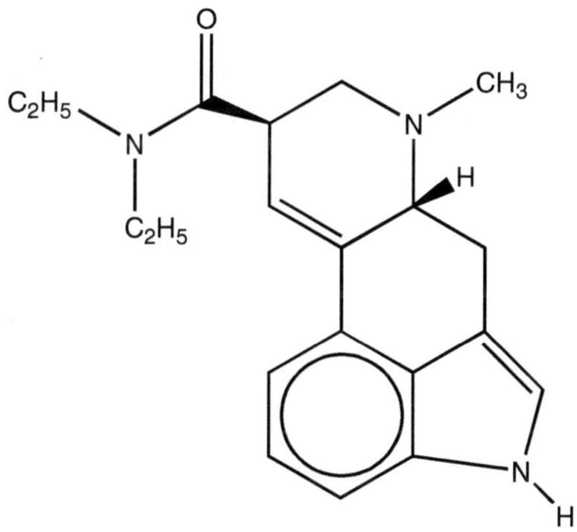

Molecular weight = 323
Base peak = 221
Prominant peaks:
72, 167, 181, 196, 207

Experiment # 13

Name ______________________

IR Spectroscopy

Reference: Chapter 11
Objective: The student will gain experience in the interpretation of infrared spectra.

Introduction:

Spectroscopy is the study of the interaction of atoms and molecules with electromagnetic radiation. These interactions often involve the emission or absorption of discrete amounts of energy, which is detected using analytical instrumentation. When energy, in the form of infrared radiation (IR), is absorbed by a particular substance, it produces measurable affects that are detected using an *infrared spectrometer.* Infrared radiation is a low energy form of radiation that is best described as "heat." When we feel the warmth of the sun, we are actually responding to infrared radiation. The heat emitted from the sun is absorbed by our bodies and stimulates molecules in our skin to vibrate. We interpret this vibrational energy as "warmth." A similar activity is observed in molecules. When infrared radiation is absorbed by a molecule, the chemical bonds convert the energy into molecular vibrations. IR spectroscopy measures the frequency of the radiation absorbed by a particular bond and records this as a band on an *infrared spectrum.* Tables containing bond vibrational frequencies are used to identify the bond from the absorption bands observed on the spectrum. The location of absorption bands on spectra are commonly represented in units of *wavenumbers* or *wavelengths.* The *wavenumber unit* (cm^{-1}, reciprocal centimeters) is used most often because it is directly proportional to vibrational energy and most modern spectrometers are linear in the wavenumber scale. Identification of a molecule based solely on an infrared spectrum is rare because IR spectroscopy is used to *identify specific bonds in a molecule.* For example, a carbonyl group (carbon–oxygen double bond) is easily identified by a characteristic absorption band present on the spectrum; however, ketones, aldehydes, carboxylic acids, and esters all contain a carbonyl group. Therefore, absolute confirmation of identity will require information from other analytical methods. Nonetheless, IR spectroscopy is often an important source of supporting evidence in the identification process.

Part A:

The IR spectrum of cocaine base and cocaine HCl are provided. Compare the two spectra and identify absorption bands common to both and those distinct to each spectrum (uncommon bands). Identify each band using wavenumber region and record your observations in the table below.

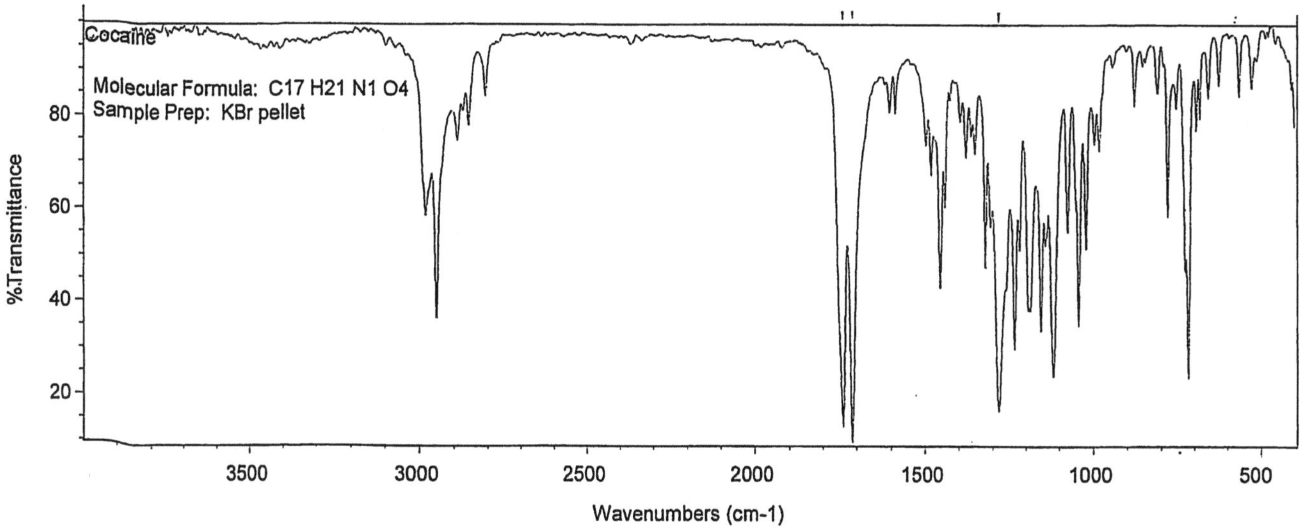

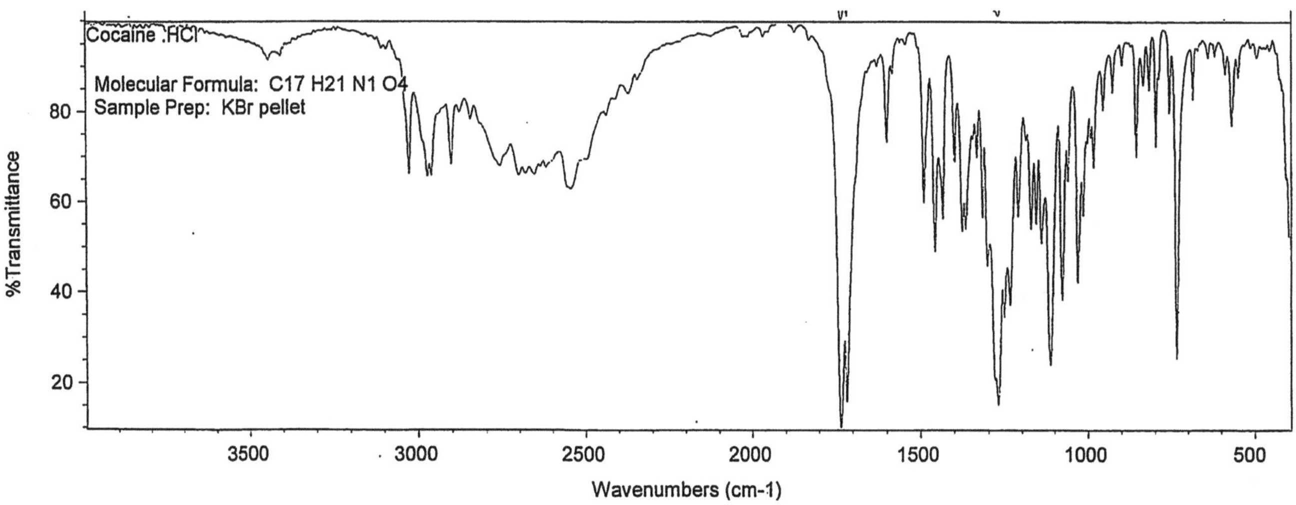

Wavenumbers (cm^{-1})

Common Bands

Distinct Bands

Cocaine HCl

Cocaine Base

Part B:

The IR spectrum of ephedrine and pseudoephedrine are provided. Compare the two spectra and identify absorption bands common to both and those distinct to each spectrum (uncommon bands). Identify each band using wavenumber region and record your observations in the table below.

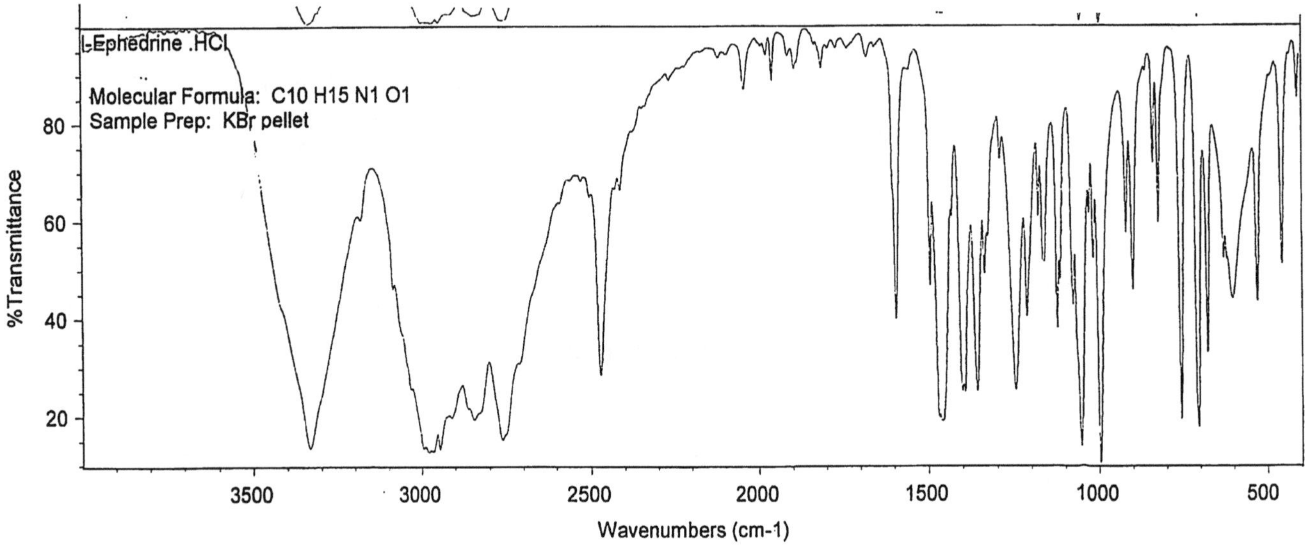

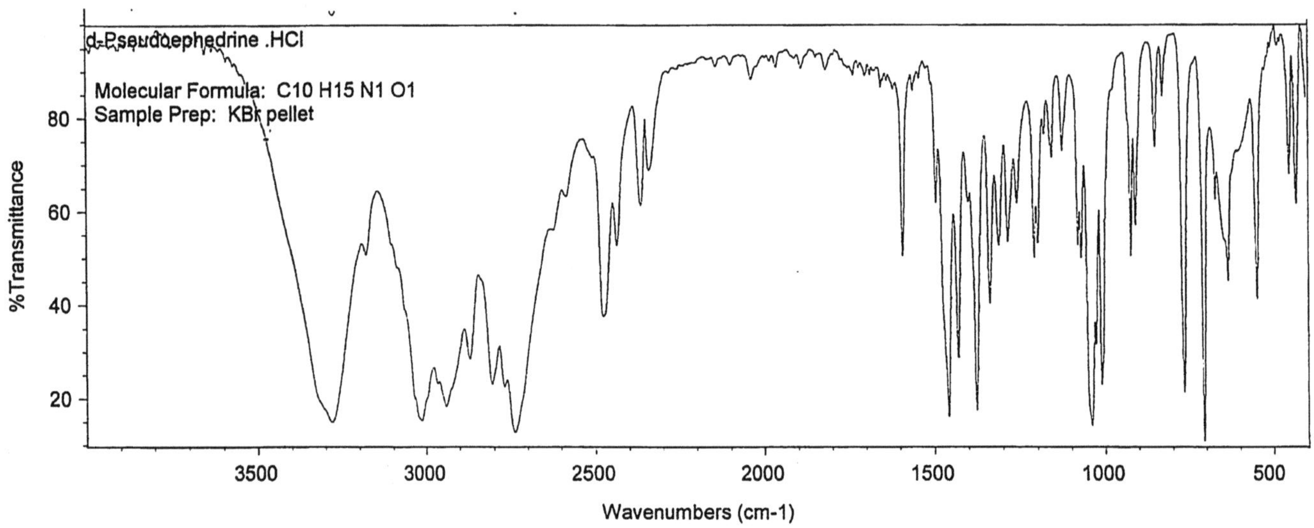

Wavenumbers (cm^{-1})

Common Bands

Distinct Bands

Ephedrine

Pseudoephedrine

Experiment # 14

Name ______________________

Examination of Marijuana (moot)

Reference: Chapter 12

Objectives: The student will gain experience in the forensic identification of marijuana plants. A report of findings will also be written and presented.

A plant material case submission contains following information/evidence.

Visual Inspection:

Microscopic Investigation:

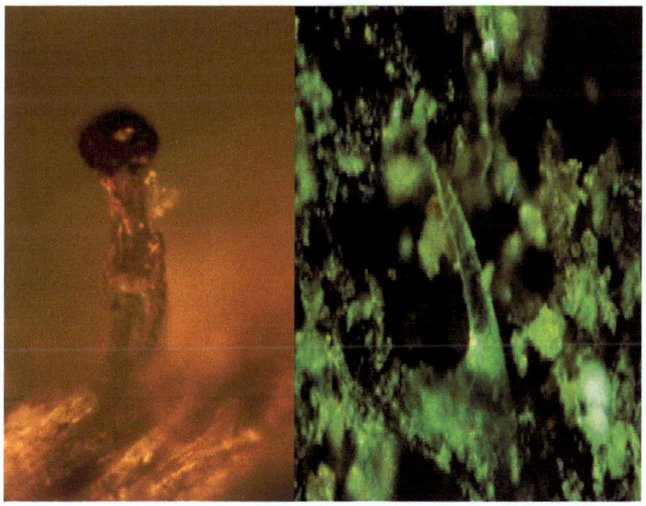

Chemical Investigation: Duquenois-Levine test results

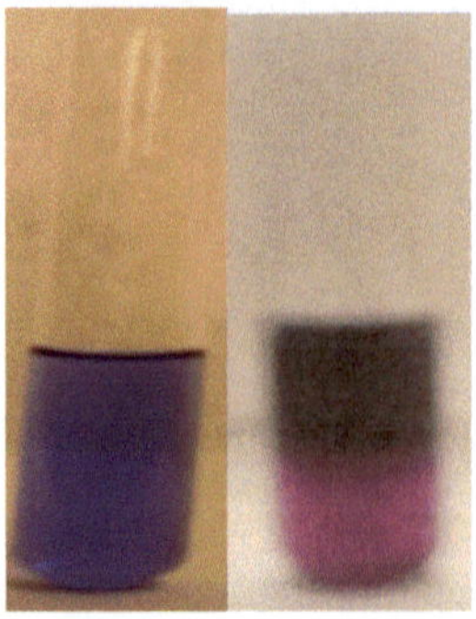

Questions:

1. Describe the physical characteristics of the leaves based on visual inspection.

2. Describe the physical characteristics of the seeds based on visual inspection.

3. Describe your observations from microscopic investigation. What types of hairs are visible?

4. The Duquenois-Levine test was performed as part of the chemical investigation.
   (a) What reagents are found in the top layer of the Duquenois-Levine test?

   (b) What reagents are found in the bottom layer of the Duquenois-Levine test?

   (c) What is the result of the Duquenois-Levine test?

5. What conclusions can be made based on your results in this investigation?

Experiment # 15

Name ____________________

Examination of Controlled Substances: Primary and Secondary Amines (moot)

Reference: Chapter 13

Objective: Students will gain experience in the forensic identification of controlled substances classified as primary and secondary amines using case data.

You are provided forensic data for three separate examination scenarios. Carefully study the information and predict the controlled substance(s) in each profile. Write a report of your findings.

Part A:

White Crystals:

Mass Spectrum:

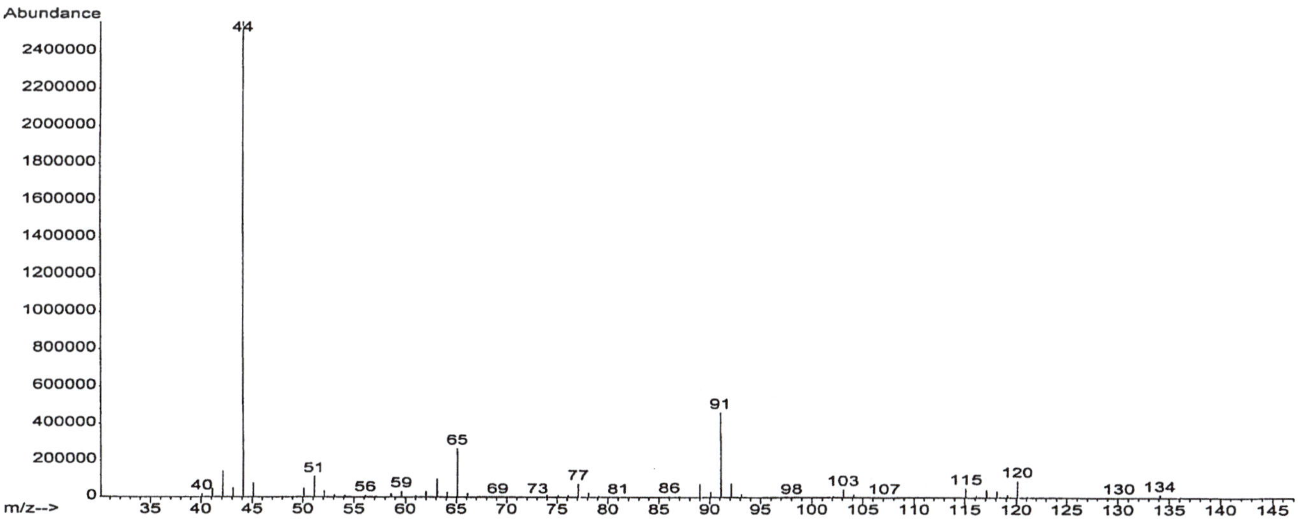

Questions:

1. Name the color-screening tests that you would perform in this case.

2. Predict the results of the screening tests in question 1.

3. Describe the method you would use as a confirmatory examination.

4. What type of extraction would be performed to prepare a sample for GCMS analysis?

5. Identify the controlled substance and provide supporting data for your conclusion.

Part B:

Red Tablets

Mass Spectrum

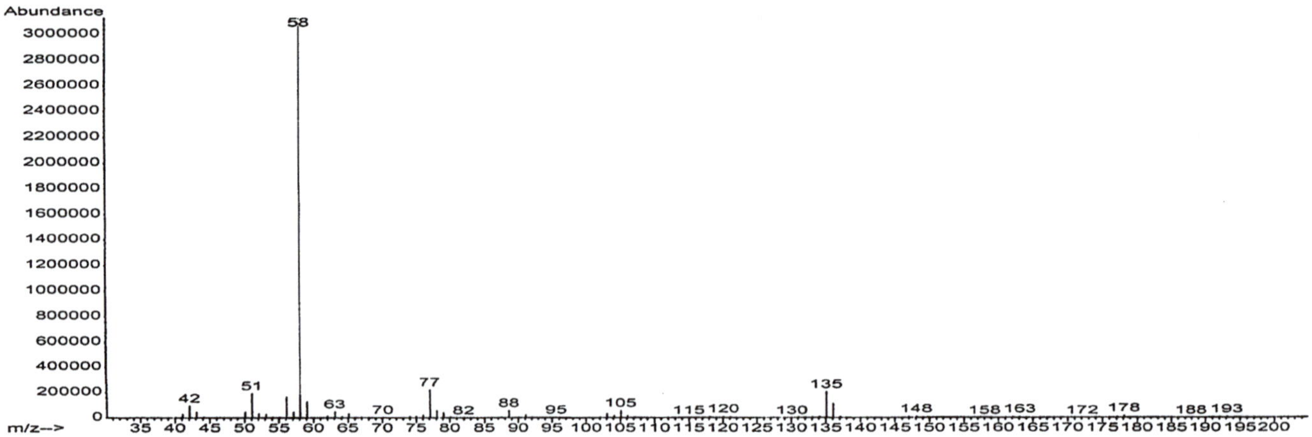

Questions:

1. Name the color-screening tests that you would perform in this case.

2. Predict the results of the screening tests in question 1.

3. Describe the confirmatory method that you would use to identify this substance.

4. Identify the controlled substance and provide supporting data for your conclusion.

Part C:
Light Brown Powder

Mass Spectrum

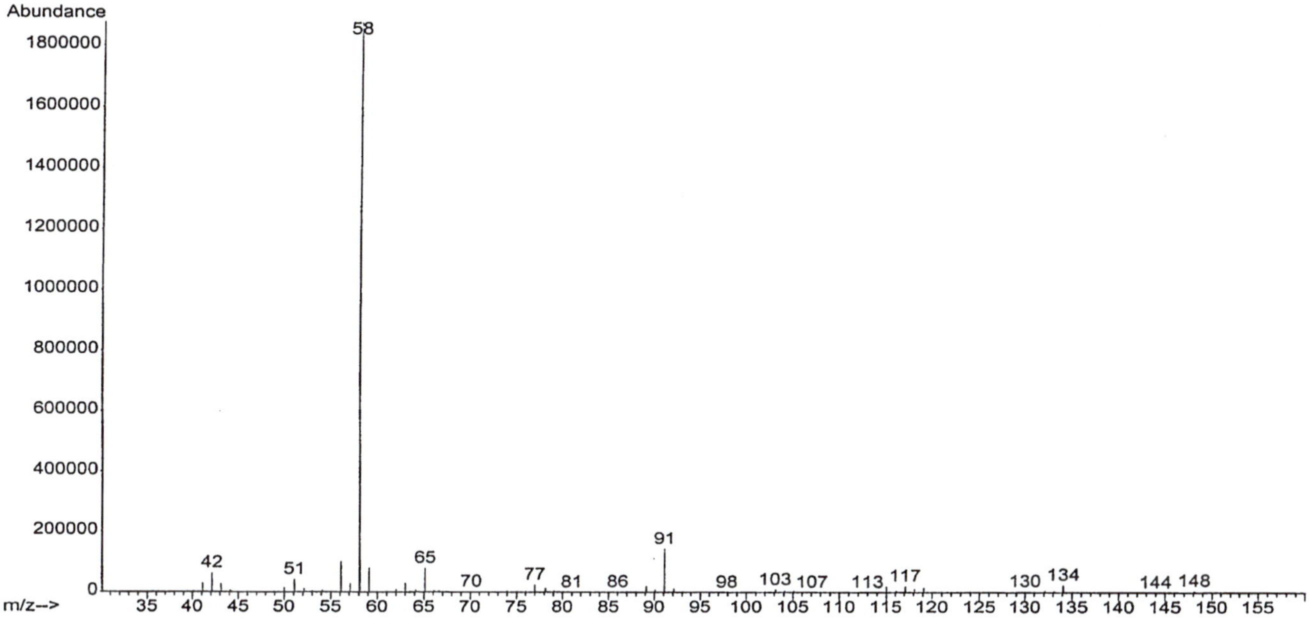

Questions:

1. What color-screening tests would you perform in this case?
2. Predict the results of the screening tests in question 1.
3. Describe the confirmatory method that you would use to identify this substance.
4. What type of extraction would you perform to prepare a sample for GCMS analysis?
5. Identify the controlled substance and provide supporting data for your conclusion.

Experiment # 16

Name ______________________________

Examination of Controlled Substances: Tertiary Amines and Opiates (moot)

Reference: Chapter 14

Objective: Students will gain experience in the forensic identification of the tertiary amine class of controlled substances using case data provided.

You are provided forensic data for three separate examination scenarios. Carefully study the information and predict the controlled substance(s) in each profile. Write a report of your findings.

Part A:

Chunks of Tan/White Substance

IR Spectrum

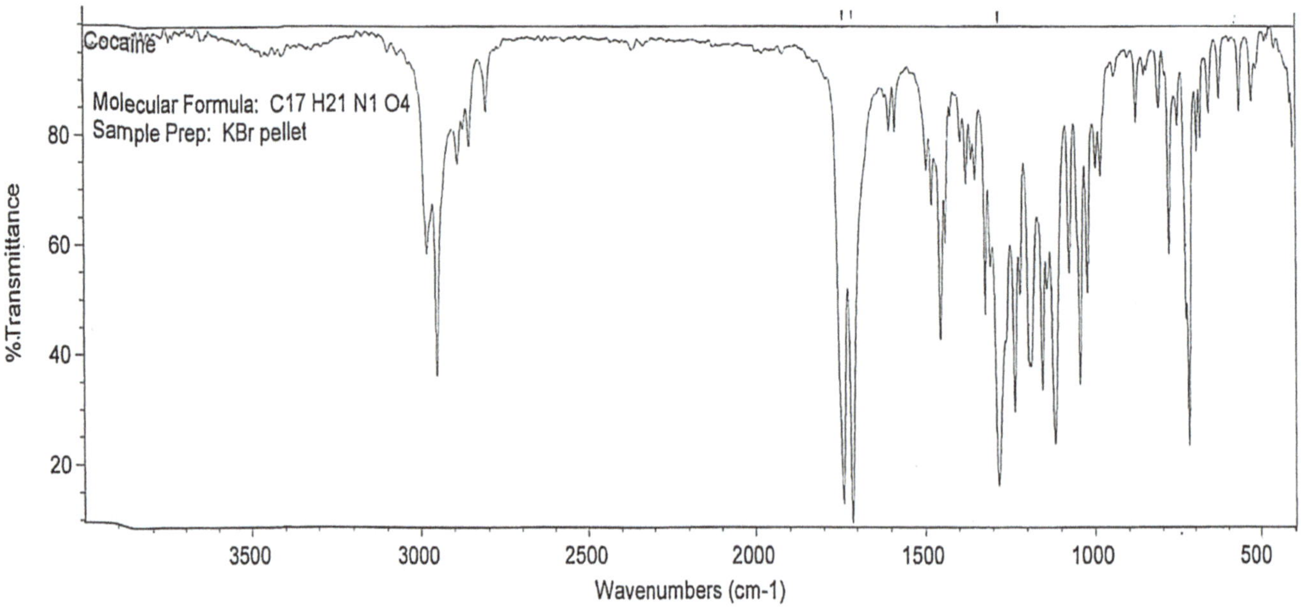

Questions:

1. What color-screening tests would you perform in this case?

2. Predict the results of the screening tests in question 1.

3. Describe the confirmatory method that you would use to identify this substance.

4. Identify the controlled substance and provide supporting data for your conclusion.

Part B:

Black Chunk

Mass Spectrum

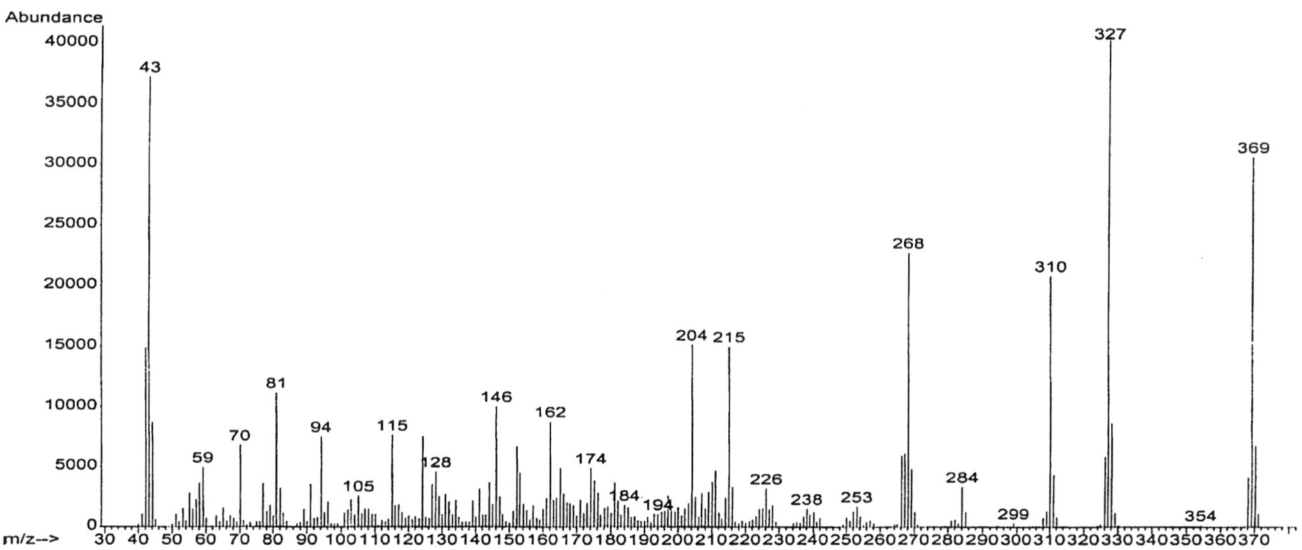

Questions:

1. What color-screening tests would you perform in this case?
2. Predict the results of the screening tests in question 1.
3. Describe the confirmatory method that you would use to identify this substance.
4. Identify the controlled substance and provide supporting data for your conclusion.

Part C:

Stained Cigarette and Yellow Oil

GCMS spectra

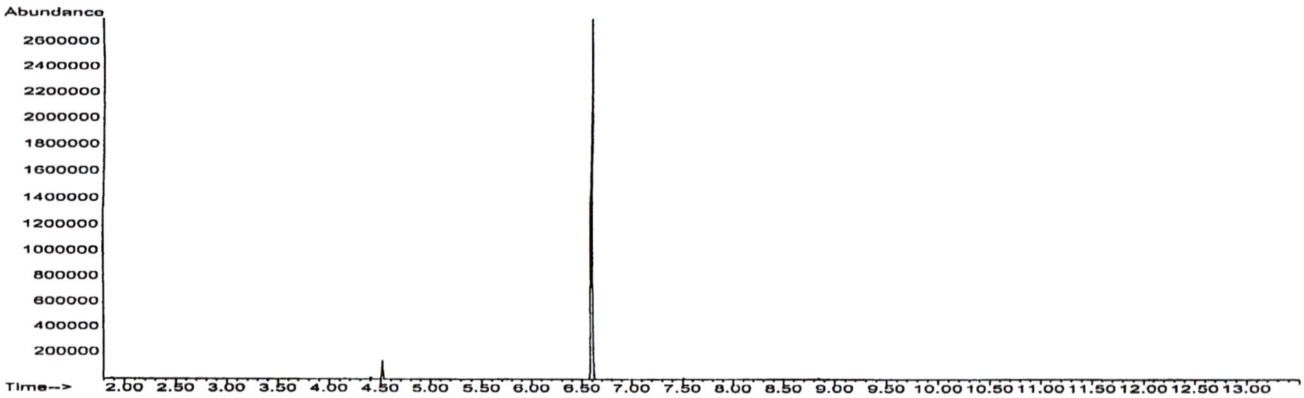

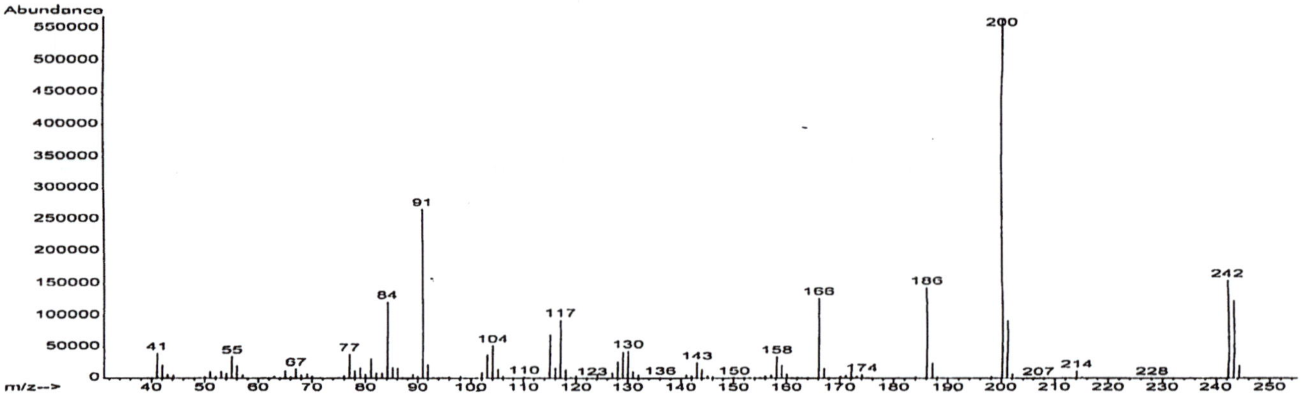

Questions:

1. What color-screening tests would you perform in this case?

2. Predict the results of the screening tests in question 1.

3. Describe the confirmatory method that you would use to identify this substance.

4. Identify the controlled substance and provide supporting data for your conclusion.

Experiment # 17

Name ____________________

Examination of Controlled Substances: Tryptamines (moot)

Reference: Chapter 15

Objective: Students will gain experience in the forensic identification of the tryptamine class of controlled substances.

You are provided forensic data for three separate examination scenarios. Carefully study the information and predict the controlled substance(s) in each profile. Write a report of your findings.

Part A:
Mushrooms

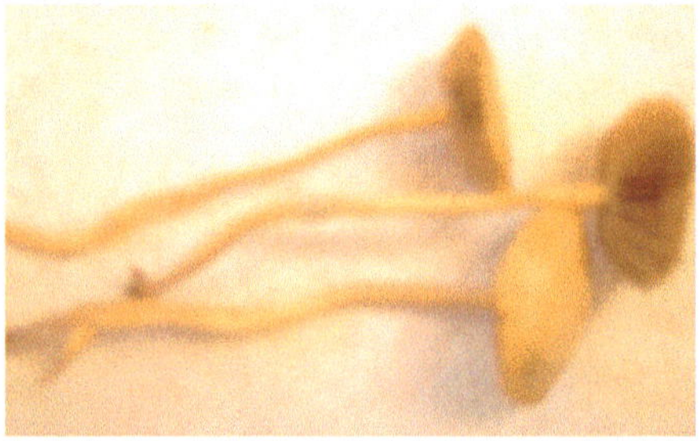

Results of Thin-Layer Chromatography (TLC)

Mass Spectrum:

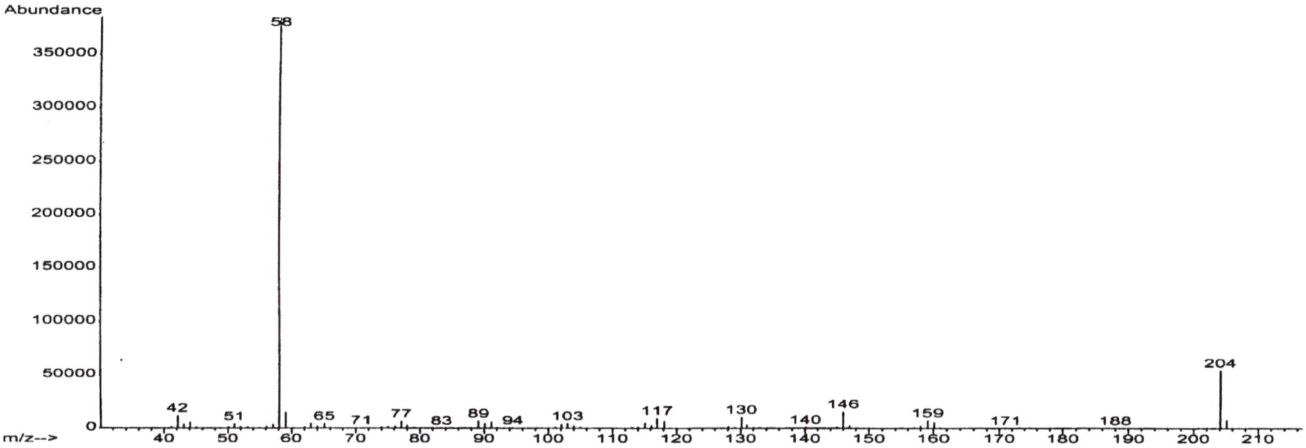

Questions:

1. What color-screening tests would you perform in this case?

2. Predict the results of the screening tests in question 1.

3. Describe the confirmatory method that you would use to identify this substance.

4. Identify the controlled substance and provide supporting data for your conclusion.

Part B:

Plant Material

Mass Spectrum:

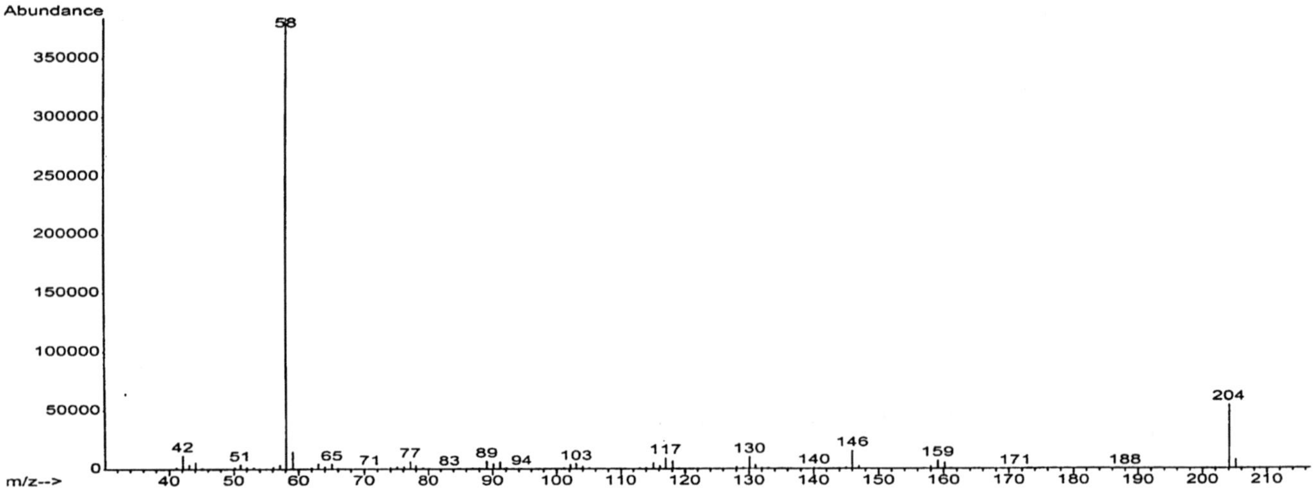

Questions:

1. What color-screening tests would you perform in this case?

2. Predict the results of the screening tests in question 1.

3. Describe the confirmatory method that you would use to identify this substance.

4. Identify the controlled substance and provide supporting data for your conclusion.

Experiment # 18

Name ____________________

Examination of Anabolic Steroids (moot)

Reference: Chapter 16

Objective: Students will be exposed to the forensic identification of anabolic steroids.

You are provided forensic data for three separate examination scenarios. Carefully study the information and predict the controlled substance(s) in each profile. Write a report of your findings.

Part A:

Red Capsules

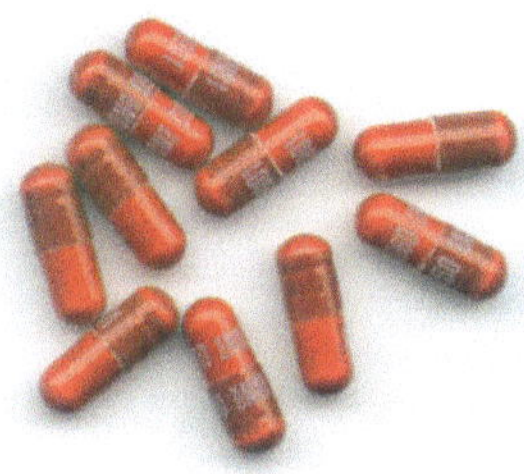

Mass Spectrum

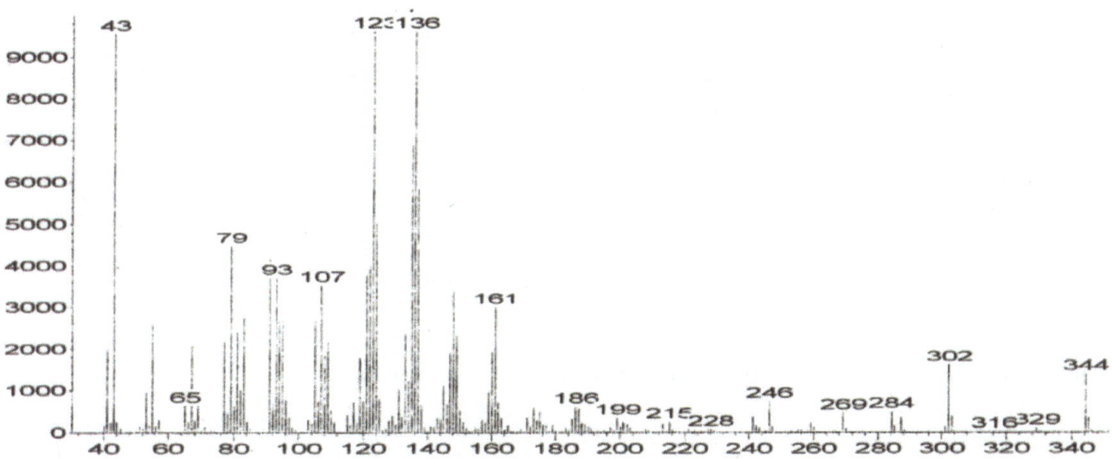

Questions:

1. What color-screening tests would you perform in this case?

2. Predict the results of the screening tests in question 1.

3. Describe the confirmatory method that you would use to identify this substance.

4. Identify the controlled substance and provide supporting data for your conclusion.

Part B:

Injection Vial

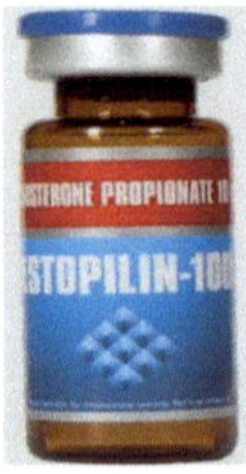

Mass Spectrum:

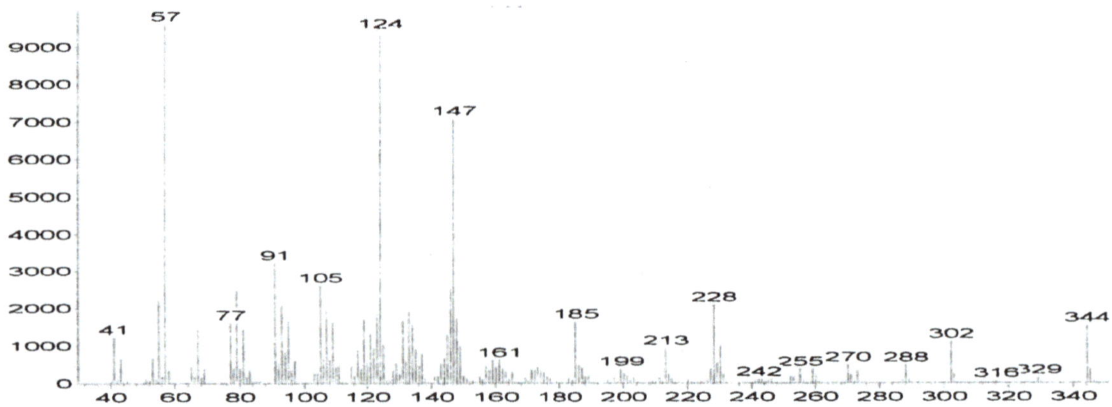

Questions:

1. What color-screening tests would you perform in this case?

2. Predict the results of the screening tests in question 1.

3. Describe the confirmatory method that you would use to identify this substance.

4. Identify the controlled substance and provide supporting data for your conclusion.

Part C:

Tablets

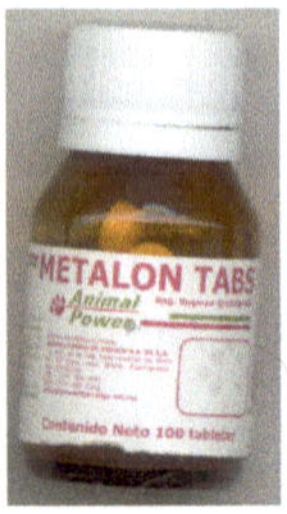

Mass Spectrum:

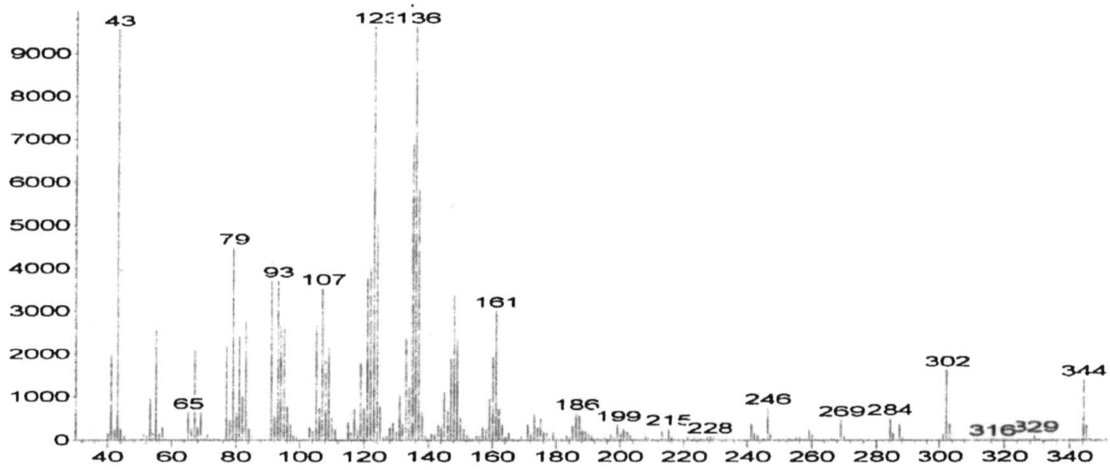

Questions:

1. What color-screening tests would you perform in this case?

2. Predict the results of the screening tests in question 1.

3. Describe the confirmatory method that you would use to identify this substance.

4. Identify the controlled substance and provide supporting data for your conclusion.

Experiment # 19

Name ______________________________

Examination of Miscellaneous Controlled Substances (moot)

Reference: Chapter 17

Objective: Students will gain experience in the forensic identification of various functional groups present in different controlled substances.

You are provided forensic data for three separate examination scenarios. Carefully study the information and predict the controlled substance(s) in each profile. Write a report of your findings.

Part A:

Paper Tabs

GSMS Spectra

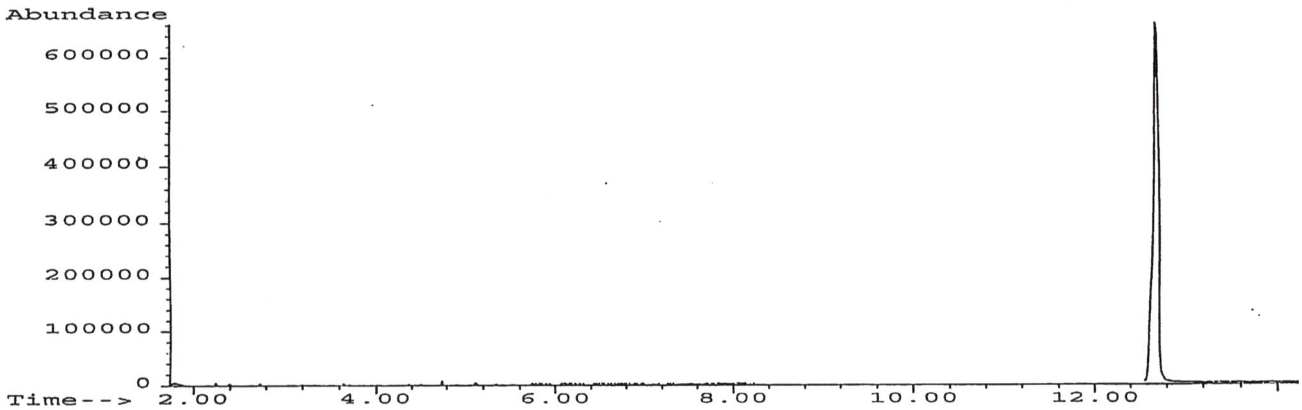

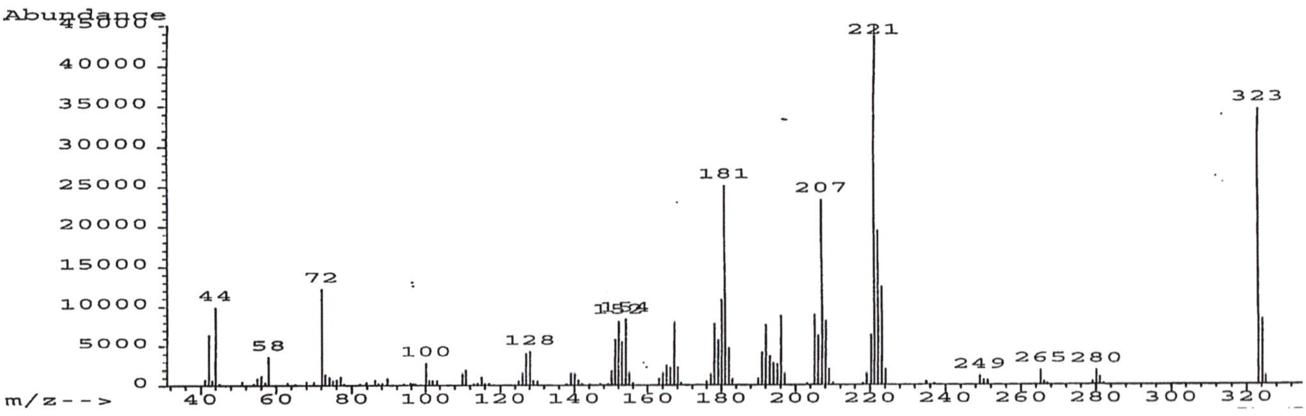

Questions:

1. What color-screening tests would you perform in this case?
2. Predict the results of the screening tests in question 1.
3. Describe the confirmatory method that you would use to identify this substance.
4. Identify the controlled substance and provide supporting data for your conclusion.

Part B:

Capsules

Questions:

1. What color-screening tests would you perform in this case?
2. Predict the results of the screening tests in question 1.
3. Describe the confirmatory method that you would use to identify this substance.
4. Identify the controlled substance and provide supporting data for your conclusion.

Part C:

Tablets

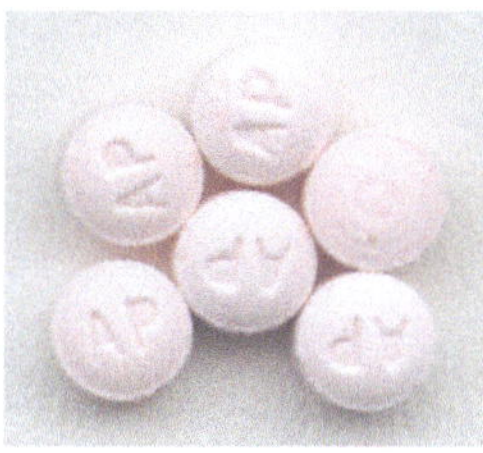

Mass Spectrum:

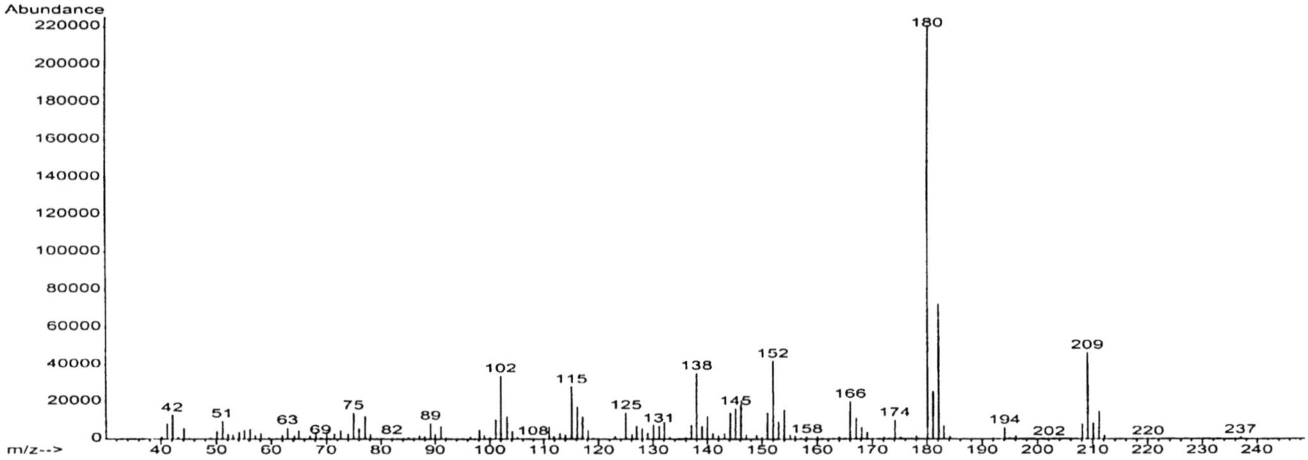

Questions:

1. What color-screening tests would you perform in this case?

2. Predict the results of the screening tests in question 1.

3. Describe the confirmatory method that you would use to identify this substance.

4. Identify the controlled substance and provide supporting data for your conclusion.

Experiment # 20

Name ____________________

Clandestine Manufacturing of Methamphetamine (moot)

Reference: Chapters 18, 19, and 20

Objectives: Students will gain experience in the forensic identification of case evidence collected at clandestine lab sites. This will be extended to include the association of distinctive evidence to specific synthetic steps used in the illegal production of methamphetamine.

Carefully study the photographs below and answer the following questions based on your observations. You will need to refer to your text.

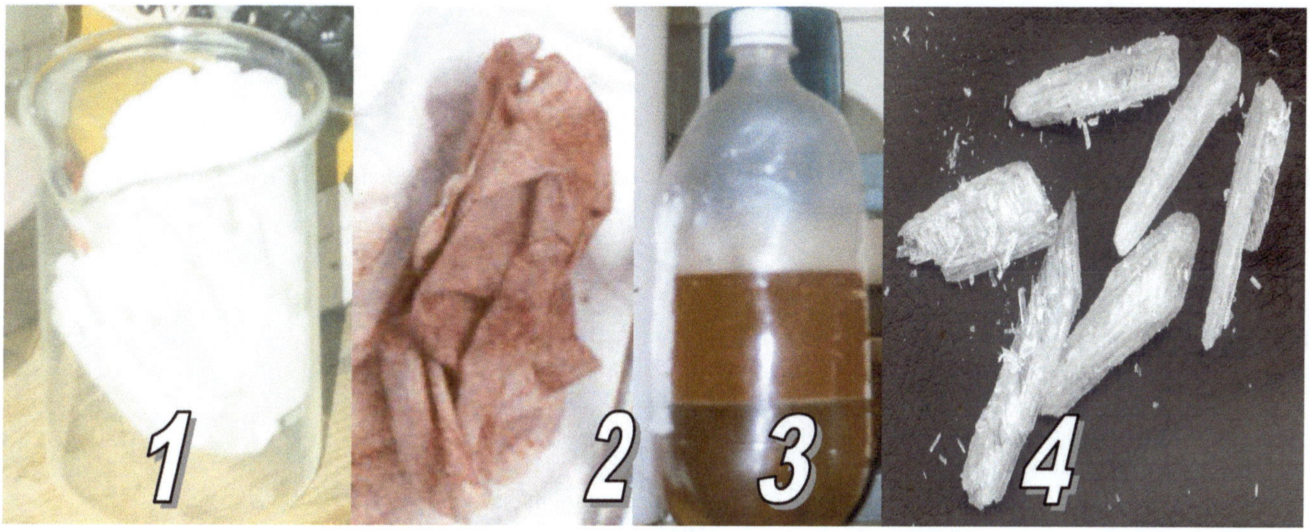

1. Identify the method used in the production of methamphetamine. Justify your answer.

2. What chemicals are required for the method in question 1?

3. List the synthetic steps used in the production of methamphetamine using this method.

4. What synthetic step is indicated by the evidence shown in photograph 1?

5. List the chemical(s) required to prove this step.

6. Briefly outline the procedure you would use in the forensic examination of the evidence shown in photograph 1.

7. Summarize your conclusion on this piece of evidence.

8. What synthetic step is indicated by the evidence shown in photograph 2?

9. List the chemical(s) required to prove this step.

10. Briefly outline the procedure you would use in the forensic examination of the evidence shown in photograph 2.

11. Summarize your conclusion on this piece of evidence.

12. What synthetic step is indicated by the evidence shown in photograph 3?

13. List the chemical(s) required to prove this step.

14. Briefly outline the procedure you would use in the forensic examination of the evidence shown in photograph 3.

15. Summarize your conclusion on this piece of evidence.

16. What synthetic step is indicated by the evidence shown in photograph 4?

17. List the chemical(s) required to prove this step.

18. Briefly outline the procedure you would use in the forensic examination of the evidence shown in photograph 4.

19. Summarize your conclusion on this piece of evidence.

Index

A
Acid and anionic test techniques, 95
Acid–base extraction, 101–102
AET. *See* Alpha (α)-ethyltryptamine
Alcohols
  bufotenine and psilocin, 43
  butanol, 43
  classification, 42
  conversion, ergot alkaloid, 245
  denatured, 258, 261, 262
  designer drugs, 43
  dimethyltryptamine (DMT), 43, 44
  ephedrine and pseudoephedrine, 44
  ethanol, 41
  fire hazard, 253
  glycerol, 42
  hydroxyl functional group (-OH), 41
  clandestine manufacturing, controlled substances, 43
  polar organic solvents, 43
  pseudoephedrine, 261–262
  Red Devil Lye, 257
Aldehydes
  acetaldehyde, 47, 86, 87
  chemical and physical properties, 47
  formaldehyde, 47, 84
Alkanes
  alkyl groups, 36
  butane, 32, 33
  carbon tetravalency, 33
  chemical formulas, 32
  2-chloropentane, 35
  cycloalkanes, 36–37
  definition, 32
  methane, 34
  naming rules, 35–36
  physical properties, 36
  skeletal structure, 34
  stereochemistry, 34
  tetrahedral geometry, 34
Alkenes
  butene, 38
  chemical formulas, 38
  condensed structural formula, 37
  cycloalkenes
    benzene, 41–42
    PCP, 39
  hexene, 37
Alkynes
  chemical formulas, 39
  C_nH_{2n-2}, 39
  cycloalkynes, 40
  hexyne, 39
  linear acetylene, 39
Alpha (α)-ethyltryptamine (AET), 197
Alpha (α)-methylfentanyl (AMF), 226
The American Society of Crime Laboratory Directors (ASCLD), 3–4
AMF. *See* Alpha (α)-methylfentanyl
Amines
  functional groups, 109
  ketamine (*see* Ketamine)
  lysergic acid diethylamide (LSD), 52
  3,4-methylenedioxyamphetamine (MDA), 52
  organic derivatives, ammonia, 51–52
  phenethyl (*see* Phenethylamines)
  primary, 52, 86, 95, 119
  secondary, 52–53, 86, 95, 119
  tertiary, 53–54, 86
  tryptamines (*see* Tryptamines)
Amphetamine
  addictive and tolerance, 159
  α-carbon, 159
  forms and street names, 159
  history, 158–159
  short-term and long-term
    physical effects, 159
    psychological effects, 159
Anabolic steroids
  administration methods, 211
  analytical methods
    GCMS, 213–214
    mass spectra, commonly encountered steroids, 214–221
    visual inspections, 213
  description, 207
  frequently encountered steroids, 212
  general structure, 210–211
  naturally occurring steroid hormones
    androgen, 209
    classes, 208
    mineralocorticoids and glucocorticoids, 209
    oral contraceptives, 209
  nomenclature
    androst, parts, 211
    beta (β) and alpha (α) substitutions, 212

Anabolic steroids (*continued*)
ketone functional groups, 212
saturated and unsaturated hydrocarbons, 211
physical and psychological effects
damaging side effects, 211
mood disturbances/disorders, 211
skeletal muscle and bone tissue, 211
Anadenanthera peregrina. See Yopo seeds
Androgen, 209
Aqueous test reagents
gold chloride test, 95
mercuric chloride test, 96
mercuric iodide test, 96
phosphoric acid test, 95
platinum chloride test, 96
potassium permanganate test, 96
sodium acetate test, 96
Aromatic compounds
benzene, C_6H_6, 40
naphthalene, 41
unusual stability, 41
ASCLD. *See* The American Society of Crime Laboratory Directors
Atomic structure
electron configurations
aufbau principle and triangle, 13–14
calculation, ions, 15
description, valence electrons, 15
distribution, atomic orbital, 17
ground state *vs.* most stable state, 16
group I, II and IV elements, 15
negative *vs.* positive ions, 15
neon (Ne) and helium (He), 16
neutral Na, 14
octet rule, 16
principle energy level, orbitals, 16
stable state, valence electrons, 16
valence electrons, Na, 14
electrons arrangement
emission, high-intensity light, 12
energy, 13
experimentation, lightning bugs, 12, 13
orbitals, 12
principle energy levels, 13
probability, 13
elements
and atoms, 10
forensic interest, 20
isotopes
description, 18
instability, 18
mole and molar mass, 20
nuclear radiation, 19–20
nucleus, subatomic particles, 11
periodic table, 9–10
periodic trends
atomic radius, 17
electronegativity, 17
elements, atomic radii, 17, 18
radioactivity
description, 18
emission, sample material, 18, 19
types, nuclear radiation, 18
subatomic particles, 11–12
types, radioactive decay, 19
Aufbau principle, 13–14

B
Barbiturates
capsules and tablets, 224
common medical practice, 225
derivation, 223
schedule IV long-acting, 224
ultra-short acting, 223
use, 223
Biphasic solutions examination
acidic layers, 275
basic layers, 275
neutral layers, 275, 276
Bufotenin
bufotoxin, 194
description, 193
hallucinogenic dose, 194
isolation, 194
species, Bufo toads, 194
Yopo seeds *(Anadenanthera peregrina)*, 194–195

C
Cannabis
description, Marijuana and tetrahydrocannabinol (THC), 145
documentation, 154, 156
forensic identification, Marijuana
botanical identification, 149
description, 149
Duquenois-Levine test, 151–153
macroscopic properties, 149, 150
microscopic identification, 151, 152
forms
Marijuana and Hashish, 147
types, 147, 148
Marijuana history
classification, Controlled Substances Act, 146
DEA, 146
as euphoric drug, 145
regional names, 146
USP, 146
packaging
decayed plants, 147
use, paper bags/envelopes, 147
programs, tetrahydrocannabinol (THC), 145
psychoactive ingredient
description, 147
structure, *Cannabinoids*, 147, 148
structure, *trans*-D^8-isomer and *trans*-D^9-isomer, 147, 149
terpenes and phenol, 147
trans-D^9-and *trans*-D^8-tetrahydrocannabinol, 147
trans-D^9-isomer, 145
visual and preliminary examination, 145
Carboxylic acids
gamma-hydroxybutyric acid (GHB), 49
protinated and deprotinated forms, 48
Case notes
content, 74
dissemination, 75
format, 75
purpose, 74
types
ambiguity, 72, 73
cocaine pricks samples, 72
correction chain, 72, 73
unacceptable corrections, 72, 74

Case report
  examples, 76
  format and content, 75
  purpose, 75
Chain of custody
  mass spectrometry, 71
  theoretical basis, instrument's operation, 71
Chemical bonding
  covalent, 24–25
  hydrogen, 26
  ionic, 23–24
  multiple, 26–27
  polar, 25
Chemical color tests
  Chen's test, 82–83
  Dille–Koppanyi's test, 83
  Froehde's test, 87
  Janovsky test, 87
  limitations, 81
  Marquis' test, 84–85
  Mecke's test, 83–84
  methods
    positive and negative controls, 81
    spot plate, 81, 82
  nitric acid test, 85
  primary amine test, 86
  secondary amine test, 86
  tertiary amine test, 86
  Van-Urk's test, 86–87
  Weber test, 88
Chemical extractions
  acid–base extraction infrared modification I and II, 105
  dry extraction gas chromatography modification, 104
  gas chromatography modifications, 104
  GCMS, 103
  infrared (IR) spectroscopy, 104
  methanol
    GCMS, 105
    screening methods, 105
  sample preparation, 103
  techniques
    acid-base, 101–102
    liquid-liquid, 100–101
    neutral compound, 102–103
    solid-liquid, 99–100
  temperature, 99
Chemical properties
  aldehydes, 47
  alkanes, 37
  burning, 7
  description, 6
  enantiomers, 55, 92
  instrumental analysis, 92
  silver nitrate test, 6–7
  solid formation, 7
  sublimation, iodine crystals, 5, 6
  substance reactivity, 6–7
Chemical screening
  color formation, 79–81
  color tests(*see* Chemical color tests)
  color transitions and instabilities, 79
  documentation, 82
  limitations, color tests, 81
  methods, 79
  screening tests results, 88, 89
  tertiary amines
    flow chart, 183, 185
    testing method, 183
Chirality, 55–56, 65, 92
Chromatography
  chromatogram peaks, 115
  column
    packing process, 110
    TLC, 110
  GC (*see* Gas chromatography)
  GCMS-advantages and disadvantages, 124
  HPLC, 111–112
  IEC, 110–111
  limitations, 115
  mass spectrometry (*see* Mass spectrometry)
  paper, 107–109
  physical method, separation, 107
  thin-layer (TLC), 109
  types, 107
Clandestine operations
  cocaine synthesis
    native coca leaves, 242
    potassium permanganate, 242
  collection, washes
    methanol, 266
    remnants, clandestine operations, 266
  drug enforcement agency (DEA), 241
  drug synthesis, 241
  encountered solutions, 260, 262, 263
  evidence collection
    bilayer liquids, 263, 265
    heating mantles, 263, 265
    icing stage, 264, 266
    methamphetamine production, 263, 265
  examination, washes, 278
  fentanyl synthesis, 242, 243
  forensic analogy, 257
  forensic chemist role
    advisory, 254
    evidence collection, 254
  γ-hydroxybutyric acid (GHB), 242–244
  heroin, 244–245
  identification, related evidence
    clandestine drug manufacturing, 258, 259
    discovery, catalysts, 258, 259
    equipment, 258, 260
    illicit drug production, 258, 260
    pseudoephedrine, 258, 261
    solvents, 258, 259
  LSD synthesis, 245
  MDMA, 246–247
  methamphetamine production
    confirmatory examination, 274
    Ephedra plant, precursor, 279–281
    evidence type, 269–270
    examination, 269–270
    extraction, pseudoephedrine, 262–263
    HI-red phosphorus method, 279, 280
    icing, methamphetamine, 263–264
    manufacturing methamphetamine, 261, 262
    processing methamphetamine, 263, 264
    signature GCMS, Nazi method, 279
  methcathinone, 279
  N,N-dimethyltryptamine (DMT), 248–249
  opium poppy, morphine, 245
  PCP synthesis, 249
  potential hazard, 253

Clandestine operations (*continued*)
  prove extraction
    evidence type, 269–270
    examination, 217–280
  prove icing, 275, 276
  prove processing, methamphetamine
    biphasic solutions, 275
    evidence type, 274
  safety equipment, 253
  signs
    operators, 257, 258
    red phosphorus, 257, 258
    stains, 257
  stains
    examination, 277
    methamphetamine, 227
Cocaine, tertiary amines natural
  Amazon and Trujillo coca, 179
  cause, excessive use, 178
  characterization, 178
  chewing coca counters, 179
  Colombian coca, 179
  description, 177
  Huanuco/Bolivian coca, 179
  novocaine and xylocaine, 178
  octopamine, 179
  origin, 178
  scientific classification, 179
  short-and long-term effects, 178
  structure, 178
Color formation
  Chen's test, 81
  electrons orientation, 79
  3,4-methylenedioxymethamphetamine (MDMA), 81
  visible light, 79
Column chromatography, 110
Controlled substances
  analytical methods
    chemical screening tests, 230, 231
    GCMS, 191, 230, 232–236
    visual identification, 229–230
  barbiturates, 223–225
  fentanyl, 225–226
  GHB, 226–227
  ketamine, 227–228
  LSD, 228–229
Controlled Substances Act (CSA), 66
Covalent bonds
  H_2 and NaCl models, 25
  hydrogen atoms, 25
  polarity, 24–25
Cycloalkanes, 36–37
Cycloalkenes, 38–39
Cycloalkynes, 40

D

DEA. *See* Drug Enforcement Administration
DET. *See* Diethyltryptamine
Deuterated triglycine sulfate (DTGS), 133
Diethyltryptamine (DET)
  structure, 194
  synthetic analogs, 197
Dispersive infrared spectrometer
  components
    monochromator and slits, 131
    optical layout, 129, 131
    radiation sources, 131
    thermal and photon detectors, 131
  design
    double-beam, 131
    radiation path, 131
  limitations, 132
Drug Enforcement Administration (DEA), 146, 241
DTGS. *See* Deuterated triglycine sulfate
Duquenois–Levine test
  proposed reaction mechanism, 151
  reagents, 152

E

Electron configurations. *See* Atomic structure
Ephedra plant
  appetite and metabolism, 164
  ma huang (Ephedra sinica), 164
  use, clandestine manufacture, 164
Ephedrine/pseudoephedrine
  description and production, 163
  dextrose fermentation, 163
  Ephedra plants, 163
  methyl and hydroxyl groups, 163
  physical and psychological effects, 163–164
Esters
  alkoxy group (-OR), 49
  systematic naming, 49

F

Fentanyl
  AMF, 226
  description, 225
  heroin and cocaine, 226
  "lollypop" form and duragesic, 225
  safety measures and clinical effects, 225
  tablets, patches, lollypops, and injections, 225, 226
Forensic chemist, 71, 74
Forensic chemistry
  ASCLD, 3–4
  chemical properties, 6–7
  chirality, 55–56
  crime-scene investigation and forensic analysis, 3
  description, 3
  esters, 49–50
  gas chromatograph, 4
  physical properties, 5–6
  properties, matter
    elements and compounds, 5
    homogeneous and heterogeneous mixtures, 5
    mass and weight, 5
    solid, liquid and gas, 5
  scientific investigation, 4
  space programs, 4
  technical procedures and lab quality manual, 3
Forensic documentation
  case note, 72–75
  case report, 75
  chain of custody, 71
  replication, and research advancement, 71
Forensic identification, Marijuana

botanical identification
Cannabis sativa, 149
physical transformation, 149, 150
scientific classification, 149
description, 149
Duquenois-Levine test
proposed reaction mechanism, 151
reagents, 152
technique, 152–153
GCMS, 154, 155
macroscopic properties
flowers and stem, 151
seeds and leaves, 149, 150
microscopic identification
glandular and cystolith hairs, 151, 152
low-power magnification, 151
TLC
description, 153
reagents, 153
visualization, 154
Forensic investigation
charges and offenses, controlled substance, 67, 68
controlled substance laws
Schedule I, 66
Schedule II, 66–67
Schedule III, 67
Schedule IV, 67
Schedule V, 67
controlled substance submission, 67
court testimony, 69
CSA, 66
definition, drugs, 61
drug abuse, 63
drug cases, crime laboratories, 68
examination, controlled substances, 67, 68
narcotics
natural drugs, 61
psychotropic drugs, 61–62
synthetic drugs, 61
physical dependence, 62
psychological dependence, 62
structural relationship
analogs, 64
designer drugs, 64–65
isomers, 65
usable quantity, 69
Fourier transform infrared spectroscopy (FTIR)
advantages, 134
bufotenin, 200, 203
components
beam path, radiation, 133
detector signal, 133
DTGS and MCT, 133
interferometer, 133
design, 133–134
methoxy/5OHDMT, 200, 204
modern design, 132
phenethylamines, 172, 174
psilocin/psilocy bin, 200, 201
sample preparation techniques
liquid/vapor phase, 134
solid-sample, 134–135
synthetic tryptamines, 200, 205
tertiary amine
C-H and C-N stretching bands, 185
description, 185
spectra, bands, 185, 186
FTIR. *See* Fourier transform infrared spectroscopy
Functional groups
alcohols, 41–44
aldehydes, 46–47
alkanes, 32–37
alkenes, 37–39
alkynes, 39–40
amines, 51–54
aromatic compounds, 40–41
carboxylic acids, 47–49
classification, 31
esters, 49–50
ketones, 44–46
methyl group, 54
multiple, 54–55
nitro compounds, 50–51

G

Gamma (γ)-hydroxybutyric acid (GHB)
activity, 226–227
"date-rape" drug, 227
injections and tablets, 227
neuroprotective nutrient, 227
street names, 227
Gas chromatography (GC)
analogy, 114
automated gas chromatographs, 112, 113
capillary, 113
coin-separating machines, 114
interpretation, 115–116
packed-column, 113
vapor-phase chromatography (VPC), 112
Gas chromatography mass spectrometry (GCMS)
advantages, 124
anabolic steroids, 213–214
analysis, 103
bufotenin and synthetic tryptamines
analogs, plant species, 200, 202
and FTIR spectrum, spectral data, 200, 201
mass spectrum and FTIR spectrum, 200, 203
toadstools, 200, 202
controlled substances, identification
allyl-cyclopentenyl-barbiteric acid, 230, 235
barbital, 230, 234
demerol and GBL/GHB, 230, 233
description, 230
ketamine, 230, 232
LSD and fentanyl, 230, 236
secobarbital, 230, 235
disadvantages, 124
dry extraction, 104
methanol, 105
mushrooms examination, tryptamines
description, 199
spectral data, 200, 201
phenethylamines, 172, 173
tertiary amine
cocaine HCL and base, 186, 187
description, 186
heroin, 186, 188
PCP, 186, 189
GCMS. *See* Gas chromatography mass spectrometry
Glucocorticoids, 109

H

High-performance liquid chromatography (HPLC)
- liquid chromatograph, 111
- normal phase, 112
- retention time, 112
- reverse-phase, 112
- structural isomers differentiation, 111

HPLC. *See* High-performance liquid chromatography
Hydrogen bonding, 27, 45, 52, 85

I

Infrared (IR) spectroscopy
- forensic identification, instrument selection, 137–138
- FTIR spectrometer
  - advantages, 134
  - components, 133
  - design, 133–134
  - sample preparation techniques, 134–135
- FTIR spectrophotometer, 127
- FTIR spectroscopy
  - advantages and disadvantages, 137
  - spectra, ephedrine and pseudoephedrine, 137
- inorganic analysis
  - anions and absorption wave numbers, 139
  - sample preparation and peak identification, 138
  - stand-alone technique, 138
- instrumentation
  - dispersive, 129
  - integrated computer workstations, 129, 130
- organic analysis
  - extensive purification, 139
  - free-base cocaine *vs.* cocaine hydrochloride, 139, 140
  - free-base forms, 139
  - HCl and methamphetamine, 139, 140
- reflectance, 137
- sampling techniques
  - cast film A, 136
  - cast film B, 136
  - Nujol Mull, 136
  - pellets, 136
  - synthetic membrane sample cards, 137
- spectrum
  - absorbance and transmittance formats, 129, 130
  - description, 129
  - transmittance, 129
  - use and conversion, absorbance, 129
- theory
  - bond activity, molecules, 128
  - frequency ranges, 128
  - observed absorption bands, 128, 129
  - radiation and spectrum, 127
  - wave numbers and frequency, 128
- transverse waves and wavelength, 127
- and ultraviolet (UV) regions, 127

Ion-exchange chromatography (IEC)
- automated computer workstation, 110, 111
- cation-exchange, 110

Ionic bonds
- crystal lattice, 24
- description, 23
- electron transfer, 23, 24

Ion trap mass analyzers
- applications, 124
- benefits, 123
- limitations, 124

IR spectroscopy. *See* Infrared (IR) spectroscopy
Isotopes, 18

K

Ketamine
- chronic use, 228
- description, 227
- dissociative anesthetic, 227–228
- effects and cause, 228
- injections and tablets, 228
- schedule III controlled substance, 228
- street names, 228

Ketones
- acetone, 45
- carbonyl group (R-CO-R), 44–45
- cathinone
  - effects, 165
  - occurrence and isolation, 164–165
  - reduction, 165
- description, 164
- functional groups, anabolic steroids, 212
- khat
  - classification, 165, 166
  - consumption, 165
  - harvested and packaged form, 165, 166
  - marijuana and alkaloids, 166
- methcathinone, 46, 165
- oxidation, 164

L

Liquid–liquid extraction, 100–101
LSD. *See* Lysergic acid diethylamide
Lysergic acid diethylamide (LSD)
- alkaloid ergonovine, 229
- clandestine production, 241
- description, 228–229
- occurrence, 229
- sclerotium, 229
- synthesis, rye ergot, 245–246
- tablet and capsule, 229

M

Magnetic sector mass analyzers
- applications, 123
- benefits, 123
- limitations, 123

MAOIs. *See* Monoamine oxidase inhibitors
Mass spectra, anabolic steroids
- comparison and parent ion peak (M^+) identification, 214, 216–221
- description, 214

Mass spectrometry
- analyzers, 119
- chemical ionization (CI)
  - advantages, 118
  - electron-impact, 118
- drugs identification, 117
- electron impact, 117–118
- GCMS, 116–117
- ionization, 117
- ion trap analyzers, 123
- magnetic sector analyzers, 123
- quadrupole analyzers, 119, 122
- spectral fragmentation

amphetamine, 119–121
identification process, 118
magnetic field strength, 123
peaks, drugs, 119, 122
phentermine and methamphetamine, 119–121
primary amines, 118–119
MCT. *See* Mercury cadmium telluride
MDMA. *See* 3,4-Methylenedioxymethamphetamine
5MeODIPT. *See* 5-Methoxy-*N,N*-diisopropyltryptamine
5MeODMT. *See* 5-Methoxy-*N,N*-dimethyltryptamine
Mercury cadmium telluride (MCT), 133
Mescaline
classification, 169
hallucinogen, 169
occurrence, 168–169
peyote plants, 169
side effects, 169
Methamphetamine
clandestine
extraction, pseudoephedrine, 261, 262
icing, methamphetamine, 263, 264
manufacturing methamphetamine, 262–263
processing methamphetamine, 263, 264
pseudoephedrine-containing cold tablets, 261, 262
Red-Devil Lye, 263, 264
Red Phosphorus-HI Method, 261
round-bottomed flask, 262–263
cold method, 249–251
element identification, 271
flame-test, red phosphorus, 271–272
history, 159–160
hot method, 251–252
hydrogen gas, 251
iodine crystals, 272
iodine identification, 272–273
as medication, depression and obesity, 160
Nazi method, 272
physical and psychological effects
description, 160
oral ingestion/snorting, 160
short-term and long-term abuse, 160
street names, 161
tolerance, 160–161
red phosphorus, 271–272
regulation, household items, 249, 250
silver nitrate test, 273
Methanol extraction, 105
Methoxy derivatives, tryptamines
5MeODIPT and 5MeODMT, 195
plants list, analogs
acanthaceae and agaricaceae, 195
aizoaceae and gramineae, 195–196
leguminosae, 196
malpighiaceae and myristicaceae, 196
ochnaceae and polygonaceae, 196
rubiaceae and rutaceae, 197
snuff preparation, 195
5-Methoxy-*N,N*-diisopropyltryptamine (5MeODIPT), 195
5-Methoxy-*N,N*-dimethyltryptamine (5MeODMT), 195
3,4-Methylenedioxyamphetamine (MDA). *See* Methylenedioxy derivatives
Methylenedioxy derivatives
MDA
capsule/pill, 167
production, 167
profound relaxation, 167
psychedelic stimulant and empathogen-entactogen, 167
MDMA
description, 167
"ecstasy", 168
relaxation state and use, 168
tablet, 168
rings and bond angles, 166
3,4-Methylenedioxymethamphetamine (MDMA) *See also* Methylenedioxy derivatives
isosafrole, 246
synthesis (ecstasy), 246–247
Microcrystal techniques
acid and anionic, 95
advantages, 91–92
aqueous, 94
critical considerations, 96–97
disadvantages
compounds identification, 92
thin-layer chromatography, 92
documentation, 92–93
GCMS/Fourier transform infrared (FTIR) spectroscopy, 91
test reagents (*see* Aqueous test reagents)
volatility, 95
Mineralocorticoids, 209
Molarity, 28
Molar mass, 27–28
Mole and molar mass, 20
Molecules
bond types prediction
hydrogen, 27
ionic, 27
nonpolar covalent, 27
polar covalent, 27
chemical bonding, 23–27
chemical reactions, 28–29
compounds, 23
molarity, 28
molar mass, 27–28
Monoamine oxidase inhibitors (MAOIs), 197
Multiple bonds
double, 26
single, 26
triple, 27

N

Narcotics
natural drugs, 61
psychotropic drugs, 61–62
synthetic drugs, 61
Natural drugs, 61
Neutral compound extraction, 102–103
Nitro compounds
chemical formula, R-NO_2, 50
illegal drugs, 50–51
Nonpolar covalent bonds, 27
Nuclear radiation, 19–20

O

Opiates, natural tertiary amines
codeine
description, 180
use, 180
description, 179
heroin

Opiates, natural tertiary amines (*continued*)
characterization, 181
description, 181
inhalation and injection, 181
short-and long-term side effects, 181
morphine
description, 179–180
inhalation and symptoms, 180
poppy
corn, 182
description, 181
flowers and seed capsules, 181
latex and seeds, 182
significance, 182
use, medicinal purposes, 182

P

Paper chromatography
capillary action, 108
samples separation, 108
solid–liquid, 109
PCP. *See* Phenylcyclohexylpiperidine
Periodic table, atomic structure
elements, 9, 10
groups and periods, 9
law, 9
metalloids/semi-metallics, 10
symbols, 9
Phenethylamines
analytical methods
chemical screening, 170, 171
confirmatory examination, 172–174
extraction techniques, 171
FTIR, 172, 174
GCMS, 172, 173
mescaline extraction, 172
microcrystal tests, 170, 171
visual inspection, 170
bronchodilators, 157
description, 157
hydroxyl derivatives
ephedra plant, 164
ephedrine/pseudoephedrine, 163–164
phenylpropanolamine, 162–163
innovative techniques, 157
ketone derivatives
cathinone, 164–165
description and oxidation, 164
khat, 165–166
methcathinone, 165
lists, 157, 158
methoxy derivatives
description, 168
mescaline, 168–169
methyl derivatives
addition, 157
amphetamine, 158–159
methamphetamine, 159–161
phentermine, 161–162
methylenedioxy derivatives
MDA, 167
MDMA, 167–168
rings and bond angles, 166
α-/β-position, 1-amino-2-phenylethane, 157
Phentermine
history
fenfluramine/dexfenfluramine, 161
fen-phen and dexfen-phen, 161
salt form, 161
physical and psychological effects, 162
side effects, 162
Phenylcyclohexylpiperidine (PCP)
contaminates, 183
description, 182
as drug of abuse, 183
effects and causes, 183
structure, 182
Phenylpropanolamine
α and β carbons, 162
physical and psychological effects, 163–164
substitution, methyl and hydroxyl group, 162
use, clandestine drug manufacture, 164
Physical properties
aldehydes, 46–47
alkanes, 36
description, 5–6
enantiomers, 55, 92
instrumental analysis, 92
melting and boiling point, 5
Polar bonds, 25
Polar covalent bonds, 27
Psilocin and psilocybin
classification, psychoactive mushrooms, 192, 193
effects, serotonin, 193
occurrence, 192
preparations, dried/brewed mushrooms, 193
psychoactive "magic" mushrooms, 192
schedule I hallucinogens, 192
Psychotropic drugs, 61–62

Q

Quadrupole mass analyzers
applications, 123
benefits, 123
limitations, 123
mass-to-charge, 119, 122

R

Radioactivity. *See* Atomic structure

S

Scientific investigation
experimentation and conclusion/theory, 4
observation and hypothesis, 4
Solid–liquid extraction, 99–100
Solid-sample FTIR spectrometer
cocaine comparison, 135
crystal-lattice effect, 134
molecule vibration and polymorphism, 135
structure limits, 134
variations, IR spectra, 135
Steroids. *See* Anabolic steroids
Subatomic particles, atomic structure
charges, protons and electrons, 11
electrons, 11
isotopes, 11

mass number, 11
net charge, protons and electrons, 11, 12
nucleus, 11
protons and neutrons, 11
Synthetic drugs, 61

T

Tertiary amines
analytical methods
chemical screening, 183, 185
confirmatory examination, 185–189
visual inspections, 183, 184
description, 177
natural
cocaine, 177–179
opiates, 179–182
synthetic, PCP
contaminates, 183
description, 182
as drug of abuse, 183
effects and causes, 183
structure, 182, 183
Thin-layer chromatography (TLC)
Cannabis
description, 153
interpretation, 154
plate and sample preparation, 153
reagents, 154
separation, 154–155
visualization, 154
tryptamines
description, 199
interpretation, 199
procedure, 199
TLC. *See* Thin-layer chromatography
Tryptamines
analogs, 191
analytical methods
chemical screening tests, Weber test, 198
GCMS, 199–205
psilocin and psilocybin extraction, 199
TLC, 199
visual identification, 197–198
description, 191
indole derivative, 191
natural
bufotenin, 193–195
methoxy derivatives, 195–197
psilocin and psilocybin, 192–193
occurrence, 191
synthetic
DET, Foxy Methoxy and AET, 197
MAOIs, 197
scientific research purpose, 197

U

United States Pharmacopoeia (USP), 146
USP. *See* United States Pharmacopoeia

V

Visual identification
controlled substances
color-screening test, 230, 231
common forms, 229, 230
description, 229
tryptamines
description, 197
identification, psychoactive mushrooms, 197–198
Visual inspections
anabolic steroids
description, 213
representative forms, 213, 214
tertiary amines
description, 183
procedure, 183, 184
Volatility test technique, 95

W

Weber test, tryptamines
methanol extraction, 198
reagent 1 and 2, 198

Y

Yopo seeds
description, 195
leaf and seeds, 194, 195
use, *cohoba snuff*, 194